Studies in Computational Intelligence

Volume 963

Series Editor

Janusz Kacprzyk, Polish Academy of Sciences, Warsaw, Poland

The series "Studies in Computational Intelligence" (SCI) publishes new developments and advances in the various areas of computational intelligence—quickly and with a high quality. The intent is to cover the theory, applications, and design methods of computational intelligence, as embedded in the fields of engineering, computer science, physics and life sciences, as well as the methodologies behind them. The series contains monographs, lecture notes and edited volumes in computational intelligence spanning the areas of neural networks, connectionist systems, genetic algorithms, evolutionary computation, artificial intelligence, cellular automata, self-organizing systems, soft computing, fuzzy systems, and hybrid intelligent systems. Of particular value to both the contributors and the readership are the short publication timeframe and the world-wide distribution, which enable both wide and rapid dissemination of research output.

Indexed by SCOPUS, DBLP, WTI Frankfurt eG, zbMATH, SCImago.

All books published in the series are submitted for consideration in Web of Science.

More information about this series at http://www.springer.com/series/7092

Janmenjoy Nayak · Bighnaraj Naik ·
Ajith Abraham
Editors

Understanding COVID-19: The Role of Computational Intelligence

Springer

Editors
Janmenjoy Nayak
Department of Computer Science
& Engineering
Aditya Institute of Technology
and Management (AITAM)
(An Autonomous Institution)
Srikakulam, India

Bighnaraj Naik
Department of Computer Application
Veer Surendra Sai University of Technology
Burla, India

Ajith Abraham
Scientific Network for Innovation
and Research Excellence
Machine Intelligence Research Labs
(MIR Labs)
Auburn, WA, USA

ISSN 1860-949X ISSN 1860-9503 (electronic)
Studies in Computational Intelligence
ISBN 978-3-030-74760-2 ISBN 978-3-030-74761-9 (eBook)
https://doi.org/10.1007/978-3-030-74761-9

This Springer imprint is published by the registered company Springer Nature Switzerland AG
The registered company address is: Gewerbestrasse 11, 6330 Cham, Switzerland

Preface

COVID-19 has emerged as an urgent emergency throughout the world for the last one year. Since the evolvement of COVID-19, several predictive models (mathematical, statistical, and machine learning-based) are developed to predict the spread of this pandemic. However, the major problem in the identification of COVID-19 is the diagnosis. Although reverse transcription polymerase chain reaction (RT-PCR) method is used for the diagnosis, still these are time-consuming, less sensitive, and short in supply because of the pandemic. The usage of medical images such as X-rays, computer tomography, and scanners can overcome the limitations of RT-PCR. Computational intelligence techniques could help the physicians for the accurate prediction and rapid detection of disease. Therefore, the present world health emergency is also seeking the support of these techniques for the efficient identification of COVID-19. Even though mathematical and statistical tools are upto some extent in the prediction and forecasting of COVID-19, still many drawbacks such as complete utilization of biomedical data, chronological dependency, dealing high-dimensionality, sparsity, heterogeneity, irregularity, and novel patterns cannot be identified with supervised methods. The advantage of the computational intelligence approach is the ability to build predictive models even when the understanding of the underlying mechanism is limited or when the computational simulations are too expensive to be carried on the basic physical models. Hence, computational intelligence techniques can be rapidly used in various healthcare applications such as identification of diseases and applying corrective diagnosis measures, drug discovery and production, medical image diagnosis, managing of smart health records, clinical trials and research, crowd-sourced data collection, better radiotherapy, and outbreak prediction. In recent times, computational intelligence-based computer-aided diagnosis system has shown major significance in solving this issue. The application of computational intelligence for the betterment of the healthcare situation is booming with each passing day. The features such as admirable performance, endpoint learning scheme with combined feature-based learning, capacity to handle the complex and multi-modal data, etc. made the computational intelligence models the best accurate models for handling medical datasets. Due to some interesting features such as high performance, absence of human intervention in the process of feature extraction and identification, not making the use of engineering advantage in the training phase, computational intelligence models are extensively used in the prediction and forecasting of COVID-19 instead of traditional computing methods.

Currently, the researchers have shown strong interest in using computational intelligence to advance diagnosis, supervision, and improved healing options of numerous diseases, particularly COVID-19. This book is a dedicated platform for the ongoing and upcoming solutions of COVID-19 using CI techniques. Also, a special emphasis will be focused on collections of useful advanced mechanisms, automated tools, and possible directions for solving this pandemic complex problem. Currently, computational intelligence techniques already acquired a key place in the healthcare system. There is a growing concern in the applications of computational intelligence in the healthcare domain to advance disease diagnosis, supervision, and the development of effective therapies. This edited volume presents fundamental research contributions from a methodological/application viewpoint in understanding the application of various computational intelligence techniques in solving COVID-19. The book contains volumes of information about advanced computational intelligence approaches spanning the areas of neural networks, connectionist systems, self-organizing systems, and hybrid intelligent systems in solving COVID-19. The book also covers the novel computational intelligence and technological models for effective COVID-19 at par with the standards laid down by the World Health Organization. Undoubtedly, various innovative technologies have helped in fighting this pandemic on all fronts. They will find this book instrumental for understanding the aspects (opportunities and challenges) of processing and handling this novel virus perspective. The book will also help them to design, develop, and integrate futuristic technologies for such pandemic. It will help to advance the exchange of ideas and lucidity among various CI methods and will augment the significance and utilization of said application areas for end-user as well as novice user applications. This volume comprises of 3 parts and 23 chapters and is organized as follows:

Part 1: Learning from various modalities of COVID-19 data

Chapter 1 presents a clear and concise idea of the nature of COVID-19, its classification, and its characteristics. Ghosh et al. have discussed in detail the causes and symptoms related to COVID-19 disease and modes of transmission and infection. The classification refers to the hierarchical naming and the reason behind the protocol accepted and the characteristics refer to the nature of the virus, sources of spreading, its strengths and weaknesses. Further, the latest and the most accurate data and findings are listed and common misconceptions and misinformation are examined, analyzed, and debunked. A systematic review of after-effects and the preventive, as well as curative measures of the COVID-19 infection is also presented. Finally, the current and future challenges are enumerated and summarized in the chapter.

In Chapter 2, Mallick et al. developed an approach that uses both graph theory and an evolutionary algorithm for achieving extractive summarization. Initially, types of text data are pre-processed by means of sentence tokenization, then removal of stop words, and finally stemming operations. Consequently, a bag-of-words is generated to represent each report in a structured form and a bi-objective strength Pareto evolutionary algorithm-based text summarization technique is concerned to produce the summary of the report. The bi-objective strength Pareto evolutionary algorithm uses the clustering coefficient of the generated graph for defining one objective function and the length of the summary for other objective functions. The method is applied on COVID-19 biomedical reports publicly available in CORD-19 dataset for generating a summary and its performance is finally evaluated by comparing it with a few of the related state-of-the-art methods.

Chapter 3 revises the basics of radiomics with a short practical and easy guide for radiologists. The main radiological findings of COVID-19 pneumonia with the most relevant information that is assessed to evaluate the extent of the disease have been presented by Federica et al. Finally, the main current literature on potential clinical application of radiomics and artificial intelligence for COVID-19 is presented together with limitations and perspectives.

Chapter 4 discussed about the usage of existing deep learning-based frameworks on an augmented dataset consisting of pixel scaled images and the diagnosis of COVID-19 infection on the original CT images. In this chapter, Sushant et al. have used convolutional neural networks to classify CT images into infected and not infected categories. A few tweaks to the hyperparameters produce good results on the available datasets. Finally, the efficiency of existing models such as ResNet, DenseNet, and VGG using pixel scaled images shows better performance in terms of accuracy, sensitivity, specificity, and F1 score when compared with results of the existing models on the original dataset.

In Chapter 5, Aliye and Hurrem investigated about the antibody-based treatments such as monoclonal antibodies, convalescent plasma therapies, and vaccine studies in order to combat the continuing pandemic. Although it may not be as troublesome as in the dengue fever, clinicians should be alert about the possible risk of antibody-dependent enhancement. Non-neutralizing antibodies (NAbs) or sub-protective levels of NAbs may trigger immunopathogenic events via immune complex-mediated or Fc receptor-dependent endocytosis, resulting in enhanced viral infection. Further, literature findings performed revealing the factors with an impact on eliciting NAbs, possible mechanisms, and risks of ADE, and safety concerns for SARS-CoV-2 treatment interventions are outlined.

Chapter 6 is about developing a system to model the spread of COVID-19. The system proposed by Behera et al. is helpful in working at post lockdown conditions when the only mode of travel is by road. The impact measures and the severity of potential disease spread in a specific area are calculated based on existing hotspots and are clustered into regions of varying danger levels, using a graph clustering algorithm. Using this method, it can be predicted where more lenient measures may be taken and which areas are less prone to the virus spread. Specifically, it focuses on preventing the virus spread, from a geographical point of view. The impact measure tells the impact of a region, on nearby regions, and the danger value tells the transmission possibility, between separate regions. Using this data, potential hotspots are easily identified. This data will help administrative bodies, to make more fine-tuned lockdown restrictions, based on the impact measures.

Chapter 7 developed a CNN model that takes input as chest X-ray images and predicts whether a person needs to go for COVID-19 testing or not. The recommendation system developed by Chittaranjan et al. takes around 35 minutes to give the result of a sample and curbs down the chances of virus spread through contact while testing, unlike the presently used methods. The experimental results reveal that proposed system attains better accuracy using the chest X-ray scans and requires less computational time. The proposed model will decrease the fear of being COVID-19 infected and can be easily implemented even in rural areas, which will also decrease the load of the COVID-19 testing team.

Chapter 8 is about the development of a concept mapping-based research framework. The framework proposed by Rajan et al. is used to review the usage of deep learning techniques for the management of the COVID-19 pandemic using multiple steps. Multiple deep learning techniques like long short-term memory, convolutional neural network, recurrent neural network, and generative adversarial network that are found to be prominently used modeling techniques for growth curve fitting and image analysis are studied in the second step. In the next step, various epidemics from the past are learned from the deep learning perspective. In the fourth step, multiple use cases for COVID-19 management are covered up, like trend forecasting, virus and drug discovery, patient diagnosis, clinical analysis, and monitoring for maintaining social distancing and face masks. The last step presents the various evaluation parameters available for analyzing the modeling techniques performance under deep learning.

Part 2: Prediction and risk analysis of COVID-19 susceptibility

Chapter 9 explores the detailed study of the mobile application requirements in the pandemic. In this chapter, Saravanan et al. have narrated different mobile apps utilized in the pandemics in the past and current with its specification. Also, an integrated mobile network platform is presented with all the actors involved in the COVID-19 pandemic management. This chapter also provides a statistical insight on the usage of mobile apps in percentage and its impact in tackling COVID-19. It also elaborates the challenges faced by various countries while using mobile apps and gives an elucidate outline for application developers.

Chapter 10 introduced the concept of deep learning and its potentials for combating the current spread COVID-19 pandemic and mitigating future outbreaks. Two DL algorithms namely long short-term memory and convolutional neural network are employed by Sarumi et al. to develop a mechanism to predict future cases of COVID-19 using a real-life dataset of China and Canada as case studies. Also, some other ongoing efforts on the use of DL as one of the non-clinical approaches in alleviating the effects and curtailing the spread of the COVID-19 pandemic are discussed in the study. Finally, some of the challenges of using deep learning in combating the effects of this pandemic are presented, and future research directions on the use of DL for mitigating the challenges of COVID-19 are deliberated.

Chapter 11 has established a novel AI-based architecture to obtain effective drugs for the COVID-19 treatment. Soumen and Banerjee have taken the COVID-19 (MT050943 and MT012098) data from the NCBI database, submitted by the Government of India to build tests and find drugs. A detailed analysis of these sequences has been performed to find the noncoding percentage, gene density, and abundances in functional gene categories using Virus Amplicon Sequencing Assembly Pipeline. This analysis finds the unique regions in the virus genome as a target region. The result of the proposed methodology obtains higher accuracy than the other existing methods and identified effective drugs like Dexamethasone, Remdesivir, etc. for the study of the COVID-19 vaccine.

Chapter 12 proposed a diffusion model by dividing the total population of constant size into four classes such as susceptible population, infected population, quarantined population, and recovered population. The disease transmission factor for both infected and quarantined population into the susceptible population is in more general form. The model developed by Reza et al. can give the wave speed at which the disease 2019-nCoV spread. Additionally, it can be predicted whether the total population in forward time will become susceptible or not in the absence of vaccine.

Chapter 13 proposed a machine learning-based anxiety prediction model to find the patterns between popular tweets and its effect on the public's sentiments, especially their fear. So, tweets of three different countries namely USA, Federative Republic of Brazil, and Republic of India have been collected by Swarup et al. to perform the analysis. Initially, predominant analysis of tweets within past two months has been done. Then a model based on machine learning algorithms is built to predict the future reaction of the general public based on the crisis level in the country. The machine learning algorithms such as logistic regression, multinomial Naïve Bayes, and support vector machine are used for classification purpose preceded by the pre-processing steps of raw data from each country, and the results reveal the efficiency of models in predicting the classes of tweets. Moreover, the chapter also provides insights to the fear and sentiments and tweets with negative sentiment and emotion indicates the cases for the pandemic outbreak.

Chapter 14 illustrated various novel methods developed by using technological inventions all around the globe, along with their merits, and demerits. The framework suggested by RamaKrishna and Sirajuddin assists policymakers in the detection and prevention of COVID-19 by using machine learning algorithms and other scalable technologies. The proposed framework includes both manual best practices with an adaption of technology change, such a hybrid solution which significantly reduces time and efforts. This framework can also help the policymakers in taking the decisions. It will trace, track, and predicts the impact with the help of novel technologies. Technologies included in this framework will help in identifying the COVID-19 personal, preparation of contact tracing, identifying the containment zones, generation of schedules of the quarantine periods, the work schedule of frontline warriors, and assessing the required materials in controlling the outspread of virus infection.

Chapter 15 proposed a Gaussian regression model for the outbreak prediction. For outbreak prediction using proposed model, Özkaya and Öztürk have analyzed pandemic data for eight countries in the European region between January 22, 2020, and April 30, 2020. These countries are Belgium, the Netherlands, Germany, UK, Spain, France, Italy, and Turkey. The number of cases is taken into consideration in the selection of countries. A comparative analysis is performed using Ard-Exponential kernel function, rational-quadratic kernel function, and squared exponential kernel functions. The quasi-Newton algorithm is used to optimize regression models. Confirmed death and recovered case data are processed in the scope of the analysis. The first 90 days of the data are used for the training of Gaussian regression models. The last nine days are evaluated for prediction. Mean absolute error, median error, mean squared error, and root mean squared error are used as performance metrics to evaluate the efficacy of the model. In general, it is seen that the Ard-Exponential kernel with Gaussian Regression method obtained high metric performances for all cases and countries.

Chapter 16 aims to examine the appropriate predictive model for COVID-19 epidemiological data in the Indian population using computational intelligence-based hyperparameter tuned regression techniques. The development of the COVID-19 epidemiological data prediction model for India is proposed by Rathipriya and Dhamodharavadhan using hyperparameter tuned regression technique which makes use of models such as Gaussian process regression model and shallow neural network models. This analysis uses a time series dataset that includes daily cumulative COVID-19 epidemiological data. In order to increase the forecast accuracy, hybrid models are constructed by merging hyperparameter tuned regression models with nonlinear autoregressive neural network. Experiential research in this work reveals significant differences between the different types of CI-hyperparameter tuned regression models for the COVID-19 fatality prediction. The results reveal the superiority of the hybrid GPR prediction model when compared with other models for the COVID-19 prediction.

Part 3: Combating COVID-19 with advanced computational intelligence techniques

Chapter 17 deliberated about the various computational intelligence, machine learning, and deep learning approaches employed in solving the problems related to COVID-19. The chapter deals with the review of the different CI approaches such as artificial neural networks, fuzzy system, and evolutionary computing that have been applied for COVID-19 screening, prediction, and vaccination of drugs. In the chapter, Aditya and Sanjay have discussed the application and suitability of computational intelligence, machine learning, and deep learning techniques for screening and treating the patients, tracing the contacts along with COVID-19 forecasting in detail. A discussion on the reason of using such techniques for the COVID-19 prediction is also presented. Finally, the chapter enumerates the limitations and future challenges of these approaches to assist the researchers and policymakers in tackling COVID-19 pandemic more effectively.

Chapter 18 described about the development of a new COVID-19-specific clinical score known as the general assessment of SARS-CoV-2 patients score. The score builds on the clinical experience gained during the first phase of the pandemic, and it is based on both clinical and laboratory data. The model developed by Passaro et al. is aimed in predicting the 30-day mortality outcome of hospitalized COVID-19 patients and showed markedly better accuracy than the Charlson comorbidity index. Furthermore, an additional predictive model based on a classical computational intelligence method is also used. Specifically, it describes the development and validation of a feedforward artificial neural network that automatically maps patients' clinical and laboratory data to a 30-day mortality risk score. However, the intrinsic black-box nature of the NN-based method makes it hard to reach an intuitive understanding of the internal computations underlying its decision process.

Chapter 19 promoted a multi-disciplinary approach to care for patients with COVID-19, including people who have high chances of getting infected, patients with mild, moderate, severe, and critical disease. The main objective of the author Jayashree is to construct an interactive decision-making environment using a Restricted Boltzmann Machine with Dynamic Estimation of Distribution Algorithm that helps in the classification of patients based on their symptoms and to identify people with a high chance of getting infected by a coronavirus. Further, the Mann—Whitney U test is accomplished on the sample dataset and obtained a better confidence interval which proves the efficiency of the proposed algorithm. Finally, the deep convolution neural network method is used on the sample chest CT images after the preliminary test to categorize patients as COVID-19 disease-ridden or not using chest CT images.

Chapter 20 developed an integrated GIS/GPS-based e-Governance platform which is effectively used by all those agencies involved in controlling the spread of the disease. The model developed by Meghna and Mitesh is useful to monitor and manage tracking of human movement, coordinate to health department and sanitary services, prepare proper scientific and socioeconomic analysis and progressive flow related to COVID-19 situation/spread in real-time basis. The platform is having different modules for different agencies including patients for effective and timely management of the crisis with minimum overheads and ensuring minimum suffering to all agencies. Data privacy and security can be achieved by deploying security at different levels like web security, network security, application security, and data encryption. This kind of approach also brings transparency in the system.

Chapter 21 focused on development of an efficient AI-based medical imaging solution for COVID-19 by leveraging the easily available COVID-19 X-ray images. Initially, Swati and Mane have developed and evaluated different variants of convolutional neural networks. Among these different variants, convolutional neural network with ten convolutional layers and 15 epochs has given the best performance. The chapter also highlights the importance of transfer learning and the utilization of data augmentation and dropout techniques in avoiding the overfitting problem.

Chapter 22 gives a thorough discussion on the role of deep learning in diagnosis of COVID-19 infection. The fundamentals of convolutional networks are elaborated in length which includes the conventional set of layers in the design of convolutional neural networks. Vidyadevi et al. have enlightened the chapter with important concepts such as augmentation, transfer learning, and various metrics that must be used in evaluation of deep learning models as applicable to coronavirus infection diagnosis. It provides a good reference to various COVID-19 dataset, and details are discussed which helps any researchers to understand the possible publicly available datasets along with the future developmental aspects. Various convolutional neural networks are discussed with an emphasis on exploring the ensemble methods and give the guidelines on selection of pretrained models for devising ensemble methods.

Chapter 23 focused on distinct novel deep learning architectures used for solving COVID-19 related problems such as identification, progression, and diagnosis based on several key performance factors. In this chapter, Prashanth et al. have also presented a comparative analysis of the different datasets used for evaluating the deep learning architectures. The chapter presents the need of utilizing deep learning approaches instead of conventional neural network architectures in solving the COVID-19-related problems with a focus on the tools and techniques for visualizing and summarizing COVID-19 related data, which greatly helps in understanding the trend to facilitate informed decision making.

We would like to thank all the contributors and the reviewers for their contributions and dedicated efforts for the successful completion of this book. We want to specially thank the editorial team of Springer for their valuable technical support and superior efforts. We hope that the work reported in this volume will motivate further research and development efforts in the performance evaluation of COVID-19 and its allied domains.

Janmenjoy Nayak
Bighnaraj Naik
Ajith Abraham

Contents

Learning from Various Modalities of COVID-19 Data

COVID-19 Pandemic: Theory, Concepts and Challenges

Anindya Banerjee, Raj krishan Ghosh, Surajit Basak, and Arijit Dey

Abstract Coronavirus is a single-stranded RNA virus, with a size of about 120 nm. Severe Acute Respiratory Syndrome Coronavirus 2 (SARS-CoV2) is an infectious disease caused by a recently discovered coronavirus Disease 2019 (COVID-19) novel coronavirus and is a major global pandemic that has already taken more than 600 thousand lives. This disease is spreading rapidly across the globe and already has more than 14 million confirmed cases worldwide. It can be latent with low or no symptoms at all for initial days after transmission. The exponential increase in new cases each day calls for swift, responsive measures and rapid control at the individual, societal, and governmental levels. Lack of awareness about COVID-19 and negligence in implementation of protective measures poses a greater threat to human lives in this pandemic. There is a shortage in hospital wards as well as the absence of proper medicine for recovery. Necessary consumer products like masks and sanitizers are short in supply and unable to quench the demand. This chapter aims to present a clear and concise idea of the nature of COVID-19, its classification, and its characteristics. Classification refers to the hierarchical naming and the reason behind the protocol accepted. The characteristics refer to the nature of the virus, sources of spreading, its strengths and weaknesses, just to name a few. Causes

A. Banerjee
Department of Electronics and Communication Engineering, Kalyani Government Engineering College, Kalyani, West Bengal, India
e-mail: anindya.banerje@gmail.com

R. Ghosh (✉)
Center of Excellence in Artificial Intelligence, Indian Institute of Technology Kharagpur, Kharagpur, West Bengal, India
e-mail: rajkrishanghosh@gmail.com

S. Basak (✉)
Department of Electronics and Communication Engineering, Guru Nanak Institute of Technology, Kolkata, India
e-mail: surajit.basak2007@gmail.com

A. Dey
Friedrich-Alexander-Universität Erlangen-Nürnberg, Bavaria, Germany
e-mail: arijit.dey@fau.de

J. Nayak et al. (eds.), *Understanding COVID-19: The Role of Computational Intelligence*, Studies in Computational Intelligence 963, https://doi.org/10.1007/978-3-030-74761-9_1

and symptoms related to this disease are explained in detail. Modes of transmission and infection are discussed. The latest and the most accurate data and findings are listed and described. Common misconceptions and misinformation are examined, analyzed, and debunked. The after-effects and steps to take are reviewed. The preventive, as well as curative measures, are conferred. Current and future challenges are also enumerated and summarized.

Keywords COVID-19 · Mode of Transmission · Detailed statistics · Preventive measures · Curative measures

1 Introduction

When a new viral disease encounters an outbreak, there are three things to be decided against it: the species, the disease, and the virus. The International Committee on Taxonomy of Viruses (ICTV) is apprehensive with the description and naming for virus taxa (i.e. family, genus, species, etc.) rather than the description of names of virus or disease. Severe Acute Respiratory Syndrome Coronavirus 2 (SARS-CoV-2) gets spread person to person via close communities. The disease name (which in numerous scenarios varies from the virus name) has been chosen as COVID-19 by WHO. The number '19' used in COVID-19 stands for the year - 2019, when the virus was seen for the first time [1].

When people with coronavirus Disease 2019 (COVID-19) cough or breath out, they eject miniscule droplets containing the virus. If these droplets enter through mouth or nose, it can cause the infection. The most common method that this illness gets spread is through close proximity with someone having the infection. It then quickly advances deeper in the body. It tries to reach the intestines, the lungs or spleen, where it often has the most dangerous effect. COVID-19 can manifest differently in people. On average five to six days are needed for symptoms to show, though in some cases it can take as long as 14 days. People having minor symptoms who otherwise are healthy should manage their symptoms at home. While COVID-19 is spreading rapidly, most people will experience only mild or moderate symptoms. Coronavirus can cause severe disease in some people [2]. Most people who have contracted the new coronavirus to date have recovered or are recovering. However, a pandemic can result in devastation when number of deaths is high, even if the percentage of death from infection is low. COVID-19 spreads more rapidly than diseases like SARS, MERS. It is unfortunate that there exist people who still believe that the virus to be a natural one alike Ebola, Rabies, and HIV, etc. while researches are still in progress to find out the exact origin of this virus.

The chances of getting affected, spreading the virus can be minimized and could be in control if we maintain basic hygiene, wear masks when in close proximities, wash hands frequently for 20 s with preferably alcohol-based sanitizer, maintain at least 6 feet distance while interacting with someone, avoid physical touch with someone, like handshakes, etc. Patients who have minimal symptoms or asymptomatic but detected positive, or patients detected positive, but no irregularities or difficulties can be considered to be treated at their home and with self-care and maintaining isolation.

The prime nature of COVID-19 is still unpredictable and tracking the mutation of the virus is under observation. During a pandemic like COVID-19, the most important aspect is to prioritize public healthcare over anything. Hence, a suitable planning as well as availability of resources for its implementation is essential to prevent and fight a pandemic like COVID-19.

COVID-19 pandemic has already affected global economy and caused significant number of deaths across the globe. Publicly available database from Johns Hopkins University was collected, analyzed and several graphical representations including number of active cases with time, number of active cases with recovered cases etc. were illustrated. Symptoms, ways of transmission of COVID-19 etc. are discussed briefly with pictorial representations and a survey based on 580 people of age group 18–65 was conducted about the information, awareness and concept about COVID-19. The chapter is organized as follows: Sect. 2 briefly describes several related works, characteristics of COVID-19 and also compares several state of art models of convolutional neural network (CNN) for COVID19 diagnosis. Section 3 describes the classification and modes of transmission of COVID-19, while Sect. 4 and 5 briefly illustrates how coronavirus affects human body and the symptoms of COVID-19. Section 6 briefly drafts the severity of COVID-19. Section 7 discusses the COVID-19 statistics and graphical representation of the same. Section 8 summarizes the common misinformation and misconception about COVID-19 among common people. Section 9 and 10 drafts the preventive and curative measures to stop the spreading of COVID-19. Section 11 briefs the current and future challenges of COVID-19. Section 12 concludes with short summary of the topics introduced in the chapter.

2 Related Works

Li et al. [3] proposed a study of the influence of COVID-19 on mental health of people and consequences on social life in China. Total 17,865 active Weibo user's data of emotion indicators including depression, oxford happiness, anxiety, depression, etc. were collected utilizing online Ecological Recognition (OER) focusing on some predictive models using machine learning. The results were compared with before declaration of COVID-19 dated 20th January, 2020. This study result illustrates that positive emotions (Oxford happiness) have been decreased significantly while negative emotions like depression, anxiety etc. have been increased. It is also evident that common people were more concerned about their loved ones, families, and health than short term leisure.

Adhikari et al. [4] introduced a review study related to COVID-19 exploring prevention, causes, clinical manifestation, diagnosis of coronavirus. It is noteworthy to say research on prevention and control measures of coronavirus has increased and several government organizations have initiated to incorporate the research findings in the community to prevent the spread of this disease.

Zhou et al. [5] confirmed the fact that COVID-19 conducts the same cell penetration receptor: angiotensin (a peptide hormone) converting enzyme II (ACE 2) alike SARS-CoV2. However, this study shows that the genome level of COVID-19 is 96% similar to bat coronavirus. This study also reports the characteristics and classification of COVID-19.

Zhou et al. [6] provided an overview of COVID-19 characteristics, pathological, epidemiological advancements. This also shows that one of the major reasons behind the spread of this virus is due to close contact of people and human to human droplet transmission.

Zeng et al. [7] concludes a serious effect of COVID-19 that the patients could suffer such as lung followed by heart. As this is considered as a viral infection, it is evident to harm the immune system drastically. It also states that the lung is the primary target organ of this virus.

Chakraborty et al. [8] highlights the effects of COVID-19 and possible ways to curb the disease. This study focuses on the use of existing drugs with minimal side effects to control the spread of the virus. Negative impacts due to COVID-19 such as economy decrease, global deaths, lack of knowledge about this novel virus etc. are described in this study, and on the contrary, it's also evident that a significant amount of global environmental changes are observed.

Qun et al. [9] collected a number of initial cases of COVID-19 infected pneumonia (NCIP) in Wuhan, China in the period between December 2019 to January 2020. Furthermore, analyzing the information of initial 425 confirmed cases with NCIP, 55% cases before January 1,2020 were directly connected with Huanan Wholesale Market for Seafood. The average period of incubation for the virus was observed to be in 5.2 days. It has also been observed that this epidemic multiplied by two times every 7.4 days. This study also shows that enough evidence has been observed that human-human transmission is one of the primary reason for spreading of this virus. It has also observed that amongst initial 425 cases, almost half of the cases were in the age group of 60 years or above and children are likely to be less prone to be infected and might exhibit mild symptoms.

Guan et al. [10] has analyzed 1099 COVID-19 positive patient datasets from 552 hospitals in mainland China. Surprisingly, the death rates caused by COVID-19 are lower than SARS-CoV as well as MERS-CoV. Notably, out of 1099 confirmed COVID-19 patients, 43.8% patients developed fever firstly and later it increased to 88.7% after hospitalization. It has also observed that some COVID-19 patients did not have fever or any common syndromes initially which has further tangled the diagnosis.

Benvenuto et al. [11] provides a molecular epidemiological and preliminary analysis of SARS-CoV-2. Phylogenetic tree associated with 12 genome sequence of SARS-CoV-2, 15 available whole genome sequence of SARS-CoV-2 and 12 identical genome sequences of SARS-CoV-2. It showed that SARS-CoV-2 certainly batched with Bat SARS-like coronavirus WIV1 while the structural analysis proves mutation of spike Glycoprotein as well as nucleocapsid protein. The results indicates that the novel coronavirus or SARS-CoV-2 is different than SARS virus.

Andrikopoulou et al. [12] characterizes the symptoms and seriousness of COVID-19 disease among pregnant women. Data collected from 13th March to 19th April 2020 of two affiliated hospitals in New York city. Out of 158 COVID-19 infected pregnant women, 22% showed moderate to severe disease while 78% had asymptomatic disease. Lukopenia and asthma seemed to be common symptoms for these 22% women. Also, nine pregnant women received ICU support and two women undergo early delivery. It concludes that one in five pregnant women diagnosed with COVID-19 would develop moderate to severe disease.

Lai et al. [13] concludes that proper vaccine and related therapies are still under evaluation as appropriate knowledge of COVID-19 remains limited and novel to the researchers and scientists. It suggests measures to curb the spread of this disease by implementing the infection control rules.

El-Aziz et al. [14] demonstrates the recent advancement and progress of drug development against coronavirus. Although, researchers across the world, governments are still struggling to discover an efficient drug to prevent this virus. Hydroxychloroquine and chloroquine drugs are generally used for the treatment of arthritis and malaria, however these are prescribed by health council of China for the treatment of coronavirus. Favilavir drug used for common influenza in Japan is now authorized to use in China for the treatment of coronavirus. This study also states that the proper vaccine for COVID-19 is expected to be prepared for clinical trial after the conclusion of December 2020 and suggests to follow the standard guidelines of WHO until the availability of proper vaccine to stop the spreading of COVID-19.

Li et al. [15] compares the CT images of COVID-19 for 51 confirmed patients with CT images of other viral diseases including viral pneumonia. This study also demonstrates the misdiagnosis rate by radiologists for COVID-19 using CT images and shows lower misdiagnosis rate of 3.9%; thus it could be effective for swift treatment of COVID-19.

Pascarella et al. [16] briefs an overview and treatment of COVID-19 and indicates that the infection spreads through human to human transmission and also from contaminated surfaces. Hand hygiene, proper personal sanitization is essential to stop the spreading of COVID-19. Antiviral drugs, respiratory therapy, hydroxychloroquine/chloroquine are being used in treatment until the discovery of dedicated vaccine for COVID-19.

Hani et al. [17] states that computed tomography (CT) could be utilized for COVID-19 diagnosis in this current pandemic situation. Polymerase chain reaction (RT-PCR) tests or sequencing could be beneficial for accurate detection of COVID-19 but these tests could not be useful in a pandemic situation to a large no of people with limited time scope. This study also suggests that while CT might be one of the best tool for early detection of COVID-19 but RT-PCR still needed for ultimate conclusion.

Nour et al. [18] proposed a CNN based model for determining features of chest X-ray images and then fed into algorithms like K-nearest neighbor algorithm, support vector machine (SVM) as well as decision trees. This model provides accuracy of 98.97%, specificity of 99.75%, F-score of 95.75% and a sensitivity of 89.39%. It is also suitable for reducing misdiagnosis rate as well as early detection of COVID-19.

Several proposed computational models of convolutional neural network (CNN) architecture, focused to detect COVID19, are compared and mentioned in Table 1. The performances of each model in terms of accuracy, sensitivity, accuracy are quite

Table 1 Comparison of different CNN models

Methods	Reference	Sensitivity (%)	Accuracy (%)	Specificity (%)
nCOVnet, transfer learning, deep CNN [19]	Panwar, Harsh, et al. [19]	97.62	88.10	89.13
COVID-Net [20]	Wang et al. [20]	91.0	93.3	–
COVIDiagnosis-Net [21]	Ucar et al. [21]	99.13	98.26	–
Deep CNN, Inception, transfer learning [22]	Wang, et al. [20]	87	89.50	88
The pretrained CNNs [23]	Apostolopoulos et al. [23]	92.85	93.48	98.75
A novel CNN model, training from scratch strategy, deep feature extraction, SVM [18]	Nour et al. [18]	89.39	98.97	99.75

convincing, although the database used in each models are restricted due to limited availability of publicly available database. The performance analysis parameters (accuracy, sensitivity, specificity) are defined in terms of true positive (TP), True negative (TN), False positive (FP) and False negative (FN) and represented by the following formulae,

$$Sensitivity = TP/(TP + FN) \tag{1}$$

$$Accuracy = (TP + TN)/(TP + FP + TN + FN) \tag{2}$$

$$Specificity = TN/(TN + FP) \tag{3}$$

It is also to be noted, X-ray images are more effective and significant than CT scan images for COVID19 diagnosis.

3 Classification and Modes of Transmission of COVID-19

For any kind of contagious disease, it is very important to know the classification and modes of transmission in order to prevent or cure it before it costs millions of lives. Classification gives a good idea about the nature and type of disease, easing the analysis of its strengths and weaknesses. The modes of transmission need to be studied extensively in order to keep the spread of a disease under control. The classification and modes of transmission for COVID-19 is given in the following sub-sections.

3.1 Classification of Coronavirus and COVID-19

When a new virus disease outbreak occurs, three names namely the species, the disease and the virus, are to be decided. The nomenclature of virus (i.e. family, genus, species etc.) is done by ICTV. The name of the disease is decided by the World Health Organization (WHO). Expert virologists are given the responsibility of naming the virus [24].

For quite a while, virus taxonomists have classified the viruses by focusing on the grouping of comparatively closely associated viruses, often in the framework of a model that stresses the fast virus evolution that causes human diseases, animals, and crops [25]. However, lately, the shotgun metagenomic sequencing has exposed the existence of a massive diversity of viruses found in samples from the environment and organisms seemingly healthy from all partitions of life [26]. As a consequence of the latest advances and progress made, ICTV has decided that virus taxonomy can be conveniently extended for inclusion of the base evolutionary relations among remotely related viruses. Subsequently, the ICTV has improved its code, allowing a hierarchy of classification with fifteen ranks that closely line up with the Linnaean taxonomic structure, thereby making it consistent with the scientific naming convention of other organisms. The novel system can embrace the entire range of genomic divergence in this virosphere [27]. The 15-rank structure is compared with the previously used 5-ranked structure, as seen in Fig. 1.

It is also predicted that the new rank hierarchy will encourage additional research on virus ancestries and evolution and aid to line up taxonomy of viruses with the classifications of cellular creatures. The ICTV Coronaviridae Study Group has studied the new virus taxonomy [24, 28]. The classification and naming of coronavirus is shown in Fig. 2.

Viruses, and then the illnesses caused by them, every so frequently have different names. For instance, HIV is the virus causing AIDS. People, most of the time, know the name of an illness, but, on the other hand, they do not know the name of the virus responsible. There are various reasons and processes in providing relevant names of viruses and the diseases they spread [7, 30]. Diseases are named with discussion on prevention of disease, transmission, spread, treatment and severity in mind. Human disease alertness and reaction is WHO's part, so illnesses are officially termed by WHO in the International Classification of Diseases (ICD). The origin for naming of the coronavirus, its species and the disease it causes, is illustrated in Fig. 3.

Viruses are termed depending on genetic arrangement to simplify the development of analytical tests, effective vaccines and medications. Virologists as well as the wider scientific groups do this task, so viruses are termed by the ICTV. Given that the Coronaviridae Study Group from ICTV is specialists in this group of viruses they also added their knowledge to the designation of the virus. The virus name is "severe

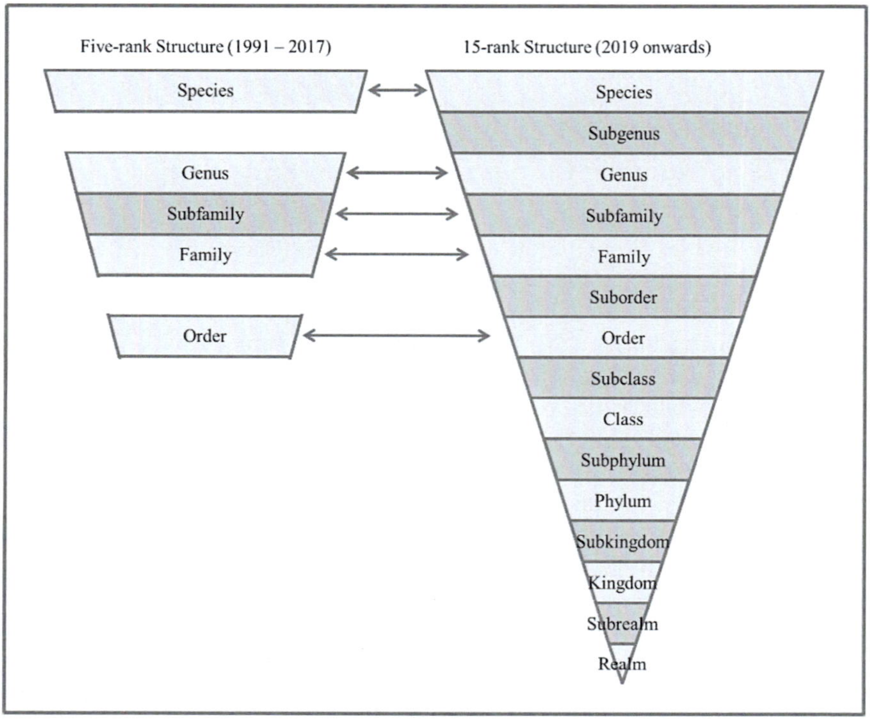

Fig. 1 Five-rank structure (1991–2017) vs. 15-rank structure (2019 onwards) [24]

acute respiratory syndrome coronavirus 2" (SARS-CoV-2). ICTV announced this name on 11th February, 2020. This term was selected since the virus is genetically linked to coronavirus accountable for the SARS outbreak which happened in 2003. Though related, these two kinds of viruses are not the same. The species where SARS-CoV-2 fits is called Severe acute respiratory syndrome-related coronavirus. The SARS-CoV-2 is called coronavirus in a day to day sense, and in the context of this chapter, unless explicitly explained, the word coronavirus is synonymous with SARS-CoV-2 [24].

3.2 Modes of Transmission of COVID-19

SARS-CoV-2 extents from one person towards another person with close vicinities. When persons with COVID-19 cough or exhale, they eject little droplets containing virus. These little droplets enter through the nose or mouth of someone devoid of virus, which results in occurrence of the infection [31]. The most common source through which this disease spreads is nearby contact with somebody having the infection. In this context, close contact is defined as the area within approximately

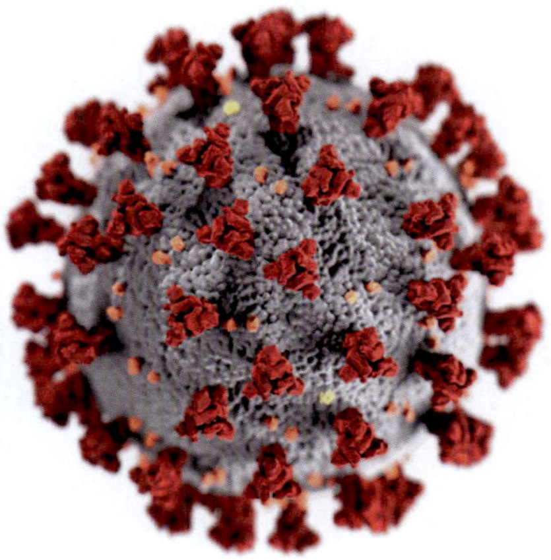

Realm:	Riboviria
Kingdom:	Orthornavirae
Phylum:	Pisuviricota
Class:	Pisoniviricetes
Order:	Nidovirales
Family:	Coronaviridae
Genus:	Betacoronavirus
Subgenus:	Sarbecovirus
Species:	Severe acute respiratory syndrome–related coronavirus

Fig. 2 Illustration and classification of SARS-CoV-2 [29]

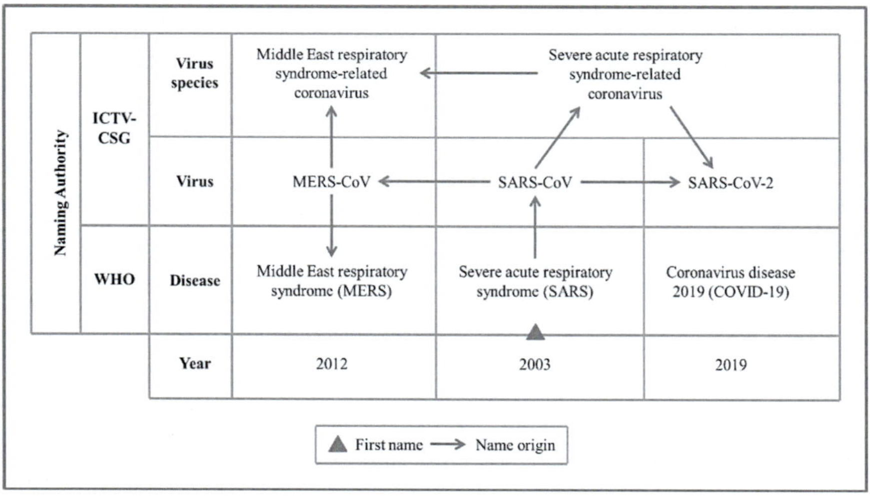

Fig. 3 Naming origin of COVID-19 [24]

six feet. The infection is most communicable when an person's symptoms are at highest. On the other hand, it is also probable for someone without indications to transmit the virus. This infection can also be spread from people having the virus but exhibiting little to no indications [32].

Fig. 4 **a** Mass gathering may be a source of infection **b** Touching nose, mouth or face may result in infection, if the virus is there in forms of Droplets on the hand surface

Droplets comprising the virus likewise land upon close articles or surfaces. People also acquire the virus through touching the infected surfaces or entities. Infection is very much likely when the individual then touches their mouth, nose, or rub their eyes, as shown in Fig. 4 [33].

It is essential to keep in mind that COVID-19 is relatively new, and the investigation is still continuing. There might be other means by which the new coronavirus can be transmitted [34].

4 How COVID-19 Affects the Body

The virus first clings onto droplets when infected individuals cough, or by touching someone who is ill, and then reach the face, say by rubbing our nose or eyes. It then travels deeper into the body into the spleen, intestine, or lungs, where it has the most intense consequence. Even only small entities of coronaviruses may ultimately cause rather a dramatic situation. The lungs has epithelial cells in billions, which line our organs and mucosa, and are infected first [35]. Coronavirus binds to a definite receptor on its target's membranes to introduce its genetic material. As the cells cannot distinguish the genetic material of the virus from the target's original genetic material, starts making virus proteins the same way it would have made the targets needed proteins. The cells bulge up with more and more copies of the virus until it can hold itself no longer, and sort of liquefies away, freeing new virus particles, ready to infect more cells [36]. As a result, there is an exponential growth in the infected cells, and consequently the virus particles. After about 10 days from the start of the infection, millions of cells of the body are infected, and billions of viruses occupy the lungs [37]. As cells of the immune system come into the lungs to fight the virus as an immune response, coronavirus contaminates several of them and generates confusion. While protecting us from harmful pathogens is the function of immune system, it can actually be pretty hazardous to us if not regulated properly. The cells communicate mostly through tiny information proteins known as cytokines. Coronavirus makes immune cells that are infected to overreact and so the immune system goes into an agitated state. Immune system then sends way more immune cells than it should, causing damage and wasting resources.

Two categories of immune cells in particular cause the most damage. First are the neutrophils, which are like killing machines, and can kill cells, our own cells included. They reach in thousands, ejecting enzymes that abolish as many healthy cells as virus cells. The other type of significant cells that causes much damage are T-cells, typically ordering infected cells to end themselves. When they get confused, they start commanding healthy cells telling to kill themselves too. More immune cells coming to the site results in more harm, and as a result, kill healthy lung tissue. This may eventually lead to irreversible injury and disabilities for life. In most of the situations, immune system recovers control slowly, killing infected cells, interrupting the viruses demanding to infect new cells, and cleans the infected sites. Consequently, the recovery period starts. Most of the population infected by coronavirus will get through it with fairly minor indications. But many circumstances become severe or critical. In these cases, millions of epithelial cells are destroyed, and as a result, the protective lining of the lungs is gone. Therefore, the little air sacs with which breathing occurs, called alveoli, can be infested by bacteria that otherwise would be harmless. This causes pneumonia in patients, resulting in difficulty in breathing. In some cases, patients need ventilators to survive [38]. At this point, the immune system is already fatigued because of such large-scale production of immune cells and thus is extremely weak. If thousands of bacteria multiply at a high rate, they affect the overall immune system [39]. As a result, they enter the blood and swarm through the body, in which case, fatality is most likely [40].

5 Symptoms of COVID-19

COVID-19 can affect people in many different ways. It takes five to six days since someone is diseased with the virus for signs to show on average, though it may take up to fourteen days. People having minor symptoms and otherwise fit should manage the symptoms at their residence [41]. Most infected individuals develop minor to moderate infection and recover with no hospitalization. The most common symptoms, as shown in Fig. 5, are as follows:

- tiredness
- dry cough

- fever

Symptoms which are relatively less common, but still can indicate COVID-19 are as follows [42], as illustrated in Fig. 6:

- headache
- sore throat
- diarrhea
- aches and pains
- conjunctivitis
- losing taste or smell
- rashes on skin, discoloration/pale color of toes or fingers

The disease, if left unattended, can develop into a serious or even fatal illness. In the worst cases, death is almost certain. The serious symptoms are as follows, as shown in Fig. 7.

- chest pressure or pain
- speech loss or movement loss
- difficulty in breathing or shortness of breath

One should consult immediate medical treatment if that person has serious symptoms. It is important to call before visiting the doctor or health facility [42].

Fig. 5 Most common symptoms of COVID-19: tired, fever, dry cough

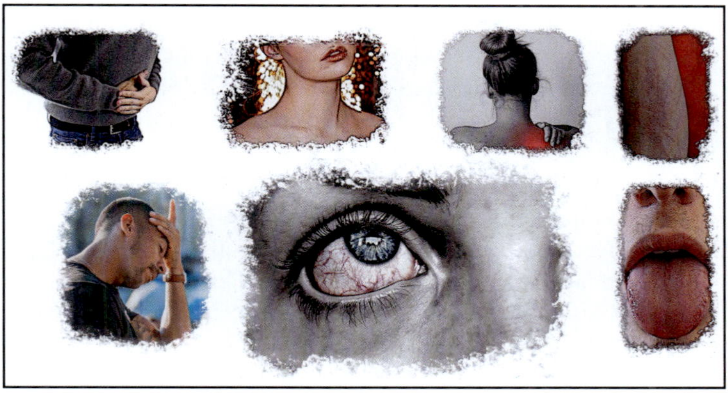

Fig. 6 Less common symptoms of COVID-19

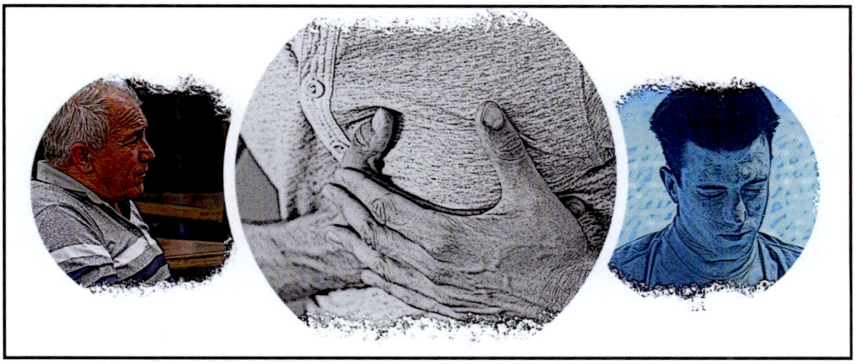

Fig. 7 Serious symptoms of COVID-19: loss of speech or movement, chest pain and difficulty breathing

6 Severity of COVID-19

While COVID-19 is increasing very quickly, most people will experience only minor or modest indications. That said, coronavirus can result in severe disease in some individuals. Mostly, individuals who have contracted the new coronavirus have recovered or are improving. However, even a disease which has a good percentage of individuals are improving from can cause a noteworthy amount of deaths if it infects sufficient individuals. There have been comparatively few COVID-19 contaminations among children. Older age groups, particularly those having underlying health conditions, are at risk. There is a thing which particularly worried about the impact of COVID-19 on the most vulnerable of the people, namely, health workers, people over 60, and those with underlying health conditions. These communities need protection [43]. COVID-19 can affect anybody, and the illness can result in indications extending from slight to severe. For other diseases triggered by influenza-like respiratory viruses, some individuals might be more probable for having the severe infection than others as they have features or medical complexities that escalate the risk, usually called 'risk factors'. Some examples of risk factors include old age and having some underlying medical problems [44].

Researchers are directing disease investigations and field surveys to better comprehend why some individuals are more susceptible to mature severe COVID-19 disease, one of the topmost priorities in combatting COVID-19. Acquirements from these struggles will deliver important statistics to assist scientists as well as officials related to public health to take better judgments in order to defend our topmost susceptible population [45]. Researchers work together with state, local and territorial health departments; vital statistics offices; public health, commercial, clinical laboratories; emergency departments; health-care providers; and academic as well as private sector associates for the analysis and observations related to this virus [46].

Individuals having risk factors are more probable to be in need of hospitalization and intensive care in case they have COVID-19. We should take protections as we go about our daily life and be present at different events and proceedings. It is very important to understand the way medical problems could disturb our own health if we get infected with COVID-19 [47]. As COVID-19 is comparatively new, more of work is required for the improved understanding of the risk factors with respect to severe infection or complications. Potential risk factors include [48, 49]:

- Age
- Race/Ethnicity
- Some medical problems
- Gender
- Pregnancy
- Usage of certain medications
- Certain occupations
- Crowing and Poverty

Additional research is definitely required and may help us in the future to identify if these are truly the risk factors for COVID-19 infection and decide if there exist other factors increasing a person's risk [50, 51].

7 COVID-19 Statistics

COVID-19 spreads more rapidly than diseases like SARS, MERS. Figure 8 shows the graphical representation of COVID-19 Active Cases with respect to time. All our observations are analyzed in the top 8 most affected countries including the United States, Brazil, India, Mexico, Peru, Russia, South Africa, United Kingdom. All the data were collected from open-source data repository of **John Hopkins University Center for Systems Science and Engineering** *(JHU CSSE)*, as of 25th August 2020 [52, 53].

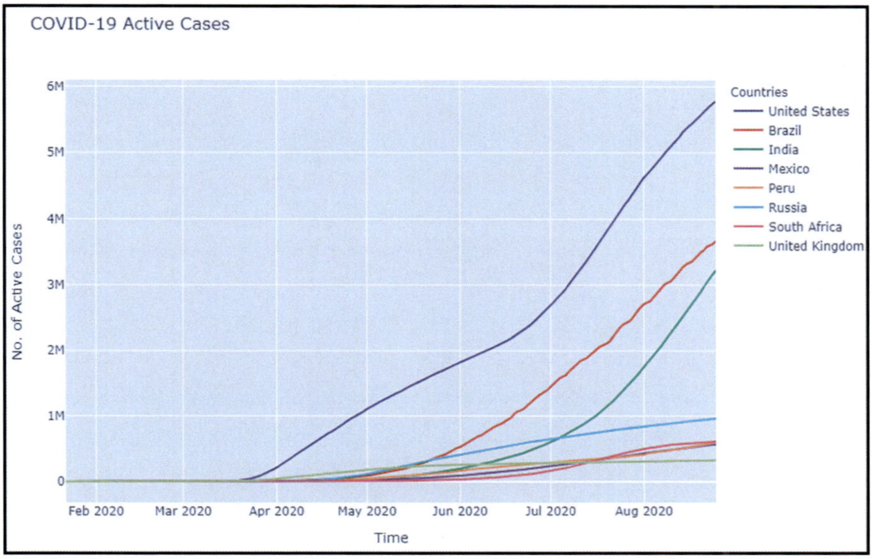

Fig. 8 Distribution of COVID-19 active cases

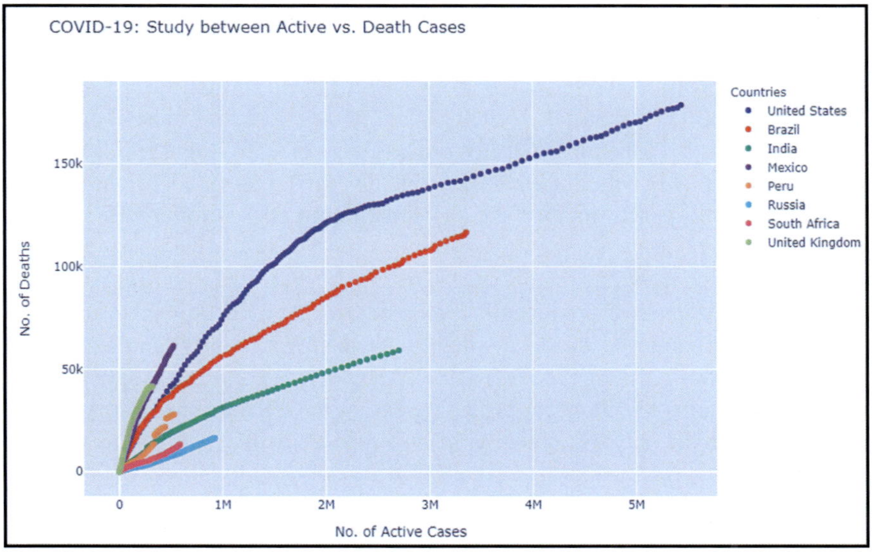

Fig. 9 Distribution of Worldwide COVID-19 death rates *with respect to* no. of active cases

Figure 9 shows the variation of the count of Deaths due to COVID-19 with No. of Active cases of above mentioned 8 countries. It is to be noted that the United states having the highest mortality rate of 1,78,486 among all countries and 54,38,325 active cases as on 25th August, 2020. A steeper graph indicates more number of

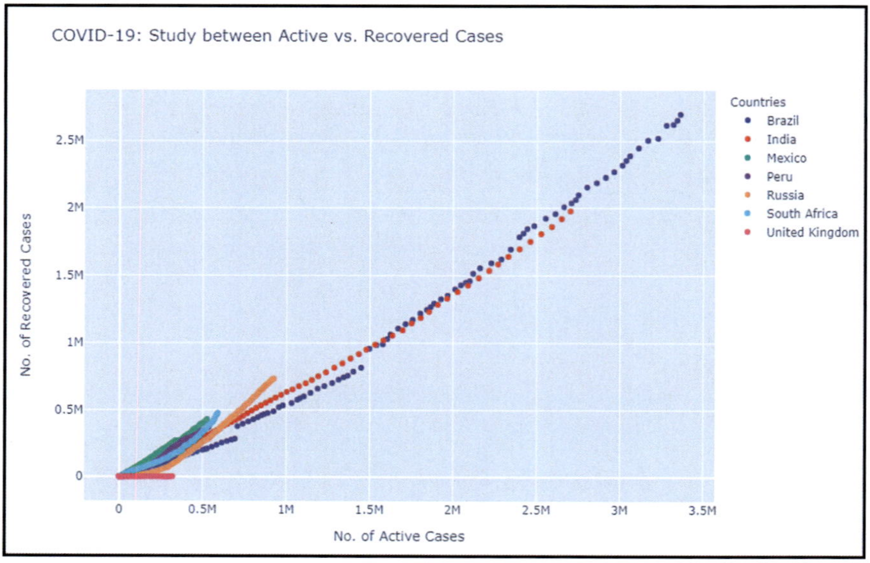

Fig. 10 Distribution of COVID-19 recoveries *versus* active cases

deaths per active case, and thus reflects more loss to life. Graphs that are less steep correspond to countries doing better in this pandemic.

Figure 10 illustrates the correlation between recoveries and active case of COVID-19. Among all the 8 countries, United Kingdom was observed to have only 319 k active cases with 1490 million recovered cases as of 25th August, 2020. Steeper graphs correspond to better performance with respect to recovery. From the plot, we see that all the countries topping the charts are doing fairly similar with respect to recovery.

Data collected from 22nd January, 2020 to 25th August, 2020 has been analyzed in Fig. 11. Death rates have significantly increased from April 2020 in all the countries. Amongst other countries, the death rate of United Kingdom has been flattened.

Figure 12 shows the recoveries from 22nd January, 2020 to 25th August, 2020 of the top 8 most affected countries excluding US, as the US data of recovery rate was not available in the public domain by the data repository of JHU CSSE.

8 Common Misconceptions and Misinformation

Based on the 580 responses collected from age groups 18-65 focused on Indian people, 65.8% people believe that the coronavirus was made in the laboratory (Table 2) and 21.5% of people still believes that the virus to be natural one alike Ebola, Rabis, and HIV etc. While researches are still in progress to find out the exact origin of this virus. 58% of people believe that COVID-19 is a man-made virus that

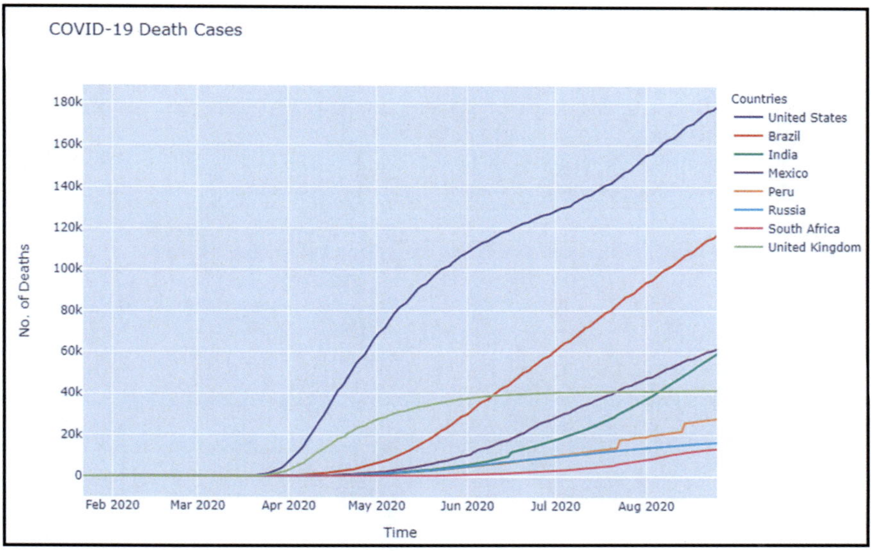

Fig. 11 Distribution of COVID-19 death rate over time

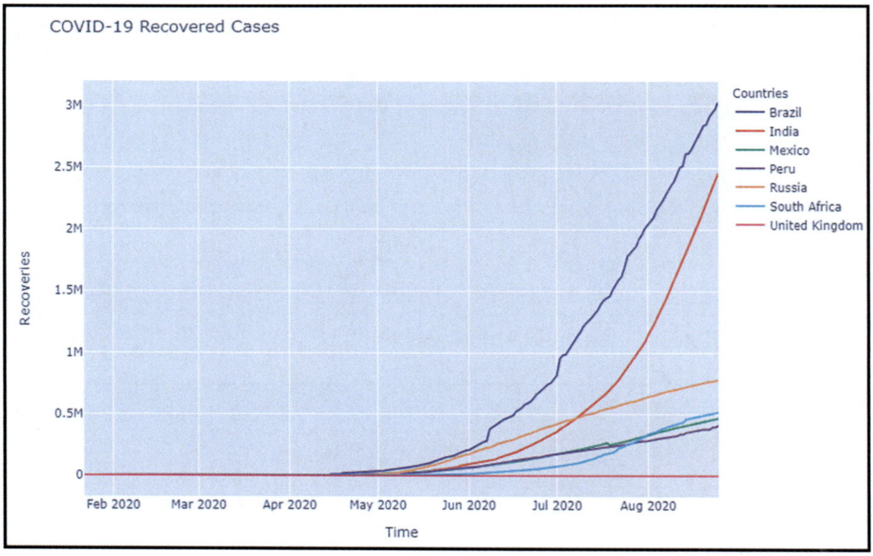

Fig. 12 COVID-19 recovery rate *versus* time

Table 2 COVID-19 related misconceptions

Questions	Yes (%)	No (%)	Maybe (%)
Do you believe, COVID-19 as created in some laboratory (i.e. it is man-made/synthetic) or it is natural virus like Ebola, Rabies, HIV etc.?	65.8	21.5	12.7

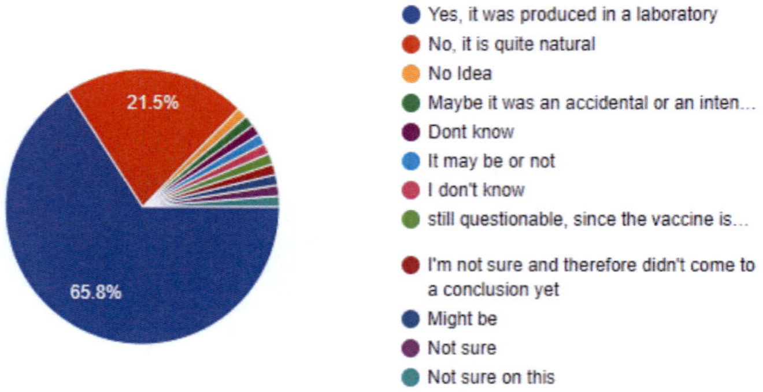

Questions	Yes (%)	No (%)	Maybe (%)
Do you think, COVID-19 was created intentionally to disrupt Human Lives and Economy around the world?	58.2	30.4	11.4

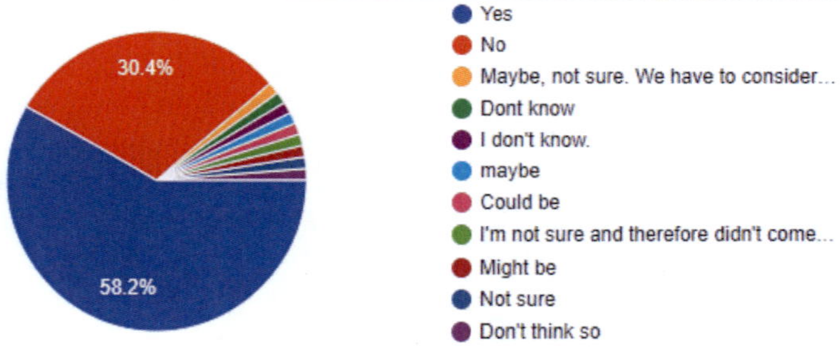

Questions	Yes (%)	No (%)	Maybe (%)
Do you really think, the world will be entirely COVID-free if we get an appropriate vaccine?	31.6	20.3	44.3

Table 3 COVID-19 related information sources

Question	TV broadcasting channel *(National, International)* (%)	Internet surfing *(Google Search)* (%)	Social media *(Excluding WhatsApp)* (%)	Others
What are the major source(s) you have gathered information about COVID-19 from?	35.4	24.1	20.3	20.2

intends to disrupt the global economy and human lives across the world, while 11% of people have no idea about the origin of the virus (Table 2). Surprisingly 31% of people think that the world will be entirely corona free after the invention of the proper vaccines, but 44% of people are still in dilemma about the same.

On the contrary, about 58% of people opted for 'Yes' when asked, "Do you think Youth and Healthy people are not taking 'Social Distancing' seriously as a necessary precaution against "COVID-19?" (See Table 2). While the terminology 'Social Distancing', 'Contact Tracing' are well versed among every citizen, still more than 35% of people said 'No' that they have not heard the term. It is evident that 87% of people said that they will try to follow all the rules strictly for Quarantine (whether it is for 14 or 28 Days) if detected positive for COVID-19 (Table 4).

In the era of fake news and in a pandemic situation more than 20% of people said they gain information about COVID-19 from Facebook, Twitter, etc. About 25% people said that they rely upon normal google search (without verifying the authenticity of the source). According to our survey top 7 sources of information related to COVID-19 are: *TV broadcasting channel (National, International) > Internet surfing (Google Search) > Social media (Excluding WhatsApp) > WhatsApp > YouTube > Local Health Workers > Books*, which is depicted in the Pie Chart and (Table 3) below:

The situation in a major pandemic like COVID-19 worsens if the information circulating among the general public is incorrect. In addition, people often deduce self-made facts about the pandemic and impose those facts on others, creating unnecessary confusion. As a result, misinformed decisions are made, resulting in more harm than good. Arguments and debates erupt, often without proper bases for the words spoken, and as such, the correct information gets buried in the talks. This might also lead to reduced sense of danger and irresponsible attitude towards adhering to necessary protocols. It is the duty of government organizations to implement campaigns so as to increase awareness among the public. In addition, the people should adopt the protocols and inform others about the possible dangers and the ways to prevent and cure the disease. The population plays a major role in the path taken by the pandemic. With responsible steps taken in our day-to-day life, it is possible to break the chain and prevent the virus from taking any more lives.

Table 4 COVID-19 related information and conception among common people

<table>
<tr><th>Questions</th><th>Yes (%)</th><th>No (%)</th><th>Maybe (%)</th></tr>
<tr><td>Do you think, Youth and Healthy people are not taking ‘Social Distancing’ seriously as a necessary precaution against COVID-19?</td><td>58.2</td><td>11.4</td><td>27.8</td></tr>
<tr><td colspan="4">

</td></tr>
<tr><td>Have you encountered the terminology 'Contact Tracing' in context of COVID-19?</td><td>50.6</td><td>36.7</td><td>12.7</td></tr>
<tr><td colspan="4">
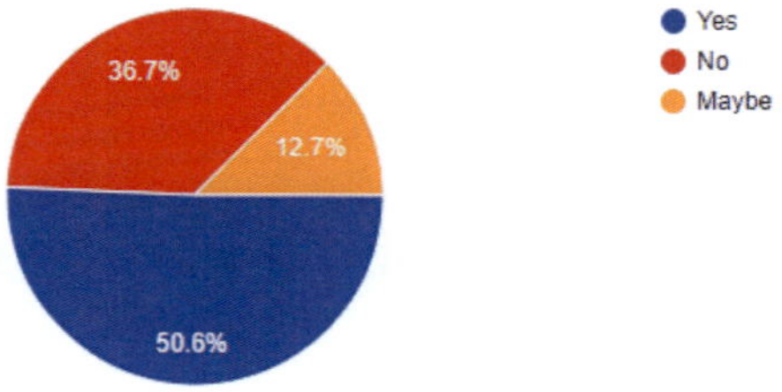

</td></tr>
<tr><td>What extent can you maintain Quarantine if you get COVID-19 Positive, somehow?</td><td>87.3</td><td>7.6</td><td>5.1</td></tr>
<tr><td colspan="4">

</td></tr>
</table>

9 Preventive Measures

COVID-19 is a deadly virus, highly contagious in nature. According to WHO, 810,492 confirmed deaths have been reported (as on 25th August, 2020) and already overturn the human lives of 216 countries/areas/territories. Figure 13 demonstrates

Fig. 13 Necessary preventive measures for COVID-19

the preventive measures for COVID-19. Meanwhile chances of getting affected, spreading the virus can be minimized and could be in control by the following preventive measures and precautions:

- People should wash their hands thoroughly, especially after traveling from public transport, visiting congested areas (banks, hospitals etc.) for at least 20 s with proper hand wash or an alcohol-based sanitizer.
- One should not touch their eyes, mouth, nose unnecessarily as these are primary entering node of this virus.
- People should make sure when they cough or sneeze, they would bend their elbow or use a tissue for the same.
- It is recommended and advised by almost every health experts to maintain proper hygiene in their surroundings.
- One should be very cautious when visiting a crowded place while keeping a distance of minimum of 3 feet. COVID-19 can also be spread from droplets
- If having symptoms like fever, difficulty in breathing, dry cough, sore throat, chest pain or pressure; one should immediately contact their local health personnel and isolate themselves. Telemedicine at initial phase is an enabling tool to provide remote virtual care to patients.

10 Curative Measures

While death rates and confirmed cases of COVID-19 are daily increasing rapidly, it is to be noted that not all patients need special hospital treatment. Patients with the minimal syndrome but detected positive, patients detected positive, but no irregularities or difficulties can be considered to be treated at their respective home and with utmost self-care. For the people having maximum symptoms of COVID-19 and aged above 50 years they need to immediately contact the nearby hospitals and admit themselves. To curb the spread of this deadly virus, following curative measures are described below:

- According to WHO, around 80% of people having COVID-19 syndromes do not require hospital treatment and can take help from local medical authorities and isolate themselves and also avoid public gathering for at least 15 days.
- People having detected positive for COVID19 should drink plenty of water to keep them dehydrated.
- Drinking hot liquids like tea, mild-warm water to get rid of sore throat. However, researchers have not confirmed yet if warm water could completely kill the virus and prevent entering the lungs.
- During the quarantine period, one should eat proper food with a balanced diet and maintain a good amount of sleep every day. Ensuring that would help to boost the immune system.

The maturation period of coronavirus is estimated to be five to six days [54], but it could also be 14 days according to the latest research findings [55, 56]. So proper medical attention and special care should need to be maintained.

11 Current and Future Challenges

COVID-19 is very contagious, and as such, has already proven to be a deadly pandemic if not dealt with proper and systematic protocols. In spite of the tremendous amount of research going on in various research facilities across the globe, a vaccine for COVID-19 is yet to be accessible in the market. Huge losses to life and economy have been observed, and are rising still. It is, thus of utmost importance that these challenges are addressed and proper implementation is done. The following sub-sections elaborate on the most important challenges that are currently faced, and others that are expected to surface as the pandemic progresses.

11.1 Current Challenges

Coronavirus (COVID-19) has now become a global threat and researchers and scientists of all over the globe are trying for a proper vaccine and medicine to curb this deadly virus. The prime nature of COVID-19 is still unpredictable and tracking the mutation of the virus is under observation. The common challenges to prevent the virus from spreading are as follows:

- This virus has a tremendous nature of human to human spread or community spread. Thus, it becomes a challenging task for scientists to find out a way to stop the transmission or reduce its effect on the human body.
- Over more than 100 studies have been going on to find a vaccine for COVID-19. But according to Wellcome trust [57], it could around 10 years or more time to develop a proper vaccine that has minimal side effects on human bodies.
- The global market has been remarkably affected and the companies all over the world are generating very little revenue; so greater consequences are on job markets and the economy.
- Healthcare challenges such as patients with other serious diseases are bound to have a routine check-up that has been severely affected and retarded medicine supply [58, 59].

11.2 Future Challenges

During a pandemic, the most important aspect is to prioritize public healthcare over any problems. Hence, a suitable contingency plan as well as a stable economy is essential to develop to prevent and fight a pandemic like COVID-19. The impacts caused by such a pandemic should be treated as a lesson to build a better future society. It is very important how the emergency measures are taken into consideration to deal with the surge [60]. The current problem initiates from an obvious failure to anticipate the occurrence of a deadly pandemic. Several aspects that can be considered as future challenges could be:

- **Implementation of Proper Telemedicine Facility:** France is a great example of implementing numerous telemedicine and telemonitoring platforms rapidly amidst COVID-19 growing at a rapid pace. However, there have been observed deficiency of management between operatives which resulted in issues like a lack of responsibility to diagnose a particular disease. Hence more transparency in communication as well as more comparability have to be incorporated for quick and better solutions. Digital initiatives to fight against COVID-19 should be feasible to both public healthcare systems and hospitals, adopted to population, and capable of understanding the newly evolved characteristics and growth of the SARS-CoV2 Virus [61].
- **Usage of IoT Based of Smart Treatment:** With advanced technologies, ongoing research and development in the field of AI, it is possible to bring IoT in medical fields, diagnosing an ailment and providing complete treatment, remotely. But to implement such technology and automated facility, we have to think of its consequences and side-effects as it is associated with healthcare, which is very crucial to us.
- **Economic Slowdown and Recession in Global Market:** With COVID-19, many people have lost their job because of complement lockdown in technical and industrial field, which results in a severe Economic Slowdown. Our prime focus

should be enhancing global economy, finding suitable alternative, starting from individual level to government in post COVID-19 era.

12 Conclusion

COVID-19 is one of the major pandemics in recent years, and has resulted in huge loss to lives all across the globe. Researchers are collaborating at massive scales in order to tackle this on-going pandemic as soon as possible. As of August 2020, a vaccine is yet to be tested with success and brought to the market, even after more than 8 months since the first cases were documented. The preventive measures are being implemented in all parts of the planet, and they are updated as and when new research provides better information about the pandemic and its prevention. The governments have implemented and campaigned about the protocols and steps to be taken with mixed success. In addition to the government, it is also the responsibility of the citizens to make sure that proper protocols are followed and personal hygiene is maintained. The misinformation needs to be tackled and addressed to as well. People should make sure as to adopt only the correct facts and protocols from the social networking platforms in order to minimize the amount of confusion going on. The basic protocols issued by WHO are easy to implement, and as such should be taken seriously and followed rigorously. In case of being affected with COVID-19, the first step is not to panic. For mild symptoms, care should be taken and it is recommended to be treated at home. In case of severe symptoms, proper hospital care should be taken. It is to be noted that COVID-19 is a very contagious disease, however the mortality rate is quite low and as such being affected should not be confused with fatality. The world has the resources and capabilities to fight a situation like this, as we should stay united in our fight against this pandemic.

While it has been observed from Figs. 11 and 12 that, mortality rates have significantly increased from April 2020 to August 2020 in the top 8 COVID19 affected countries, except the UK, and the recovered case in Brazil is the highest amount other countries in the said interval of time. The data collected from the publicly available JHU CSSE database and analyzed in this chapter are up to August of 2020 as this is the latest data available at the time of preparing the manuscript. In the upcoming months, the availability of more data enables future analysis of the progression of COVID-19. Pandemic like these often has multiple recurring peaks, with the first and the second peak being the most dangerous and deadly. As such, the second peak could be properly analyzed and the models used for prediction of the path taken by COVID-19 could be tweaked accordingly. The progression of the second peak and the response from different countries should also be analyzed to observe the best and worst effective practices and regulations and a prediction model could be developed from the inferences.

References

1. V.V. Hamparian, M.R. Hilleman, A. Ketler, Contributions to characterization and classification of animal viruses. Proc. Soc. Exper. Biol. Med. **112**(4), 1040–1050 (1963)
2. Naming the coronavirus disease (COVID-19) and the virus that causes it, https://www.who.int/emergencies/diseases/novel-coronavirus-2019/technical-guidance/naming-the-coronavirus-disease-(covid-2019)-and-the-virus-that-causes-it/. Accessed 24 Aug 2020
3. S. Li, Y. Wang, J. Xue, N. Zhao, T. Zhu, The impact of COVID-19 epidemic declaration on psychological consequences: a study on active Weibo users. Int. J. Environ. Res. Public Health **17**(6), 2032 (2020)
4. S.P. Adhikari, S. Meng, Y.J. Wu, Y.P. Mao, R.X. Ye, Q.Z. Wang, C. Sun, S. Sylvia, S. Rozelle, H. Raat, H. Zhou, Epidemiology, causes, clinical manifestation and diagnosis, prevention and control of coronavirus disease (COVID-19) during the early outbreak period: a scoping review. Infect. Dis. Poverty **9**(1), 1–12 (2020)
5. P. Zhou, X.L. Yang, X.G. Wang, B. Hu, L. Zhang, W. Zhang, H.R. Si, Y. Zhu, B. Li, C.L. Huang, H.D. Chen, A pneumonia outbreak associated with a new coronavirus of probable bat origin. Nature, **579**(7798) 270–273 (2020)
6. M. Zhou, X. Zhang, J. Qu, Coronavirus disease 2019 (COVID-19): a clinical update. Front. Med. 1–10 (2020)
7. J.H. Zeng, Y.X. Liu, J. Yuan, F.X. Wang, W.B. Wu, J.X. Li, L.F. Wang, H. Gao, Y. Wang, C.F. Dong, Y.J Li, First case of COVID-19 complicated with fulminant myocarditis: a case report and insights. Infection, 1 (2020)
8. I. Chakraborty, P. Maity, COVID-19 outbreak: migration, effects on society, global environment and prevention. Sci. Total Environ. 138882 (2020)
9. Q. Li, X. Guan, P. Wu, X. Wang, L. Zhou, Y. Tong, R. Ren, K. Leung, E. Lau, J.Y. Wong, X. Xing, N. Xiang, Y. Wu, C. Li, Q. Chen, D. Li, T. Liu, J. Zhao, M. Li, Z. Feng, Early Transmission dynamics in Wuhan, China, of novel coronavirus-infected Pneumonia. New Engl. J. Med. **382**(13), 1199–1207 (2020). https://doi.org/10.1056/NEJMoa2001316
10. W.J. Guan, Z.Y. Ni, Y. Hu, W.H. Liang, C.Q. Ou, J.X. He, L. Liu, H. Shan, C.L. Lei, D.S. Hui, B. Du, Clinical characteristics of coronavirus disease 2019 in China. New Engl. J. Med. **382**(18), 1708–1720 (2020)
11. D. Benvenuto, M. Giovanetti, A. Ciccozzi, S. Spoto, S. Angeletti, M. Ciccozzi, The 2019-new coronavirus epidemic: evidence for virus evolution. J. Med. Virol. **92**(4), 455–459 (2020)
12. M. Andrikopoulou, N. Madden, T. Wen, J.J. Aubey, A. Aziz, C.D. Baptiste, N. Breslin, M.E. D'Alton, K.M. Fuchs, D. Goffman, C. Gyamfi-Bannerman, Symptoms and critical illness among obstetric patients with coronavirus disease 2019 (COVID-19) infection. Obstet. Gynecol. **136**(2), 291–299 (2020)
13. C.C. Lai, T.P. Shih, W.C. Ko, H.J. Tang, P.R. Hsueh, Severe acute respiratory syndrome coronavirus 2 (SARS-CoV-2) and corona virus disease-2019 (COVID-19): the epidemic and the challenges. Int. J. Antimicrob. Agents, 105924 (2020)
14. T.M. Abd El-Aziz, J.D. Stockand, Recent progress and challenges in drug development against COVID-19 coronavirus (SARS-CoV-2)-an update on the status. Infect. Gen. Evol. 104327 (2020)
15. Y. Li, L. Xia, Coronavirus disease 2019 (COVID-19): role of chest CT in diagnosis and management. Am. J. Roentgenol. **214**(6), 1280–1286 (2020)
16. G. Pascarella, A. Strumia, C. Piliego, F. Bruno, R. Del Buono, F. Costa, S. Scarlata, F.E. Agrò, COVID-19 diagnosis and management: a comprehensive review. J. Intern. Med. **288**(2), 192–206 (2020). https://doi.org/10.1111/joim.13091
17. C. Hani, N.H. Trieu, I. Saab, S. Dangeard, S. Bennani, G. Chassagnon, M.P. Revel, COVID-19 pneumonia: a review of typical CT findings and differential diagnosis. Diagn. Interv. Imaging **101**(5), 263–268 (2020)
18. M. Nour, Z. Cömert, K. Polat, A novel medical diagnosis model for COVID-19 infection detection based on deep features and Bayesian optimization. Appl. Soft Comput. 106580 (2020)

19. H. Panwar, R.K. Gupta, M.K. Siddiqui, R. Morales-Menendez, V. Singh, Application of deep learning for fast detection of COVID-19 in X-Rays using nCOVnet. Chaos, Solitons Fractals, 109944 (2020)
20. L. Wang, Z.Q. Lin, A. Wong, Covid-net: a tailored deep convolutional neural network design for detection of covid-19 cases from chest x-ray images. Sci. Rep. **10**(1), 1–12 (2020)
21. F. Ucar, D. Korkmaz, COVIDiagnosis-Net: deep bayes-squeezenet based diagnostic of the coronavirus disease 2019 (COVID-19) from X-Ray images. Med. Hypotheses, 109761 (2020)
22. S. Wang, B. Kang, J. Ma, X. Zeng, M. Xiao, J. Guo, M. Cai, J. Yang, Y. Li, X. Meng, B. Xu, A deep learning algorithm using CT images to screen for Corona Virus Disease (COVID-19). medRxiv. Published online 24 April, 2020–2002 (2020)
23. I.D. Apostolopoulos, T.A. Mpesiana, Covid-19: automatic detection from x-ray images utilizing transfer learning with convolutional neural networks. Phys. Eng. Sci. Med. 1 (2020)
24. International Committee on Taxonomy of Viruses (ICTV), https://talk.ictvonline.org/. Accessed 24 Aug 2020
25. International Committee on Taxonomy of Viruses Executive Committee, The new scope of virus taxonomy: partitioning the virosphere into 15 hierarchical ranks. Nat. Microbiol. **5**(5), 668 (2020)
26. T. Ando, J.S. Noel, R.L. Fankhauser, Genetic classification of norwalk-like viruses. J. Infect. Dis. **181**(Supplement_2), S336-S348 (2020)
27. E.O. Caul, H. Appleton, The electron microscopical and physical characteristics of small round human fecal viruses: an interim scheme for classification. J. Med. Virol. **9**(4), 257–265 (1982)
28. The new scope of virus taxonomy: partitioning the virosphere into 15 hierarchical ranks. Nat. Microbiol. **5**(5), 668–674 (2020). https://doi.org/10.1038/s41564-020-0709-x
29. Severe acute respiratory syndrome coronavirus 2 – Wikipedia, https://en.wikipedia.org/wiki/Severe_acute_respiratory_syndrome_coronavirus_2. Accessed 01 Sept 2020
30. D.G. Diel, L.H. da Silva, H. Liu, Z. Wang, P.J. Miller, C.L. Afonso, Genetic diversity of avian paramyxovirus type 1: proposal for a unified nomenclature and classification system of newcastle disease virus genotypes. Infect. Gen. Evol. **12**(8), 1770–1779 (2012)
31. J.E. Wong, Y.S. Leo, C.C. Tan, COVID-19 in Singapore—current experience: critical global issues that require attention and action. JAMA **323**(13), 1243–1244 (2020)
32. Y. Bai, L. Yao, T. Wei, F. Tian, D.Y. Jin, L. Chen, M. Wang, Presumed asymptomatic carrier transmission of COVID-19. JAMA **323**(14), 1406–1407 (2020)
33. C. Del Rio, P.N. Malani, COVID-19—new insights on a rapidly changing epidemic. JAMA **323**(14), 1339–1340 (2020)
34. L. Heller, C.R. Mota, D.B. Greco, COVID-19 faecal-oral transmission: are we asking the right questions?. Sci. Total Environ. 138919 (2020)
35. S. Bhutani, J.A. Cooper, COVID-19 related home confinement in adults: weight gain risks and opportunities. Obesity, **28**(9) (2020). https://doi.org/10.1002/oby.22904
36. S. Cuschieri, S. Grech, COVID-19 and diabetes: the why, the what and the how. J. Diab. Complications, Sep, 34(9): 107637. https://doi.org/10.1016/j.jdiacomp.2020.107637
37. E. Mahase, Covid-19: what treatments are being investigated? BMJ (Clin. Res. ed.) **368**, (2020). https://doi.org/10.1136/bmj.m1252
38. S.R. Baker, R.A. Farrokhnia, S. Meyer, M. Pagel, C. Yannelis, How does household spending respond to an epidemic? consumption during the 2020 covid-19 pandemic (No. w26949). National Bureau of Economic Research (2020)
39. M. Nicola, Z. Alsafi, C. Sohrabi, A. Kerwan, A. Al-Jabir, C. Iosifidis, M. Agha, R. Agha, The socio-economic implications of the coronavirus pandemic (COVID-19): a review. Int. J. Surg. (London, Engl.) **78**, 185 (2020)
40. Sources Corona, https://sites.google.com/view/sourcescorona. Accessed 25 Aug 2020
41. Y. Liu, L.M. Yan, L. Wan, T.X. Xiang, A. Le, J.M. Liu, M. Peiris, L.L. Poon, W. Zhang, Viral dynamics in mild and severe cases of COVID-19. Lancet Infect. Dis. **20**(6), 656–657 (2020). https://doi.org/10.1016/S1473-3099-2030232-2
42. Coronavirus, https://www.who.int/health-topics/coronavirus#tab=tab_3. Accessed 24 Aug 2020

43. COVID-19 advice - Know the facts _ WHO Western Pacific, https://www.who.int/westernpacific/emergencies/covid-19/information/severity. Accessed 25 Aug 2020
44. X. Xu, M. Han, T. Li, W. Sun, D. Wang, B. Fu, Y. Zhou, X. Zheng, Y. Yang, X. Li, X. Zhang, Effective treatment of severe COVID-19 patients with tocilizumab. Proc. Natl. Acad. Sci. **117**(20), 10970–10975 (2020)
45. C. Qin, L. Zhou, Z. Hu, S. Zhang, S. Yang, Y. Tao, C. Xie, K. Ma, K. Shang, W. Wang, D.S. Tian, Dysregulation of immune response in patients with COVID-19 in Wuhan, China clinical infectious diseases: an official publication of the infectious diseases. Soc. Am. **71**(15), 762–768 (2020). https://doi.org/10.1093/cid/ciaa248
46. J.B. Moore, C.H. June, Cytokine release syndrome in severe COVID-19. Science **368**(6490), 473–474 (2020)
47. R.E. Jordan, P. Adab, K.K. Cheng, Covid-19: risk factors for severe disease and death, BMJ,368:m1198 (2020)
48. Y. Dong, X. Mo, Y. Hu et al., Epidemiological characteristics of 2143 pediatric patients with 2019 coronavirus disease in China. Pediatrics (2020). https://doi.org/10.1542/peds.2020-0702
49. Y. Gao, T. Li, M. Han, X. Li, D. Wu, Y. Xu, Y. Zhu, Y. Liu, X. Wang, L. Wang, Diagnostic utility of clinical laboratory data determinations for patients with the severe COVID-19. J. Med. Virol. **92**(7), 791–796 (2020). https://doi.org/10.1002/jmv.25770
50. Assessing Risk Factors for Severe COVID-19 Illness _ CDC, https://www.cdc.gov/coronavirus/2019-ncov/covid-data/investigations-discovery/assessing-risk-factors.html. Accessed 25 Aug 2020
51. www.github.com/CSSEGISandData/COVID-19. Accessed 25 Aug 2020
52. E. Dong, H. Du, L. Gardner, An interactive web-based dashboard to track COVID-19 in real time. Lancet Infect. Dis. **20**(5), 533–534 (2020)
53. L. Meng, F. Hua, Z. Bian, Coronavirus disease 2019 (COVID-19): emerging and future challenges for dental and oral medicine. J. Dental Res. **99**(5), 481–487 (2020). https://doi.org/10.1177/0022034520914246
54. J.A. Backer, D. Klinkenberg, J. Wallinga, Incubation period of 2019 novel coronavirus (2019-nCoV) infections among travellers from Wuhan, China, 20–28 January 2020. Euro Surveill. **25**(5) (2020). https://doi.org/10.2807/1560-7917.ES.2020.25.5.2000062
55. Z.Y. Li, L.Y. Meng, Zhonghua Kou Qiang Yi Xue Za Zhi, 55(0):E001 (2020). https://doi.org/10.3760/cma.j.issn.1002-0098.2020.0001
56. A. Haleem, M. Javaid, R. Vaishya, Effects of COVID-19 pandemic in daily life. Curr. Med. Res. Pract. (2020). https://doi.org/10.1016/j.cmrp.2020.03.011
57. Wellcome trust, https://wellcome.org/news/how-can-we-develop-covid-19-vaccine-quickly. Accessed 19 Oct 2020
58. C. Huang, Y. Wang, X. Li, L. Ren, J. Zhao, Y. Hu, L. Zhang, G. Fan, J. Xu, X. Gu, Z. Cheng, Clinical features of patients infected with 2019 novel coronavirus in Wuhan. China. Lancet **395**(10223), 497–506 (2020)
59. M. Chinazzi, J.T. Davis, M. Ajelli, C. Gioannini, M. Litvinova, S. Merler, A.P. Piontti, K. Mu, L. Rossi, K. Sun, C. Viboud, The effect of travel restrictions on the spread of the 2019 novel coronavirus (COVID-19) outbreak. Science, **368**(6489), 395–400 (2020)
60. J.J. Klemeš, Y. Van Fan, R.R. Tan, P. Jiang, Minimising the present and future plastic waste, energy and environmental footprints related to COVID-19. Renew. Sustain. Energy Rev. **127**, (2020)
61. G. Fagherazzi, C. Goetzinger, M.A. Rashid, G.A. Aguayo, L. Huiart, Digital health strategies to fight COVID-19 worldwide: challenges, recommendations, and a call for papers. J. Med. Internet Res. **22**(6), (2020)

Evolutionary Algorithm Based Summarization for Analyzing COVID-19 Medical Reports

Chirantana Mallick, Sunanda Das, and Asit Kumar Das

Abstract The COVID-19 pandemic emerged in the month of December from the very popular city Wuhan of China and spread rapidly all over the country. The present situation depicts the progress in various less-explored areas to strengthen the global knowledge base for preventing and addressing the COVID-19 pandemic. Novel coronavirus disease became a major health concern due to its pathogenicity and world widespread distribution. Many good scientific research papers related to COVID-19 have recently published using the dataset CORD-19. Analysis of these massive data and identifying meaningful information is a crucial and very challenging task in health care and bio-medical science research. In this chapter, we try to extract important information about COVID-19 from the available text documents, such as research papers, articles, journals, reports, and other publications. Initially, in the chapter, these types of text data are preprocessed by means of sentence tokenization, then removal of stopwords, and finally stemming operations. Consequently, a bag-of-words is generated to represent each report in a structured form and a bi-objective strength Pareto evolutionary algorithm based text summarization technique is concerned to produce the summary of the report. The method has two objective functions: one based on the length of the summary, a predetermined value, and the other based on the clustering coefficient concept of graph theory. After convergence of the evolutionary algorithm, the fittest chromosome in the final population provides a summary of the report. The method is applied on COVID bio-medical reports publicly available in CORD-19 dataset for generating a summary and its performance is finally evaluated by comparing it with a few of the related state-of-the-art methods.

Keywords Medical report · Summarization · COVID-19 · Pandemic · Bi-objective evolutionary algorithm · Natural language processing · Bio-medical science

---

C. Mallick (✉) · A. K. Das
Department of Computer Science and Technology, Indian Institute of Engineering Science and Technology, Shibpur 711103 Howrah, India
e-mail: akdas@cs.iiests.ac.in

S. Das
Department of Computer Science and Engineering, SVCET, RVS Nagar, Chittoor 517127, AP, India

J. Nayak et al. (eds.), *Understanding COVID-19: The Role of Computational Intelligence*, Studies in Computational Intelligence 963, https://doi.org/10.1007/978-3-030-74761-9_2

1 Introduction

A series of unknown pneumonia cases have emerged in early December 2019 from Wuhan [1], a popular city in China, and spread rapidly throughout the world. The outbreak of this new coronavirus, SARS-CoV-2, which is now known as COVID-19 had been monitored by the Centers for Disease Control and Prevention (CDC). The unknown virus was first identified in Wuhan, China. Nine countries, namely China, Japan, Thailand, South Korea, Vietnam, Singapore, the United States, Taiwan, and Nepal had diagnosed at least 830 cases on 24th January 2020. A huge increasing number of cases appeared due to human-to-human interaction and many unknown details like the origin of the emergence of this virus and its ability to spread among humans. On 2nd March 2020, 85650 number of confirmed cases of COVID-19 with death counts of over 132 had been reported from 67 territories outside mainland China and several other countries including Iran and Italy. On 11th March, COVID-19 was declared as a worldwide pandemic by the World Health Organization (WHO) [2, 3]. As the COVID-19 pandemic unfolds, the attention of the global scientific community had been attracted by this novel coronavirus disease. Researchers from all fields are contributing their knowledge together to strengthen the global knowledge base for preventing and addressing the COVID-19 pandemic. To fight the current and future pandemics, multidisciplinary open collaboration has proved its significance and several initiatives [4, 5] has been initiate to share COVID-19-related scientific research to a wide variety of people. As some of the publicly available information regarding the outbreak is not reliable an 'infodemic' has also added with the COVID-19 outbreak [6]. Due to the lack of accurate information researchers and clinicians are facing several critical problems to provide proper health care. So, in this present scenario health crises are also recognized as information crises [7–10]. A daily updated list of relevant research and a portal is maintained by the WHO to provide information publicly [11], like the European Commission [12]. As a response to this pandemic, a weekly updated and growing dataset (CORD-19) [13] of COVID-19 publications, capturing all the past and present researches on COVID-19 and its family of viruses has been released by the White House along with the leading research groups to help and accelerate the research by various researchers and the global research community. To develop a new perception on the ongoing fight against this infectious disease, this freely available dataset is provided publicly for applying different advanced Natural Language Processing (NLP) and other Artificial Intelligence (AI) methods. Various researchers are already trying to develop a search engine [14] to reduce the valuable time for finding COVID-19-related information. A set of challenges is released with CORD-19 which is hosted by Kaggle for extraction of meaningful and structured information from such an enormous set of publications. Analyzing these rapidly growing data and identifying meaningful important information is a crucial and very challenging task. Various summarizing techniques have been implemented to reduce the difficulties of extracting important information from the massive data.

Summarization means the exclusion of extraneous information keeping the prominent idea of the document. Automatic text summarization diminishes the number of texts of the document for generating a summary that contains the most important information of the actual text. The various applications like a summary of newspaper, magazine, article, book, and film increases the value of text summarization methods. Automatic summarization builds some mathematical models to get a summary of the document. Automatic document summarization [15] is currently a very important research field because of the impotence and complexity of the task. Automatic summarization dates back to Luhn's work [16] where it was assumed that the words which are repeated many times are more important than the other words in the document and to indicate that importance, he used word frequency in his proposed work [16]. Many methods [17, 18] have been implemented to extract relevant information from the target document for representing the intermediate representation of the document. Early methods [19, 20] mainly used the traditional statistical methods like term frequency, word frequency, inverse document frequency, etc. for identifying the key concepts within a document. In general, Automatic text summarization approaches can mainly be divided into two types, extractive summarization, and abstractive summarization. Extractive summary [21] is a subset of the sentences in the original document; whereas abstractive summary [21] conveys the ideas in the source document using different sentences and words. In this paper, we have used an extractive summarization based approach in our proposed methodology. It contains three different and independent tasks: (a) Intermediate representation (b) Sentence scoring and (c) Summary selection strategy.

1.1 Intermediate Representation

Identify the hidden information in the target report is the main task of this step. It is also known as the "Topic Representation" approach [22] as the approach is widely used for identifying the topic words pertaining to the particular report. The approach provides an impressive variation in sophistication and representation power. The frequency-based approach, topic signature or word approach [22], Term Frequency-Inverse Document Frequency(TF-IDF) approach, etc. are included in this representation technique.

1.2 Sentence Scoring

After intermediate representation, it is very much required to score the sentences based on their importance in the original report. It is quite obvious that the sentences with relevant information be given a higher score value than the others. Different statistical measures like vertex similarity or distance between vertices, different centrality measures, etc. are used to score the sentences in the graph-based approach. The technique of scoring sentences differs for different intermediate representations.

1.3 Summary Selection Strategy

For generating an appropriate summary the last and final step is to select the summary sentences. In this step, the sentences with key information of the report are selected and arranged in a concise form. The genre of the report is important for controlling what sentences should go into the summary.

2 Related Work

Topic representation approaches (i.e. "Topic Word" and "Topic Signature") [23] are very traditional methods in automatic text summarization. To classify the topic words from the report, a frequency threshold has been considered in these approaches. Frequency driven approaches like Term Frequency (TF) and Inverse Document Frequency(IDF) is mainly used to measure the depth of topic words. To compute the salience of sentences centroid summarization [24] uses a given set of extracted features. In general, extractive summarization creates the summary from the extracted sentences and phrases in the target document(s) whereas abstractive summarization conveys the key concepts of the target document by reconstructing the important sentences from the document(s). Abstractive summarization can retrieve relevant information from multiple documents more precisely and efficiently than extractive summarization. Text summarization can also be categorized as Indicative [25] and informative [17]. An indicative summary points out the main idea of the given document whereas an informative summary conveys the key topics with a proper elaboration which also includes some suggestions as per the reader's requirement. For making generic [18] and query-oriented [26] summary, informative summarization technique has been used. Generic summaries can present the generic concept of the given document by covering maximum information but Query-oriented summaries provide the answer with a minimal set of sentences from the document to cover a given set of queries. Just-the-news summary [27] yields the most updated news based on the reader's interest. Single-document summary [28] creates summary by taking sentences from only single document, whereas in multi-document summarization [29, 30], summary is created by combining sentences from the multiple documents. Topic-oriented summary [31] is generated based on the reader's topic of interest and the extracted information from some given topics. In centrality based summarization [32], the centrality of the sentences is calculated from the word centrality which is described based on the centroid of that clustered document. The central sentences of the document are those sentences that are closer to the centroid. In current years, various graph-based clustering algorithms have been considered for automatic text summarization. In Markov Clustering Algorithm (MCL) [33], the graph is formed by clustering the highly connected vertices with the help of the flow simulation technique. It is a more natural unsupervised clustering algorithm, where a mathematical bootstrapping procedure [34] has been used to detect the cluster structure of the graph. Here, the probabilities of random walks are calculated using stochastic matrices (Markov matrices) to capture the mathematical concept of random walks on a graph. The paper is focused [35] on the algorithmic challenges and formalized the Bounded Correlation Clustering problem and illustrated the intractable problem. In the paper [36], the semantic information is summarized where a folksonomy data model is utilized to signify spatial and semantic relationships among tags, places, and users. The paper [37] is concentrated on developing a structure integrating a series of data mining methods including clustering techniques, pattern changing discovery algorithm, and analysis technique.

Evolutionary Algorithms (EAs) [38] with the single objective as well as multi-objective are widely used for solving any optimization problem. But in real life, most of the problems in optimization are naturally multi-objective as they require multiple objectives that are conflicting in nature. In multi-objective optimization problems, a set of Pareto optimal solutions [39] is considered instead of a single best one. Many multi-objective evolutionary algorithms (MOEAs), like NSGA [40], NSGA-II [41], MOEA/D [42] etc. efficiently handle the multi-objective optimization problems in high dimensional space. The remaining work is arranged into other three sections: Sect. 3 describes the proposed methodology for summarization and the experimental results with its performance evaluation for the publicly available CORD-19 dataset are elaborately discussed in Sect. 4. Finally, this work is concluded in Sect. 5.

3 Proposed Methodology

In this chapter, we have presented a graph theory and evolutionary algorithm-based approach to find the summary of a text report. For calculating the similarity between a pair of sentences, cosine similarity [43] is used. The multi-objective Strength Pareto Evolutionary Algorithm (SPEA) [44] is applied to the pre-processed report. The report is transformed into a matrix where each row represents a sentence and columns represent the bag-of-words identified from the report after pre-processing and it also indicates the significant words from the given report. At first, a random solution from the search space has been selected. Then the better solution has been searched based on the quality of the defined objective functions. Finally, the optimal solution from the search space has been found by achieving the convergence of the algorithm iteratively. Figure 1 represents the flow diagram of the proposed methodology.

3.1 Preprocessing

In the proposed method, the input reports are processed to create bag of significant words, required for summarization method. The following steps are used to preprocess the dataset.

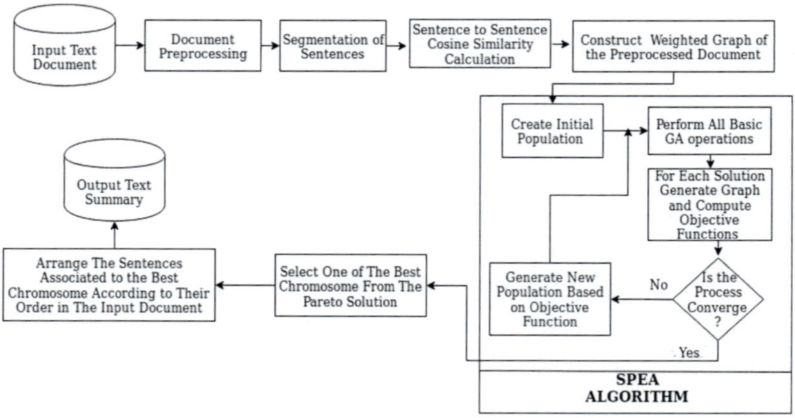

Fig. 1 Work flow of the Evolutionary Algorithm based summarization

3.1.1 Removal of Special Characters and Sentence Tokenization

The input report is transferred into sentences using the Natural Language Tool Kit (NLTK) package [45]. The indices for each of the sentences are assigned according to the input sequence for making a meaningful summary. Then Python word tokenizer has been used for word tokenization of each sentence of the input report.

3.1.2 Removal of Articles and Stopwords

Some words are found irrelevant in order to convey the main context of the input report or to evaluate the importance of each sentence of the given report. That words can be considered as the stopwords (such as "the", "a", "an", "in" etc.) which are more commonly used words in any report. So, we remove the stopwords and other articles which are irrelevant for conveying information regarding the sentence using NLTK (Natural Language Toolkit). For 16 different languages, the stopwords lists are available in Python NLTK and we have accessed it via the NLTK corpus.

3.1.3 Stemming

The procedure of identifying the root words by removing prefixes and suffixes from the words is known as Stemming. For example, 'play' and 'playing' should be considered as play in both cases. There are different algorithms for stemming but among all other algorithms, the Porter Stemming Algorithm [46] is a widely used stemming algorithm for optimal representation of the original document. A rule-based approach has been applied for separating the affix from a word. Lovin stemmer [47] uses comparatively a large set of suffixes which contains multiple morphemes. All though this mechanism is more conventional than Porter stemmer, the bottleneck is it's over diversity. In Krovetz stemmer [48], the word stemming rules are restructured by the non-iterative stripping mechanism which also preserves the meaning for the current string before removing a suffix. Due to its high dependency on the dictionary, the main drawback of this algorithm is its conservative tendency. The Xu and Croft stemmer algorithm [49] defines equivalence classes of words instead of removing suffixes though it is computationally expensive. A normalized co-occurrence method has been used in this algorithm to make conclusions about different word forms with the same meaning are rooted in the same words. A set of affix removal rules is applied in *P*-stemmer using Latent Semantic Analysis of the stemming algorithm [50] and multiple possible "parses" of the word are generated efficiently. In this chapter, the Krovetz stemmer method is used to select the bag-of-words.

A word should have a higher importance in a report if it appears many times in a sentence but very few sentences of the report. A sentence will be important if it contains many words with higher importance. Term-Frequency (TF) computes the frequency of a word that appears in the sentence. Sentence frequency (SF) of a word is the number of sentences containing the word. Inverse sentence Frequency [51] (ISF) of a word is defined using Eq. (1), where T is the size of the given report.

$$ISF = \log_{10} \frac{T}{SF} \tag{1}$$

Certain words may appear several times but have little importance in a report if it appears in many sentences. Thus, the rare less frequent terms are needed to scale up while the frequent terms are needed to bring down to reflect their relative importance by computing the ISF value. Hence, popular weighting schema, Term Frequency-Inverse Sentence Frequency(TFISF) is defined using Eq. (2).

$$TFISF = TF \times \log_{10} \frac{T}{SF} = TF \times ISF \tag{2}$$

The TFISF is used to build the vector matrix where the sentences are represented in rows and the columns show the occurrence of the words in the sentences (i.e., BoW model) [52]. TSISF of j-th word in the i-th sentence is represented by the (i, j)th entry in the matrix. This matrix will be used as input of the proposed SPEA based summarization algorithm.

3.2 Initial Population

In evolutionary algorithms [53], initial population generation is a vital task. Here, an initial population of size M of binary strings is generated randomly where each string size has been equal to the target size (S) of the summary. Each row is an individual chromosome with length S and the population contains M chromosomes. Each chromosome signifies the sentences by which the report is characterized, where

'0' or '1' in a position of the chromosome represents the absence or presence of the corresponding sentence, respectively. Initially, these chromosomes have been considered as the possible solutions for this proposed summarization technique. The size of the population has been determined by making a trade-off between the efficiency and effectiveness of the proposed method. For searching appropriate solution, the population size should not be too small as in that case the sufficient dimension is not available and similarly, the size should not be too large as it reduces the efficiency of the process. All the '1's in a chromosome represent the sentences considered for a summary of the report and only these have been considered for calculating the strength of the chromosomes for that specific iteration. After convergence of the process, the optimal solution is the best-fit chromosomal solution of the proposed algorithm. The sentences corresponding to the best chromosome provides a summary of the report.

3.3 Fitness Function

The proposed work is based on the bi-objective genetic algorithm, which uses two objective functions to measure the strength of a possible solution. Two objective functions(i.e., fitness functions) are described below.

a. The first fitness function (f_{1i}) for i-th chromosome is defined using Eq. (3), where S is the target size of the summary and ch_i (n_1) denotes how many '1' presents in the i-th chromosome (ch_i).

$$f_{1i} = S - ch_i(n_1) : ch_i(n_1) \leq S \tag{3}$$

The minimum value of f_{1i} is required because it represents the candidate summary size near to the model summary size, which is desired in our summarization method.

b. The second fitness function (f_{2i}) for i-th chromosome is defined using Eq. (4).

$$f_{2i} = CC(G) - CC(G_i) \tag{4}$$

Here, $CC(G)$ is the cluster co-efficient of graph G constructed for the whole dataset, and $CC(G_i)$ is the cluster co-efficient of the graph G_i built by removing the edges from the vertices of G based on the i-th chromosome (ch_i), as explained below. Obviously, the maximum value of f_{2i} is required for determining the better quality chromosome.

The following steps describe the second fitness function computation method.

i) Cosine similarity matrix is obtained from the computed cosine similarity for every pair of sentences using Eq. (5). A high value of this similarity implies the sentences are extremely alike to each other and vice versa.

$$SIM(X, Y) = \frac{\sum_{i=1}^{n} X_{wi} Y_{wi}}{\sqrt{\sum_{i=1}^{n}(X_{wi})^2} \times \sqrt{\sum_{i=1}^{n}(Y_{wi})^2}} \tag{5}$$

Where, X and Y are two sentence vectors of pre-processed report. The X_{wi} and Y_{wi} represent the i-th word of sentences X and Y, respectively and n is the number of distinct words in the pre-processed report.

ii) A similarity graph G representing the dataset is built from the cosine similarity matrix. Here, each vertex is an n-dimensional sentence vector and the edges represent relationship between the sentences. The weight of an edge is the cosine similarity value between two sentences representing terminal vertices of the edge. The edges are established between vertices, if the similarity measure between the vertices is greater than the defined threshold value (δ), which is experimentally set. That means, if edge weight is less than δ then the edge is removed. Thus the graph is modified and denoted by the same name G.

iii) Now, the graph G_i is constructed corresponding to the chromosome ch_i from the graph G. If i-th chromosome contains n_1 number of 1's then the associated n_1 sentences are the n_1 vertices in graph G. The edges between every pair of these n_1 vertices (if exist) are eliminated from original graph and so, a new graph G_i has been obtained from G for chromosome ch_i.

The clustering coefficient of an un-weighted graph G is the mean of the clustering coefficients of all the vertices of the graph, where the clustering coefficient (CC) of a vertex v is defined using Eq. (6).

$$CC_G(v) = \frac{2 \times T(v)}{d(v) \times (d(v) - 1)} \tag{6}$$

Where, $T(v)$ is the number of triangles through vertex v which is the number of links between neighbors of v and $d(v)$ is the degree of vertex v. But as the constructed graph is the weighted graph, so the weights of the edges are considered during calculation of clustering coefficient of a vertex v_i in G, as defined in Eq. (7), where $d(v_i)$ is the degree of vertex v_i, n_{ij} is the normalized value of weight w_{ij} of edge (v_i, v_j) in G, computed as $n_{ij} = w_{ij}/\max\{ w_{lm} \forall w_{lm}$ = weight assigned to edge in G } and summation is taken over all the triangles formed considering all the adjacent vertices v_j of v_i.

$$CC_G(v_i) = \frac{1}{d(v_i) \times (d(v_i) - 1)} \Sigma_{v_i v_j} (n_{ij} n_{jk} n_{ki})^{1/3} \tag{7}$$

Then the clustering coefficient [54] $CC(G)$ of the graph G is computed as the average clustering coefficient of all the vertices in the graph G, as defined in Eq. (8), where V is the set of vertices in G.

$$CC(G) = \frac{1}{|V|} \Sigma_{v_i \in V} CC_G(v_i) \tag{8}$$

Similarly, clustering coefficient $CC(G_i)$ of the graph G_i for chromosome ch_i is calculated using Eq. (7) and Eq. (8).

3.4 External Population

In GA, to store and maintain the generated non-dominated solutions external population (or archive) is used at every generation. This ensures the elitism property and allows the non-dominated best chromosome(s) to preserve from the current generation to the next generation directly without reducing the quality of the solutions generated by The GA from one to the next generation. At each generation, both the external population (P') and the current population (P) are generated. The fitness values are computed and assigned to each chromosome of both the populations at

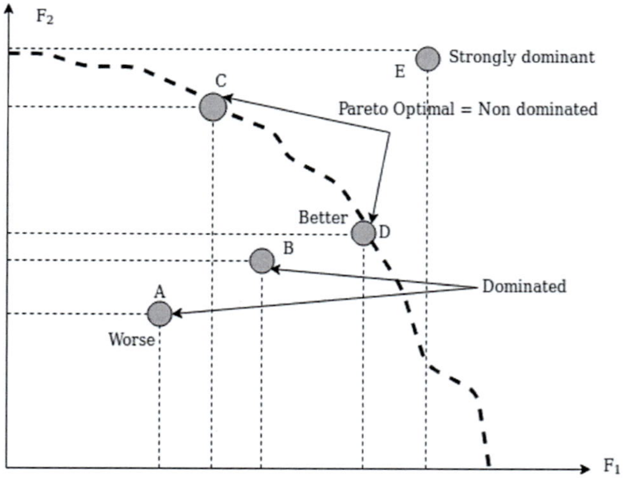

Fig. 2 Pareto optimal concept

each generation. The size of the external population has been set to one-third of the initial population size. The chromosomes of the external population are selected from the non-dominated Pareto optimal front obtained from the current population based on the fitness functions and this is demonstrated in Fig. 2.

In Fig. 2, F_1 and F_2 are two objective functions. This is the solution space for a bi-objective optimization problem. The solution represented by the point D is better and the solution with the point A is worse with respect to the solution represented by the point B as the solution of B is better in at least one objective function and not worse in any objective. C is not better than that of D and vice versa because one objective function value of each point is higher than the other. These two points form a Pareto optimal front or Non-dominated set represented by a dotted line. E is the strongly dominant solution compared to all other solutions with respect to two objective functions.

After selection, the selected members are feed in an Agglomerative hierarchical clustering algorithm [55] if the size of the external population exceeds and this process is continued until the number of clusters obtained by merging is less or equal to the external population size. From each cluster, the chromosome nearest to the centroid of the cluster is considered as the member of the external population. In external population P', the assigned strength for each chromosome is computed using Eq. (9), where, M is the size of the population and n_i is the size of the current population dominated by chromosome ch_i.

$$SE(ch_i) = \frac{n_i}{M+1} \tag{9}$$

This implies that the strength of a chromosome is directly proportional to the number of chromosomes dominated by it. In the internal population, the strength of each chromosome ch_j is defined using Eq. (10).

$$SI(ch_j) = 1 + \Sigma_{ch_i \in P' \wedge i \neq j} SE(ch_i) \tag{10}$$

In Eq. (10), condition $i \neq j$ indicates that one more than the sum of the strength of all chromosomes (ch_i) in an external population that weakly dominates the chromosome ch_j is assigned to the strength of the chromosome ch_i of the internal population. So, in SPEA algorithm the chromosome with less strength is the best suited for the selection of the mating pool.

3.5 Selection

In GA, the selection is the first genetic operation. The objective of this operation is to select better chromosomes from the population, which will be used as the parents, known as a mating pool for generating offspring for the next generation. The target of selection is to give preference to better individuals by allowing them to pass on their genes to the next generation. The quality of an individual is measured by a fitness value. The fitness value is determined by one or more objective functions. During the creation of the mating pool various popular selection methods like rank selection, roulette wheel selection, tournament selection, steady-state selection, and many more other selection methods are widely used for selecting the best chromosomes. Among all those popular selection methods tournament selection is preferable due to its acceptability of negative fitness values. In our proposed algorithm, based on the strength assigned to the individuals as per the SPEA algorithm a binary tournament selection method [56] has been used. Here, two chromosomes are selected from the population at random and the best one (i.e., less strength, determined using Eqs. (9), (10) is selected as a parent. The same process is repeated for selecting another parent. Until the mating pool is filled this selection method is continued.

3.6 Crossover

In GA [57] is much similar to biological crossover. Crossover operation affects the performance of GA. To help in searching the solutions towards the local optima and to get better solutions more than one parent are selected for producing one or more chromosomes by crossover operation. From various types of crossover such as single-point crossover, uniform crossover, multi-point crossover, etc. single-point crossover is used in this work. In single-point crossover, the crossover point is selected and a new chromosome is created in form of a binary string from the beginning of the chromosome to the crossover point is copied from one parent, and the rest is copied from the second parent to the offspring. Crossover probability indicates how often Crossover is performed between two parents to form two new offspring. If crossover probability is 100%, that means that all offspring are made by crossover and if crossover probability is 0%, that means that all offspring are an exact copy of the parents. A randomly generated value between 0 and 1 is compared to the crossover probability, and the crossover operation is performed if the randomly generated value is less than the crossover probability, otherwise, the offspring have remained unchanged which is identical to the parents. In the proposed work, the crossover probability for offspring generation is considered as 0.8 (i.e. 80%) for experimental purposes.

3.7 Mutation

To maintain genetic diversity, mutation [57] is used normally after crossover operation in GA. Just like the biological mutation it also alters the gene values of a chromosome. It is necessary to discover a better optimal space for some chromosomes and break out of local optimum space. It affects less number of chromosomes in the population as it occurs less frequently. For different genome types, different mutation types are suitable. The proposed method is used bit string mutation, which flips the bits in the randomly selected position of the chromosome. Mutation probability indicates a measure of the likeness that random bits of a single chromosome will be flipped. If mutation probability is 100%, that means that no mutation is performed and all the offspring are an exact copy of the parent. Mutation probability 0% means all the bits of the chromosome are flipped to create new offspring. The mutation operation is performed if a randomly selected floating-point value for every bit of the chromosome is less than the mutation probability; otherwise, no mutation happens (i.e. the offspring are just alike to the parents). The proposed algorithm uses 0.02 (i.e. 20%) mutation probability for experimental purposes.

3.8 Elitism

In GA there is a possibility of removing the better solution from the population which might make it difficult to achieve it by slowing down the process of finding an optimal solution. In general, elitism [58, 59] is the process of memorizing the better solution. In elitism, better chromosomes are moved to the next generation without any change and it improves the overall performance as it does not consider the previously discarded partial solutions. In this method, to generate the external population where the best non-dominated solutions are kept elitism is applied. The flip of elitism helps GA to converge on local maxima instead of the global maximum. The high number of elite individuals causes population degeneration. For this reason, the high number of elite individuals should not be preferable.

After the method converges, the best chromosome (with respect to the objective functions) of the final population is considered. The 1's in this chromosome indicates the positions of the sentences in the original report and these sentences are selected as a summary of the report. The pseudo code of the proposed evolutionary algorithm based text summarization (EATS) method is described below.

Algorithm 1: EVOLUTIONARY ALGORITHM BASED TEXT SUMMARIZATION (EATS)

Input: P= Population of size m, D = Text report, s = predetermined Summary size, Arc_s = External Population Size, J_{max} = Maximum number of generation, c_p = Crossover probability, m_p = Mutation probability, Ch_{best} = Best Chromosome and δ = user defined similarity threshold.

Output: Summary S of report D

Individual sentences of the report D is identified by sentence segmenter;
Index value is assigned to each sentence in the report;
Perform preprocessing and build sentence vector of each sentence;
Compute cosine similarity of each pair of sentence vectors using Eq. (5);
Construct a graph $G = (V, E, W)$, where V is the set of vertices represent sentence vectors, E is the set of edges, and W is the set of weights assigned to the edges such that weight of an edge is the similarity between to terminal vertices (i.e., sentences);
Modify graph G by removing the edges with weight $< \delta$;
Calculate clustering coefficient of graph G using Eqs. (7) and (8);
Generate a binary population P having size m randomly;
$Generation\_Count = 0$;
while *(Generation_Count =* J_{max}*)* **do**
    for *each chromosome* $ch_i \in P$ **do**
        V_i = subset of vertices of V representing sentences corresponding to 1 in ch_i;
        Generate graph G_i by removing edges between any pair of vertices in V_i from G;
        Compute clustering coefficient of G_i using Eqs. (7) and (8);
        Compute first fitness function (f_{1i}) of ch_i using Eq. (3);
        Compute second fitness function (f_{2i}) of ch_i using Eq. (4);
        Determine C = set of non-dominated solutions;
        if $|C| \leq Arc_s$ **then**
            C is moved to the external population P';
        else
            C is replaced according to SPEA algorithm described in section 3.4 and stored in P';
        Compute strength of ch_i in P' using Eq. (9);
        Compute strength of ch_i in P using Eq. (10);
    while $(|P| = |P'_{new}|)$ **do**
        Apply binary tournament selection on P and P' to create P_{new} and P'_{new}, respectively;
    while *(Perform crossover operation* $|P_{new}|/2$ *times)* **do**
        Select chromosomes ch_1 and ch_2 randomly from P_{new};
        Single-point crossover is applied on ch_1 and ch_2 with Probability c_p;
        Compute their fitness values for the generated two offspring of the parents;
        Select the best two among offspring and parents;
    for *each* $ch_i \in P_{new}$ **do**
        Mutation operation is applied on ch_i with probability m_p;
        Compute the fitness values for the new mutated offspring;
        if *Fitness values increases* **then**
            Offspring is replaced by mutated one;
        $Generation\_Count = Generation\_Count + 1$;
Find best chromosome Ch_{best} of M;
Select sentences associated with Ch_{best};
Arrange sentences according to their original order using indices and store in S;
return S.

4 Experimental Results

Experiments are carried out and performances are measured on publicly available newly release CORD-19 dataset [13]. Over 47000 reports, of which 36000 are with full text (April 4, 2020) are contained in the CORD-19 publication dataset. COVID-19 research reports from the WHO [3] and pre-prints reports from bioRxiv [60] and medRxivs [61] and also reports from PubMed's PMC open access corpus [62] are collected together to prepare this dataset. The reports on "COVID-19" OR "Corona virus" OR "2019-nCoV" OR "SARS-CoV" OR "MERS-CoV" OR "Severe Acute Respiratory Syndrome" OR "Middle East Respiratory Syndrome" have been considered to prepare this dataset for COVID-19 research. All these publications are from a few popular areas which include coronaviruses (primarily COVID-19, MERS, SARS); the molecular biology of viruses; public health and viral epidemics; immunology and antivirals; influenza and other families of viruses; methodology (testing, clinical trials, diagnosing). Kaggle [63] provides the extracted metadata and the extracted text from CORD-19 dataset in a json format that is easily accessible openly. In this proposed work, we have used the NLTK library [45] of Python 2.7 [64] for dataset preprocessing. To build a corpus for our experiments, we have randomly selected 1000 bio-medical published reports of each of five different types, namely Corona virus (i.e. including "COVID-19", "Corona virus" and "2019-nCoV"), SARS-CoV, MERS-CoV, Severe Acute Respiratory Syndrome and Middle East Respiratory Syndrome from CORD-19 database.

4.1 Experimental Setup

The parameters of genetic algorithm used in the proposed EATS method are listed in Table 1. These listed parameters of our proposed algorithm have been selected after evaluating the algorithm on the CORD-19 dataset several time. The proposed summarization algorithm is implemented by us using python programming language and runs for 50 times and 500 generations for each time on HP Pentium core i5 processor PC.

Table 1 Required parameters of proposed algorithm

Parameter	Value
Population size (P)	400
Crossover probability (C_P)	0.80
Mutation probability (M_P)	0.02
Maximum generation (J_{max})	500
Similarity threshold Value (δ)	0.2

Target size (S) of the summary is considered to be 10, 15, 20, and 25% of the original report size (i.e., the number of sentences presents in it). The comparative analysis between some state-of-art summarization techniques and the proposed technique by using some statistical measures helps to show the performance evaluation of our proposed method, which has been discussed in the following subsections.

4.2 Performance Analysis Using ROUGE

The most commonly used performance evaluation of summarization technique is Recall-Oriented Understudy for Gisting Evaluation (ROUGE) [54, 65]. Among all the ROUGE methods, the ROUGE_N, ROUGE_L, ROUGE_W, and ROUGES_U [54] are universally used to measure the performances. In the proposed work, we have considered ROUGE_N and ROUGE_L for measuring the performance of the proposed summarization method.

4.2.1 ROUGE_N

ROUGE_N method is the N-gram ($N \geq 1$) recall measurement between a set of model summaries and a system summary. It estimates how much the summaries are fluent and it is defined using Eq. (11).

$$ROUGE\_N = \frac{\Sigma_{s \epsilon refsum} \Sigma_{gram_N \epsilon S} count_{match}(gram_N)}{\Sigma_{s \epsilon refsum} \Sigma_{gram_N \epsilon S} count(gram_N)} \quad (11)$$

Where the size/length of the N-gram is symbolized by N and the maximum co-occurrence number of N-gram in a set of model summaries and a system summary is symbolized by $count(gram_N)$. Here, only ROUGE_1 and ROUGE_2 have been considered for evaluation as the value of N is taken as 1 and 2. ROUGE_1 and ROUGE_2 indicate the 1-gram and bi-grams overlapping between the system and model summaries, respectively.

4.2.2 ROUGE_L

The ROUGE_L [54] is the longest sub-sequence related statistics which is employed automatically to obtain the longest co-occurring in sequence N-grams. Suppose, $S = (s_1, s_2, s_3, \ldots s_R)$ and $M = (m_1, m_2, m_3, \ldots m_L)$ are the set of sentences of the system summary and the model summary, respectively. These summaries are represented by the sequence of words and LCS (Longest Common Sequence) based F-score (F_{lcs}) represents the similarity between S(of length R) and M(of length L) according to Eqs. (12), (13), and (14), respectively.

$$R_{lcs} = \frac{LCS(S, M)}{L} \tag{12}$$

$$P_{lcs} = \frac{LCS(S, M)}{R} \tag{13}$$

$$F_{lcs} = \frac{(1 + \beta^2) R_{lcs} P_{lcs}}{(R_{lcs} + \beta^2 P_{lcs})} \tag{14}$$

where, LCS(S, M) stands for the length of the LCS of S and M and $\beta = P_{lcs}/R_{lcs}$.

4.2.3 Performance Measure:

The proposed method has been thoroughly tested on the CORD-19 database of biomedical and life sciences journal citations. CORD-19 is the dataset of the COVID report that contains more than 47000 references to journal reports in life sciences with a concentration on COVID pandemic research. To build a corpus for our experiments, we have randomly selected 1000 biomedical published reports for each dataset based on Corona virus, SARS-CoV, MERS-CoV, Severe Acute Respiratory Syndrome, and Middle East Respiratory Syndrome from CORD-19 dataset.

As per our knowledge, there does not exist a specific corpus for biomedical summarization. The authors' written abstracts of any scientific reports have been often considered as the golden summary [66] of that report. So, the authors' written abstracts have been considered as the model summaries of that report in order to quality evaluation of the proposed method. Also, we have considered the summary constructed by the domain expert as the model summary of that report for evaluating our proposed method. Similarly, the performance of the summaries are evaluated using Eq. (12), (13), and (14) individually for all 1000 CORD-19 biomedical published reports and the average values of performance measurement metrics are listed in Table 2 considering abstract as the model summary and Table 3 considering domain expert constructed summary as the model summary.

From Tables 2 and 3, it is observed that the Recall and Precision ROUGE values are more or less the same for any considering summary size but the summary size containing 20% sentences of the reports provides better f-score ROUGE values compare to other summary sizes. The importance of the f-score ROUGE value is more than that of the individual Recall and Precision ROUGE value as the f-score provides the combined weighted harmonic mean of Recall and Precision. This indicates that the proposed EATS technique generates the summary almost closed to the ground truth summary for summary size 20% sentences of the original report.

4.3 Performance Comparison with Other Summarization Techniques

To measure the efficiency of the proposed EATS method, some existing text summarization methods like Textrank summarization (TR) [67], Lex-Rank summarisation (LR) [68], Graph-of-words-based summarization (GOW) [69], ClusterRank summarization (CR) [70], FreqRank summarization (FR) [71], Latent Semantic Analysis for summarization (LSA) [72], and Correlation SumBasic summarization (SB) [73] are considered. These methods are used to generate automatic text summary for the biomedical reports and compared with the proposed EATS method with the help of ROUGE values and Receiver operating Curve (ROC) plotting.

Table 2 ROUGE values of summary generated by proposed EATS method for selected top 1000 CORD-19 biomedical published reports where abstract is considered as the model summary

Length	10%			15%		
ROUGE value	Recall	Precision	f-score	Recall	Precision	f-score
N = 1	0.5120	0.5301	0.5220	0.4080	0.5010	0.4406
N = 2	0.3990	**0.5001**	0.4303	0.3900	0.4210	0.4030
LCS	0.3250	0.2009	0.2304	0.3021	0.2130	0.2360
Length	20%			25%		
ROUGE value	Recall	Precision	f-score	Recall	Precision	f-score
N = 1	**0.5781**	**0.5900**	**0.5837**	0.5010	0.5200	0.5090
N = 2	**0.4020**	0.4901	**0.4335**	0.4000	0.4910	0.4320
LCS	0.3310	**0.3010**	**0.3102**	**0.3402**	0.2400	0.2001

Table 3 ROUGE values of summary generated by proposed EATS method for selected top 1000 CORD-19 biomedical published reports where domain expert is constructed summary considered as the model summary

Length	10%			15%		
ROUGE value	Recall	Precision	f-score	Recall	Precision	f-score
N = 1	**0.7721**	0.6421	0.7011	0.6880	**0.7894**	**0.7352**
N = 2	0.6794	0.5638	0.6062	0.6296	0.7234	0.6732
LCS	**0.7549**	0.6325	0.6895	0.5981	0.6881	0.6400
Length	20%			25%		
ROUGE value	Recall	Precision	f-score	Recall	Precision	f-score
N = 1	0.7321	0.7284	0.7302	0.7060	0.6350	0.6648
N = 2	**0.7152**	0.7028	**0.7088**	0.6824	**0.7107**	0.6960
LCS	0.7232	**0.7463**	**0.7328**	0.6741	0.7039	0.6880

4.3.1 Performance Comparison Using ROUGE

The ROUGE values of the results acquired by the other methods are listed in Tables 4, 5, 6, 7 and 8 considering summary size 20% of the size of the original report and abstract as the model summary as it provides a better result for the proposed method, shown in Table 2. The ROUGE values for different summarizers are shown in Tables 9, 10, 11, 12 and 13 considering summary size 20% of the size of the original report and domain expert constructed summary as the model summary as it provides a better result for the proposed method, shown in Table 3. The values are obtained from the average result of selected 1000 biomedical published reports based on SARS-CoV, Corona virus, MERS-CoV, Middle East Respiratory Syndrome from CORD-19 and Severe Acute Respiratory Syndrome, as listed in Tables 4, 5, 6, 7, 8, 9, 10, 11, 12 and 13, respectively.

Table 4 Comparison of EATS with other methods using CORD-19 biomedical selected published reports based on Corona virus where abstract is considered as the model summary

ROUGE value		Summarization techniques							
		TR [67]	LR [68]	GOW [69]	CR [70]	FR [71]	LSA [72]	SB [73]	EATS
Recall	N = 1	0.5600	0.5485	0.3800	0.4550	0.4100	0.4830	0.3850	**0.5781**
	N = 2	0.3928	0.3896	0.4120	0.4010	0.4020	**0.4252**	0.3700	0.4020
	LCS	0.3300	0.3294	0.2646	0.3120	0.2900	0.3110	0.3004	**0.3310**
Precision	N = 1	0.5893	0.5796	0.5670	0.5460	0.5630	0.5640	0.5200	**0.5900**
	N = 2	0.4801	**0.4991**	0.4730	0.4570	0.4240	0.4120	0.3670	0.4901
	LCS	0.2998	0.2983	0.2787	0.2378	0.2830	0.2039	0.2120	**0.3010**
f-score	N = 1	0.5600	0.5702	0.4984	0.5288	0.5159	0.5078	0.4976	**0.5837**
	N = 2	0.4237	0.4239	0.4030	0.4150	0.3890	0.4006	0.3750	**0.4335**
	LCS	0.3010	0.3070	0.2890	0.2310	0.2270	0.2130	0.2290	**0.3102**

Table 5 Comparison of EATS with other methods using CORD-19 biomedical selected published reports based on SARS-CoV where abstract is considered as the model summary

ROUGE value		Summarization techniques							
		TR [67]	LR [68]	GOW [69]	CR [70]	FR [71]	LSA [72]	SB [73]	EATS
Recall	N = 1	0.5600	0.5485	0.3800	0.4550	0.4100	0.4830	0.3850	**0.5581**
	N = 2	0.3928	0.3896	0.4120	0.4010	0.4020	**0.4252**	0.3700	0.4011
	LCS	0.3300	0.3294	0.2646	0.3120	0.2900	0.3110	0.3004	**0.3300**
Precision	N = 1	0.5893	0.5796	0.5670	0.5460	0.5630	0.5640	0.5200	**0.5960**
	N = 2	0.4801	**0.4991**	0.4730	0.4570	0.4240	0.4120	0.3670	0.4923
	LCS	0.2998	0.2983	0.2787	0.2378	0.2830	0.2039	0.2120	**0.3102**
f-score	N = 1	0.5600	0.5702	0.4984	0.5288	0.5159	0.5078	0.4976	**0.5807**
	N = 2	0.4237	0.4239	0.4030	0.4150	0.3890	0.4006	0.3750	**0.4435**
	LCS	0.3010	0.3070	0.2890	0.2310	0.2270	0.2130	0.2290	**0.3312**

Table 6 Comparison of EATS with other methods using CORD-19 biomedical selected published reports based on MERS-CoV where abstract is considered as the model summary

ROUGE value		Summarization techniques							
		TR [67]	LR [68]	GOW [69]	CR [70]	FR [71]	LSA [72]	SB [73]	EATS
Recall	N = 1	0.5600	0.5485	0.3800	0.4550	0.4100	0.4830	0.3850	**0.5790**
	N = 2	0.3928	0.3896	0.4120	0.4010	0.4020	**0.4252**	0.3700	0.4005
	LCS	0.3300	0.3294	0.2646	0.3120	0.2900	0.3110	0.3004	**0.3299**
Precision	N = 1	0.5893	0.5796	0.5670	0.5460	0.5630	0.5640	0.5200	**0.5910**
	N = 2	0.4801	**0.4991**	0.4730	0.4570	0.4240	0.4120	0.3670	0.4891
	LCS	0.2998	0.2983	0.2787	0.2378	0.2830	0.2039	0.2120	**0.3012**
f-score	N = 1	0.5600	0.5702	0.4984	0.5288	0.5159	0.5078	0.4976	**0.5876**
	N = 2	0.4237	0.4239	0.4030	0.4150	0.3890	0.4006	0.3750	**0.4340**
	LCS	0.3010	0.3070	0.2890	0.2310	0.2270	0.2130	0.2290	**0.3111**

Table 7 Comparison of EATS with other methods using CORD-19 biomedical selected published reports based on Severe Acute Respiratory Syndrome where abstract is considered as the model summary

ROUGE value		Summarization techniques							
		TR [67]	LR [68]	GOW [69]	CR [70]	FR [71]	LSA [72]	SB [73]	EATS
Recall	N=11	0.5600	0.5485	0.3800	0.4550	0.4100	0.4830	0.3850	**0.5777**
	N = 2	0.3928	0.3896	0.4120	0.4010	0.4020	**0.4252**	0.3700	0.4009
	LCS	0.3300	0.3294	0.2646	0.3120	0.2900	0.3110	0.3004	**0.3340**
Precision	N=1	0.5893	0.5796	0.5670	0.5460	0.5630	0.5640	0.5200	**0.5911**
	N = 2	0.4801	**0.4991**	0.4730	0.4570	0.4240	0.4120	0.3670	0.4894
	LCS	0.2998	0.2983	0.2787	0.2378	0.2830	0.2039	0.2120	**0.3201**
f-score	N = 1	0.5600	0.5702	0.4984	0.5288	0.5159	0.5078	0.4976	**0.5840**
	N = 2	0.4237	0.4239	0.4030	0.4150	0.3890	0.4006	0.3750	**0.4325**
	LCS	0.3010	0.3070	0.2890	0.2310	0.2270	0.2130	0.2290	**0.3111**

The Tables 4, 5, 6, 7, 8, 9, 10, 11, 12 and 13 show that, our proposed EATS method provides better result than other listed summarization methods in most of the summarization techniques and it signifies the efficiency of the proposed method. In time of evaluation, several statistical measurement metrics such as, Recall, Precision, and F-score are evaluated for accuracy measurement.

4.3.2 Performance Comparison Using Receiver Operating Characteristic Curve:

The methods are also compared by constructing Receiver Operating Characteristics (ROC) curve for evaluating generated summary obtained by various summarization techniques with reference to the model summary as the ground truth. The evaluation is based on an evaluation matrix given in Table 14, similar to the confusion matrix [74] used for evaluation of the classifiers. In Table 14, 'Agree' stands for the number of sentences of the report present in the summary and 'Disagree' means the number of sentences of the report not present in the summary. Here, u indicates the total number of sentences present in both the generated and the model summaries, v indicates the total number of sentences present in the model summary but not in the generated

Table 8 Comparison of EATS with other methods using CORD-19 biomedical selected published reports based on Middle East Respiratory Syndrome where abstract is considered as the model summary

ROUGE value		Summarization techniques							
		TR [67]	LR [68]	GOW [69]	CR [70]	FR [71]	LSA [72]	SB [73]	EATS
Recall	N=1	0.5600	0.5485	0.3800	0.4550	0.4100	0.4830	0.3850	**0.5672**
	N=2	0.3928	0.3896	0.4120	0.4010	0.4020	**0.4252**	0.3700	0.4166
	LCS	0.3300	0.3294	0.2646	0.3120	0.2900	0.3110	0.3004	**0.3325**
Precision	N=1	0.5893	0.5796	0.5670	0.5460	0.5630	0.5640	0.5200	**0.5906**
	N=2	0.4801	**0.4991**	0.4730	0.4570	0.4240	0.4120	0.3670	0.4961
	LCS	0.2998	0.2983	0.2787	0.2378	0.2830	0.2039	0.2120	**0.3210**
f-score	N=1	0.5600	0.5702	0.4984	0.5288	0.5159	0.5078	0.4976	**0.5840**
	N=2	0.4237	0.4239	0.4030	0.4150	0.3890	0.4006	0.3750	**0.4329**
	LCS	0.3010	0.3070	0.2890	0.2310	0.2270	0.2130	0.2290	**0.3302**

Table 9 Comparison of EATS with other methods using CORD-19 biomedical selected published reports based on Corona virus where domain expert is constructed summary considered as the model summary

ROUGE value		Summarization techniques							
		TR [67]	LR [68]	GOW [69]	CR [70]	FR [71]	LSA [72]	SB [73]	EATS
Recall	N = 1	0.5600	0.5485	0.3800	0.4550	0.4100	0.4830	0.3850	**0.7321**
	N = 2	0.3928	0.3896	0.4120	0.4010	0.4020	0.4252	0.3700	**0.7152**
	LCS	0.3300	0.3294	0.2646	0.3120	0.2900	0.3110	0.3004	**0.7232**
Precision	N = 1	0.5893	0.5796	0.5670	0.5460	0.5630	0.5640	0.5200	**0.7284**
	N = 2	0.4801	0.4991	0.4730	0.4570	0.4240	0.4120	0.3670	**0.7028**
	LCS	0.2998	0.2983	0.2787	0.2378	0.2830	0.2039	0.2120	**0.7463**
f-score	N = 1	0.5600	0.5702	0.4984	0.5288	0.5159	0.5078	0.4976	**0.7302**
	N = 2	0.4237	0.4239	0.4030	0.4150	0.3890	0.4006	0.3750	**0.7088**
	LCS	0.3010	0.3070	0.2890	0.2310	0.2270	0.2130	0.2290	**0.7328**

Table 10 Comparison of EATS with other methods using CORD-19 biomedical selected published reports based on SARS-CoV where domain expert is constructed summary considered as the model summary

ROUGE value		Summarization techniques							
		TR [67]	LR [68]	GOW [69]	CR [70]	FR [71]	LSA [72]	SB [73]	EATS
Recall	N = 1	0.5600	0.5485	0.3800	0.4550	0.4100	0.4830	0.3850	**0.6881**
	N = 2	0.3928	0.3896	0.4120	0.4010	0.4020	0.4252	0.3700	**0.6393**
	LCS	0.3300	0.3294	0.2646	0.3120	0.2900	0.3110	0.3004	**0.5804**
Precision	N = 1	0.5893	0.5796	0.5670	0.5460	0.5630	0.5640	0.5200	**0.7798**
	N = 2	0.4801	0.4991	0.4730	0.4570	0.4240	0.4120	0.3670	**0.7112**
	LCS	0.2998	0.2983	0.2787	0.2378	0.2830	0.2039	0.2120	**0.6978**
f-score	N = 1	0.5600	0.5702	0.4984	0.5288	0.5159	0.5078	0.4976	**0.7012**
	N = 2	0.4237	0.4239	0.4030	0.4150	0.3890	0.4006	0.3750	**0.6005**
	LCS	0.3010	0.3070	0.2890	0.2310	0.2270	0.2130	0.2290	**0.6223**

Table 11 Comparison of EATS with other methods using CORD-19 biomedical selected published reports based on MERS-CoV where domain expert is constructed summary considered as the model summary

ROUGE value		Summarization techniques							
		TR [67]	LR [68]	GOW [69]	CR [70]	FR [71]	LSA [72]	SB [73]	EATS
Recall	N = 1	0.5600	0.5485	0.3800	0.4550	0.4100	0.4830	0.3850	**0.6432**
	N = 2	0.3928	0.3896	0.4120	0.4010	0.4020	0.4252	0.3700	**0.5767**
	LCS	0.3300	0.3294	0.2646	0.3120	0.2900	0.3110	0.3004	**0.6342**
Precision	N = 1	0.5893	0.5796	0.5670	0.5460	0.5630	0.5640	0.5200	**0.7022**
	N = 2	0.4801	0.4991	0.4730	0.4570	0.4240	0.4120	0.3670	**0.6011**
	LCS	0.2998	0.2983	0.2787	0.2378	0.2830	0.2039	0.2120	**0.6670**
f-score	N = 1	0.5600	0.5702	0.4984	0.5288	0.5159	0.5078	0.4976	**0.6645**
	N = 2	0.4237	0.4239	0.4030	0.4150	0.3890	0.4006	0.3750	**0.6978**
	LCS	0.3010	0.3070	0.2890	0.2310	0.2270	0.2130	0.2290	**0.6883**

summary, w is the total number of sentences present in the generated summary but not in the model summary and x is the total number of sentences not present in both the generated and the model summary. From Table 14, we can calculate the True Agree Rate (TAR) and False Agree Rate (FAR) for the proposed summarizer using Eq. (15) and (16)

$$TAR = \frac{u}{u + v} \tag{15}$$

$$FAR = \frac{w}{w + x} \tag{16}$$

Table 12 Comparison of EATS with other methods using CORD-19 biomedical selected published reports based on Severe Acute Respiratory Syndrome where domain expert is constructed summary considered as the model summary

ROUGE value		Summarization techniques							
		TR [67]	LR [68]	GOW [69]	CR [70]	FR [71]	LSA [72]	SB [73]	EATS
Recall	N = 1	0.5600	0.5485	0.3800	0.4550	0.4100	0.4830	0.3850	**0.7302**
	N = 2	0.3928	0.3896	0.4120	0.4010	0.4020	0.4252	0.3700	**0.7122**
	LCS	0.3300	0.3294	0.2646	0.3120	0.2900	0.3110	0.3004	**0.7312**
Precision	N = 1	0.5893	0.5796	0.5670	0.5460	0.5630	0.5640	0.5200	**0.7277**
	N = 2	0.4801	0.4991	0.4730	0.4570	0.4240	0.4120	0.3670	**0.7327**
	LCS	0.2998	0.2983	0.2787	0.2378	0.2830	0.2039	0.2120	**0.7021**
f-score	N = 1	0.5600	0.5702	0.4984	0.5288	0.5159	0.5078	0.4976	**0.7906**
	N = 2	0.4237	0.4239	0.4030	0.4150	0.3890	0.4006	0.3750	**0.7098**
	LCS	0.3010	0.3070	0.2890	0.2310	0.2270	0.2130	0.2290	**0.7314**

Table 13 Comparison of EATS with other methods using CORD-19 biomedical selected published reports based on Middle East Respiratory Syndrome where domain expert is constructed summary considered as the model summary

ROUGE value		Summarization techniques							
		TR [67]	LR [68]	GOW [69]	CR [70]	FR [71]	LSA [72]	SB [73]	EATS
Recall	N = 1	0.5600	0.5485	0.3800	0.4550	0.4100	0.4830	0.3850	**0.7106**
	N = 2	0.3928	0.3896	0.4120	0.4010	0.4020	0.4252	0.3700	**0.6743**
	LCS	0.3300	0.3294	0.2646	0.3120	0.2900	0.3110	0.3004	**0.6593**
Precision	N = 1	0.5893	0.5796	0.5670	0.5460	0.5630	0.5640	0.5200	**0.6153**
	N = 2	0.4801	0.4991	0.4730	0.4570	0.4240	0.4120	0.3670	**0.6543**
	LCS	0.2998	0.2983	0.2787	0.2378	0.2830	0.2039	0.2120	**0.6781**
f-score	N = 1	0.5600	0.5702	0.4984	0.5288	0.5159	0.5078	0.4976	**0.7121**
	N = 2	0.4237	0.4239	0.4030	0.4150	0.3890	0.4006	0.3750	**0.7069**
	LCS	0.3010	0.3070	0.2890	0.2310	0.2270	0.2130	0.2290	**0.7354**

In a ROC curve, the X-axis shows False Agree Rate (FAR) and Y axis represents the True Agree Rate (TAR). The top left point where the TAR is one and FAR is zero is known as the 'ideal' point of the ROC plotting. It denotes that the higher the accuracy of the summarizer as larger the area under the curve and vice versa. It is ideal to minimize the FAR while maximizing the TAR. So, the steepness of the ROC plotting is considered a crucial aspect. ROC curve for Summarization of CORD-19 datasets are demonstrated in Figs. 3 and 4. From Figs. 3 and 4, it is observed that the Area Under Curve (AUC) for the proposed EATS method is higher than other considered summarization techniques, though the AUC for LSA, TR, and LR are very closed to that of EATS.

Table 14 Evaluation matrix for summarizer

		Predicted	
		Agree	Disagree
Actual	Agree	u	v
	Disagree	w	x

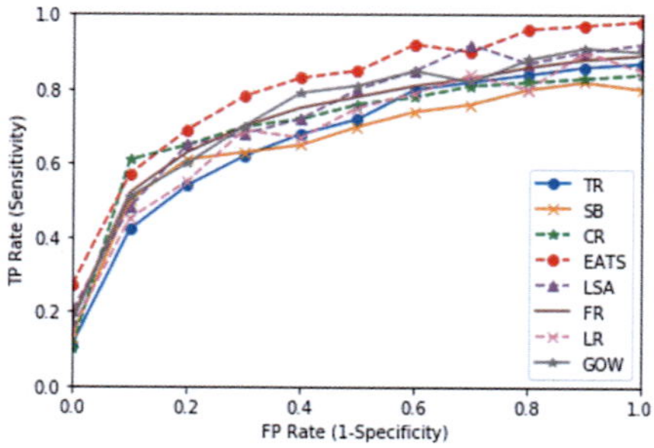

Fig. 3 ROC plotting for the CORD-19 biomedical report with Paper-ID 32389849

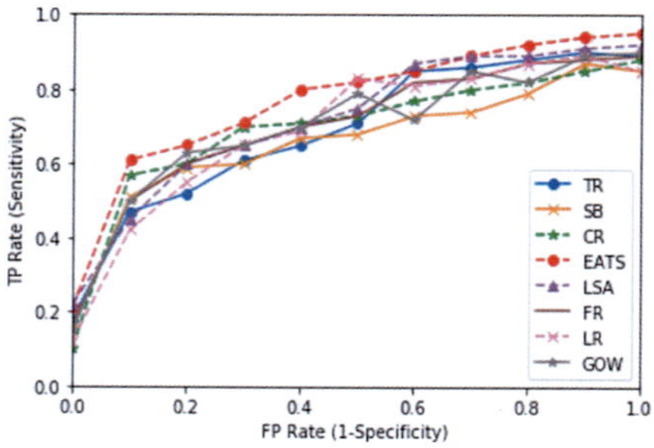

Fig. 4 ROC plotting for the extracted 1000 CORD-19 reports

5 Conclusion and Future Direction

The proposed method presents an integration of graph theory and an evolutionary algorithm-based approach for achieving extractive summarization. The proposed bi-objective strength Pareto evolutionary algorithm uses the clustering coefficient of the generated graph for defining one objective function and the length of the summary for other objective functions. The method is applied for CORD-19 biomedical reports and achieved promising results. In the method, each sentence of the text report is vectorized using the bag-of-words concept, but in the bio-medical domain, each sentence can be represented only by the medical terms of the sentences to extract more relevant information. Also SPEA2 (i.e., improved SPEA) [75] can be incorporated with other compact fitness assignment strategy, which considers density information. We can also consider some lexical relationships between the medical terms in order to summarize the bio-medical reports and an ensemble summarization technique may be proposed to increase the accuracy. The proposed automated summarizer may be helpful to the researchers and scientists of healthcare to make themselves updated about the new research information by devoting a minimum amount of time. Thus the clinicians and clinical researchers can improve the quality of their laboratories and develop a smart healthcare system.

References

1. H. Lu, C.W. Stratton, Y.-W. Tang, Outbreak of pneumonia of unknown etiology in Wuhan, China: the mystery and the miracle. J. Med. Virol. **92**(4), 401–402 (2020)
2. L.Y. Hsu, P.Y. Chia, J.F.Y. Lim, The novel coronavirus (SARS-COV-2) pandemic (2020)
3. C. Sohrabi, Z. Alsafi, N. O'Neill, M. Khan, A. Kerwan, A. Al-Jabir, C. Iosifidis, R. Agha, World health organization declares global emergency: a review of the 2019 novel coronavirus (covid-19). Int. J. Surg. (2020)
4. J. Bedford, J. Farrar, C. Ihekweazu, G. Kang, M. Koopmans, J. Nkengasong, A new twenty-first century science for effective epidemic response. Nature **575**(7781), 130–136 (2019)
5. C. Birkle, D.A. Pendlebury, J. Schnell, J. Adams, Web of science as a data source for research on scientific and scholarly activity. Quan. Sci. Stud. **1**(1), 363–376 (2020)
6. F.B. Hamzah, C. Lau, H. Nazri, D.V. Ligot, G. Lee, C.L. Tan, M.K.B.M. Shaib, U.H.B. Zaidon, A.B. Abdullah, M.H. Chung, et al, Coronatracker: worldwide covid-19 outbreak data analysis and .prediction. Bull. World Health Organ. **1**(32) (2020)
7. B. Xie, D. He, T. Mercer, Y. Wang, D. Wu, K.R. Fleischmann, Y. Zhang, L.H. Yoder, K.K. Stephens, M. Mackert, et al. Global health crises are also information crises: a call to action. J. Assoc. Inf. Sci. Technol. **71**(21), 1419–1423 (2020)
8. M. Cinelli, W. Quattrociocchi, A. Galeazzi, C.M. Valensise, E. Brugnoli, A.L. Schmidt, P. Zola, F. Zollo, A. Scala, The covid-19 social media infodemic. arXiv preprint arXiv:2003.05004 (2020)
9. B. Swire-Thompson, D. Lazer, Public health and online misinformation: challenges and recommendations. Ann. Rev. Public Health **41**, 433–451 (2020)
10. J.P.A Ioannidis, Coronavirus disease 2019: the harms of exaggerated information and non-evidence-based measures. Euro. J. Clin. Invest. **50**(4), e13222 (2020)
11. World Health Organization et al, Epi-win: Who information network for epidemics. World Health Organization (2020)
12. Pamment, J, The EU's role in fighting disinformation: taking back the initiative. Future (2020)
13. L.L. Wang, K. Lo, Y. Chandrasekhar, R. Reas, J. Yang, D. Eide, K. Funk, R. Kinney, Z. Liu, W. Merrill, et al, Cord-19: The Covid-19 open research dataset. ArXiv (2020)
14. H. Taub Tabib, M. Shlain, S. Sadde, D. Lahav, M. Eyal, Y. Cohen, Y. Goldberg, Interactive extractive search over biomedical corpora, in *Proceedings of the 19th SIGBioMed Workshop on Biomedical Language Processing* (2020), pp. 28–37
15. F. Mehta, Machine learning techniques for document summarization: a survey (2016)
16. C. Shekhar Yadav, A. Sharan, Hybrid approach for single text document summarization using statistical and sentiment features. Int. J. Inf. Retrieval Res. (IJIRR), **5**(4), 46–70 (2015)
17. H. Saggion, G. Lapalme, Generating indicative-informative summaries with sumum. Comput. Linguist. **28**(4), 497–526 (2002)
18. Y. Gong, X. Liu, Generic text summarization using relevance measure and latent semantic analysis, in *Proceedings of the 24th Annual International ACM SIGIR Conference on Research and Development in Information Retrieval* (ACM, 2001), pp. 19–25
19. T. Dunning, Accurate methods for the statistics of surprise and coincidence. Comput. Linguistics **19**(1), 61–74 (1993)
20. H. Eduard, C.-Y. Lin, Automated text summarization and the summarist system, in *Proceedings of a Workshop on Held at Baltimore, Maryland:*, October 13-15, 1998 (Association for Computational Linguistics, 1998), pp. 197–214
21. S. Dhingra, S. Sharma, P. Kaur, C. Dabas, Fault tolerant streaming of live news using multi-node cassandra. in *2017 Tenth International Conference on Contemporary Computing (IC3)* (IEEE, 2017)
22. C.-Y. Lin., E. Hovy, The automated acquisition of topic signatures for text summarization. in *Proceedings of the 18th Conference on Computational Linguistics-Volume 1* (Association for Computational Linguistics, 2000), pp. 495–501
23. A. Khan, N. Salim, A review on abstractive summarization methods. J. Theor. Appl. Inf. Technol. **59**(1), 64–72 (2014)

24. K.-F. Wong, M. Wu, W. Li, Extractive summarization using supervised and semi-supervised learning. in *Proceedings of the 22nd International Conference on Computational Linguistics-Volume 1* (Association for Computational Linguistics, 2008), pp. 985–992
25. M.-Y. Kan, K.R. McKeown, J.L. Klavans, Applying natural language generation to indicative summarization. in *Proceedings of the 8th European workshop on Natural Language Generation-Volume 8* (Association for Computational Linguistics, 2001), pp. 1–9
26. J. Tang, L. Yao, D. Chen, Multi-topic based query-oriented summarization. in *Proceedings of the 2009 SIAM International Conference on Data Mining*, SIAM (2009), pp. 1148–1159
27. J.-Y. Yeh, H.-R. Ke, W.-P. Yang, I.-H. Meng, Text summarization using a trainable summarizer and latent semantic analysis. Inf. Process. Manage. **41**(1), 75–95 (2005)
28. M. Litvak, M. Last, Graph-based keyword extraction for single-document summarization. In *Proceedings of the workshop on Multi-source Multilingual Information Extraction and Summarization* (Association for Computational Linguistics, 2008), pp. 17–24
29. J. Goldstein, V. Mittal, J. Carbonell, M. Kantrowitz, Multi-document summarization by sentence extraction. in *Proceedings of the 2000 NAACL-ANLP Workshop on Automatic summarization* (Association for Computational Linguistics, 2000), pp. 40–48
30. X. Wan, J. Yang, J. Xiao, Using cross-document random walks for topic-focused multi-document. in *2006 IEEE/WIC/ACM International Conference on Web Intelligence (WI 2006 Main Conference Proceedings)(WI 2006)* (IEEE, 2006), pp. 1012–1018
31. S. Harabagiu, F. Lacatusu, Topic themes for multi-document summarization. in *Proceedings of the 28th Annual International ACM SIGIR Conference on Research and Development in Information Rtrieval* (ACM 2005), pp. 202–209
32. A. Nenkova, K. McKeown, A survey of text summarization techniques. in *Mining text data* (Springer, 2012), pp. 43–76
33. L. Szilágyi, S.M. Szilágyi, Efficient Markov clustering algorithm for protein sequence grouping. in *Engineering in Medicine and Biology Society (EMBC), 2013 35th Annual International Conference of the IEEE* (IEEE, 2013), pp. 639–642
34. T.Di. Battista, D.D. Spalatro, A bootstrap method for adaptive cluster sampling. in *Classification and Data Analysis* (Springer, 1999), pp. 19–26
35. F. Geerts, R. Ndindi, Bounded correlation clustering. Int. J. Data Sci. Anal. **1**(1), 17–35 (2016)
36. S. Mohamed, A.I. Abdelmoty, Spatio-semantic user profiles in location-based social networks. Int. J. Data Sci. Anal. **4**(2), 127–142 (2017)
37. S. Wang, C.F. Eick, A data mining framework for environmental and geo-spatial data analysis. Int. J. Data Sci. Anal. **5**(2–3), 83–98 (2018)
38. M. Xiong, W. Li, J. Zhao, L. Jin, E. Boerwinkle, Feature (gene) selection in gene expression-based tumor classification. Mol. Gen. Metab. **73**(3), 239–247 (2001)
39. C. Lazar, J. Taminau, S. Meganck, D. Steenhoff, A. Coletta, C. Molter, V. de Schaetzen, R. Duque, H. Bersini, A. Nowe, A survey on filter techniques for feature selection in gene expression microarray analysis. IEEE/ACM Trans. Comput. Biol. Bioinform. **9**(4), 1106–1119 (2012)
40. I. Inza, P. Larrañaga, Y. Saeys, A review of feature selection techniques in bioinformatics. Bioinformatics **23**(19), 2507–2517 (2007)
41. P. Mitra, C.A. Murthy, S.K. Pal, Unsupervised feature selection using feature similarity. IEEE Trans. Pattern Anal. Mach. Intell. **24**(3), 301–312 (2002)
42. L. Song, A. Smola, A. Gretton, K.M. Borgwardt, J. Bed, Supervised feature selection via dependence estimation. in *Proceedings of the 24th International Conference on Machine Learning*, ICML 2007, New York, NY, USA (ACM, 2007), pp. 823–830
43. D.T. Nguyen, L. Chen, C.K. Chan et al., Clustering with multiviewpoint-based similarity measure. IEEE Trans. Knowl. Data Eng. **24**(6), 988–1001 (2012)
44. E. Zitzler, L. Thiele, An evolutionary algorithm for multiobjective optimization: the strength pareto approach. TIK-report, **43** (1998)
45. S. Bird, E. Klein, E. Loper, Natural Language Processing with Python, O'Reilly (2009)
46. P. Willett, The porter stemming algorithm: then and now. Program **40**(3), 219–223 (2006)
47. M.F. Porter, Snowball: a language for stemming algorithms (2001)

48. R. Krovetz, Viewing morphology as an inference process. in *Proceedings of the 16th Annual International ACM SIGIR Conference on Research and Development in Information Retrieval* (ACM, 1993), pp. 191–202
49. J. Xu, W.B. Croft, Corpus-based stemming using cooccurrence of word variants. ACM Trans. Inf. Syst. (TOIS) (1998)
50. G. Kou, Y. Peng, An application of latent semantic analysis for text categorization. Int. J. Comput. Commun. Control **10**(3), 357–369 (2015)
51. A. Kanapala, S. Jannu, R. Pamula, Summarization of legal judgments using gravitational search algorithm. Neural Comput. Appl. **31**(12), 8631–8639 (2019)
52. K. Soumya George, S. Joseph, Text classification by augmenting bag of words (bow) representation with co-occurrence feature. IOSR J. Comput. Eng. **16**(1), 34–38 (2014)
53. Soumen Kumar Pati and Asit Kumar Das, Ensemble classifier design selecting important genes based on extracted features. Int. J. Data Min. Bioinform. **19**(2), 117–149 (2017)
54. C. Mallick, A.K. Das, M. Dutta, A.K. Das, A. Sarkar, Graph-based text summarization using modified textrank. in *Soft Computing in Data Analytics* (Springer, 2019), pp. 137–146
55. H.S. Behera, J. Nayak, B. Naik, A. Abraham, in *Computational Intelligence in Data Mining: Proceedings of the International Conference on CIDM 2017*, vol 711 (Springer, 2018)
56. N. Hou, F. He, Y. Zhou, Y. Chen, X. Yan, A parallel genetic algorithm with dispersion correction for HW/SW partitioning on multi-core CPU and many-core GPU. IEEE Access **6**, 883–898 (2018)
57. S.K. Pati, A.K. Das, A. Ghosh, Gene selection using multi-objective genetic algorithm integrating cellular automata and rough set theory. in *International Conference on Swarm, Evolutionary, and Memetic Computing* (Springer, 2013), pp. 144–155
58. C.W. Ahn, R.S. Ramakrishna, Elitism-based compact genetic algorithms. IEEE Trans. Evol. Comput. **7**(4), 367–385 (2003)
59. S. Yang. Genetic algorithms with elitism-based immigrants for changing optimization problems. in *Workshops on Applications of Evolutionary Computation* (Springer, 2007), pp. 627–636
60. J.R. Inglis, R. Sever, biorxiv: a progress report. *ASAPbio blog.* (2016). http://asapbio.org/biorxiv. Accessed 14 Jan 2019
61. medrxivs, https://www.medrxiv.org, note = . Accessed 13 Oct 2020
62. Pubmed medline dataset, https://www.nlm.nih.gov/databases/download/pubmed_medline.html, note = . Accessed 13 Oct 2020
63. kaggle cord-19, https://www.kaggle.com/allen-institute-for-ai/CORD-19-research-challenge, note = . Accessed 13 Oct 2020
64. Python 2.7.14 documentation, https://docs.python.org/2/index.html. Accessed 13 Oct 2020
65. M.M. Haque, S. Pervin, Z. Begum (2016), Enhancement of keyphrase-based approach of automatic Bangla text summarization. in *Region 10 Conference (TENCON), 2016 IEEE*, (IEEE, 2016)
66. L.H. Reeve, H. Han, A. Brooks, The use of domain-specific concepts in biomedical text summarization. Inf. Process. Manage. **43**, 1765–1776 (2007)
67. R. Mihalcea, P. Tarau, Textrank: bringing order into text. in *Proceedings of the 2004 Conference on Empirical Methods in Natural Language Processing* (2004)
68. G. Erkan, D.R. Radev, Lexrank graph-based lexical centrality as salience in text summarization. J. Artif. Intell. Res. **22**, 457–479 (2004)
69. A. Tixier, K. Skianis, M. Vazirgiannis. Gowvis: a web application for graph-of-words-based text visualization and summarization. *Proceedings of ACL-2016 System Demonstrations* (2016), pp. 151–156
70. N. Garg, B. Favre, K. Reidhammer, D. Hakkani-Tür. Clusterrank: a graph based method for meeting summarization. in *Tenth Annual Conference of the International Speech Communication Association* (2009)
71. A.V. Luong, N.-T. Tran, V.-G. Ung, M.-Q. Nghiem, Word graph-based multi-sentence compression: Re-ranking candidates using frequent words. in *Seventh International Conference on Knowledge and Systems Engineering (KSE)* (IEEE, 2015), pp. 55–60

72. J. Steinberger, K. Jezek, Using latent semantic analysis in text summarization and summary evaluation. Proc. ISIM **4**, 93–100 (2004)
73. A. Nenkova, L. Vanderwende. The impact of frequency on summarization. *Microsoft Research, Redmond, Washington, Tech. Rep. MSR-TR-2005*, 101 (2005)
74. G.M. Foody. Status of land cover classiffication accuracy assessment. Remote Sens. Environ. **80**(1), 185–201 (2002)
75. E. Zitzler, M. Laumanns, L. Thiele, Spea2: improving the strength pareto evolutionary algorithm. *TIK-report*, 103 (2001)

Chest CT in COVID-19 Pneumonia: Potentials and Limitations of Radiomics and Artificial Intelligence

Federica Vernuccio, Giuseppe Cutaia, Roberto Cannella, Laura Vernuccio, Roberto Lagalla, and Massimo Midiri

Abstract SARS-CoV-2 epidemics has resulted in an unprecedented global health crisis causing a high number of deaths with pneumonia being the most common manifestation. Chest CT is the best imaging modality to identify pulmonary involvement, but unfortunately there are no pathognomonic features for COVID-19 pneumonia, since many other infectious and non-infectious diseases may cause similar alterations. The adoption of artificial intelligence in biomedical imaging has the potential to revolutionize the identification, management, and the patient's outcome. If adequately validated, it could be used as a support with predictive and prognostic purposes in symptomatic patients but also as a screening test in asymptomatic patients in COVID-19 epidemics. Some studies have already shown the potential adoption of artificial intelligence for detection of COVID-19 infection, or even to differentiate from community-acquired pneumonia, but at present artificial intelligence cannot routinely applied for COVID-19 due to several limitations. This book chapter will first revise the basics of radiomics with a short practical and easy guide for radiologists; then, the main radiological findings of COVID-19 pneumonia will be presented with the most relevant information that are assessed to evaluate extent of the disease; finally, the main current literature on potential clinical application of radiomics and artificial intelligence for COVID-19 will be presented together with limitations and perspectives.

F. Vernuccio (✉) · G. Cutaia · R. Cannella · R. Lagalla · M. Midiri
Department of Biomedicine, Neuroscience and Advanced Diagnostics (Bi.N.D.), University of Palermo, Via del Vespro 129, 90127 Palermo, Italy

R. Lagalla
e-mail: roberto.lagalla@unipa.it

M. Midiri
e-mail: massimo.midiri@unipa.it

G. Cutaia · R. Cannella
Department of Health Promotion, Mother and Child Care, Internal Medicine and Medical Specialties (PROMISE), University of Palermo, Palermo, Italy

L. Vernuccio
U.O.C. Geriatria E Lungodegenza, AOUP University Hospital Palermo, CDCD Geriatria, University Hospital Palermo, Palermo, Italy

J. Nayak et al. (eds.), *Understanding COVID-19: The Role of Computational Intelligence*, Studies in Computational Intelligence 963, https://doi.org/10.1007/978-3-030-74761-9_3

Keywords COVID-19 · Radiomics · Artificial intelligence · GLCM · GLRLM · GLSZM · GLDM

1 Introduction

Severe acute respiratory syndrome coronavirus 2 (SARS-CoV-2) is part of the Coronaviridae family that in the past caused serious epidemics, although with geographically less extensive pandemic damage, such as SARS and MERS, respectively in 2003 and 2012 [1]. SARS-CoV-2 has a genome consisting of a single strand of positive sense RNA of about 30 kb and the transmission occurs mainly by means of respiratory droplets, although the mode of infection with skin contact mechanisms has been also demonstrated [2, 3]. The virus within the droplets penetrates the broncho-alveolar mucosa using the angiotensin-converting enzyme 2 (ACE2) enzyme receptor, thus first down-regulating ACE2, and then causing the release of inflammatory cytokines, which leads to a diffuse alveolar damage (DAD) [4–7]. As consequence, the first expression of the disease is the DAD in which the alveoli are coated with eosinophilic material, composed of necrotic pneumocytes and plasma proteins which, in the most serious cases, are followed by a complete filling of the alveolus, by exudate, often hemorrhagic, up to the onset of an acute respiratory distress syndrome. The expression of the ACE2 gene in the nasal mucosa of adults seems to be greater than that of children and this could explain why the disease has a higher frequency in the adult population [8].

COVID-19 may involve many organs, but the lung is the most common, with clinical presentation ranging from asymptomatic patients to respiratory insufficiency requiring mechanical ventilation and admission to intensive care units (ICU) [7, 9, 10]. The histological and imaging changes induced by SARS-CoV-2 are very similar to those caused by SARS-CoV-1 and MERS and this is not surprising as they belong to the same virus family [11–13]. However, the type of functional, anatomopathological and radiological evolution in COVID-19 pneumonia is yet not clear due to paucity of date on long-term follow-up. In the near future, it will be important to also assess the respiratory function, complications, and quality of life of subjects discharged from the ICUs as these patients are expected to likely have the worst lung the long-term compromise.

The radiological assessment of lung parenchyma in COVID-19 patients can be performed by ultrasound (US), chest x-ray (CXR), and computed tomography (CT) [9, 10]. Chest CT is the best imaging modality to identify pulmonary involvement in COVID-19, but unfortunately there are no pathognomonic features for COVID-19 pneumonia, since many other diseases, infectious and non-infectious, may cause similar alterations [9, 10, 14]. As chest CT has been widely performed during this pandemic, a large amount of radiological data on this specific disease are now available and this has increased the interest on the application of radiomics and artificial intelligence (AI) for COVID-19 pneumonia, with promising, but yet to be validated, results [15]. This book chapter will first revise the basics of radiomics with a short practical and easy guide for radiologists; then, radiological findings of COVID-19 pneumonia will be presented with the most relevant information that are assessed to evaluate extent of the disease; finally, the main current literature on potential clinical application of radiomics and AI for COVID-19 will be presented together with limitations and perspectives.

2 Literature Study and Methodology

The following section describes the systematic review and workflow of the radiomics and artificial intelligence in biomedical imaging.

2.1 Radiomics and Artificial Intelligence

Photographies are similar to medical images: they depict a 3D or 4D real world in a bidimensional space and, in addition to what we can see, they contain not-visible information such as emotions and feelings of both the photographer and the photographed subjects. In a similar way, medical images contain more information than those that human's eye can appreciate. Thanks to the technological advances this information – i.e., radiomics features – may be unveiled, extracted, quantified, and analyzed through radiomics tool [16]. The analysis of this information represents one of the basis of AI.

AI is currently used in many settings of everyday life. Among the many possibilities related to the adoption of AI there is the possibility to recognize specific objects and individuals by simply uploading images into a web search engine. This is possible thanks to the meticulous labeling of millions of images and application of image recognition algorithms for machine learning. Similarly, it is expected that the routine adoption of AI in CT/MR imaging will help to increase radiologists' confidence and diagnostic accuracy in many fields, but the potential application of AI in radiology can be further extended to the improvement of CT/MR protocols and in reporting workflow, especially in the emergency department [17]. Anyhow, it must be acknowledged that the clinical applicability is still under investigation and the reasons are multifold. Among these, the clinical impact of current radiomics studies has been lowered because of the differences among patient populations, CT/MRI acquisition parameters, reconstruction, and segmentation techniques [18–21]. All scientists working with radiomics and AI – e.g., engineers, clinicians, radiologists, technician, hospital managers - should be aware of the theoretical background behind these solutions.

2.2 *The Workflow of Radiomics: A Short Practical and Easy Guide for Radiologists*

Radiomics refers to the process of extracting and quantifying mineable high dimensional features from radiological images and to post-process them using advanced feature analysis algorithms with the aim of building a predictive model. The overall workflow of radiomics is shown in Fig. 1 and each of the steps has a relevant role to ensure the robustness and reproducibility of radiomics and AI algorithms [17].

First, the clinical chest CT images are acquired and collected. Chest CT protocol has been standardized by the Fleischner Society for the management of lung nodules, but the same protocol (thin sections of 1.5 mm or less, low radiation technique adjusting exposure factors to body habitus) is used for lung imaging in a broader sense, including the assessment of pneumonia [10, 22]. An important factor is the use of comparable acquisition technique as it has been demonstrated that about 11.3% of radiomic features vary based on different technical parameters [19]. Therefore, the adoption of consistent technical parameters allows for obtaining more reproducible data in the long term. Whenever images with variable technical parameters are processed, it is suggested to improve reproducibility of radiomic analysis by choosing the most reproducible radiomic features. The second step of the radiomic process is represented by image segmentation [17, 23]. Image segmentation consists in the delineation of the regions of interest, which are represented in COVID-19 pneumonia by areas of ground glass opacities (GGOs) that may coalesce into consolidations. The segmentation process may include different approaches such as the use of dimensional (2D) or three-dimensional (3D) areas or volumes of interest, the segmentation of an area or volume with a predefined value in terms of width

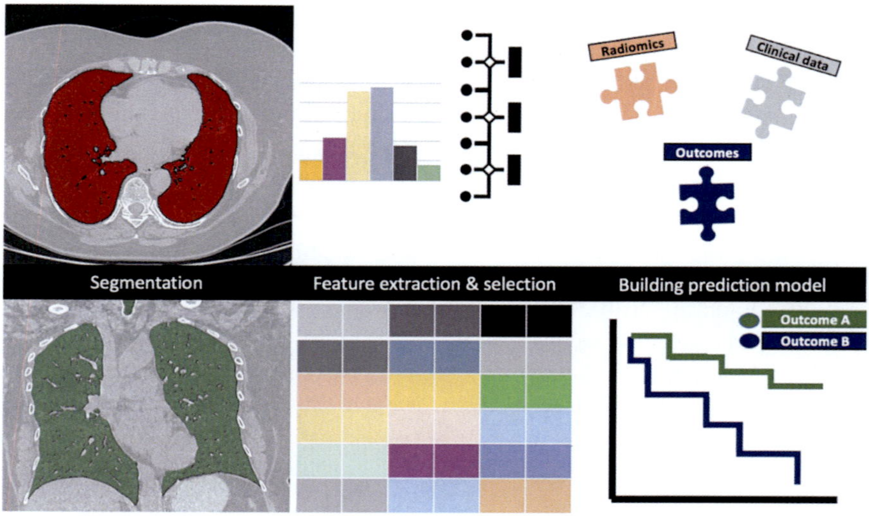

Fig. 1 Workflow of radiomics

or the contouring of an area or volume, the inclusion of the affected area with or without inclusion of normal parenchyma or tissue, as well as the manual, semi-automated or automated process for contouring. This implies that the information obtained from the different segmented area or volumes may be potentially different and operator-dependent [20, 23].

Texture features are then extracted through an algorithm that allows the characterization of the segmented area or volume of interests [17, 23]. Imaging features may be included within many texture matrices, such as "grey level co-occurrence" (GLCM), "grey level run-length" (GLRLM), "grey level size-zone" (GLSZM), "gray level dependence" (GLDM) and "neighborhood grey tone difference" (NGTDM) [17, 23–26]:

- GLCM assesses every combination of gray level pairs at distances and angles that may be chosen;
- GLRLM considers runs of pixels and the calculation of the texture features resulting from the matrix GLRLM can differ between different radiomics software;
- GLSZM looks at size of homogeneous zones for each grey-level in 3 dimensions;
- GLDM quantifies the number of connected voxels within a predefined distance;
- NGTDM looks at the difference between a certain gray value and the average gray value of its neighbors.

Finally, a filtering process- i.e., feature selection from a dataset - is performed and a multi-feature signature is then built; this latter should be high efficient and accurate in the identification of the disease and correlation with different outcomes [17, 27, 28].

2.3 From Radiomics to Artificial Intelligence: A Short Practical and Easy Guide for Radiologists

Nowadays, both radiomics and AI have been getting attention in the scientific literature, but there is still difficulties on the differences between them. Radiomics allows extraction of quantitative information from CT/MRI images [16, 17]. AI is broad term to indicate a set of advanced computational algorithms that basically learn the patterns from input data provided to make predictions (output) on unseen data sets. These algorithms may be used to either analyze provided numeric data or directly analyze the medical images in order to automatically design its own radiomic features, with a process called deep learning [29]. Deep learning solutions may be used in the radiological field not only for diagnosis or outcome prediction for a specific disease but also for image preprocessing, reconstruction and denoising, for lesion detection, segmentation and classification, or to enhance diagnostic workflows [29]. Deep learning adopts multi-layered neural networks to reach these outcomes [30]. Neural networks were first inspired by the human brain itself, although the network design used now is very loosely connected to it. These networks consist of neurons where the connections between them is modelled with weights, that are adjusted as learning proceeds. The major learning paradigms are supervised, semi-supervised, unsupervised and reinforcement learning, each corresponding to a particular learning task [31–33]:

- In supervised learning – i.e., the most common type used in radiology research – the algorithm is trained from a set of images or data where the output labels are already known and are always based on the reference standard;
- Semi-supervised learning combines labeled data based on reference standard etih unlabeled data;
- In unsupervised learning – used to draw inferences from unlabeled input data without reference standard – the algorithm is left on its own to find structure in the input data. Tasks that fall within this type of learning are in general estimation problems.
- Reinforcement learning consists in learning the associations between stimuli, actions, and the occurrence of rewards or penalties for the performed actions.

3 Radiological Findings in COVID-19 Pneumonia – A Clinical Approach

The current most common radiological approach in worldwide clinical routine is based on the radiologists' assessment of lung opacities in terms of types and extension rather than on radiomics or AI approach. The choice of imaging technique must take into account international recommendations but also depends on the experience of radiologists and the type of available equipment [10, 34]. Currently, the most used imaging techniques for assessment of lung parenchyma are x-ray (CXR) and CT [9]. Knowledge of the pulmonary changes in both imaging techniques is essential since COVID-19 pneumonia imaging findings change according to the stage of the disease [35].

3.1 Chest X-ray

CXR is an inexpensive and easily performed technique. In patients COVID-19 patients its use has not yet been completely defined with sensitivity ranging between 68 and 90% [36, 37] and increasing with time interval between symptom onset and exam acquisition [9]. Therefore, CXR is currently performed as a first-line investigation exclusively in geographic areas with a lack of real time PCR assays or unavailability of CT scanners [38]. CXR also has a fundamental role in hospitalized patients to monitor the progression of the disease and highlight any complications (e.g., pleural effusion, bacterial superinfection, pneumothorax, pneumomediastinum, and subcutaneous emphysema) [10, 39, 40]. In such patients, the Fleischner Society does not recommend performing a CXR every 24 h but the examination should be requested taking into account only the clinical and laboratory parameters of the patients [10, 41]. It is possible that in future clinical practice, portable CXR will replace most CT exams for imaging follow-ups in hospitalized patients as the mobile x-ray machine allows to acquire exams directly at the patient's bed, reducing the risk of cross-infection and avoiding to suspend the acquisition of secondary exams to the decontamination of the CT room requested after the scan of COVID-19 patients [42].

The study by Cozzi et al. [37] performed between the second and fifteenth day from the onset of symptoms on 234 patients with COVID-19 infection showed irregular or bilateral opacities and reticular-nodular consolidation, with basal and peripheral predominance in most cases. In this study, 66.6% of patients had reticulo-nodular opacities, 62.8% areas of ground glass (GGO), 57.7% areas of parenchymal consolidation, and 23.5% of pulmonary nodules [37]. In addition, 35.5% of patients in the study had both GGO and parenchymal consolidation areas at the same time and 69.2% had bilateral involvement, while when the alteration was unilateral the right lung was more frequently affected than the left (58 vs 42%) [37]. Bilateral involvement in COVID-19 pneumonia is useful as an important criterion for the differential diagnosis with community-acquired bacterial pneumonia (CAP, usually unilateral and involving only one lobe) [43]. To speed up the reporting system and make disease monitoring easier, some authors have developed CXR severity scoring systems [44]. Among these, the radiographic evaluation of pulmonary edema score is based on the extent of the CXR findings and each lung is assigned a score of 0–4 depending on percentage of involvement [36, 45].

3.2 Chest CT

As suggested by the World Health Organization guide, chest imaging should be done in symptomatic patients in case of unavailability or potential false negative real time PCR [46]. Many studies demonstrated the high sensitivity (>90%) of chest CT, with the drawback of a low specificity (up to 56%) due to the overlap of imaging features with other pneumonia [47–50].

In COVID-19 epidemics, chest CT is used not only as a diagnostic tool in patients with high clinical suspicion (i.e., recent exposure, severe-moderate clinical features), but also to monitor the evolution of the disease or any complications, to identify other causes that could justify patient respiratory symptoms (e.g., pulmonary edema), as well as in long-term follow-up programs in patients at high risk of fibrotic evolution [9, 10, 51, 52]. The assessment of lung parenchyma on CT requires a high resolution technique (HRCT), usually without injection of contrast medium, unless there is a high clinical suspicion of pulmonary embolism (PE), one of the potential complications of COVID-19 pneumonia in severe forms of the disease [9, 53, 54].

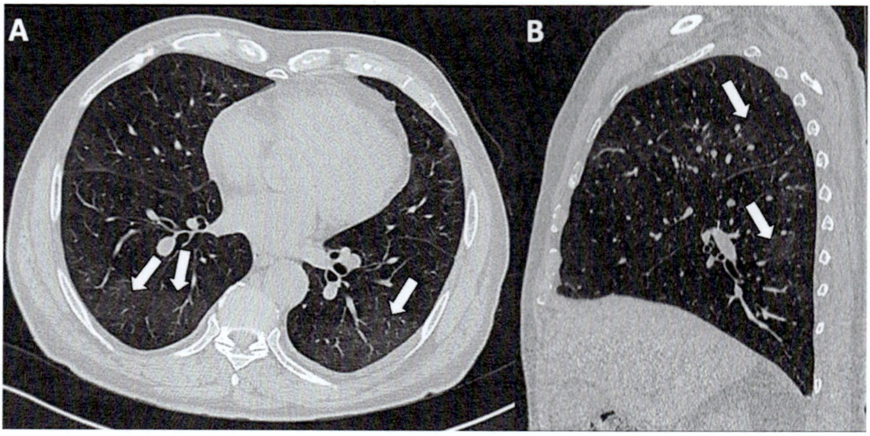

Fig. 2 CT images (axial, a – sagittal, **b** in a patient with COVID-19 pneumonia show bilateral ground glass opacities (arrows) with peripheral/subpleural distribution

In COVID-19 pneumonia, chest CT can be normal in 56% of cases within the first 3 days of the onset of symptoms [54–56]. In most cases, COVID-19 pneumonia has a fairly characteristic presentation on CT [57, 58] which includes the presence of bilateral, multilobar, subpleural/peripheral GGOs, with a posterior with apico-basal gradient (Fig. 2) [55, 59, 60].

With evolution of the disease, GGOs tend to evolve, both in extension and in attenuation value, thus leading to the crazy paving pattern (Fig. 3) - i.e., GGO with thickening of the interlobular septa that generally show an irregular subpleural and peribroncovascular distribution - and even parenchymal consolidation [53, 55, 61] with a peak around day 10, usually followed by a progressive decrease in extension and attenuation value of the lung damage [53, 55, 61–63].

The temporal evolution of CT findings of this pneumonia allows radiologists and clinicians to trace the period of infection with the virus with good approximation [35]. In the early stages of the disease, it is possible to observe an enlargement (diameter < 3 mm) of the subsegmental pulmonary vessels within GGO in up to 89% of cases [9]. This sign is more easily depicted with maximum intensity projection (MIP) and resolves when the GGOs evolve into parenchymal consolidation. This could be helpful to distinguish between the GGOs of patients with COVID-19 and the GGOs observed in other pneumonia viral [9], although this sign may be present also in non-COVID-19 pneumonia.

After 2–3 weeks from the symptom's onset (average 10 days), the peak of the disease is highlighted with the possible CT appearance of the reversed halo sign (i.e., atoll sign), reticular opacities (i.e., "strip-like opacities"), and bronchial wall thickening (Fig. 4) [59, 64].

Furthermore, bronchiectasis and traction broncholectasis can be observed within the parenchymal consolidation areas (Fig. 5) [65].

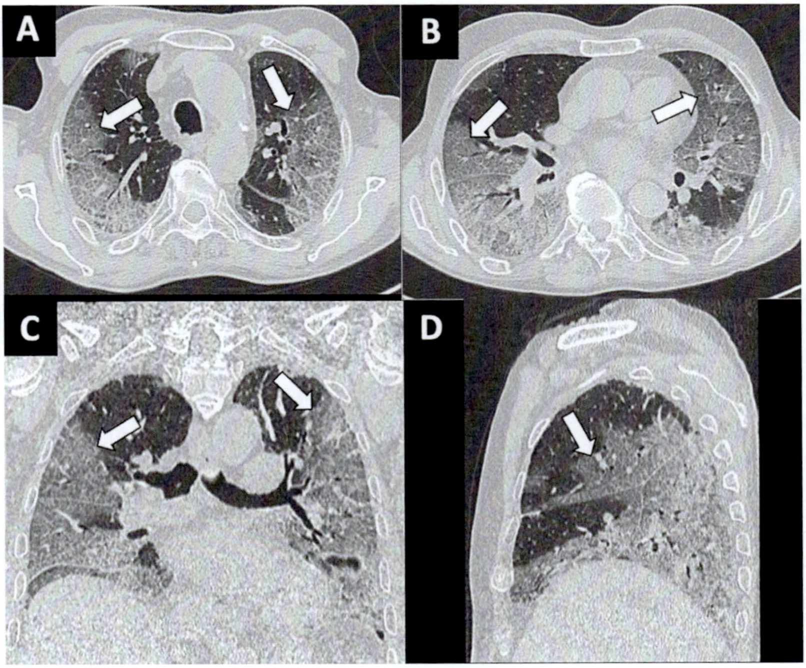

Fig. 3 CT images (axial, a, b – coronal, c – sagittal, **d** images show multiple and bilateral areas of crazy paving (arrows)

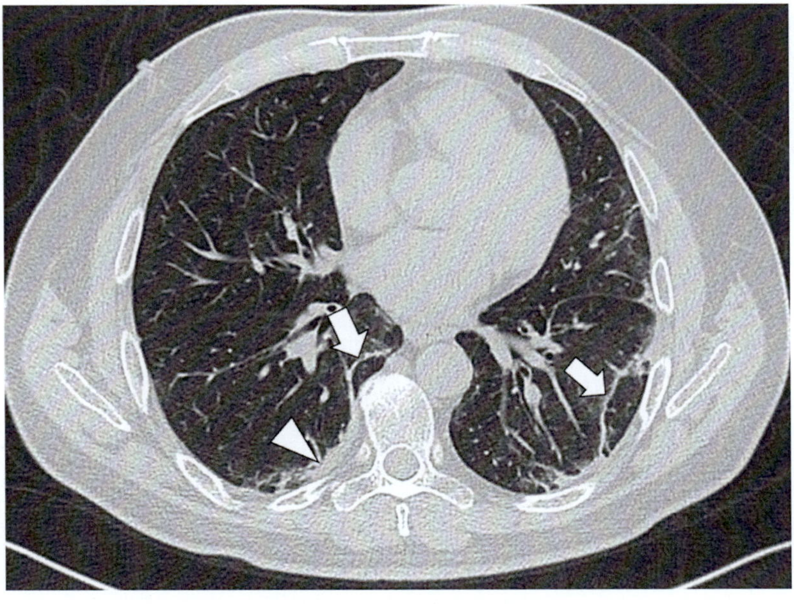

Fig. 4 Axial HRCT image demonstrate strip like opacities (arrows) with subpleural consolidation (arrowhead)

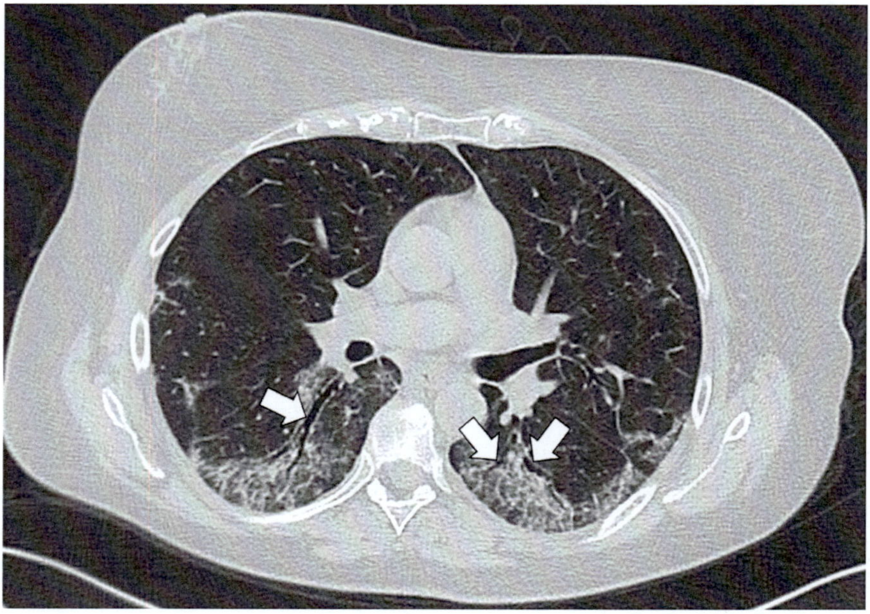

Fig. 5 CT image show traction bronchiectasis (arrows) within the parenchymal consolidation areas

After disease peak, there is a slow and gradual resolution of the HRCT findings, but it is still possible to notice the GGO areas, some parenchymal consolidations, and strip-like opacities even after 1 month [66]. Despite the young history of this epidemic, some studies showed a complete resolution of CT findings in cases of mild pneumonia, and a fibrotic evolution in patients with severe disease (e.g., patients recovered after a full-blown ARDS) although the proportion is not known yet [11, 53, 65, 67].

In addition to the aforementioned typical radiographic findings of COVID-19 pneumonia, some patients - mainly young or asymptomatic - may present atypical radiological findings including unilateral, unilobar, apical opacities that can involve the anterior and peribroncovascular areas and the target sign (Fig. 6) – which has been recently described [56, 60, 61, 68, 69].

In case of unilateral involvement, the right lung is more commonly affected [37]. Pleural effusion, pericardial effusion and enlargement of the mediastinal lymph nodes are atypical findings found in 5, 3.6 and 5.4% of cases, respectively, and usually associated with severe disease [61, 70–72]. The study by Jalaber et al. [53] states that 15 to 30% of hospitalized patients progress to ARDS, the leading cause of death from COVID-19. Other complications include pneumothorax – generally secondary to the rupture of a subpleural bulla –, pneumomediastinum and subcutaneous emphysema – generally secondary to the Macklin effect (Fig. 7) [73–75].

With the aim to facilitate the radiologist's job in the routine practice, the COVID-19 Reporting and Data System (CO-RADS) – which seeks to define the degree of

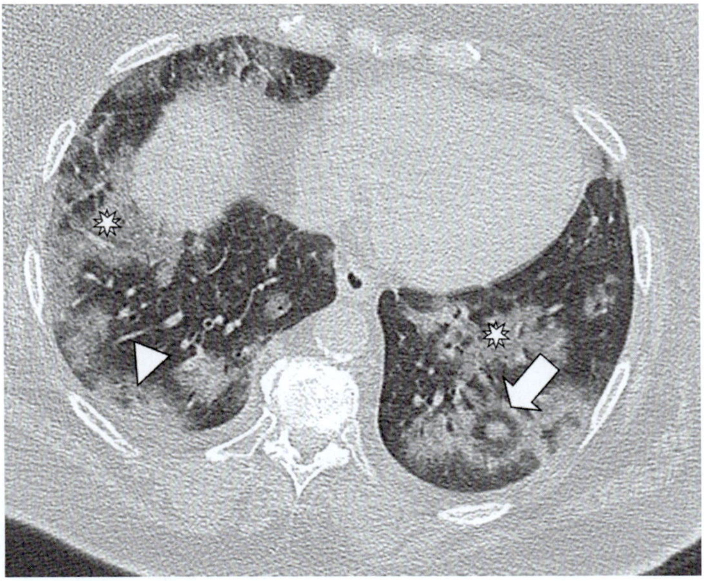

Fig. 6 Axial HRCT image depicts "target sign" (arrow) with extensive areas of crazy paving (asterisks) alternating with parenchymal consolidations (arrowhead)

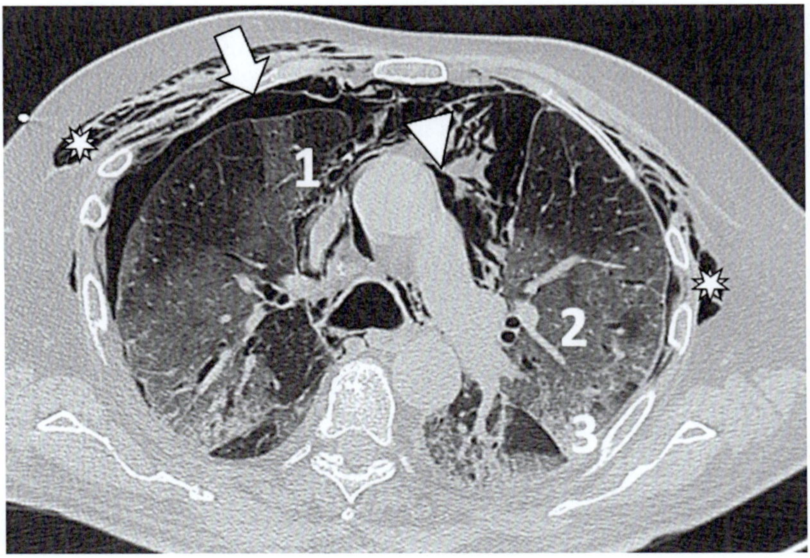

Fig. 7 CT image shows pneumomediastinum (arrowheads), right pneumothorax (arrows), bilateral chest subcutaneous emphysema (asterisks), as well as ground glass opacities (1), crazy paving areas (2), and subpleural consolidations (3)

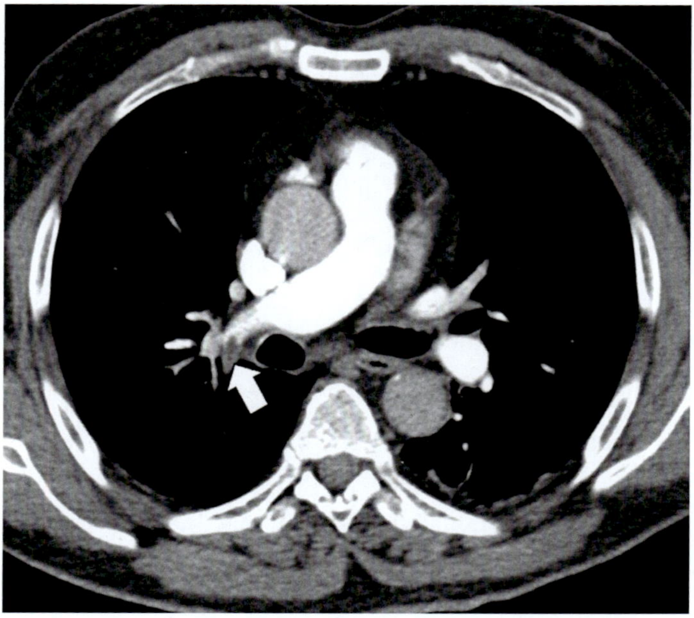

Fig. 8 Axial CT image with contrast medium injection in a COVID-19 patient with pulmonary embolism shows a filling defect in the right pulmonary artery

probability of COVID-19 pneumonia on radiological images – allows to assign a score from 1 to 5 to each exam according to a progressive scale of probability of positivity, while the CO-RADS 6 score is reserved for patients with positive RT-PCR [76].

Pulmonary coinfections and pulmonary embolism deserve a separate discussion. The study by Koehler et al. [77] demonstrates the overlap of pulmonary aspergillosis in 5 of 19 patients with ARDS associated with COVID-19, while viral co-infections are more frequent in pediatric age [78]. Pulmonary embolism is now considered an epiphenomenon of the disease rather than a real complication [54]; it occurs in 24–30% of patients and is secondary to the hypercoagulable state of critically ill patients with exacerbated systemic inflammatory response (Fig. 8) [54, 79, 80].

4 Clinical Applications of Radiomics

The main way to contain COVID-19 pandemic is to speed up the process of identifying infected patients. Several radiomics and AI tools are currently being developed, in particular based on deep learning. These studies focus on several topics but have as their main objectives the diagnostic, prognostic, and predictive evaluation of patients affected by COVID-19.

Xie et al. [81] assessed 474 radiomic features in 301 patients with GGOs on chest CT, including 33 patients with COVID-19, 136 with lung infections from other causes, 48 with lung adenocarcinoma, and 84 with benign indeterminate lesions and developed a non-invasive radiomic model for the diagnosis of COVID-19 achieving a sensitivity of 83% and a specificity of 90%. COVID-19 needs to be differentiated from influenza A pneumonia, although CT characteristics of these two types of pneumonia are similar [82]. Zeng et al. [83] have recently proposed a nomogram, developed on the basis of 8 out of 56 radiomic features was able to distinguish between patients with COVID-19 and influenza A pneumonia with an AUC of 0.87. Other radiomics studies on COVID-19 were focused on investigated how to predict outcome.

In addition, prediction of outcomes in patients with COVID-19 is highly relevant [84–87]. Wu et al. [84] assessed CT images of 492 positive RT-PCR patients acquired in four different hospitals including 40 patients (8.1%) with negative outcome were analyzed and, after automatic segmentation of lung lesions, the extraction of the features was made with the creation and validation of the radiomic signature (i.e., RadScore) to predict unfavorable results [84]. This RadScore created by authors was capable of stratifying COVID-19 patients into low- or high-RadScore groups having different survival time [84].

Clinical and radiomics information were also combined for outcome prediction in the study by Chao et al. [85]. These authors included CT datasets from Iran, USA, and Italy proposing a model that combines the radiomic features of lung lesions extracted through Pyradiomics package with clinical and laboratory results to predict patient need of admission to ICUs, achieving AUC values across the three datasets of up to 0.777 [85]. This study highlighted the variability accuracy of the proposed model based on scanner technologies and study site, thus underscoring that the most relevant results from radiomics studies need a regional/site finetuning of the models [85].

CT texture analysis may be also helpful for staging COVID-19. Wei et al. [86] created a radiomics signature with 92% sensitivity and 80% specificity for the differentiation between mild-to moderated and severe disease.

Another interesting outcome assessed by Cai et al. [88] using radiomics was the prediction of negativity of RT-PCR during treatment using chest CT images. The objective of this study is clinically relevant because until now, guidelines require a negative RT-PCR to discharge a COVID-19 patient. The model proposed in this study allowed the prediction of negativity of RT-PCR with sensitivity and specificity respectively of 76.5 and 62.5% in the training dataset and 78.4 and 60% in the testing dataset, thus being potentially a good indicator of the proper time for repeating RT-PCR test. Yue et al. [89] performed a multicenter study to predict length of hospitalization in COVID-19 pneumonia.

5 Clinical Applications of Artificial Intelligence

There are many ongoing studies on the clinical applications of AI on both CXR and CT images of COVID-19 pneumonia, including assessment of diagnosis, severity and outcome prediction. Apostolopoulos et al. [90] recently published a study on CXR images of more than 1400 patients (i.e. both normal CXR and CXR in patients with COVID-19 and non-COVID-19 pneumonia) and showed that deep learning may allow automatic detection and automatic extraction of essential characteristics for diagnosis of COVID-19. Li et al. [91] have proposed, tested, and approved the fully automatic COVID-19 detection neural network (COVNet) with to differentiate COVID-19 pneumonia from CAP and other lung diseases using 4352 chest CT scans, achieving a sensitivity and specificities of 90 and 96% for COVID-19 detection, and 87 and 92%, respectively, for CAP. Huang et al. [92] developed a deep learning method, using serial CT, to quantify and monitor the severity of the disease evaluating lung changes in the entire lung and five lobes.

From a clinical standpoint it is also of utmost importance to try to predict the time of progression towards the severe stage of the disease, in order to plan the most appropriate treatments in advance and this objective has been the basis for the study by Zhu et al. [60] that analyzed 408 chest CT studies and created a method that predicts sever cases with an accuracy of 77%.

6 Conclusion

Pneumonia is the main manifestation of COVID-19 with a wide range of clinical and radiological presentations. Considering the worldwide adoption of chest CT in this pandemic for diagnostic and prognostic purposes, a large amount of radiological data on this disease has become available, thus increasing the interest on the potential application of radiomics and AI. Many studies have been published so far for adoption of radiomics and AI in COVID-19 pneumonia with promising, but yet to be validated, results.

References

1. N. Petrosillo, G. Viceconte, O. Ergonul, G. Ippolito, E. Petersen, COVID-19, SARS and MERS: are they closely related? Clin. Microbiol. Infect. **26**, 729–734 (2020)
2. K. Yuki, M. Fujiogi, S. Koutsogiannaki, COVID-19 pathophysiology: a review. Clin. Immunol. **215**, 108427 (2020)
3. Q. Li, X. Guan, P. Wu et al., Early transmission dynamics in Wuhan, China, of novel coronavirus-infected pneumonia. N Engl. J. Med. **382**, 1199–1207 (2020)
4. F. Silhol, G. Sarlon, J.C. Deharo, B. Vaïsse, Downregulation of ACE2 induces overstimulation of the renin-angiotensin system in COVID-19: should we block the renin-angiotensin system? Hypertens Res. **43**, 854–885 (2020)
5. M. Mahmudpour, J. Roozbeh, M. Keshavarz, S. Farrokhi, I. Nabipour, COVID-19 cytokine storm: the anger of inflammation. Cytokine **133**, 155151 (2020)
6. Q. Ye, B. Wang, J. Mao, The pathogenesis and treatment of the 'Cytokine Storm' in COVID-19. J. Infect. **80**, 607–613 (2020)

7. N. Lopes, F. Vernuccio, C. Costantino et al., An Italian guidance model for the management of suspected or confirmed COVID-19 patients in the primary care setting. Front. Public Health (2020). https://doi.org/10.3389/fpubh.2020.572042
8. S. Bunyavanich, A. Do, A. Vicencio, Nasal gene expression of angiotensin-converting enzyme 2 in children and adults. JAMA **323**(23), 2427–2429 (2020)
9. A.R. Larici, G. Cicchetti, R. Marano et al., Multimodality imaging of COVID-19 pneumonia: from diagnosis to follow-up. A Compr. Rev. Eur. J. Radiol. **131**, 109217 (2020)
10. G.D. Rubin, C.J. Ryerson, L.B. Haramati et al., The role of chest imaging in patient management during the COVID-19 pandemic: a multinational consensus statement from the Fleischner society. Radiology **296**, 172–180 (2020)
11. T. Schaller, K. Hirschbühl, K. Burkhardt et al., Postmortem examination of patients with COVID-19. JAMA **323**, 2518–2520 (2020)
12. L.M. Barton, E.J. Duval, E. Stroberg, S. Ghosh, S. Mukhopadhyay, COVID-19 Autopsies, Oklahoma, USA. Am. J. Clin. Pathol. **153**, 725–733 (2020)
13. H.J. Koo, S. Lim, J. Choe, S.H. Choi, H. Sung, K.H. Do, Radiographic and CT features of viral pneumonia. RadioGraphics **38**, 719–739 (2018)
14. G. Soldati, A. Smargiassi, R. Inchingolo et al., Is there a role for lung ultrasound during the COVID-19 pandemic? J. Ultrasound Med. **39**, 1459–1462 (2020)
15. F. Homayounieh, S. Ebrahimian, R. Babaei et al., CT radiomics, radiologists and clinical information in predicting outcome of patients with COVID-19 Pneumonia. Radiol. Cardiothoracic Imaging **2**(4) (2020). https://doi.org/10.1148/ryct.2020200322
16. R.J. Gillies, P.E. Kinahan, H. Hricak, Radiomics: images are more than pictures, they are data. Radiology **278**, 563–577 (2016). https://doi.org/10.1148/radiol.2015151169
17. F. Vernuccio, R. Cannella, A. Comelli, G. Salvaggio, R. Lagalla, M. Midiri, Radiomica e intelligenza artificiale: nuove frontiere in medicina [Radiomics and artificial intelligence: new frontiers in medicine.]. Recenti. Prog. Med. **111**, 130–135 (2020)
18. A. Traverso, L. Wee, A. Dekker, R. Gillies, Repeatability and reproducibility of radiomic features: a systematic review. Int. J. Radiat. Oncol. Biol. Phys. **102**, 1143–1158 (2018)
19. M. Meyer, J. Ronald, F. Vernuccio et al., Reproducibility of CT radiomic features within the same patient: influence of radiation dose and CT reconstruction settings. Radiology **293**, 583–591 (2019)
20. C. Haarburger, G. Müller-Franzes, L. Weninger, C. Kuhl, D. Truhn, D. Merhof, Radiomics feature reproducibility under inter-rater variability in segmentations of CT images. Sci. Rep. **10**, 12688 (2020)
21. F. Prayer, J. Hofmanninger, M. Weber et al., Variability of computed tomography radiomics features of fibrosing interstitial lung disease: a test-retest study. Methods **S1046–2023**(20), 30194–30198 (2020)
22. H. MacMahon, D.P. Naidich, J.M. Goo et al., Guidelines for management of incidental pulmonary nodules detected on CT images: from the Fleischner society 2017. Radiology **284**, 228–243 (2017)
23. J.E. van Timmeren, D. Cester, S. Tanadini-Lang, H. Alkadhi, B. Baessler, Radiomics in medical imaging-"how-to" guide and critical reflection. Insights Imaging **11**, 91 (2020)
24. D. Giambelluca, R. Cannella, F. Vernuccio et al., PI-RADS 3 lesions: role of prostate MRI texture analysis in the identification of prostate cancer. Curr. Probl. Diagn. Radiol. **S0363–0188**(19), 30222–30231 (2019)
25. Radiomic Features: https://pyradiomics.readthedocs.io/en/latest/features.html Accessed 28 Nov 2020
26. J.J.M. van Griethuysen, A. Fedorov, C. Parmar et al., Computational radiomics system to decode the radiographic phenotype. Can. Res. **77**, e104–e107 (2017)
27. A. Bommert, X. Sun, B. Bischl, J. Rahnenführer, M. Lang, Benchmark for filter methods for feature selection in high-dimensional classification data. Comput. Stat. Data Anal. **143**(106839) (2020). https://doi.org/10.1016/j.csda.2019.106839
28. A. Comelli, A. Stefano, C. Coronnello, et al.: Radiomics: a new biomedical workflow to create a predictive model. Commun. Comput. Inf. Sci. **1248** CCIS, 280–293 (2020). https://doi.org/10.1007/978-3-030-52791-4_22

29. V.S. Parekh, M.A. Jacobs, Deep learning and radiomics in precision medicine. Expert Rev. Precis. Med. Drug. Dev. **4**, 59–72 (2019). https://doi.org/10.1080/23808993.2019
30. R. Laudicella, A. Comelli, A. Stefano et al., Artificial neural networks in cardiovascular diseases and its potential for clinical application in molecular imaging. Curr. Radiopharm (2020). https://doi.org/10.2174/1874471013666200621191259
31. E. Montagnon, M. Cerny, A. Cadrin-Chênevert et al., Deep learning workflow in radiology: a primer. Insights Imaging **11**, 22 (2020). https://doi.org/10.1186/s13244-019-0832-5
32. L. Agnello, A. Comelli, E. Ardizzone, S. Vitabile, Unsupervised tissue classification of brain MR images for voxel-based morphometry analysis. Int. J. Imaging Syst. Technol. **26**(2), 136–150 (2016). https://doi.org/10.1002/ima.22168
33. H.H. Tseng, Y. Luo, S. Cui, J.T. Chien, R.K. Ten Haken, I.E. Naqa, Deep reinforcement learning for automated radiation adaptation in lung cancer. Med. Phys. **44**, 6690–6705 (2017). https://doi.org/10.1002/mp.12625
34. Q.Y. Peng, X.T. Wang, L.N. Zhang, Chinese critical care ultrasound study group (CCUSG) 2020. Findings of lung ultrasonography of novel corona virus pneumonia during the 2019–2020 epidemic. Intensive Care Med. **46**, 849–850 (2020)
35. F. Vernuccio, D. Giambelluca, R. Cannella et al., Radiographic and chest CT imaging presentation and follow-up of COVID-19 pneumonia: a multicenter experience from an endemic area. Emerg. Radiol. **27**, 623–632 (2020). https://doi.org/10.1007/s10140-020-01817-x
36. H.Y.F. Wong, H.Y.S. Lam, A.H. Fong et al., Frequency and distribution of chest radiographic findings in patients positive for COVID-19. Radiology **296**, E72–E78 (2020)
37. D. Cozzi, M. Albanesi, E. Cavigli et al., Chest X-ray in new Coronavirus Disease 2019 (COVID-19) infection: findings and correlation with clinical outcome. Radiol. Med. **125**, 730–737 (2020)
38. S. Schiaffino, S. Tritella, A. Cozzi et al., Diagnostic performance of chest X-Ray for COVID-19 pneumonia during the SARS-CoV-2 pandemic in lombardy Italy. J. Thorac. Imaging **35**, W105–W106 (2020)
39. N. Chen, M. Zhou, X. Dong et al., Epidemiological and clinical characteristics of 99 cases of 2019 novel coronavirus pneumonia in Wuhan, China: a descriptive study. Lancet **395**, 507–513 (2020). https://doi.org/10.1016/S0140-6736(20)30211-7
40. G. McGuinness, C. Zhan, N. Rosenberg et al., Increased incidence of barotrauma in patients with COVID-19 on invasive mechanical ventilation. Radiology **297**(2), E252–E262 (2020)
41. G. Fichera, R. Stramare, G. De Conti, R. Motta, C. Giraudo, It's not over until it's over: the chameleonic behavior of COVID-19 over a six-day period. Radiol. Med. **125**, 514–516 (2020)
42. ACR recommendations for the use of chest radiography and computed tomography (CT) for suspected COVID-19 infection. American College of Radiology https://www.acr.org/Advocacy-and-Economics/ACR-Position-Statements/Recommendations-for-Chest-Radiography-and-CT-for-Suspected-COVID19-Infection. Accessed 28 Nov 2020
43. J. Vilar, M.L. Domingo, C. Soto et al., Radiology of bacterial pneumonia. Eur. J. Radiol. **51**, 102–113 (2004)
44. A. Borghesi, R. Maroldi, COVID-19 outbreak in Italy: experimental chest X-ray scoring system for quantifying and monitoring disease progression. Radiol. Med. **125**, 509–513 (2020)
45. M.A. Warren, Z. Zhao, T. Koyama et al., Severity scoring of lung oedema on the chest radiograph is associated with clinical outcomes in ARDS. Thorax **73**, 840–846 (2018)
46. E.A. Akl, I. Blazic, S. Yaacoub et al., Use of chest Imaging in the diagnosis and management of COVID-19: a WHO rapid advice guide. Radiology **30**, 203173 (2020). https://doi.org/10.1148/radiol.2020203173
47. T. Ai, Z. Yang, H. Hou et al., Correlation of chest CT and RT-PCR testing for coronavirus disease 2019 (COVID-19) in China: a report of 1014 cases. Radiology **296**, E32–E40 (2020)
48. D. Caruso, M. Zerunian, M. Polici et al., Chest CT Features of COVID-19 in Rome Italy. Radiology **296**, E79–E85 (2020)
49. M. Hosseiny, S. Kooraki, A. Gholamrezanezhad, S. Reddy, L. Myers, Radiology perspective of coronavirus disease 2019 (COVID-19): lessons from severe acute respiratory syndrome and middle East respiratory syndrome. AJR Am. J. Roentgenol. **214**, 1078–1082 (2020)

50. Y.H. Jin, L. Cai, Z.S. Cheng, H. Cheng et al., A rapid advice guideline for the diagnosis and treatment of 2019 novel coronavirus (2019-nCoV) infected pneumonia (standard version). Mil. Med. Res. **7**, 4 (2020)
51. W. Dai, H. Zhang, J. Yu et al., CT Imaging and differential diagnosis of COVID- 19. Can Assoc. Radiol. J. **71**(2), 195–200 (2020)
52. Y. Li, L. Xia, Coronavirus disease 2019 (COVID-19): role of chest CT in diagnosis and management. AJR Am. J. Roentgenol. **214**, 1280–1286 (2020)
53. C. Jalaber, T. Lapotre, T. Morcet-Delattre, F. Ribet, S. Jouneau, M. Lederlin, Chest CT in COVID-19 pneumonia: a review of current knowledge. Diagn. Interv. Imaging **101**, 431–443 (2020)
54. F. Vernuccio, F.P. Lombardo, R. Cannella et al., Thromboembolic complications of COVID-19: the combined effect of a pro-coagulant pattern and an endothelial thrombo-inflammatory syndrome. Clin. Radiol. **75**, 804–810 (2020)
55. A. Bernheim, X. Mei, M. Huang et al., Chest CT Findings in coronavirus disease-19 (COVID-19): relationship to duration of infection. Radiology **295**, 200463 (2020)
56. S. Inui, A. Fujikawa, M. Jitsu et al., Chest CT findings in cases from the cruise ship "Diamond Princess" with coronavirus disease 2019 (COVID-19). Radiol. Cardiothoracic Imaging **2**(2), e200110 (2020)
57. F. Pan, T. Ye, P. Sun et al., Time course of lung changes at chest CT during recovery from coronavirus disease 2019 (COVID-19). Radiology **295**, 715–721 (2020)
58. C. Hani, N.H. Trieu, I. Saab et al., COVID-19 pneumonia: a review of typical CT findings and differential diagnosis. Diagn. Interv. Imagin **101**, 263–268 (2020)
59. S. Salehi, A. Abedi, S. Balakrishnan, A. Gholamrezanezhad, Coronavirus disease 2019 (COVID-19): a systematic review of imaging findings in 919 patients. AJR Am. J. Roentgenol. **215**, 87–93 (2020)
60. J. Zhu, Z. Zhong, H. Li et al., CT imaging features of 4121 patients with COVID-19: a meta-analysis. J. Med. Virol. **92**, 891–902 (2020)
61. H.J.A. Adams, T.C. Kwee, D. Yakar, M.D. Hope, R.M. Kwee, Chest CT imaging signature of coronavirus disease 2019 infection: in pursuit of the scientific evidence. Chest. **158**, 1885–1895 (2020)
62. S. Zhou, T. Zhu, Y. Wang, L. Xia, Imaging features and evolution on CT in 100 COVID-19 pneumonia patients in Wuhan. China Eur. Radiol. **30**, 5446–5454 (2020)
63. H.X. Bai, B. Hsieh, Z. Xiong et al., Performance of radiologists in differentiating COVID-19 from non-COVID-19 viral pneumonia at chest CT. Radiology **296**, E46–E54 (2020)
64. D.M. Hansell, A.A. Bankier, H. MacMahon, T.C. McLoud, N.L. Müller, J. Remy, Fleischner society: glossary of terms for thoracic imaging. Radiology **246**, 697–722 (2008)
65. W. Zhao, Z. Zhong, X. Xie, Q. Yu, J. Liu, Relation between chest CT findings and clinical conditions of coronavirus disease (COVID-19) pneumonia: a multicenter study. Am. J. Roentgenol. **214**, 1072–1077 (2020)
66. K. Ichikado, High-resolution computed tomography findings of acute respiratory distress syndrome, acute interstitial pneumonia, and acute exacerbation of idiopathic pulmonary fibrosis. Semin. Ultrasound CT MR **35**, 39–46 (2014)
67. Y. Wang, C. Dong, Y. Hu et al., Temporal changes of CT findings in 90 patients with COVID-19 pneumonia: a longitudinal study. Radiology **296**, E55–E64 (2020)
68. M. Chung, A. Bernheim, X. Mei et al., CT imaging features of 2019 novel coronavirus (2019-nCoV). Radiology **295**, 202–207 (2020)
69. H. Shi, X. Han, N. Jiang et al., Radiological findings from 81 patients with COVID-19 pneumonia in Wuhan, China: a descriptive study. Lancet Infect Dis. **20**, 425–434 (2020)
70. V. Ojha, A. Mani, N.N. Pandey, S. Sharma, S. Kumar, CT in coronavirus disease 2019 (COVID-19): a systematic review of chest CT findings in 4410 adult patients. Eur. Radiol. **30**, 6129–6138 (2020)
71. F. Sardanelli, A. Cozzi, L. Monfardini, C. Bnà, R.A. Foà, A. Spinazzola, S. Tresoldi, M. Cariati, F. Secchi, S. Schiaffino, Association of mediastinal lymphadenopathy with COVID-19 prognosis. Lancet Infect Dis. **20**, 1230–1231 (2020)

72. K. Li, J. Wu, F. Wu et al., The clinical and chest CT features associated with severe and critical COVID-19 pneumonia. Invest Radiol. **55**, 327–331 (2020)
73. G. McGuinness, C. Zhan, N. Rosenberg et al., Increased incidence of barotrauma in patients with COVID-19 on invasive mechanical ventilation. Radiology **297**, E252–E262 (2020)
74. C.C. Macklin, Transport of air along sheaths of pulmonic blood vessels from alveoli to mediastinum. Clin. Implications. Arch. Intern. Med. Chic. (Chic) **64**, 913–926 (1939)
75. A. Bruno, R. Gargano, D. Albano et al., Pulmonary complications after cocaine abuse: a case series. Minerva Pneumologica **58**, 36–44 (2019)
76. M. Prokop, W. van Everdingen, T. van Rees Vellinga, et al., COVID-19 standardized reporting working group of the dutch radiological society. CO-RADS: a categorical CT assessment scheme for patients suspected of having COVID-19-definition and evaluation. Radiology **296**(2), E97–E104 (2020)
77. P. Koehler, O.A. Cornely, B.W. Böttiger et al., COVID-19 associated pulmonary aspergillosis. Mycoses **63**, 528–534 (2020)
78. W. Xia, J. Shao, Y. Guo, X. Peng, Z. Li, D. Hu, Clinical and CT features in pediatric patients with COVID-19 infection: different points from adults. Pediatr. Pulmonol. **55**, 1169–1174 (2020)
79. B. Bikdeli, M.V. Madhavan, D. Jimenez et al., COVID-19 and thrombotic or thromboembolic disease: implications for prevention, antithrombotic therapy, and follow-Up: JACC state-of-the-art review. J. Am. Coll. Cardiol. **75**, 2950–2973 (2020)
80. F. Grillet, J. Behr, P. Calame, S. Aubry, E. Delabrousse, Acute pulmonary embolism associated with COVID-19 pneumonia detected with pulmonary CT angiography. Radiology **296**, E186–E188 (2020)
81. C. Xie, M.Y. Ng, J. Ding et al., Discrimination of pulmonary ground-glass opacity changes in COVID-19 and non-COVID-19 patients using CT radiomics analysis. Eur. J. Radiol. Open. **7**, 100271 (2020)
82. E.A. Kim, K.S. Lee, S.L. Primack, et al., Viral pneumonias in adults: radiologic and pathologic findings. Radiographics **22**(Spec), S137–S149 (2002)
83. Q.Q. Zeng, K.I. Zheng, J. Chen, et al., Radiomics-based model for accurately distinguishing between severe acute respiratory syndrome associated coronavirus 2 (SARS-CoV-2) and influenza a infected pneumonia. MedComm (Beijing) (2020). https://doi.org/10.1002/mco2.14.
84. Q. Wu, S. Wang, L. Li et al., Radiomics analysis of computed tomography helps predict poor prognostic outcome in COVID-19. Theranostics **10**, 7231–7244 (2020)
85. H. Chao, X. Fang, J. Zhang et al., Integrative analysis for COVID-19 patient outcome prediction. Med. Image Anal. **67**, 101844 (2020)
86. W. Wei, X.W. Hu, Q. Cheng, Y.M. Zhao, Y.Q. Ge, Identification of common and severe COVID-19: the value of CT texture analysis and correlation with clinical characteristics. Eur. Radiol. **30**, 6788–6796 (2020)
87. F. Homayounieh, R. Babaei, Karimi et al., Computed tomography radiomics can predict disease severity and outcome in coronavirus disease 2019 pneumonia. J. Comput. Assist. Tomogr. **44**, 640–646 (2020)
88. Q. Cai, S.Y. Du, S. Gao et al., A model based on CT radiomic features for predicting RT-PCR becoming negative in coronavirus disease 2019 (COVID-19) patients. BMC Med. Imaging **20**, 118 (2020)
89. H. Yue, Q. Yu, C. Liu et al., Machine learning-based CT radiomics method for predicting hospital stay in patients with pneumonia associated with SARS-CoV-2 infection: a multicenter study. Ann. Transl. Med. **8**, 859 (2020)
90. I.D. Apostolopoulos, T.A. Mpesiana, Covid-19: automatic detection from X-ray images utilizing transfer learning with convolutional neural networks. Phys. Eng. Sci. Med. **43**, 635–640 (2020)
91. L. Li, L. Qin, Z. Xu et al., Using artificial intelligence to detect COVID-19 and community-acquired pneumonia based on pulmonary CT: evaluation of the diagnostic accuracy. Radiology **296**, E65–E71 (2020)
92. L. Huang, R. Han, T. Ai, P. Yu, H. Kang, Q. Tao, L. Xia, Serial quantitative chest CT assessment of COVID-19: deep-learning approach. Radiol. Cardiothoracic Imaging **2**(2), e200075. Radiological Society of North America (2020)

Use of Deep Learning Based Frameworks on Pixel Scaled Images of Chest CT Scans for Detection of COVID-19

Sushant Shetty, Amit Gawade, and Subodh Deolekar

Abstract As on 15th September 2020, the total cases of SARS-CoV-2 infected patients in the world has crossed 29 million, with more than 930,000 deaths occurring due to the virus. Real time RT-PCR (Reverse Transcription Polymerase Chain Reaction), which is the standard detection method for COVID-19, is likely to have low detection rate in early stages of the infection possibly due to less viral load in the patients. On the other hand, in comparison to RT-PCR, patterns obtained from radiography on chest CT scans show higher sensitivity and specificity. However, due to the sensitive nature and difficulties in publicly acquiring medical data, only 2 open-sourced COVID CT datasets with images containing clinical findings of COVID-19 could be found. Applying existing deep learning models to the limited CT scans can distinguish COVID-19 and non-COVID-19 CT images but with lesser accuracy. The paper proposes to use existing deep learning-based frameworks on an augmented dataset consisting of pixel scaled images (of the original CT images) and the original CT images to diagnose COVID-19 infection. Since this is a binary classification problem the paper proposes to use Convolutional Neural Networks (CNN) to classify CT images into infected and not infected categories. The implementation is done with Keras and Tensorflow using an 80/20 ratio for training and validation. The proposed methodology (using pixel scaled images) achieved a validation accuracy of over 90% in detecting COVID-19 with an F1-score of 0.96 compared to the best F1-score of 0.86 on the original dataset.

Keywords COVID-19 · Chest CT scan · Deep learning · Convolutional neural network · Data augmentation · Pixel scaling

---

S. Shetty (✉) · A. Gawade
REDX Innovation Laboratory, Prin. L.N. Welingkar Institute of Management Development and Research, Mumbai, India
e-mail: sushant.shetty@welingkarmail.org

A. Gawade
e-mail: amit.gawade@welingkarmail.org

S. Deolekar
Prin. L.N. Welingkar Institute of Management Development and Research, Mumbai, India
e-mail: subodh.deolekar@welingkarmail.org

J. Nayak et al. (eds.), *Understanding COVID-19: The Role of Computational Intelligence*, Studies in Computational Intelligence 963, https://doi.org/10.1007/978-3-030-74761-9_4

1 Introduction

COVID-19 or Coronavirus Disease of 2019 is an infectious disease caused by the novel coronavirus named SARS-CoV-2. Large number of cases of this disease were first reported in Wuhan, China in late 2019 which has since spread across the world making it a global pandemic. Initially, the virus was named "2019 novel coronavirus" or 2019-nCoV but was later renamed to "Severe Acute Respiratory Syndrome Coronavirus 2" or SARS-CoV-2 by the ICTV (International Committee on Taxonomy of Viruses) in early 2020 as the virus is genetically similar to the coronavirus which started the SARS outbreak in 2002 (SARS-CoV) [1]. Evidence suggests that SARS-CoV-2, like SARS and MERS has a zoonotic origin meaning that the disease first started in animals and was then transmitted from animals (mainly bats) to humans [2]. COVID-19 symptoms include cough along with fever, with some people experiencing symptoms such as headache, sore throat, diarrhoea, loss of taste and/or smell and skin rashes to name a few. The disease can even escalate to a serious case with symptoms including difficulty in breathing, pneumonia and chest pain. The most common method of infection among humans is through contact with saliva droplets or nose discharge which are released into the air during coughing or sneezing by an infected person [3].

At the time of writing (15th September 2020), more than 29.3 million COVID-19 cases have been reported worldwide, with around 19.9 million recoveries and 930,000 deaths due to complications arising from SARS-CoV-2. The current diagnosis methods of COVID-19 in suspected patients include nucleic acid testing, rapid antigen testing and analysing CT scans. Testing for nucleic acid is more reliable and suitable than testing based on symptoms since they are designed in a way to identify and target certain micro-organisms [4]. Formulation of techniques based on nucleic acid are dependent on having a complete idea about the protein and genetic composition of the micro-organism and the overall modifications which are induced in the host body during the period of infection. Based on the above-mentioned factors, several Reverse Transcription Polymerase Chain Reaction (RT-PCR) methods have been developed to identify COVID-19 virus genetically [4]. The RT-PCR chain reaction is an in vitro method which is very sensitive and has varied applications in medical science and biomaterial fields. The RT-PCR has been extensively used before in the diagnosis of RNA virus infections such as HIV, hepatitis C virus, lymphomas, leukaemia and sarcomas. For the current RT-PCR tests, respiratory samples are used which usually include taking swabs from the throat and nose. However, accurate viral load detection depends on the number of days since the start of infection. The most reliable source of SAR-CoV-2 virus, 14 days from the day of infection is the mucous secretion followed by swabs from the nasal canal. On the other hand, swabs from the throat were found to be unreliable after 8 days of symptom onset. Such irregularities in viral load could result in false negatives which could be a result of improper handling, inadequate samples, very little viral matter in the collected swabs or genetic variations in the virus [4]. It is suggested that if the RT-PCR test returns a negative result, but the patient is symptomatic, RT-PCR be repeated on the patient to eliminate false negatives. This is however, time consuming as it requires additional tests and given the number of infections happening daily and the limited availability of test kits, chances are that infected patients might be left out of testing, thereby likely increasing the infection rates.

Antibodies are the human immune system's response to fight off diseases and can be used to detect if people have been infected by a particular micro-organism. Antibodies are released in response to antigens which are pathogen specific. Antibodies bind themselves to specific parts of the pathogen depending on the antigen, thereby neutralizing or tagging it for other immune responses. Based on these bindings, antibody or antigen based diagnostic tests can be developed for disease detection [5]. In case of COVID-19 detection, antigen testing is done using a respiratory sample, preferably nose or throat of the suspected individuals to detect if the virus is present in the patient. In case the target antigen from the virus is present in a given sample in sufficient concentrations, it binds itself to specific antibodies which are fixed onto strips of paper enclosed inside a plastic container. This binding of antibodies creates a visually detectable signal (change of colour) in under 30 min [6]. Due to the quicker detection times, these tests are also called rapid antigen tests. However, rapid antigen tests are prone to lower sensitivity rates typically in the range of 34 to 80%. Owing to these numbers, the test might result in a lot of false negatives due to which more than half of the infected patients may be missed by the test [6]. In some cases, the false negative rates are greater than 40% [7]. Since the number of false negative outcomes are high, it is recommended that patients who test negative undergo a RT-PCR test to be completely sure of not being infected. However, reports suggest that in the Indian state of Maharashtra, almost 99% of the citizens have not taken the second recommended test with other states too reporting such issues [8]. Such issues combined with the false negative rates of RT-PCR can dent the efforts in efficiently containing the virus and putting an end to the pandemic.

In such a situation Chest CT scans can be looked upon as a pre-diagnostic tool to assist in the detection of possible COVID-19 cases. CT or Computerized Tomography scans of a body combine multiple X-Ray images from different angles which is then fed to a computer algorithm to create cross-sectional images of bones, tissues and blood vessels inside a human body. Chest CT scans create a 3D model of the lungs which is used to determine features associated with problems such as tumours, pleural effusion, tuberculosis, pneumonia. Features associated with COVID-19 can be immediately identified by a trained radiologist and since CT scans are more easily obtained compared to RT-PCR results they can provide doctors with much quicker information about a patient's infection. In infected patients, associated pneumonia usually shows ground-glass opacification along with bilateral abnormalities. Along with this, CT scans can also show crazy paving's, vascular dilation, traction bronchiectasis, subpleural bands and architectural distortion which can be possible indicators of COVID-19 induced pneumonia [9].

Chest CT scans seem to be a more efficient and reliable pre-diagnostic tool in detecting SARS-CoV-2 infection. However, given the work stress on healthcare professionals and the high number of daily infections around the world, routine examination of Chest CT scans by medical practitioners may not be possible. Deep Learning systems have been used as an assistive technology in detecting various diseases from medical images. Since computer systems can process CT scans, a Deep Learning system can be developed to detect or indicate possible SARS-CoV-2 infections as the images are being processed. A computer-aided diagnosis can greatly reduce diagnosis periods while reducing the burden on kits and the healthcare industry. This paper uses existing Deep Learning architectures (previously trained on the Imagenet dataset) on open-sourced COVID-19 image datasets and presents the results. These architectures can further be modified according to the medical images to better the results.

2 Literature Study

Traditional methods of medical image classification require experienced radiologists and trained medical practitioners to analyse each image and give the diagnosis. However, in recent years, the availability of medical imaging data through open-sourced datasets and extensive literature from field experts along with the advancements in Machine Learning and Deep Learning frameworks have opened up multiple avenues for computer aided diagnosis systems. However, the key challenge for computer aided diagnosis systems is the different visual characteristics in medical images: some are very distinct and in some the differences might be very subtle [10]. Support Vector Machine (SVM) has been used in medical image diagnosis for a long time. However, the performance of SVM's is far below the practical standards of medical imaging and they are more time consuming due to selective feature extraction for different objects [11]. Deep Learning has demonstrated to be more effective in solving intricate problems due to their sophisticated architectures. In Deep Learning, CNN or Convolutional Neural Network is widely used for classification of images [12]. Computer scientist Yann LeCun in the late 90's proposed the term CNN's, which have performed well in the computer vision domain. The main feature of a CNN is the extraction of visual information by the convolutional filters. Based on results from the convolutional layers the multilayer perceptron then decides on the recognition operation to be performed [13]. CNN's became a revolutionary technique during the ImageNet Large Scale Visual Recognition Challenge (ILSVRC), when AlexNet became the first CNN to win the competition [14]. The number of individual layers and the order in which they are put together has given rise to multiple CNN architectures which have performed well on natural image classification. There have been instances of CNN's performing better than trained medical practitioners in medical image classification too.

CheXNet, a CNN to detect pneumonia with a depth of 121 layers was trained on a dataset in excess of 100,000 chest X-rays. This custom-built CNN performed better compared to the mean diagnosis of 4 radiology specialists [15]. The weighted average error of a CNN trained using transfer learning on over 100,000 Optical Coherence Tomography images to classify macular degeneration and diabetic retinopathy was similar to the mean diagnosis of 6 domain specialists [16]. Another CNN on Oral cancer detection got a classification accuracy of 91.4% with sensitivity of 0.94 and specificity of 0.91 [17]. A denoising based CNN trained on a publicly available Kaggle dataset consisting of images of the retina for Diabetic Retinopathy got a score of 95% in identifying 2 classes and around 85% accuracy for 5 class classification [18].

In line with the image classification problem at hand, previous attempts have been made to develop and train a CNN to detect COVID-19 infection. A study working with Chest X-rays, trained a CNN using transfer learning technique on an open-sourced image repository called the COVIDx consisting of 13,975 chest X-ray. The developed architecture called the COVID-Net achieved a test accuracy of 93% which is better than the results obtained with VGG19 and ResNet50 [19]. Another research working on the same COVIDx open-sourced dataset used automatic selection of learning rate and progressive image resizing on a ResNet50 architecture to achieve an accuracy of 96%. However, the dataset was divided into four classes: normal, bacterial, viral and COVID-19 [20].

The examples stated in Table 1 above ascertain the fact that CNN can achieve the diagnosis levels of trained medical professionals and can be used as an assistive if not a self-diagnostic tool in the healthcare industry.

From the introduction and literature review sections we can see that COVID-19 is a serious threat to human life and development, but techniques like deep learning which have already proven to be effective in early diagnosis of certain diseases can prove

Table 1 Summary of existing CNN architectures used in medical diagnosis

Researchers	Disease	Dataset	Performance metric
Rajpurkar P, et al. [15]	Pneumonia	>100,000 chest X-rays	Better than the mean diagnosis of 4 radiology experts
Kermany DS, et al. [16]	Macular degeneration and diabetic retinopathy	>100,000 OCT images	Weighted average error similar to mean diagnosis of 6 domain specialists
Jeyaraj PR and Nadar ERS [17]	Oral cancer	>600 hyperspectral images	Accuracy–91.4%, Sensitivity–0.94, Specificity–0.91
Ghosh R, et al. [18]	Diabetic Retinopathy	>33,000 eye fundus images	2 class accuracy–95%, 5 class accuracy–85%
Wang L and Wong W [19]	COVID-19	13,975 chest X-rays	Accuracy–93%
Farooq M and Hafeez A [20]	COVID-19	13,975 chest X-rays	4 class accuracy–96%

to be useful here too. However, since the disease is fairly new, diagnosis techniques and treatment methods are still in the planning and development phase. Hence, data is very scarce to obtain and since deep learning requires a lot of data to perform relatively well, lack of suitable data could hamper the progress in the development of Deep Learning based computer aided diagnosis systems. Even with less amount of data, techniques like transfer learning and data augmentation can solve the problem to a certain extent. But known drawbacks in traditional data augmentation techniques like rotation, mirroring etc. for CT scans can lead to significant issues like incorrect feature detection in deep learning models. Hence, this research work proposes the use of a fairly new data augmentation technique called pixel scaling to be used instead of traditional augmentation. Pixel scaling refers to changing the value of individual pixels for the entire image around the median pixel value of each image. So, for individual values less than the median, the value is scaled down by a predetermined value and for pixel values above the median, the value is scaled up. The scaling values are decided based on how much it changes the new image from the original. Scaling can be done multiple times using different values to augment the final dataset. The paper proposes using scaling once to see the effectiveness of this data augmentation technique by evaluating different model parameters on the original and augmented dataset.

3 Proposed Methodology

CT scans are relatively easy to obtain and analyse for medical professionals. Figure 1 shows a sample CT scan of a patient infected with SARS-CoV-2, with the infected area circled by a trained medical examiner. However, challenges like limited data availability, patient data privacy and extensive time required to make the datasets are some of the issues which concern the Deep Learning community when it comes to developing computer assisted diagnostic solutions in the medical domain.

However, in recent years, the challenge of limited data availability has been solved by transfer learning up to a certain extent. The logic behind transfer learning is to transfer model knowledge from a larger dataset to a model working with limited data. Since transfer learning can train deep neural networks with relatively less data on a pre-trained model, it is very popular in the Deep Learning community. Transfer learning works by using the weights of the early convolutional layers as they extract general, basic features such as corners and edges, patterns and gradients which are applicable across images. It is the upper layers which identify specific features which are very specific to a dataset such as the eyes or nose in a face dataset. Since the low-level features are shared among images, an existing model can be trained on a relatively new dataset containing few images along with hyperparameter tuning to get better overall accuracy. Since CT scan images are relatively hard to come by due to reasons such as patient data privacy and extensive preparation times, open-sourced Chest CT scan datasets have been used. However, since the dataset contains a smaller

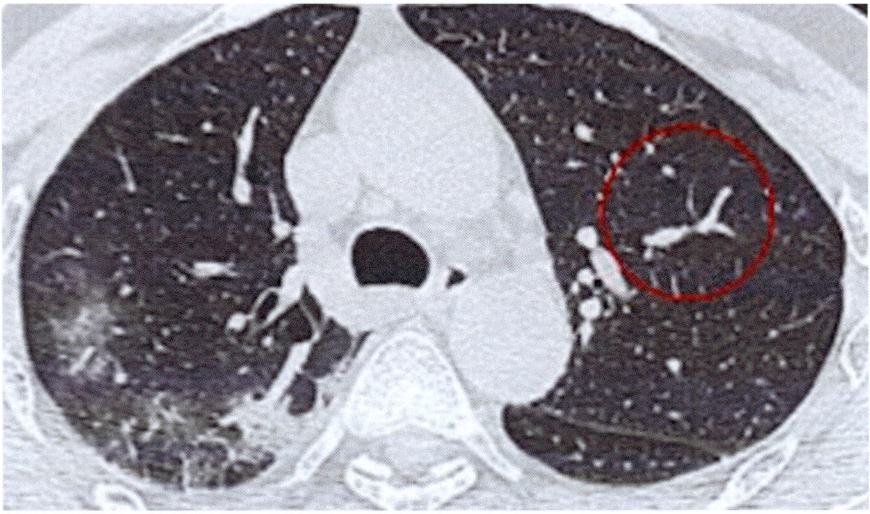

Fig. 1 Sample CT scan of a patient with COVID-19

number of images, transfer learning has applied on existing architectures which were previously trained on the ImageNet dataset.

Instead of developing a new framework/architecture for working with the CT scan images, existing architectures which have performed well on real world tasks and competitions such as the ImageNet detection, ImageNet localization, COCO detection and COCO segmentation have been used. Architectures like AlexNet, VGG, ZFNet, GoogleNet, DenseNet etc. and multiple variants of the above-mentioned architectures have been used for training on this dataset. However, due to obvious limitations, results of only a few architectures which had better results than the others have been presented in this paper.

3.1 CNN Architecture

A CNN network can be divided into 2 blocks: the first block which returns feature maps through convolution filtering operations and the second block common to all neural networks, a classification block which returns the class probabilities.

A CNN comprises of the following 4 layers:

1. The convolutional layer: The first layer and the key component of any CNN network. The goal of a convolutional layer is to identify basic characteristics from the dataset using convolutional filtering.

2. The pooling layer: Pooling refers to minimizing the size of the images while at the same time upholding the important features from the data. A pooling layer is generally placed between two convolutional layers and is often 2×2 or 3×3

in size to not lose too much information. The major advantage of using a pooling layer is that it brings down the number of features in turn reducing the number of calculations in the network thus improving on the efficiency of the network while avoiding overfitting on the data.

3. The ReLU activation layer: The ReLU layer is used to replace all negative values received from the previous layers to zero.

4. The fully-connected layer: This layer outputs a vector of size N (N = number of classification classes). Each number in the vector shows the probability for each class in the classification problem. This is done by a combination of linear values and an activation function on input values.

3.2 Dataset

For training and validation purposes, two open-source datasets of Chest CT scan images were combined and used. The first dataset [21] is available on Github and contains 349 CT scans with COVID-19 findings and 397 CT scans of class non-COVID. As per the Github description, the dataset was shown to a senior radiologist in Wuhan, China who ascertained its utility for developing automated solutions. The second dataset [22] is available on Kaggle and contains 1252 CT scan images with COVID-19 findings and 1230 CT scans which are non-COVID. After combining both the datasets, the final dataset originally contained 3228 images, however after cleaning i.e., removing corrupt and excessively marked images it was brought down to 3193 images. The augmented dataset (consisting of the original and the pixel scaled images) contains a total of 6386 images of which 3134 have COVID-19 findings and 3252 are non-COVID. Furthermore, the results section highlights the different model evaluation parameters for models trained on the original and the augmented dataset.

3.3 Preprocessing and Data Augmentation

Preprocessing techniques like Thresholding and Edge detection were implemented on the images to help the model better understand features from the images which might assist the model in differentiating Covid from Non Covid. Thresholding in image processing is an image segmentation technique, wherein the pixel values of an image are changed to make the image easier to analyse. In case of a colour image, it is first converted into a grayscale image and then into a binary or an image with just 2 colours, black and white. Using a certain threshold value, thresholding helps to target a certain part of the image, while ignoring the others.

Figure 2 demonstrates the results of the thresholding techniques which were used on an image from the CT scan dataset. From Fig. 2 it is evident that none of the above-mentioned techniques were successful in extracting features that would aid the model during training. Global thresholding failed to capture all the details in the

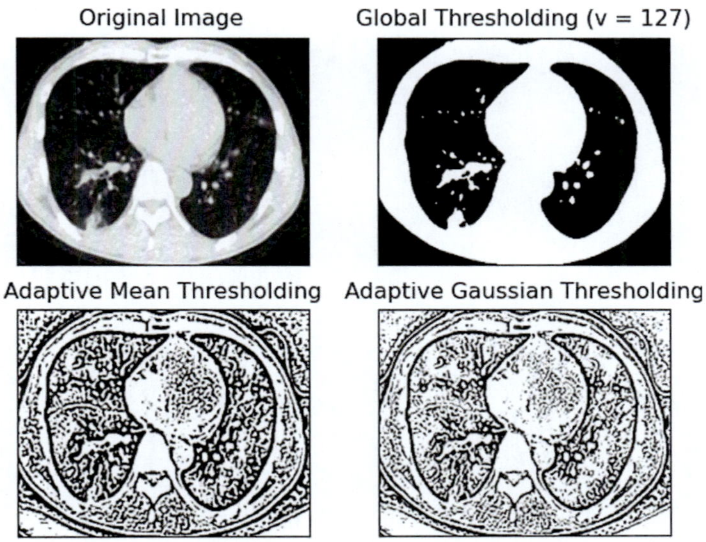

Fig. 2 Preprocessing using Thresholding techniques - Image 1

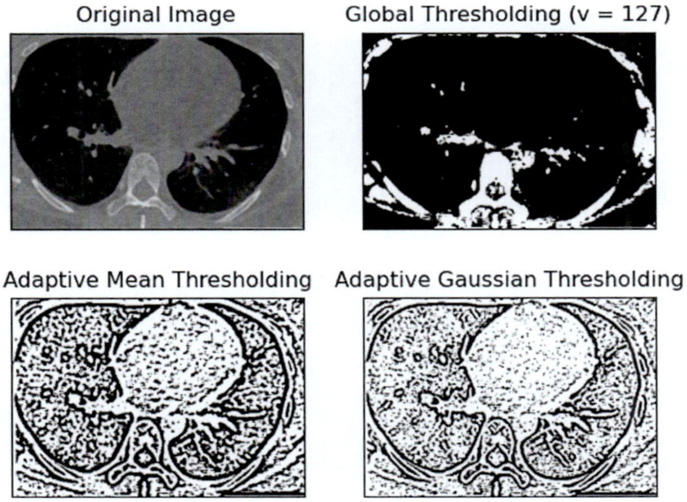

Fig. 3 Preprocessing using Thresholding techniques - Image 2

image whereas adaptive mean and adaptive gaussian thresholding introduced a lot of noise in the image. The same thresholding techniques were applied to another image (shown in Fig. 3) from the same dataset which was more towards the gray side compared to Fig. 2. It can be seen that the same thresholding of 127 gave poor results on darker images and the other two methods introduced noise in the image.

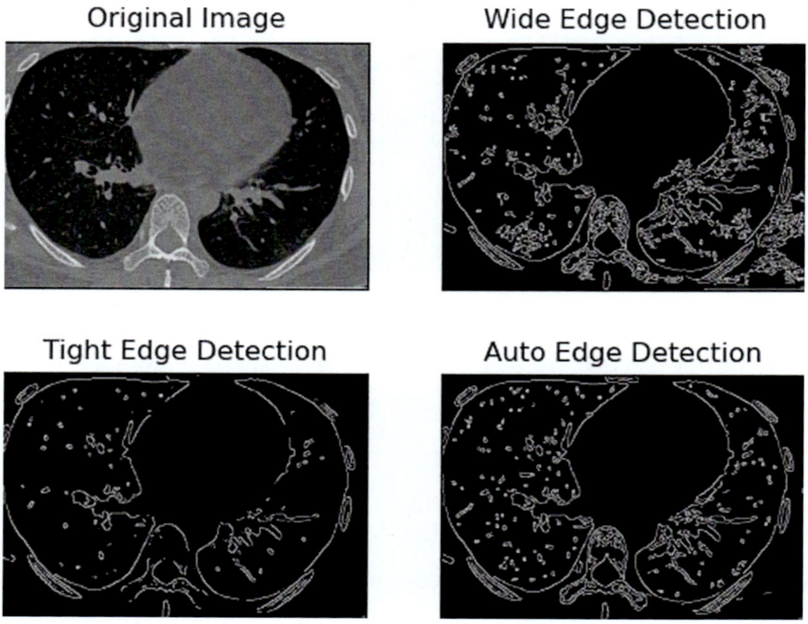

Fig. 4 Preprocessing using edge detection methods

Edge detection is another common image segmentation technique used while preprocessing a dataset. It is used to find boundaries of objects in images by looking for discontinuities in pixel brightness. The dataset was preprocessed using a canny edge detector to detect edges from a wider range. The image is first blurred using a Gaussian blur to remove noise followed by edge detection. Figure 4 shows the results for different types of edge detectors used. Due to a non-uniform pixel value range for multiple images, edge detection did not work well for the collected dataset.

As preprocessing proved to be unsuccessful in extracting the features to help while training a model, the next best option was to increase the number of images via data augmentation techniques. Simple doubling of the images by copy pasting the same images would lead to better accuracy on the training dataset but poor validation accuracy. This is known as over fitting. Image rotation or mirroring can lead to serious consequences in medical images and can be even more consequential in case of CT scans as a simple mirroring of the image can cause the heart to be seen as an anomaly by the model. In order to get more images via data augmentation, a new technique called pixel scaling was used. Firstly, the median pixel value of each image was found out. Since the median value is unique to each image the technique is more adaptive rather than being based on a static value as was the case with global thresholding. Each pixel value is then compared to the median value for changing the value. If greater than the median value, the pixel value was scaled up by 2% by multiplying it by 1.02. If the pixel value was found out to be less than the median, the corresponding pixel was scaled down by 2% by multiplying it with 0.98.

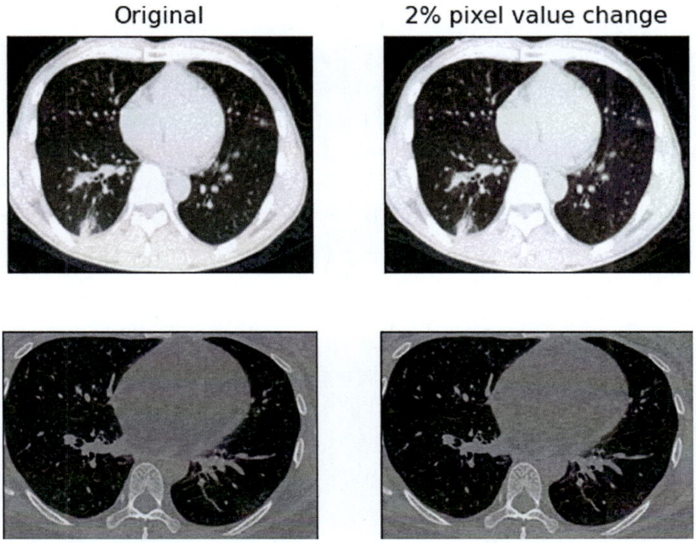

Fig. 5 Two percent pixel value change

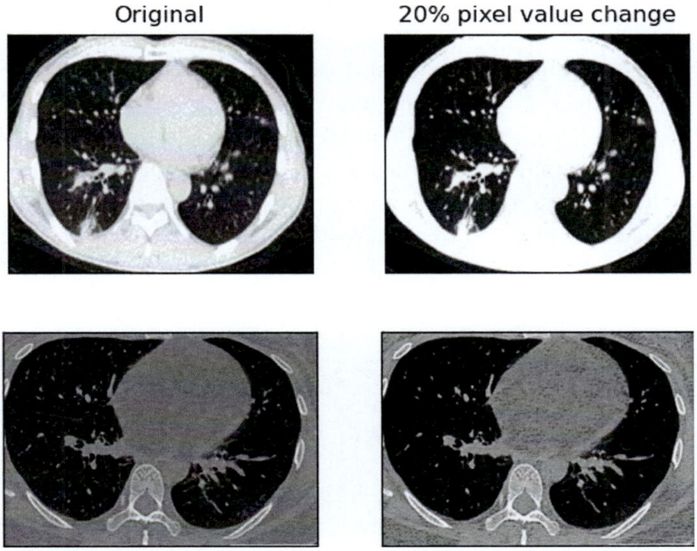

Fig. 6 Twenty percent pixel value change

In Fig. 5, the results of a 2% scaling can be seen in the right column. To the naked eye both images in the right column look identical to the ones in the left. Since CT images are multiple slices of the Chest from the same patients, such differences in pixel values might naturally exist. To ascertain the usability of this technique, both the images in the top row were shown to a radiologist in their original resolutions. The radiologist did not notice any difference in the images, meaning both of them could be used by a trained medical professional in diagnosing COVID-19 infection. However, scaling pixels by more than 3% can lead to noticeable differences in the image compared to the original image. Due to image resolution restrictions, the right column in Fig. 6 shows an image scaled by a factor of 20%. Such a drastic change can lead to loss of information. To maintain the originality of the dataset and to prevent skewed data because of pixel value scaling, the operation was performed only once and a value of 2% was chosen for scaling. Although not recommended, the technique was used as a last resort for data augmentation and was only used after consulting a trained medical professional.

3.4 Hyperparameters

Learning rate is an essential hyperparameter which requires tuning to ensure that the model converges properly to a global minimum. It controls how much the model changes with respect to the calculated error each time the weights are updated. If the learning rate value is too large, the model can quickly converge to a suboptimal minimum value, thereby failing to achieve the global minimum. If it's too small, the model might take too long to train and achieve a global minimum. The best value for learning rate can be finalized after multiple training iterations by observing the loss function values. After fine tuning, 0.001 was chosen as the learning rate for training the model on the dataset.

Neural networks either use gradient descent or back propagation to update weights. Both these methods use an error value (for weight updation) which is calculated using a subset of the train dataset. The size of this subset is referred to as the batch size. For a batch size of 8, 8 images are used to calculate the error after which the weights of the model are updated accordingly. Smaller batch sizes are generally useful as it helps to load the training data in smaller chunks into the memory and gives a better idea about the model owing to the quick updates in the error values. After trying different sizes, a batch size of 8 was selected for the training purpose.

Epochs refer to the number of the times the model trains itself over the whole dataset. The number of epochs can be decided from trial and error. Generally, the epoch value should be capped when the change in the model's loss value is considerably small or doesn't change over multiple epochs. A value of 50 epochs was chosen for training beyond which the loss and accuracy value did not change much.

Table 2 summarizes hyperparameter tunings done during the training to obtain optimal results.

Table 2 Summary of hyperparameter tunings done during training

Hyperparameter	Tuning approach	Range	Note
Learning rate	Fixed => Fine tune	0.001–0.1	Based on trial and error
Batch size	Fixed => Fine tune	1–128	Depends on the size of dataset
Epochs	Fixed => Fine tune	25–100	Depends on the learning rate and training loss value

3.5 Architectures Used for Training

The following section contains a brief description of the various architectures, namely, Inception, VGG, ResNet, DenseNet used for training purposes.

3.5.1 Inception

Deep Neural Networks learn better as the number of layers in the network increases. However, as the model size increases, the model becomes more prone to overfitting. This can be problematic in the medical domain as the dataset available for training is very less. Also, as the network deepens, the number of parameters increases thereby increasing the need for computational resources. GoogLeNet or Inception Net which topped the ImageNet Large Scale Visual Recognition Challenge (ILSVRC) solved this problem by developing a sparsely connected architecture which replaces a fully connected network. The idea of micro-architectures in Inception Net was conceptualized by Szegedy et al. in 2014 [23]. Inception V1 is a 27-layer deep network with special inception layers in between. The inception layer is a combination of 3 filters whose outputs are appended and fed to the next stage. Two major add-ons in the original inception layer are: 1. the use of a 1×1 convolutional layer at each stage which helps in dimensionality reduction and a parallel max pooling layer. Memory wise, weights for the Inception model are smaller than VGG and ResNet.

3.5.2 VGG

Simonyan and Zisserman introduced the concept of VGG in their 2014 paper [24]. This architecture won the first and second place in object localization and image classification at the ImageNet Large Scale Visual Recognition Challenge (ILSVRC) 2014. The network is characterised by the 3×3 convolutional layers with max pooling layers handling size reduction. In the end, fully connected layers each with 4096 nodes are followed by a softmax classifier. VGG has two variants: VGG16 and VGG19 where '16' and '19' stand for the number of weight layers in the network. VGG networks are very slow to train and have very large model weights in terms of disk usage. VGG16 alone is 533 MB and VGG19 is 574 MB making their deployment difficult on mobile systems and systems with low RAM.

3.5.3 ResNet

Most CNN's use more layers in neural networks to reduce the overall error value. This technique works better with less layers but as the network goes deeper, the problem of vanishing/exploding gradient comes into picture which causes the gradient to become too small or very large. Neural networks learn via a back-propagation process, which relies on gradient descent. If the architecture has too many layers, repetitive multiplication makes the gradient value very small until it 'vanishes' or makes it too large causing it to 'explode' causing the model performance to saturate with each additional layer. ResNet which was first introduced as 'Deep Residual Learning for Image Recognition' in the year 2015 [25] by researchers at Microsoft solved this problem with the help of residual blocks. The ResNet solution is to identify short cut connections. ResNet does this by using a technique called skip connections which skips training from a few layers and directly connects to the output. It stacks up identity mappings, layers that initially contribute to the vanishing gradients or which don't do anything. Skipping layers initially compresses the network into only a few layers, thus speeding up the learning process. This happens because the inputs can propagate faster through the residual connection across the layers. The model then slowly restores or expands the skipped layers as it learns the feature map from the previous layers. Data suggests that ResNet is easier to optimize and hence gains better accuracy due to the increased depth in the number of layers. In 2015, ResNet152 won in the ImageNet classification, Detection and Localization problem. It also won the COCO Detection and Segmentation challenge. The number 152 in ResNet152 specifies the number of layers in the architecture. In comparison, GoogLenet which won the ImageNet competition in 2014 had just 22 layers. Over the years, depending on the use case, multiple ResNet architectures have come into existence. ResNet18, ResNet34, ResNet50, ResNet101, ResNet152 are a few ResNet based architectures. Since ResNet uses global average pooling instead of fully connected layers, the models take up less space in memory. Compared to VGG16 which takes up 533MB of disk space, ResNet50 takes up just 102 MB due to the above-mentioned features.

3.5.4 DenseNet

A traditional Convolutional Neural Network with M layers has M connections. However, Densely Connected Convolutional Networks [26] or DenseNet have M(M+1)/2 direct connections. DenseNet consists of dense blocks, wherein layers receive information from the previous ones and then pass their features to the next layers. Each layer in DenseNet is hence gathering a collection of feature maps from all its previous layers. Since each layer is receiving all the information from previous layers, the network is compact and less wide. Due to this DenseNets have higher computational efficiency and are more efficient in memory management. Also, since all information is passed onto the future layers, the error value is propagated early on to all the layers thus all layers can rectify the error values from the final classification layers. A Dense block is connected to all the upper layers and it consists of Batch Normalization, ReLU activation and a 3×3 convolution filter. Instead of summing up the

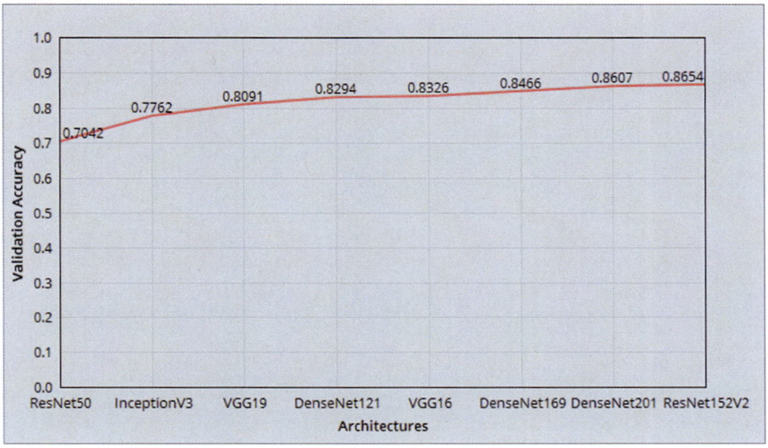

Fig. 7 Validation accuracies for various architectures - original dataset

residual values like in ResNet, DenseNet concatenates all the feature maps. Over the year's different versions of DenseNet have been developed namely DenseNet121, DenseNet169, DenseNet201 where the number specifies the number of layers in the model.

4 Result Analysis and Discussion

The models mentioned in the methodology were trained on two datasets. The first one (Original dataset) consisting of the original 3193 images and the second one (Augmented dataset) consisting of pixel scaled images of the original images together with the original ones, thus making a total of 6386 images. Both the datasets are divided into training (80%) and validation (20%) sets and contain two categories: 'Covid' and 'Not Covid'.

4.1 Results on Original Dataset

Of all the models and the different variations which were used for training and validation, the validation accuracy of 8 architectures on the original dataset is shown in Fig. 7 below.

In Fig. 8, the training and validation loss and accuracy of the better performing model i.e., ResNet152V2 is shown. In the first 30 epochs the graphs for loss and accuracy show considerable improvement, however, after around 30 epochs the curve of the loss and accuracy function starts to straighten. The model achieved a validation

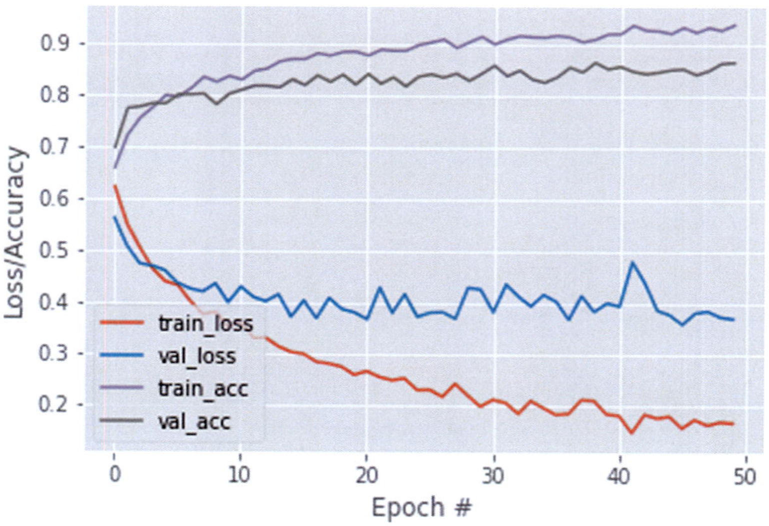

Fig. 8 Loss and accuracy graph for ResNet152V2 - original dataset

accuracy of 86.54% after 50 epochs of training with a loss value of 0.3724. The training accuracy achieved is 93.85% with a loss value of 0.1681. Also, the results mentioned in the tables for both the datasets are arranged in increasing order of the validation accuracies for each architecture.

As a performance measure for various models, the confusion matrix (shown in Fig. 9) was used. The X-axis consists of predicted labels with true labels on the Y-axis. The validation dataset forms the basis when calculating the numbers for the confusion matrix. When the samples that belong to class 'Covid' are identified correctly by the model, they are put in the True Positive category. Similarly, when it correctly identifies a 'Not Covid', they are put in the True Negative category. When the model incorrectly identifies a 'Covid' image as 'Not Covid' it is put in the False Negative category and when it mistakes a 'Not Covid' as 'Covid' it belongs to the False Positives category. Of the total 639 images used for validation, it had 269 True Positives, 284 True Negatives, 45 False Negatives and 41 False Positives. The confusion matrix values for other architectures which were tried are mentioned in Table 3.

Two other parameters used in evaluating the performance of a model are the sensitivity and specificity values on the validation dataset. Sensitivity measures how often the model correctly predicts a positive result for people with the infection (also known as the true positive rate). A model with high-sensitivity will predict almost everyone infected with the disease and not generate many false-negatives.

Sensitivity is calculated as follows:

$$Sensitivity = \frac{TruePositive}{TruePositive + FalseNegative}$$

Table 3 Confusion matrix values for different architectures - original dataset

Architecture	True positives	True negatives	False negatives	False positives
ResNet50	200	250	114	75
InceptionV3	265	252	49	73
VGG19	269	227	45	98
DenseNet121	286	255	28	70
VGG16	241	289	73	36
DenseNet169	283	249	31	76
DenseNet201	257	293	57	32
ResNet152V2	269	284	45	41

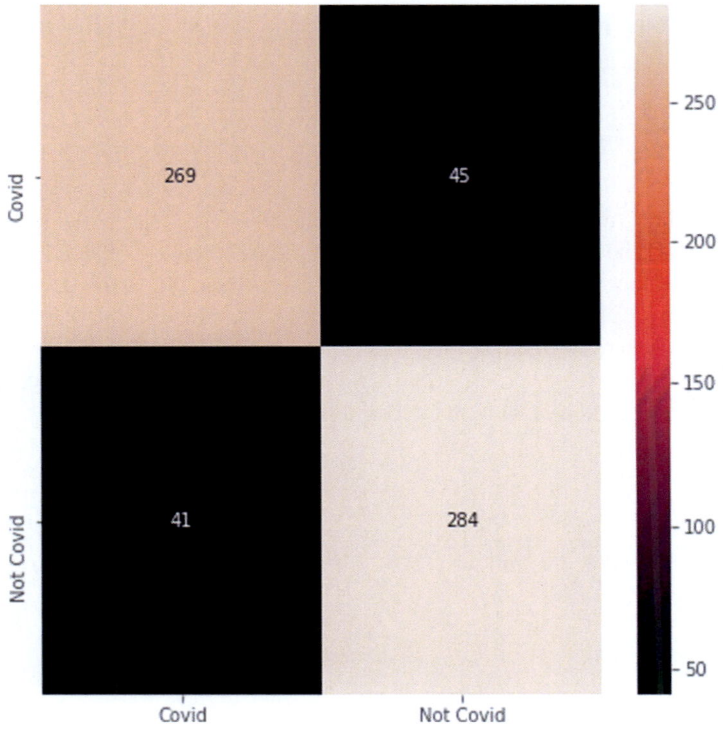

Fig. 9 Confusion matrix for ResNet152V2 - original dataset

Specificity on the other hand measures a model's ability to correctly predict a negative infection for people who don't have the disease. A model with high-specificity will correctly specify the number of people who don't have the disease while at the same time generate less false-positive results.

Table 4 Evaluation parameters for different architectures - original dataset

Architecture	F1 score	Sensitivity	Specificity
ResNet50	0.68	0.6369	0.7692
InceptionV3	0.81	0.8439	0.7754
VGG19	0.79	0.8567	0.6985
DenseNet121	0.85	0.9108	0.7846
VGG16	0.82	0.7675	0.8892
DenseNet169	0.84	0.9013	0.7662
DenseNet201	0.85	0.8185	0.9015
ResNet152V2	0.86	0.8567	0.8738

The formula for Specificity is:

$$Specificity = \frac{TrueNegatives}{TrueNegatives + FalsePositives}$$

The architectures in Table 4 have been arranged as per their validation accuracies and the F1 is related to the precision and recall values for class Covid. It also mentions other evaluation parameters for different CNN architectures on the original dataset. The F1 score is a measure of the test's accuracy and conveys the balance between the precision and the recall (sensitivity) of a model. The precision of a model is the number of correctly predicted positive cases divided by the total number of all positive predictions including the wrong ones.

$$Precision = \frac{TruePositive}{TruePositive + FalsePositives}$$

F1 score is calculated as follows:

$$F1 = \frac{2 \times Precision \times Recall}{Precision + Recall}$$

4.2 Results on Augmented Dataset

The augmented dataset was trained on the same better performing architectures as was the original dataset to maintain similarity in the training process.

Figure 10 demonstrates the validation accuracy of 8 different CNN architectures which were trained on the augmented dataset.

In Fig. 11, the training and validation loss and accuracy of the better performing model i.e., ResNet152V2 is shown. The model achieved a validation accuracy of

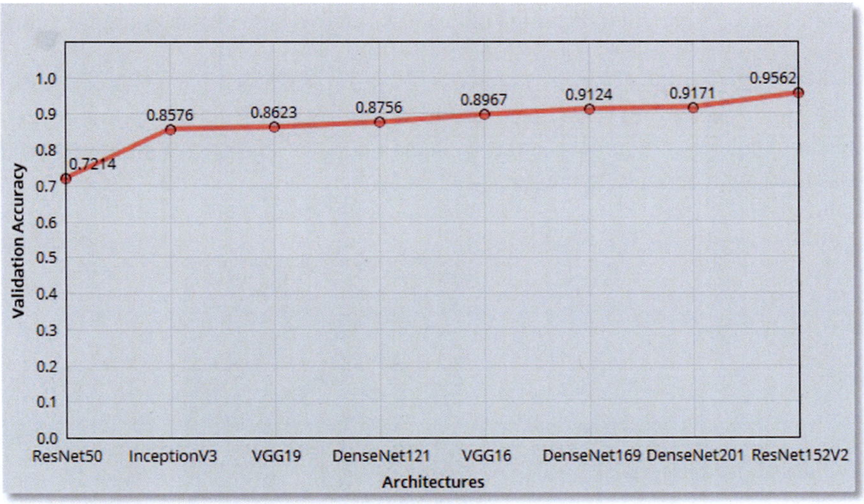

Fig. 10 Validation accuracies for various architectures - augmented dataset

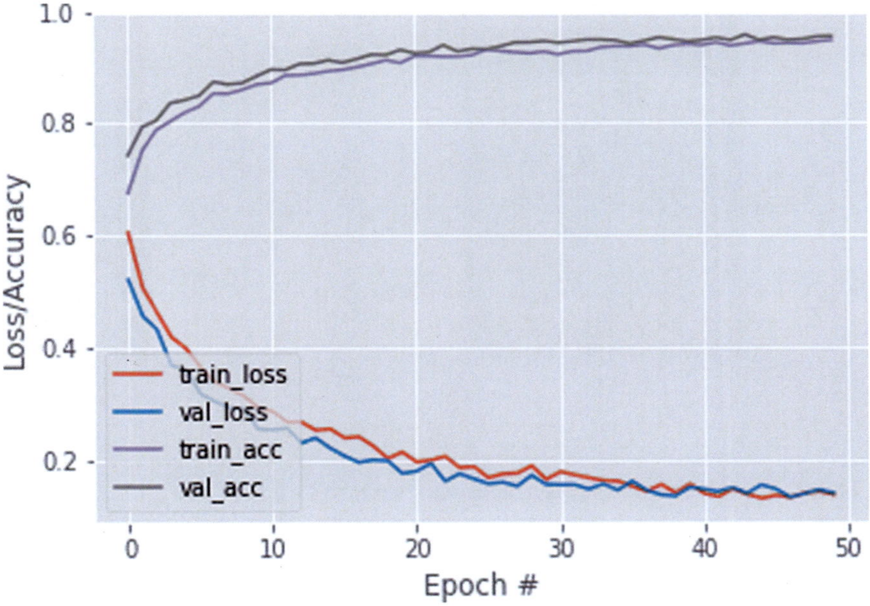

Fig. 11 Loss and accuracy graph for ResNet152V2 - augmented dataset

95.62% after 50 epochs of training with a loss value of 0.1388. The training accuracy achieved is 95.01% with a loss value of 0.1188.

For the augmented dataset, the confusion matrix (shown in Fig. 12) consists of 611 True Positives, 611 True Negatives, 16 False Negatives and 40 False Positives from a total of 1278 images in the validation dataset. The confusion matrix values for other architectures are mentioned in Table 5.

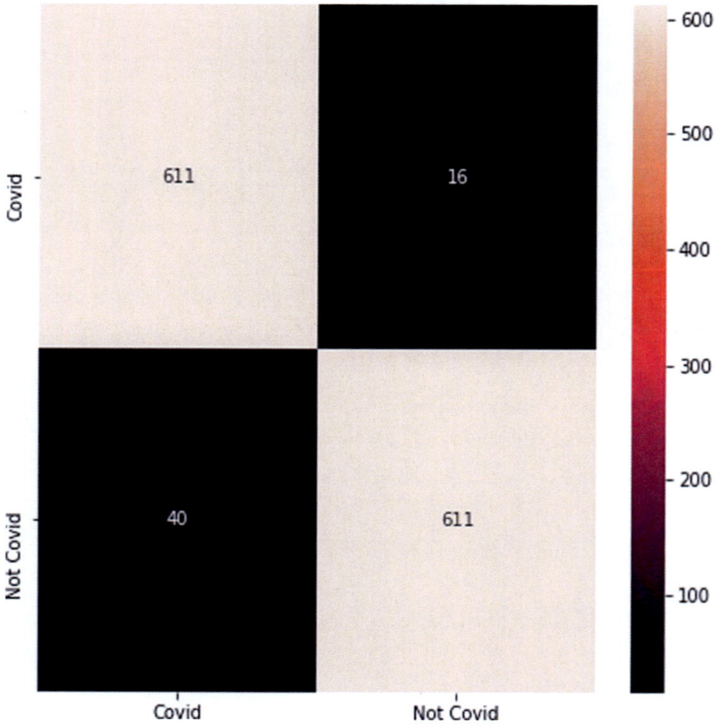

Fig. 12 Confusion matrix for ResNet152V2 - augmented dataset

Table 5 Confusion matrix values for different architectures - augmented dataset

Architecture	True positives	True negatives	False negatives	False positives
ResNet50	335	587	292	64
InceptionV3	516	580	111	71
VGG19	527	575	100	76
DenseNet121	556	563	71	88
VGG16	576	570	51	81
DenseNet169	563	603	64	48
DenseNet201	594	578	33	73
ResNet152V2	611	611	16	40

Table 6 Evaluation parameters for different architectures - augmented dataset

Architecture	F1 score	Sensitivity	Specificity
ResNet50	0.65	0.5343	0.9017
InceptionV3	0.85	0.8230	0.8909
VGG19	0.86	0.8405	0.8833
DenseNet121	0.87	0.8868	0.8648
VGG16	0.90	0.9187	0.8756
DenseNet169	0.91	0.8979	0.9263
DenseNet201	0.92	0.9474	0.8879
ResNet152V2	0.96	0.9745	0.9386

Other evaluation parameters for different CNN architectures on the augmented dataset are mentioned in Table 6 below.

5 Discussion and Conclusion

During the pandemic, the Deep Learning community has stepped up and come together to tackle the problem of COVID-19 diagnosis using techniques other than the conventional RT-PCR, Antigen and Antibody testing. As the number of daily infections are on the rise and the risk of community transmission is high, it becomes necessary to use different methods and not rely on a single method for diagnosis. Apart from the above-mentioned tests, medical images are another tool to help in early diagnosis of diseases. However, given the situation in the world, where most of the medical professionals are needed on ground to tackle the situation, diagnosis using medical images could be slow and inefficient. Convolutional Neural Networks have shown a lot of promise in developing Computer-Aided automatic diagnostic systems. Existing CNN architectures like Inception, DenseNet, ResNet have achieved accuracy levels at par with humans on object detection and classification. However, one of the problems with Neural Networks is that to perform on a level at par with medical experts, they need to be trained on a lot of data which is hard to come by in the healthcare domain due to reasons such as data privacy and dataset preparation times. Data Augmentation techniques such as pixel value scaling proposed in this research can help address the issue of limited data available for training a model. Results from this research validate the usefulness of such a technique to help counter the issues of data scarcity. CNN models perform slightly better as indicated by the different model evaluation parameters on the augmented dataset as compared to using just the original images for training. The accuracy can be further increased by creating a dataset which contains multiple pixel scaling images from the original dataset. However, care should be taken to scale it in such a way that the scaled and the original images are indistinguishable to the human eye. Hence, the utility of such methods should be verified from a trained medical expert before being incorporated into the methodology. Also, care should be taken to not use standard data augmentation techniques like rotation and mirroring to avoid certain anomalies like the heart position being mistaken for a feature. In addition to data augmentation, transfer learning is a very effective method while working with limited data. The results of this paper show the efficiency of existing models such as ResNet, DenseNet and VGG among a few others, to adapt to new data. A few tweaks to the hyperparameters can get really good results on the available datasets. Finally, the average accuracy should not be the only parameter used while determining the efficiency of a model. Other parameters like sensitivity, specificity and F1 score can provide a really good measure of how the model would perform on unseen real-world images, where the model would be required the most.

References

1. S. Vergnaud, What does COVID-19 stand for? GoodRx (2020). https://www.goodrx.com/blog/what-does-covid-19-mean-who-named-it. Accessed 10 Sept 2020
2. WHO: Coronavirus disease 2019 (COVID-19) Situation Report-94 (2019). https://www.who.int/docs/default-source/coronaviruse/situation-reports/20200423-sitrep-94-covid-19.pdf. Accessed 10 Sept 2020
3. WHO: Coronavirus (2020). https://www.who.int/health-topics/coronavirus. Accessed 10 Sept 2020
4. B. Udugama et al., Diagnosing COVID-19: the disease and tools for detection. ACS Nano **14**(4), 3822–3835 (2020). https://doi.org/10.1021/acsnano.0c02624
5. W. MacDonald, Antibody vs Antigen testing for COVID-19. Diagnostics from technology networks (2020). https://www.technologynetworks.com/diagnostics/articles/antibody-vs-antigen-testing-for-covid-19-336486. Accessed 10 Sept 2020
6. WHO: Advice on the use of point-of-care immunodiagnostic tests for COVID-19 (2020). https://www.who.int/news-room/commentaries/detail/advice-on-the-use-of-point-of-care-immunodiagnostic-tests-for-covid-19. Accessed 10 Sept 2020
7. Department of Health, Australian Government: Post-market evaluation of serology-based point of care tests (2020). https://www.tga.gov.au/post-market-evaluation-serology-based-point-care-tests. Accessed 10 Sept 2020
8. K. Mohanty, Which COVID test passes the test for accuracy? Times Of India (2020). https://timesofindia.indiatimes.com/india/which-covid-test-passes-the-test-for-accuracy/articleshow/77278075.cms. Accessed 10 Sept 2020
9. COVID-19 Imaging finding. Radiology Assistant (2020). https://radiologyassistant.nl/chest/covid-19/covid19-imaging-findings. Accessed 10 Sept 2020
10. A. Kumar, J. Kim, D. Lyndon, M. Fulham, D. Feng, An ensemble of fine-tuned convolutional neural networks for medical image classification. IEEE J. Biomed. Health Inform. **21**(1), 31–40 (2017). https://doi.org/10.1109/jbhi.2016.2635663
11. S.S. Yadav, S.M. Jadhav, Deep convolutional neural network based medical image classification for disease diagnosis. J. Big Data **6**, 113 (2019). https://doi.org/10.1186/s40537-019-0276-2
12. M.V. Valueva, N.N. Nagornov, P.A. Lyakhov, G.V. Valuev, N.I. Chervyakov, Application of the residue number system to reduce hardware costs of the convolutional neural network implementation. Math. Comput. Simul. **177**, 232–243 (2020). https://doi.org/10.1016/j.matcom.2020.04.031
13. Y. Lecun, L. Bottou, Y. Bengio, P. Haffner, Gradient-based learning applied to document recognition. Proc. IEEE **86**(11), 2278–2324 (1998). https://doi.org/10.1109/5.726791

14. K. Smeda, Understand the architecture of CNN. Medium (2019). https://towardsdatascience.com/understand-the-architecture-of-cnn-90a25e244c7. Accessed 10 Sept 2020
15. P. Rajpurkar et al., CheXNet: radiologist-level pneumonia detection on Chest X-Rays with deep learning (2017). Preprint at https://arxiv.org/abs/1711.05225v3
16. D.S. Kermany et al., Identifying medical diagnoses and treatable diseases by image-based deep learning. Cell **172**(5), 1122-1131.e9 (2018). https://doi.org/10.1016/j.cell.2018.02.010
17. P.R. Jeyaraj, E.R. Samuel Nadar, Computer-assisted medical image classification for early diagnosis of oral cancer employing deep learning algorithm. J. Cancer Res. Clin. Oncol. **145**, 829–837 (2019). https://doi.org/10.1007/s00432-018-02834-7
18. R. Ghosh, K. Ghosh, S. Maitra, in Automatic Detection and Classification of Diabetic Retinopathy Stages Using CNN. 4th International Conference on Signal Processing and Integrated Networks, pp. 550–554 (2017). https://doi.org/10.1109/spin.2017.8050011
19. L. Wang, A. Wong, COVID-Net: a tailored deep convolutional neural network design for detection of COVID-19 cases from chest X-Ray images (2020). Preprint at https://arxiv.org/abs/2003.09871
20. M. Farooq, A. Hafeez, COVID-ResNet: a deep learning framework for screening of COVID19 from radiographs (2020). Preprint at https://arxiv.org/abs/2003.14395
21. X. Yang, X. He, J. Zhao, Y. Zhang, S. Zhang, P. Xie, COVID-CT-dataset: a CT scan dataset about COVID-19. Preprint at https://arxiv.org/abs/2003.13865
22. E. Soares, P. Angelov, S. Biaso, M.H. Froes, D.K. Abe, SARS-CoV-2 CT-scan dataset: a large dataset of real patients CT scans for SARS-CoV-2 identification. medRxiv (2020). https://doi.org/10.1101/2020.04.24.20078584
23. C. Szegedy et al., Going deeper with convolutions (2014). Preprint at https://arxiv.org/abs/1409.4842
24. K. Simonyan, Zisserman, A very deep convolutional networks for large-scale image recognition (2014). Preprint at https://arxiv.org/abs/1409.1556
25. K. He, X. Zhang, S. Ren, J. Sun, *Proceedings of the IEEE Conference on Computer Vision and Pattern Recognition (CVPR)*, pp. 770-778 (2016). https://arxiv.org/abs/1512.03385
26. G. Huang, Z. Liu, L. van der Maaten, K.Q. Weinberger, Densely connected convolutional networks (2016). Preprint at https://arxiv.org/abs/1608.06993

SARS-CoV-2 Infection and Antibody-Dependent Enhancement

Aliye Bastug and Hurrem Bodur

Abstract There is still knowledge gap about the antibody response in COVID-19 patients. Factors affecting the kinetics/titers of the elicited neutralizing antibodies (NAbs), the mean persistence time of NAbs after infection, risk of reinfection, and enhanced respiratory disease via antibody-dependent enhancement (ADE) are the main issues needing to be clarified. Recently published studies have suggested an association between increased antibody titers and COVID-19 infection severity. Although there exist insufficient data, the titers of antibodies have been reported to be decreasing rapidly, suggesting a risk of reinfection. If reinfection occurs, another concern is whether it will be more severe due to the preexisting low level antibodies. To combat the continuing pandemic, antibody-based treatments, such as monoclonal antibodies (mAbs), convalescent plasma (CP) therapies, and vaccine studies, have been investigated. Although it may not be as troublesome as in the Dengue fever, clinicians should be alert about the possible risk of ADE. Non-neutralizing antibodies or sub-protective levels of NAbs may trigger immunopathogenic events via immune complex-mediated or Fc receptor-dependent endocytosis, resulting in enhanced viral infection. In this chapter, literature findings revealing the factors with an impact on eliciting NAbs, possible mechanisms and risks of ADE, and safety concerns for SARS-CoV-2 treatment interventions are outlined.

Keywords Antibody-dependent enhancement · Neutralizing antibodies · SARS-CoV-2 · COVID-19 · Reinfection

1 Introduction

SARS-CoV-2 virus is a pathogen causing global challenges for public health. It causes a wide spectrum of clinical conditions from asymptomatic infection to severe acute respiratory distress syndrome (ARDS). Several factors have an impact on disease severity such as older age, comorbidities, lymphopenia, higher levels of CRP, and

A. Bastug (✉) · H. Bodur
Department of Infectious Diseases and Clinical Microbiology, University of Health Science, Ankara City Hospital, Ankara 06230, Turkey

J. Nayak et al. (eds.), *Understanding COVID-19: The Role of Computational Intelligence*,
Studies in Computational Intelligence 963, https://doi.org/10.1007/978-3-030-74761-9_5

NLR (neutrophil lymphocyte ratio) (Bastug et al. 2020; Cloutier et al. 2020). The host hyperinflammatory response, known as cytokine storm triggered by the virus, results in poorer outcomes (Cloutier et al. 2020). In reports involving severe patients having more prominent antibody response, a raising concern is whether there is an adequate immune response elicited after primary infection to protect reinfections in all patients and how long it persists (Kang et al. 2020; Zhao et al. 2020). Another concern is the probable risk of antibody-dependent enhancement (ADE) development via antibody-based vaccines and/or therapeutics, which may exacerbate the severity of COVID-19 (Lee et al. 2020). In addition, the preexisting non-neutralizing antibodies elicited during the primary infection could facilitate viral internalization of the virus in the host cell and enhance the infection (Peron and Nakaya 2020). The current pandemic requires the need for the development of therapeutic agents and vaccines faster than ever. In this chapter, published studies on neutralizing antibody response elicited against SARS CoV-2 infection, the reinfection probability, the hypothesis and mechanisms of ADE phenomenon are discussed.

2 Methodology

Electronic databases including PubMed, Embase, and Cochrane Library were searched for studies published before December 12, 2020 on SARS CoV-2 and other coronaviruses. The following terms were used for literature search: Neutralizing antibody response in SARS CoV-2, SARS- CoV-2 reinfection, SARS CoV-2 and ADE, Coronaviruses and ADE. *BioRxiv*.org and MedRxiv.org were also searched for preprints. Two authors performed literature search and evaluated abstracts of the relevant articles. Articles yielding important evidences on the present issue were selected and discussed in the relevant subheadings. The selected 78 article that had important evidence from the 583 initial search results were evaluated and the important outcomes of the studies were discussed in the present chapter.

3 Discussion with Literature Findings

The following section presents a systematic review on neutralizing antibody response elicited against SARS CoV-2 infection, the reinfection probability, the hypothesis and mechanisms of ADE phenomenon.

3.1 Viral Characteristics and Important Viral Genome Structures for Cellular Entry

Coronaviruses are enveloped, positive stranded RNA viruses (Perlman and Netland 2009). SARS CoV-2 is the last of the seven human coronaviruses (HCoVs). SARS CoV and MERS CoV are the two highly pathogenic coronaviruses with high mortality rates (de Groot et al. 2013; Peiris et al. 2003). Sequence analysis revealed that SARS CoV-2 is very similar to SARS CoV with ~80% sequence similarity. MERS CoV has the second closest sequence identity with the percentage of 54. Other endemic HCoVs are HKU1, OC43, NL63, and 229E. The similarity in sequence between these HCoVs and SARS-CoV-2 ranges between 48 and 52% (Lee et al. 2020; Lu et al. 2020; Zhou et al. 2020).

The genome of the SARS-CoV-2 encodes both structural and non-structural proteins. The most critical structural protein for cellular entry is the viral spike (S) protein. It is responsible for binding to cellular receptor (angiotensin-converting enzyme 2 (ACE 2)) and membrane fusion. It consists of S1 and S2 subunits. The S1 subunit has a receptor binding domain (RBD), which manages the binding process, and the S2 subunit, which manages viral membrane fusion (Lee et al. 2020; Pillay 2020). After binding with the ACE receptor, conformational changes of the spike protein occur. Whereupon, the proteolytic activation of the spike protein by host cell proteases, such as transmembrane protease serine 2 (TMPRS2), arises, which is also necessary for viral entry to the host cell (Pillay 2020). Apart from the known cellular receptor (ACE2) of SARS CoV-2, CD147 (aka Basigin or EMMPRIN) has been identified recently, which acts as another receptor for SARS-CoV-2 entry to the host cells (K. 2020).

3.2 Neutralizing Antibody (NAb) Response in SARS CoV-2 Infection

NAbs are key components of humoral immune response, and they bind the viral particles preventing attachment and internalization of the virus to the host cell (Y. 2020). Wu et al. (F. 2020)reported that particular Nabs could be detected in COVID-19 patients 10–15 days after disease onset; however, one third of the patients had failed to produce high titers of NAbs. The higher NAb titers were positively correlated with older age and higher CRP levels and negatively correlated with lymphocyte levels. Although there is a knowledge gap in detailed dynamics and characteristics of the NAb response in SARS CoV-2 infections, a limited number of studies have reported that severely infected patients had prominent NAb response as compared to patients in milder condition (Y. 2020; Zhao et al. 2020). In contrary to these findings, other studies have reported no significant associations between COVID-19 intensity and SARS-CoV-2 antibodies (Moderbacher et al. 2020; To et al. 2020). In a current study investigating adaptive immune response kinetics, it was reported that asymptomatic individuals had a weaker immunity after the primary infection response, and NAb titers decreased in the early convalescent phase of the infection (within 2 months after primary infection) in most of the recovered patients of COVID-19 (Long et al. 2020). Recently, a short duration of immunity was also suggested in a mathematical model study (Kissler et al. 2020). Further studies are warranted to outline the exact and long-term kinetics of the NAbs.

3.3 The Probability of SARS CoV-2 Reinfection

Acquired immunity is anticipated via NAbs in many recovered COVID-19 patients. It is expected that the elicited NAbs via primary infection resist to reinfection with the same virus. However, how long Nabs are effective is still unclear. In a study analyzing the dynamics of NAb titers, it was reported that NAb titers decreased 6–7 weeks after disease onset in four of the eight convalescent patients (Wang et al. 2020). The lack of NAb production or decline in NAbs titers in time may give rise to the risk of reinfection. However, T-cell immunity, which may be detected at least up to three months after primary infection, is also important (T.M. 2020).

The rechallenge of the rhesus macaques with SARS-CoV-2 in an early period after the primary infection showed that NAbs have the capability of protection from subsequent reinfection (L. 2020). It is not known whether these results are still valid for the subsequent periods further. Chandrashekar et al. (Chandrashekar et al. 2020) also revealed in a study of the rhesus macaque model that SARS CoV-2 infection elicited protective immune response and reinfection did not occur after re-challenge in non-human primates. On the other hand, it is important to underscore that there are considerable differences between humans and macaques in terms of SARS-CoV-2 infection.

Although there exist reports on the re-test positivity of recovered COVID-19 patients (J. 2020), it should not be interpreted as reinfection. It may be due to a false negative previous PCR test, incorrect sampling procedures, or false positive PCR results via contamination (Leung and Xu 2020). The differentiation of re-test positivity, prolonged viral shedding, and reinfection should be made properly. Although it is rare, there are reports on prolonged viral shedding extending over one month (S. 2019; Molina et al. 2020). To et al. (K.K.W. 2019) performed a whole genome analysis and showed evidence of reinfection with different clades of the SARS-CoV-2 virus. They concluded that, reinfection may occur due to the waning immunity by time or the lack of NAb response after primary infection. In addition, SARS-CoV-2 may persist in human beings even though they have had natural infection or have been vaccinated. These factors raise concerns for the possibility of reinfection and risk of ADE due to the low level of antibodies obtained from primary infection. Further studies assessing long term results are necessary to shed light on this issue.

3.4 The Risks and Mechanisms of ADE

The main objective of therapeutic interventions (antibody-based treatments, vaccine trials, etc.) is preventing viral entry to the host cells. Therefore, S proteins are the main targets of these approaches (Lee et al. 2020). One possible risk for these interventions is the development of ADE. There are also concerns as to whether ADE will occur in a state of reinfection (if reinfection occurs) via decreased immunity, non-neutralizing antibodies, or low level titers of Nab (K.K.W. 2019). ADE has been reported previously for many virus, such as dengue virus (Sridhar et al. 2018), HIV (Willey et al. 2011), Ebola virus (Takada et al. 2003), respiratory syncytial virus (RSV) (Graham 2016), SARS CoV, and Middle East respiratory syndrome related coronavirus (MERS- CoV), etc. (Lee et al. 2020). The characteristics of antibodies such as specificity, titers, and affinity are the main factors which have an impact on the risk of ADE.

Molecular mechanisms of ADE consist of two different types: enhanced infection and exacerbated immune activation. Enhanced infection occurs via sub-neutralizing or non-neutralizing antibodies, which causes excessive uptake of the virus into macrophages or monocytes expressing Fc gamma receptor IIa (FcγIIa, receptor-mediated endocytosis). This may initiate enhanced viral replication, dissemination of infection, and lead to poorer outcomes. Dengue virus infection may be given as the best example of clinical ADE occurring via this mechanism (Lee et al. 2020; Sridhar et al. 2018).

Another possible mechanism of ADE is exacerbated immune activation (activation of complement pathways, excessive release of cytokines) which exists via immune complex formed by non-neutralizing antibodies and viral antigens. It was well defined in respiratory viral infections such as RSV and measles (Polack et al. 2002). This form of ADE phenomenon may result in airway obstruction due to the increased inflammation via excessive secretion of cytokines. Although it is not known clearly, it is more likely the possible mechanism of ADE expected to be seen in COVID-19 (Lee et al. 2020).

3.5 Evidence of ADE in Viral Infections

Dengue virus exemplifies the best-known cause of the infections which has an ADE risk. It belongs to the Flaviviridae family and has four serotypes (DENV-1 to DENV-4). Although it usually causes mild infections, secondary infections with other serotypes may be more severe, which are named as dengue hemorrhagic fever or dengue shock syndrome. Previous studies have showed that already occurring cross-reactive, non-neutralizing antibodies developed due to the primary infection cause enhanced infection via ADE (Guzman et al. 2013; Morens et al. 1987; Ulrich et al. 2020). Similarly, they may be acquired from vaccination or maternal immunity as well, and in case of secondary infection with heterologous dengue serotype, they may cause ADE development (Simmons et al. 2007; Taylor et al. 2015). After non-Nabs-bound viral epitope, the formed immune complex is internalized by immune cells bearing FcγR, which causes an enhanced illness. This phenomenon is defined as ADE, which is seen in the Dengue virus infection. The facilitating factors of ADE in Dengue virus infections are followings: it has four circulating serotypes with different antigenicity, there is a possibility of reinfections of individuals via different serotypes, and it is a macrophage tropic virus, which leads to high probability of FcγRIIa mediated endocytosis (Arvin et al. 2020). It is important to underscore that ADE is a rare phenomenon (around 0.5%) even in dengue virus endemic areas and it correlates with the percentage of pre-existing antibody levels (a peak antibody level for ADE is 1:21–1:80) (Taylor et al. 2015).

Even though the detailed underlying processes are not clearly known, ADE has also been defined for coronaviruses. Feline infectious peritonitis is an alpha coronavirus causing infections in cats. It is also macrophage tropic virus like dengue, and ADE may occur as it has the same molecular mechanism of the dengue virus (Pedersen 2014).

In regards to HCoVs, it was suggested that partial immunity may be acquired via the primary infection of HCOv 229E (Callow et al. 1990). A study conducted in Canada in 2003 reported that antibodies elicited by the nucleocapsid protein of HCoV-OC43 may cross-react with SARS CoV (Patrick et al. 2006). Khan et al. (S. 2020)reported low levels of cross-reactivity between seasonal HCoVs antibodies and S2 subunits of SARS CoV-2 strains. It has been reported previously that cross reactive antibodies against endemic HCoV have not been associated with neither severe nor milder infection of SARS CoV-2 (K. 2020; van der Hoek et al. 2006). Although HCoV NL63 uses the ACE2 receptor for viral entry, no antibody cross-reactivity has been observed since receptor binding domain (RBD) is quite dissimilar from the RBD of SARS-CoV-2 (Arvin et al. 2020). The presence of cross-reactive antibodies of the S-protein belonging SARS-CoV and MERS CoV might be expected. Wang et al. (Wang et al. 2014)reported that in vitro diluted sera including anti-spike antibodies against SARS CoV mediated ADE in cell lines. LV et al. (Lv et al. 2020)reported that cross- neutralization is rare, but cross-reactivity of antibodies bound to the spike protein is common between SARS CoV and SARS CoV-2. They concluded that whether these non-NAb responses will result in ADE has not been clarified yet. However, the limited prevalence of these infections may not allow finding out the clinical possibility of ADE risk via cross-reactive antibodies (Arvin et al. 2020; X. 2020). As for MERS CoV-related ADE, Wan et al. (Y. 2020) reported that antibody dosages, tissue specific expression of Fc receptor, dipeptidyl peptidase 4 (cell receptor of MERS CoV), and the affinity of antibodies are important factors determining the risk of ADE for MERS CoV. Even though no in vivo ADE has been described for MERS CoV, they suggested that ADE existed in vitro if intermediate dosages of mAbs are administered.

Concerns regarding the possibility of ADE for SARS CoV-2 arose when the correlation between NAb titers and severe infection was suggested in the literature (Zhao et al. 2020). There are conflicting data about this issue since some other studies have found no relationship between severity of COVID-19 and NAb titers against

SARS CoV-2 (Moderbacher et al. 2020; To et al. 2020). An unmet demand for biomarkers or immunological assays exists to differentiate whether the severity of human infection is derived from ADE or any other cause.

At this time of point, no reliable data are available regarding ADE with clinical importance for SARS-CoV-2 infection. Nonetheless, scientists need to be aware of the probable risks and mechanisms of ADE. It is not known yet whether clinically significant RBD mutations on S protein will appear in the future. In such a case, individuals with immunity due to primary infection or vaccination may encounter infections via mutated strains, and if ADE is induced, exacerbation of the clinical course of the infection may exist.

3.6 Risk of ADE for Therapeutic Interventions: Plasma Therapies, Monoclonal Antibodies, and Vaccines

There are two critical challenges for the immune system to control highly pathogenic respiratory coronavirus conditions like such as SARS- CoV, MERS- CoV, and SARS-CoV-2. In addition to rapid elimination of the virus, enabling the control of lung inflammation have a key role for preventing acute lung injury (Braciale et al. 2012; Liu et al. 2019). Currently, no highly effective antiviral therapy options for patients with COVID-19 is available. Hence, immunotherapeutic interventions, such as convalescent plasma (CP) therapies (obtained from recovered patients) and mAbs targeting S proteins of the virus, have been investigated as potential therapeutic agents in various clinical studies (Duan et al. 2020; Ju et al. 2020; Li et al. 2020; Shen et al. 2020). Although the intended use of such therapeutics is to provide clinical improvement, they could potentially induce immunopathogenic events.

Preliminary data of a small case series suggested some benefits of CP implementations, such as a decrease in viral load, an increase in titers of NAbs, the resolution of acute respiratory distress syndrome, and clinical improvement without severe adverse effects (Duan et al. 2020; Shen et al. 2020; X. 2020). On the other hand, since such studies include small case series without any control groups, conclusions should be evaluated with caution. Recently, Li et al. (Li et al. 2020)evaluated the effectiveness of CP implementation among 103 severe COVID-19 patients via a randomized clinical trial. Although they used at least 1:640 titers of S-RBD specific IgG (equivalent to a titer of 1:160 neutralizing antibodies), they were not be able to detect significant improvement via CP implementation. They determined early viral clearance from nasopharyngeal swabs via CP therapies. It is concluded that it might not be possible to identify the exact efficacy of the CP therapies since most of the patients in that study used CP after 2 weeks or later from the onset of the symptoms. Future randomized controlled studies are warranted to evaluate the efficiency of CP in early periods after infection. It is noteworthy that implementation time of plasma therapies is as important as drafting appropriate donors having ample neutralizing antibodies. The use of convalescent plasma for severe and critically ill

COVID-19 patients is approved by the US Food and Drug Administration (FDA). It is recommended measuring the titers of NAbs and using plasma with high titers if available (Tanne 2020). Despite the lack of clear evidence, if convalescent plasma including low titers of NAbs is implemented, ADE may occur theoretically (Sullivan and Roback 2020). Wang et al. (Wang et al. 2014) reported that diluted antibodies against spike protein of SARS CoV increased ADE in vitro.

To reduce the potential ADE risk in CP therapies, donors should be chosen from patients with higher NAbs titers (>1/320), and also the purification of anti-S or anti-RBD antibodies from the donated plasma should be ensured.

Monoclonal antibodies against S protein also mostly promoted ADE in SARS CoV infections in vitro (Wang et al. 2014). In opposite to this finding, the useful effects of mAb in SARS CoV infection, such as a decrease in viral load and reduced lung pathology, were shown in a macaque model study (Miyoshi-Akiyama et al. 2011). It was also reported that the implementation of mAbs against MERS CoV in macaques resulted in less pulmonary pathology (Johnson et al. 2016). Shi et al. (Shi et al. 2020)reported that CB6-a neutralizing mAb-may be a possible therapeutic agent for COVID-19. Even though there exist inadequate data obtained in clinical trials, it is suggested that mAbs could be beneficial in outpatients in the early period after infection. Bamlanivimab and imdevimab plus casirivimab are approved by FDA for emergency use in COVID-19 patients with high risk of severe infection (S.A. 2020; E. 2020; Rizk et al. 2020).

As COVID-19 has been continuing to spread globally, reliable and effective vaccines are urgently needed to prevent infection. Major challenge for vaccines and antibody-based therapeutic interventions are achieving viral clearance as well as protecting from immunopathology. Enhancement of infection after vaccination has been previously reported for RSV, pandemic influenza, and dengue hemorrhagic fever (Liu et al. 2019; Monsalvo et al. 2011; Openshaw and Tregoning 2005). Regarding inactivated vaccine for MERS CoV, it is reported that it could lead to an enhanced lung pathology via increased mononuclear infiltration as it was seen in SARS CoV vaccines (Prescott et al. 2018; Wen et al. 2020). Conversely, other studies have revealed that vaccination of macaques prevented both viral replication and lung damage (Qin et al. 2006; Zhou et al. 2005). Previous reports have revealed that immunopathology occur if vaccine-induced Th1 response is inefficient (Zhao et al. 2016; Zhao et al. 2010). As for SARS CoV-2, preliminary data suggest that several vaccine candidates have encouraging results. Although inactive vaccines are blamed for the cause of ADE, in vitro studies of purified inactivated SARS CoV-2 vaccine candidate have revealed that a high neutralization effect were obtained without leading ADE (Gao et al. 2020). Safety and efficacy data of BNT162b2 mRNA vaccine phase 3 trial showed that it is effective and safe with similar serious adverse effects as placebo (Polack et al. 2020). On December 11, 2020, FDA authorized emergency use of this vaccine for COVID-19 (J.H. 2020). However, there are still insufficient data for long term safety of the vaccine. In addition, the length of vaccine protection time and its preventive effect of the person-to-person transmission are still under question.

4 Conclusion

Reinfection is still a matter of widely unknown issue. However, it is considered it will not be an extensive problem. Lung inflammation developed via cytokine storm is an important factor that causes a severe course of COVID-19. Especially, low levels of NAbs or non-neutralizing antibodies may have an impact on increased inflammatory response via the activation of complement pathways and excessive release of cytokines that defined as ADE. There have been no clinical biomarkers being able to differentiate a severe infection from ADE, yet. Hence, it is an issue to detect the exact role of ADE. Its risk may vary depending on virus characteristics and it is a rare issue even in dengue infection, in which ADE may typically be seen. In addition to in vitro and animal studies, long term safety data for antibody-based treatments and vaccines are necessary to understand the exact development risk of ADE. The development of vaccines that induce potent T cell immunity besides neutralizing antibody response is a critical issue (Chen and Wherry 2020; Diamond and Pierson 2020; Hotez et al. 2020). Although, the risk of ADE should be kept in mind, it should not obstruct vaccine and antibody-based implementations. Antibody based therapies should be implemented early and they should include high neutralizing antibody titers. In consequence, SARS-CoV-2 virus is a new emergent pathogen causing a pandemic, there are many unknown issues to be resolved.

References

A.M. Arvin, K. Fink, M.A. Schmid, A. Cathcart, R. Spreafico, C. Havenar-Daughton, A. Lanzavecchia, D. Corti, H.W. Virgin, A perspective on potential antibody-dependent enhancement of SARS-CoV-2. Nature **584**, 353–363 (2020)

L. Bao, W. Deng, H. Gao, C. Xiao, J. Liu, J. Xue, Q. Lv, J. Liu, P. Yu, Y. Xu, Lack of reinfection in rhesus macaques infected with SARS-CoV-2. BioRxiv (2020)

A. Bastug, H. Bodur, S. Erdogan, D. Gokcinar, S. Kazancioglu, B.D. Kosovali, B.O. Ozbay, G. Gok, I.O. Turan, G. Yilmaz, Clinical and laboratory features of COVID-19: predictors of severe prognosis. Int. Immunopharmacol. **88**, 106950 (2020)

S.A. Board, G. Team, Z.V.T. Force, C.T. Force, G.W. Brief, C.B. Blog, Progress in the Treatments of COVID-19 (2020)

T.J. Braciale, J. Sun, T.S. Kim, Regulating the adaptive immune response to respiratory virus infection. Nat. Rev. Immunol. **12**, 295–305 (2012)

K. Callow, H. Parry, M. Sergeant, D. Tyrrell, The time course of the immune response to experimental coronavirus infection of man. Epidemiol. Infect. **105**, 435–446 (1990)

A. Chandrashekar, J. Liu, A.J. Martinot, K. McMahan, N.B. Mercado, L. Peter, L.H. Tostanoski, J. Yu, Z. Maliga, M. Nekorchuk, SARS-CoV-2 infection protects against rechallenge in rhesus macaques. Science **369**, 812–817 (2020)

Z. Chen, E.J. Wherry, T cell responses in patients with COVID-19. Nat. Rev. Immunol. **20**, 529–536 (2020)

M. Cloutier, M. Nandi, A.U. Ihsan, H.A. Chamard, S. Ilangumaran, S. Ramanathan, ADE and hyperinflammation in SARS-CoV2 infection-Comparison with dengue hemorrhagic fever and feline infectious peritonitis. Cytokine **136**, 155256 (2020)

R.J. de Groot, S.C. Baker, R.S. Baric, C.S. Brown, C. Drosten, L. Enjuanes, R.A. Fouchier, M. Galiano, A.E. Gorbalenya, Z.A. Memish, Commentary: Middle east respiratory syndrome coronavirus (mers-cov): announcement of the coronavirus study group. J. Virol. **87**, 7790–7792 (2013)

M.S. Diamond, T.C. Pierson, The challenges of vaccine development against a new virus during a pandemic. Cell Host Microbe **27**, 699–703 (2020)

K. Duan, B. Liu, C. Li, H. Zhang, T. Yu, J. Qu, M. Zhou, L. Chen, S. Meng, Y. Hu, Effectiveness of convalescent plasma therapy in severe COVID-19 patients. Proc. Natl. Acad. Sci. **117**, 9490–9496 (2020)

Q. Gao, L. Bao, H. Mao, L. Wang, K. Xu, M. Yang, Y. Li, L. Zhu, N. Wang, Z. Lv, Development of an inactivated vaccine candidate for SARS-CoV-2. Science **369**, 77–81 (2020)

Barney S. Graham, Vaccines against respiratory syncytial virus: the time has finally come. Vaccine **34**(30), 3535–3541 (2016)

X. Guo, Z. Guo, C. Duan, G. Wang, Y. Lu, M. Li, J. Lu, Long-Term persistence of IgG antibodies in SARS-CoV infected healthcare workers. MedRxiv (2020)

M.G. Guzman, M. Alvarez, S.B. Halstead, Secondary infection as a risk factor for dengue hemorrhagic fever/dengue shock syndrome: an historical perspective and role of antibody-dependent enhancement of infection. Adv. Virol. **158**, 1445–1459 (2013)

S. Hao, J. Lian, Y. Lu, H. Jia, J. Hu, G. Yu, X. Wang, K. Xu, Q. Ni, Y. Li, Decreased B cells on admission was associated with prolonged viral RNA shedding from respiratory tract in Coronavirus Disease 2019: a case control study. J. Infect. Dis. (2020)

P.J. Hotez, D.B. Corry, M.E. Bottazzi, COVID-19 vaccine design: the Janus face of immune enhancement. Nat. Rev. Immunol. **20**, 347–348 (2020)

J. Huang, L. Zheng, Z. Li, S. Hao, F. Ye, J. Chen, X. Yao, J. Liao, S. Wang, M. Zeng, Recurrence of SARS-CoV-2 PCR positivity in COVID-19 patients: a single center experience and potential implications. MedRxiv (2020)

R.F. Johnson, U. Bagci, L. Keith, X. Tang, D.J. Mollura, L. Zeitlin, J. Qin, L. Huzella, C.J. Bartos, N. Bohorova, 3B11-N, a monoclonal antibody against MERS-CoV, reduces lung pathology in rhesus monkeys following intratracheal inoculation of MERS-CoV Jordan-n3/2012. Virology **490**, 49–58 (2016)

B. Ju, Q. Zhang, J. Ge, R. Wang, J. Sun, X. Ge, J. Yu, S. Shan, B. Zhou, S. Song, Human neutralizing antibodies elicited by SARS-CoV-2 infection. Nature **584**, 115–119 (2020)

H. Kang, Y. Wang, Z. Tong, X. Liu, Retest positive for SARS-CoV-2 RNA of "recovered" patients with COVID-19: persistence, sampling issues, or re-infection? J. Med. Virol. **92**(11), 2263–2265 (2020)

S. Khan, R. Nakajima, A. Jain, R.R. de Assis, A. Jasinskas, J.M. Obiero, O. Adenaiye, S. Tai, F. Hong, D.K. Milton, Analysis of serologic cross-reactivity between common human coronaviruses and SARS-CoV-2 using coronavirus antigen microarray. BioRxiv (2020)

S.M. Kissler, C. Tedijanto, E. Goldstein, Y.H. Grad, M. Lipsitch, Projecting the transmission dynamics of SARS-CoV-2 through the postpandemic period. Science **368**, 860–868 (2020)

W.S. Lee, A.K. Wheatley, S.J. Kent, B.J. Dekosky, Antibody-dependent enhancement and SARS-CoV-2 vaccines and therapies. Nat. Microbiol. **5**, 1185–1191 (2020)

A.W. Leung, C. Xu, Is reinfection possible after recovery from COVID-19? Hong Kong Med. J. **26**, 264–265 (2020)

L. Li, W. Zhang, Y. Hu, X. Tong, S. Zheng, J. Yang, Y. Kong, L. Ren, Q. Wei, H. Mei, Effect of convalescent plasma therapy on time to clinical improvement in patients with severe and life-threatening COVID-19: a randomized clinical trial. Jama **324**, 460 (2020)

L. Liu, Q. Wei, Q. Lin, J. Fang, H. Wang, H. Kwok, H. Tang, K. Nishiura, J. Peng, Z. Tan, Anti–spike IgG causes severe acute lung injury by skewing macrophage responses during acute SARS-CoV infection. JCI Insight **4**, e123158 (2019)

Q.-X. Long, X.-J. Tang, Q.-L. Shi, Q. Li, H.-J. Deng, J. Yuan, J.-L. Hu, W. Xu, Y. Zhang, F.-J. Lv, Clinical and immunological assessment of asymptomatic SARS-CoV-2 infections. Nat. Med. **26**, 1200–1204 (2020)

R. Lu, X. Zhao, J. Li, P. Niu, B. Yang, H. Wu, W. Wang, H. Song, B. Huang, N. Zhu, Genomic characterisation and epidemiology of 2019 novel coronavirus: implications for virus origins and receptor binding. Lancet **395**, 565–574 (2020)

H. Lv, N.C. Wu, O.T.-Y. Tsang, M. Yuan, R.A. Perera, W.S. Leung, R.T. So, J.M.C. Chan, G.K. Yip, T.S.H. Chik, Cross-reactive antibody response between SARS-CoV-2 and SARS-CoV infections. Cell Rep. **31**, 107725 (2020)

E. Mahase, Covid-19: FDA authorises neutralising antibody bamlanivimab for non-admitted patients. British Medical Journal Publishing Group (2020)

T. Miyoshi-Akiyama, I. Ishida, M. Fukushi, K. Yamaguchi, Y. Matsuoka, T. Ishihara, M. Tsukahara, S. Hatakeyama, N. Itoh, A. Morisawa, Fully human monoclonal antibody directed to proteolytic cleavage site in severe acute respiratory syndrome (SARS) coronavirus S protein neutralizes the virus in a rhesus macaque SARS model. J. Infect. Dis. **203**, 1574–1581 (2011)

C.R. Moderbacher, S.I. Ramirez, J.M. Dan, A. Grifoni, K.M. Hastie, D. Weiskopf, S. Belanger, R.K. Abbott, C. Kim, J. Choi, Antigen-specific adaptive immunity to SARS-CoV-2 in acute COVID-19 and associations with age and disease severity. Cell **183**(996–1012), e19 (2020)

L.P. Molina, S.-K. Chow, A. Nickel, J.E. Love, Prolonged detection of severe acute respiratory syndrome coronavirus 2 (SARS-CoV-2) RNA in an obstetric patient with antibody seroconversion. Obstet. Gynecol. **136**, 838–841 (2020)

A.C. Monsalvo, J.P. Batalle, M.F. Lopez, J.C. Krause, J. Klemenc, J.Z. Hernandez, B. Maskin, J. Bugna, C. Rubinstein, L. Aguilar, Severe pandemic 2009 H1N1 influenza disease due to pathogenic immune complexes. Nat. Med. **17**, 195–199 (2011)

D.M. Morens, C. Venkateshan, S.B. Halstead, Dengue 4 virus monoclonal antibodies identify epitopes that mediate immune infection enhancement of dengue 2 viruses. J. Gen. Virol. **68**, 91–98 (1987)

K. Ng, N. Faulkner, G. Cornish, A. Rosa, C. Earl, A. Wrobel, D. Benton, C. Roustan, W. Bolland, R. Thompson, Pre-existing and de novo humoral immunity to SARS-CoV-2 in humans. BioRxiv (2020)

P.J. Openshaw, J.S. Tregoning, Immune responses and disease enhancement during respiratory syncytial virus infection. Clin. Microbiol. Rev. **18**, 541–555 (2005)

D.M. Patrick, M. Petric, D.M. Skowronski, R. Guasparini, T.F. Booth, M. Krajden, P. McGeer, N. Bastien, L. Gustafson, J. Dubord, An outbreak of human coronavirus OC43 infection and serological cross-reactivity with SARS coronavirus. Canad. J. Infect. Dis. Med. Microbiol. **17**, 330–336 (2006)

N.C. Pedersen, An update on feline infectious peritonitis: virology and immunopathogenesis. Vet. J. **201**, 123–132 (2014)

J. Peiris, S. Lai, L. Poon, Y. Guan, L. Yam, W. Lim, J. Nicholls, W. Yee, W. Yan, M. Cheung, Coronavirus as a possible cause of severe acute respiratory syndrome. Lancet **361**, 1319–1325 (2003)

S. Perlman, J. Netland, Coronaviruses post-SARS: update on replication and pathogenesis. Nat. Rev. Microbiol. **7**, 439–450 (2009)

J.P.S. Peron, H. Nakaya, Susceptibility of the elderly to SARS-CoV-2 infection: ACE-2 overexpression, shedding, and antibody-dependent enhancement (ADE). Clinics **75**, 1–6 (2020)

T.S. Pillay, Gene of the month: the 2019-nCoV/SARS-CoV-2 novel coronavirus spike protein. J. Clin. Pathol. **73**(7), 366–369 (2020)

F.P. Polack, M.N.L. Teng, P. Collins, G.A. Prince, M. Exner, H. Regele, D.D. Lirman, R. Rabold, S.J. Hoffman, C.L. Karp, A role for immune complexes in enhanced respiratory syncytial virus disease. J. Exp. Med. **196**, 859–865 (2002)

F.P. Polack, S.J. Thomas, N. Kitchin, J. Absalon, A. Gurtman, S. Lockhart, J.L. Perez, G. Pérez Marc, G., E.D. Moreira, C. Zerbini, Safety and efficacy of the BNT162b2 mRNA covid-19 vaccine. New Engl. J. Med. **389**, 2603–2615 (2020)

J. Prescott, D. Falzarano, E. de Wit, K. Hardcastle, F. Feldmann, E. Haddock, D. Scott, H. Feldmann, V.J. Munster, Pathogenicity and viral shedding of MERS-CoV in immunocompromised rhesus macaques. Front. Immunol. **9**, 205 (2018)

E. Qin, H. Shi, L. Tang, C. Wang, G. Chang, Z. Ding, K. Zhao, J. Wang, Z. Chen, M. Yu, Immunogenicity and protective efficacy in monkeys of purified inactivated Vero-cell SARS vaccine. Vaccine **24**, 1028–1034 (2006)

J.G. Rizk, D.N. Forthal, K. Kalantar-Zadeh, M.R. Mehra, C.J. Lavie, Y. Rizk, J.P. Pfeiffer, J.C. Lewin, Expanded access programs, compassionate drug use, and emergency use authorizations during the COVID-19 pandemic. Drug Disc. Today **26**, 593–603 (2020)

C. Shen, Z. Wang, F. Zhao, Y. Yang, J. Li, J. Yuan, F. Wang, D. Li, M. Yang, L. Xing, Treatment of 5 critically ill patients with COVID-19 with convalescent plasma. JAMA **323**, 1582–1589 (2020)

R. Shi, C. Shan, X. Duan, Z. Chen, P. Liu, J. Song, T. Song, X. Bi, C. Han, L. Wu, A human neutralizing antibody targets the receptor binding site of SARS-CoV-2. Nature **584**, 1–8 (2020)

C.P. Simmons, T.N.B. Chau, T.T. Thuy, N.M. Tuan, D.M. Hoang, N.T. Thien, L.B. Lien, N.T. Quy, N.T. Hieu, T.T. Hien, Maternal antibody and viral factors in the pathogenesis of dengue virus in infants. J. Infect. Dis. **196**, 416–424 (2007)

T.M. Snyder, R. Gittelman, M. Klinger, D.H. May, E.J. Osborne, R. Taniguchi, H.J. Zahid, I.M. Kaplan, J.N. Dines, M.T. Noakes, Magnitude and dynamics of the T-cell response to SARS-CoV-2 infection at both individual and population levels. MedRxiv (2020)

S. Sridhar, A. Luedtke, E. Langevin, M. Zhu, M. Bonaparte, T. Machabert, S. Savarino, B. Zambrano, A. Moureau, A. Khromava, Effect of dengue serostatus on dengue vaccine safety and efficacy. N. Engl. J. Med. **379**, 327–340 (2018)

H.C. Sullivan, J.D. Roback, Convalescent plasma: therapeutic hope or hopeless strategy in the SARS-CoV-2 pandemic. Transfusion Med. Rev. **34**, 145–150 (2020)

A. Takada, H. Feldmann, T.G. Ksiazek, Y. Kawaoka, Antibody-dependent enhancement of Ebola virus infection. J. Virol. **77**, 7539–7544 (2003)

J.H. Tanne, Covid-19: FDA approves use of convalescent plasma to treat critically ill patients. BMJ **368**, m1256 (2020a)

J.H. Tanne, Covid-19: FDA panel votes to approve Pfizer BioNTech vaccine. British Medical Journal Publishing Group (2020b)

Adam Taylor, Suan-Sin. Foo, Roberto Bruzzone, Luan Vu Dinh, Nicholas J. C. King, Suresh Mahalingam, Fc receptors in antibody-dependent enhancement of viral infections. Immunol. Rev. **268**(1), 340–364 (2015)

K.K.W. To, I.F.N. Hung, J.D. Ip, A.W.H. Chu, W.M. Chan, A.R. Tam, C.H.Y. Fong, S. Yuan, H.W. Tsoi, A.C.K. Ng, Coronavirus disease 2019 (COVID-19) re-infection by a phylogenetically distinct severe acute respiratory syndrome coronavirus 2 strain confirmed by whole genome sequencing. Clin. Infect. Dis. (2020a)

K.K.-W. To, O.T.-Y. Tsang, W.-S. Leung, A.R. Tam, T.-C. Wu, D.C. Lung, C.C.-Y. Yip, J.-P. Cai, J.M.-C. Chan, T.S.-H. Chik, Temporal profiles of viral load in posterior oropharyngeal saliva samples and serum antibody responses during infection by SARS-CoV-2: an observational cohort study. Lancet Infect. Dis. **20**, 565–574 (2020)

H. Ulrich, M.M. Pillat, A. Tárnok, Dengue fever, COVID-19 (SARS-CoV-2), and antibody-dependent enhancement (ADE): a perspective. Cytometry Part A **97**, 662–667 (2020)

L. van der Hoek, K. Pyrc, B. Berkhout, Human coronavirus NL63, a new respiratory virus. FEMS Microbiol. Rev. **30**, 760–773 (2006)

Y. Wan, J. Shang, S. Sun, W. Tai, J. Chen, Q. Geng, L. He, Y. Chen, J. Wu, Z. Shi, Molecular mechanism for antibody-dependent enhancement of coronavirus entry. J. Virol. **94** (2020)

K. Wang, W. Chen, Y.-S. Zhou, J.-Q. Lian, Z. Zhang, P. Du, L. Gong, Y. Zhang, H.-Y. Cui, J.-J. Geng, SARS-CoV-2 invades host cells via a novel route: CD147-spike protein. BioRxiv (2020a)

S.-F. Wang, S.-P. Tseng, C.-H. Yen, J.-Y. Yang, C.-H. Tsao, C.-W. Shen, K.-H. Chen, F.-T. Liu, W.-T. Liu, Y.-M.A. Chen, Antibody-dependent SARS coronavirus infection is mediated by antibodies against spike proteins. Biochem. Biophys. Res. Commun. **451**, 208–214 (2014)

X. Wang, X. Guo, Q. Xin, Y. Pan, Y. Hu, J. Li, Y. Chu, Y. Feng, Q. Wang, Neutralizing antibodies responses to SARS-CoV-2 in COVID-19 inpatients and convalescent patients. Clin. Infect. Dis. **71**, 2688–2694 (2020b)

Y. Wang, L. Zhang, L. Sang, F. Ye, S. Ruan, B. Zhong, T. Song, A.N. Alshukairi, R. Chen, Z. Zhang, Kinetics of viral load and antibody response in relation to COVID-19 severity. J. Clin. Invest. **130** (2020c)

J. Wen, Y. Cheng, R. Ling, Y. Dai, B. Huang, W. Huang, S. Zhang, Y. Jiang, Antibody-dependent enhancement of Coronavirus. Int. J. Infect. Dis. **100**, 483–489 (2020)

S. Willey, M.M. Aasa-Chapman, S. O'Farrell, P. Pellegrino, I. Williams, R.A. Weiss, S.J. Neil, Extensive complement-dependent enhancement of HIV-1 by autologous non-neutralising antibodies at early stages of infection. Retrovirology **8**, 16 (2011)

F. Wu, A. Wang, M. Liu, Q. Wang, J. Chen, S. Xia, Y. Ling, Y. Zhang, J. Xun, L. Lu Neutralizing antibody responses to SARS-CoV-2 in a COVID-19 recovered patient cohort and their implications (2020)

X. Zhang, Y. Zou, S. Zhang, Treatment with convalescent plasma for critically Ill patients With SARS (2020)

J. Zhao, Q. Yuan, H. Wang, W. Liu, X. Liao, Y. Su, X. Wang, J. Yuan, T. Li, J. Li, Antibody responses to SARS-CoV-2 in patients of novel coronavirus disease 2019. Clin. Infect. Dis. **4**, E123158 (2020)

J. Zhao, J. Zhao, A.K. Mangalam, R. Channappanavar, C. Fett, D.K. Meyerholz, S. Agnihothram, R.S. Baric, C.S. David, S. Perlman, Airway memory CD4+ T cells mediate protective immunity against emerging respiratory coronaviruses. Immunity **44**, 1379–1391 (2016)

J. Zhao, J. Zhao, S. Perlman, T cell responses are required for protection from clinical disease and for virus clearance in severe acute respiratory syndrome coronavirus-infected mice. J. Virol. **84**, 9318–9325 (2010)

J. Zhou, W. Wang, Q. Zhong, W. Hou, Z. Yang, S.-Y. Xiao, R. Zhu, Z. Tang, Y. Wang, Q. Xian, Immunogenicity, safety, and protective efficacy of an inactivated SARS-associated coronavirus vaccine in rhesus monkeys. Vaccine **23**, 3202–3209 (2005)

Y. Zhou, Y. Hou, J. Shen, Y. Huang, W. Martin, F. Cheng, Network-based drug repurposing for novel coronavirus 2019-nCoV/SARS-CoV-2. Cell Disc. **6**, 1–18 (2020)

Graph-Based Clustering Algorithm for Social Community Transmission Control of COVID-19 During Lockdown

Varun Nagesh Jolly Behera, Ashish Ranjan, and Motahar Reza

Abstract This paper proposes a system to model the spread of COVID-19. This system will work in post lockdown conditions, when the only mode of travel, is by road. It defines impact measures, that state the severity of potential disease spread, in a specific area. These impact measures are calculated based on existing hotspots, and are clustered into regions of varying danger-levels, using a graph clustering algorithm. Using this method, it can be predicted where more lenient measures may be taken, and which areas are less prone to the virus spread. There exist other methodologies to model the spread of viruses, but most overlook the spatial nature of viruses. The proposed system focuses on this limitation. Specifically, it focuses on preventing the virus spread, from a geographical point of view. Since the virus spread depends entirely on contact, regions near existing hotspots may potentially become new hotspots. The entire country is first visualized as a weighted graph of regions, at an appropriate administrative level, such as districts. The weights of the nodes are the number of active cases, and the weights of the edges are the geographical distances between those nodes. This graph is connected based on a distance threshold. The impact measure tells the impact of a region, on nearby regions, and the danger value tells the transmission possibility, between separate regions. Using this data, potential hotspots are easily identified. This data will help administrative bodies, to make more fine-tuned lockdown restrictions, based on the impact measures.

Keywords Graph · Covid · Algorithm · Transmission

V. N. J. Behera (✉) · A. Ranjan
R&D Division, Just Another Media Laboratory (JAM Lab), Mumbai, India
e-mail: varun.behera@jamlab.in

A. Ranjan
e-mail: ashish.ranjan@jamlab.in

M. Reza
Department of Mathematics, GITAM University Hyderabad Campus, Hyderabad, India
e-mail: mreza@gitam.edu

J. Nayak et al. (eds.), *Understanding COVID-19: The Role of Computational Intelligence*, Studies in Computational Intelligence 963, https://doi.org/10.1007/978-3-030-74761-9_6

1 Introduction

The 2020 pandemic, caused by the SARS-CoV-2 [1, 2], popularly known as COVID-19, has resulted in thousands of deaths and millions being infected. It was officially first identified in December 2019, in Wuhan, China [3]. In mid March, 2020, the World Health Organization declared it a pandemic, changing its previous designation as an outbreak. Over 65 million have been infected and more than 1.5 million dead worldwide, as of 4th December 2020 [4].

Similar to most flu viruses, it is spread via sneeze and cough droplets [5]. Touching the face after touching a contaminated surface may also cause infection. Those who are asymptomatic may also spread the disease, although chances of infection spreading this way are low. The virus is most contagious starting a few days before the initial symptoms, and then staying the same a week or two later [6, 7].

Cough, fever, tiredness, breathing issues, are a few common symptoms. Since this virus primarily causes pneumonia and respiratory issues, those with existing conditions, are more at risk. This is evident by the fact that almost 80% deaths in China, were aged above 60 [8], most having pre-existing health complications like diabetes and heart diseases. The ratio of the number of death to the number of cases is 2.3%, as of 4th December 2020, according to John Hopkins university.

Social distancing has been one of the best preventive measures for COVID-19, accompanied by wearing face masks, maintaining personal hygiene, regularly sanitizing one's surroundings, and self-isolating, if flu-like symptoms evolve [9]. Governments have imposed lockdowns, travel-restrictions, and mandatory safety measures to control the virus spread [10].

While the above measures prove to be very effective, they are not 100% foolproof. Restrictions should be based on local situations, rather than be imposed at a country level. It makes no sense to restrict movement in an area that has a negligible number of cases [11]. A balance is required to make sure that the economy does not completely destabilize, and people do not end up in a position where they are so financially unstable, that they may starve to death. At the same time, people need to be protected from the infection. If dynamic real-time decisions are not taken, a developing country like India will end up suffering from both a dying economy and a dying workforce [12].

Most research done regarding disease epidemiology is done on factors like total case numbers, infection rates, and primarily focus on the SEIR (susceptible - exposed - infected - recovered) model [13]. But COVID-19 is unique in the sense that after imposing travel restrictions on air and rail, the only way of virus transmission is via roads. This means the virus does not have much spatial transmission range once mass travel restrictions and lockdowns are imposed. This makes spatial epidemiology an ideal candidate for analyzing and predicting the spread of this type of disease.

A possible model for analyzing diseases like COVID-19 post-lockdown stage is visualizing administrative zones as graph nodes, and the roads connecting them as the edges. Then these nodes may be color-coded, for spread severity. This methodology will work regardless of the administrative level to be analyzed, as long as the zone-wise infection data is available, and has been accurately documented.

2 Related Work

Epidemiology is generally either observational or experimental. Experimental studies follow a controlled factor-based approach, while observational studies go for an entirely natural approach, with no interference [14]. Epidemiological studies should reveal relations with underlying health conditions. For example, it is known that cardiovascular diseases, diabetes, and other conditions, severely affect a COVID-19 patient. This paper focuses on observational studies, specifically on spatial epidemiology [15].

A lot of work has been done on spatial epidemiology research. One of them being the mapping of diseases, allowing for a better understanding of disease spread and assisting the government in making well-informed decisions on resource allocation [16]. There also exist methodologies, based on geographical demographics, including factors like race, age, etc. that may influence the spread of diseases. [17]. Disease clusters, infection groups, hotspots, are also dynamic to be analyzed. Although there are some issues to these approaches, provided the study has been done properly, they can be beneficial, especially for epidemics and pandemics [18].

There has also been a lot of recent research specific to COVID-19. More prominent studies are based in China, being the origin point of the virus. For example, a study [19], explores the spatial epidemic dynamics using Moran's I spatial statistic. This study was based in Mainland China. It concluded, that in fact, there is a spatial correlation regarding the spread of COVID-19. A study, specific to the effects of public transport, also exists, again based in China [20]. Most studies are based on the SEIR model [21], although SEIRS [22] should have been the way to go, due to WHO stating that reinfection is possible. Other recent work includes using machine learning for modelling the trend [23], estimation of reproduction number [24], and clustering analysis [25].

Studies based on the Indian scenario are more in line with the work explained in this paper. There have been a few studies with regards to spatial transmission, including multi-city modeling [26], trend analysis using geospatial approaches [27], and spatial prediction using ARIMA [28].

Other recent work with respect to geospatial analysis and spread control includes reopening strategies [29], crowds in urban areas [30], and cluster identification using location data [31]. Literature review has been summarised in Table 1.

Table 1 Summary of related work

Researchers	Focus	Methodolgy/Findings	Comments
Kang, Dayun, et al.	Spatial epidemics of COVID-19 in mainland China	Using Moran's I spatial statistic, it concludes that the disease does have spatial association	Does not explain benefits/practical use-case of these findings
Zheng, Ruizhi, et al.	Spatial transmission of COVID-19 via transportation in China	Association was found between public transport and the virus spread	Paper is poorly explained as methodologies have not been expressed
Huang, Rui, Miao Liu, and Yongmei Ding.	Spatial-temporal distribution and prediction of COVID-19 in China	Also used Moran's I spatial statistic, similar conclusions to Kang, Dayun, et al.	Does not explain benefits/practical use-case of these findings
Tuli, Shreshth, et al.	Prediction of growth and trend of COVID-19	Using Generalized Inverse Weibull distribution for predicting virus growth	Practical usecase, but does not explore spatial dynamic of the virus
Zhao, Shi, et al.	Estimation of the reproduction number of COVID-19	Using the exponential growth data, the reproduction rate of virus in the body was found	Confirms that COVID-19 will always cause outbreaks
Zarikas, Vasilios, et al.	Clustering analysis of COVID-19 affected countries	Using a 45-day time period for number of infections, clustering of countries was done	Does not explore spatial correlation while clustering countries
Pujari, Bhalchandra S., and Snehal M. Shekatkar	Multi-city modelling of COVID-19 in India	Using an SIR model with spatial dynamics, to model disease spread	Focuses on pre-lockdown condition, primarily on flights between major cities
Murugesan, Bagyaraj, et al.	Geospatial distribution and trend analysis of COVID-19 in India	Using GIS-based spatial distribution, countrywide spread was mapped out	Interpolation is used to estimate number of cases in between regions
Roy, Santanu, Gouri Sankar Bhunia, and Pravat Kumar Shit.	Spatial prediction of COVID-19 in India	Using ARIMA to forecast epidemiological patter of COVID-19 spread	There is no uniformity in the shown spread, as moderate regions are somehow unaffected by high risk regions nearby
Liu, Meng, Raphael Thomadsen, and Song Yao.	Forecasting COVID-19 spread under different reopening strategies	Using an SIR model with mobility data for virus spread	No visual representations done for the mobility case
Rader, Benjamin, et al.	Role of climate, urbanization, restrictions in COVID-19 spread and prevention	Using Lloyd's index, and correlating with number of cases	Identifies a relation, but does not perform predictions based on it
Hisada, Shohei, et al.	Identifying clusters using location-aware mobile devices	Using the confidential geolocation data, based on web search query logs, to identify user locations	Authors have themselves identified a bias after public announcement of the system

3 Goals and Objectives

The objective of this paper is to study the spatial epidemiology of diseases like COVID-19 and devise a methodology to do so. This methodology will help predict future disease hotspots and help in devising preventive measures as necessary. The modeling and prediction will be done at the district level in this paper, but the same methodology can be applied on any administrative level. Following are a few goals of this paper:

1. **Country Independent**: The methodology needs to be independent of the country. As long as the assumptions are met, the algorithm should work automatically for any and all countries.
2. **Administrative Level Independent**: It also should be independent of the administration level that needs to be analyzed. This includes state, county/district, zone, zip codes, etc. As long as both location and infection data is available, the algorithm should run flawlessly.
3. **Automatic Graph Construction**: Graph for that administrative level should be generated automatically on the fly. This includes connecting nodes using clustering, forming edges, and assigning weight to both the nodes and edges.
4. **Division Inclusive/Exclusive**: Impact of a location on itself may be included or excluded, based on user preference. This will give a slightly different result in each case.
5. **Real Time Updates**: The system should update periodically, preferably daily, so that better strategies can be made for disease spread prevention and control. The result can be visualized by creating a web app dashboard.

4 Requirements and Assumptions

The algorithm defined in this paper is robust, as long as few base assumptions are met. The reliability of this method is based on a lot of requirements and assumptions including:

1. **Restricted Travel**: This means that both air and rail travel are restricted. Also, long-distance travel by road is strictly monitored, making it difficult for the disease to have a large radius of influence.
2. **Data Quality**: The data needs to be well defined, and properly cleaned. If the geographical names do not match with the Map API, then the longitude and latitude information of those locations cannot be retrieved. This means it is necessary to make sure there are no naming discrepancies, and the data should be modified as required so that its information aligns with that of the Map API to be used. Not checking and refining the data, will result in varying amounts of information loss.
3. **Data Granularity**: This model can work on various administrative levels including at the district, state, and country level. Modeling can be done for a specific

political map level only if the data is available for it. Meaning, if the model is to be defined at the district level, both the Map API and the disease spread data should have granularity till the district level. If, either the co-ordinate data or the disease spread data of districts are missing, this model cannot be applied at that level.

4. **Data Accuracy**: The data needs to be accurate. If data is old or incorrect, the calculations made by the model will be drastically affected. The accuracy of the current number of infected is most important. This means testing more, and using better quality tests will help this model a lot.
5. **Proper Map API**: Map API that gives accurate co-ordinates of the geographical locations at the required administrative level is necessary. Only then can the graph be created with proper weight values and connections.

5 Methodology Overview

Following is a brief overview of the workflow described in this paper. It is visually represented in Fig. 1.

1. **Choosing Administrative Level**: The first step is choosing the administrative level, at which infection spread has to be analyzed and predicted. Administrative levels like countries, states, counties, districts, zones, wards, zipcodes, can be chosen for analysis. The only limitation in choosing the administrative level is the unavailability of data at that level. Otherwise, any administrative level can be chosen and analyzed.
2. **Data Acquisition**: The reliability of the algorithm is entirely based on the 'goodness' of data. This includes aspects like availability, accuracy, comprehensiveness, etc. If both, infection data and location data are available and reliable, only then can the algorithm be used. The algorithm does not have the ability to fix inaccuracies and errors in the data. It is assumed that the data is accurate enough to be used. There exist many infection data sources, depending on the disease whose transmission has to be studied. To find location data, a satisfactorily capable Map API is used.
3. **Pre-processing**: The data obtained from the previous step has to be cleaned and restructured properly. This includes dropping unnecessary columns from the infection data, and the row of unassigned infection count. It is also necessary to make sure that the latitudes and longitudes are properly mapped with their respective divisions.
4. **Graph Construction**: Based on the restructured data, individual divisions are designated as separate graph nodes. This is done to create an abstract view of the country's administrative divisions, and how they are connected. Using a clustering method inspired by k-means, the nearest nodes for every node are discovered. All pairs of nearest nodes are connected, to form graph edges. A distance metric is used to define the clustering factor. Only divisions within a

certain range will be connected to each other. This distance metric has to be changed based on the administrative level chosen, to factor in the difference in distances between division centers.

5. **Finding Danger Level**: Three different measures have been defined in this paper, to evaluate the probability of a district becoming a hotspot, termed as 'danger level'. These measures evaluate this danger level, using different parameters, giving different levels of importance to each parameter.
6. **Predicting Potential Hotspots**: The danger level value obtained can be easily color-coded. The color green represents a division having a low risk of becoming a hotspot anytime soon. Orange signifies that the division is near a few existing hotspots, and if precautions are not taken, may become one itself. Red marks regions that either already have, or will very soon become major hotspots, unless strict measures are not taken.

6 Data Acquisition

This model requires primarily two types of datasets. First, is the current number of infections at the desired administrative level. This paper explores the model's robustness at the district level, but as mentioned before, the model can be used at any administrative level. Second, is the co-ordinates of various locations at the same administrative level. This information can be extracted efficiently using a Map API.

6.1 Infection Data

This is the data of the total number of active infections, at the required administration level. A varied number of sources exist, depending on the disease that needs to be studied. Some prominent datasets include the Malaria Atlas Project [32], and Scientific Data for Dengue [33], Leishmania [34] and Ebola [35]. This paper explores the impact of COVID-19. Some datasets currently available for it are nCoV-2019 [36], CORD-19 [38], Lancet Dashboard [37] and Coronavirus Twitter Dataset [39]. Since this paper focuses on the Indian scenario at the district level, the dataset chosen is from covid19india.org [40], a crowd-sourced dataset, as seen in Fig. 2. This dataset is highly accurate and is updated daily, sourced both, from official government sources and media. There also exists an official Indian Government Dataset that can only be used for state-level analysis.

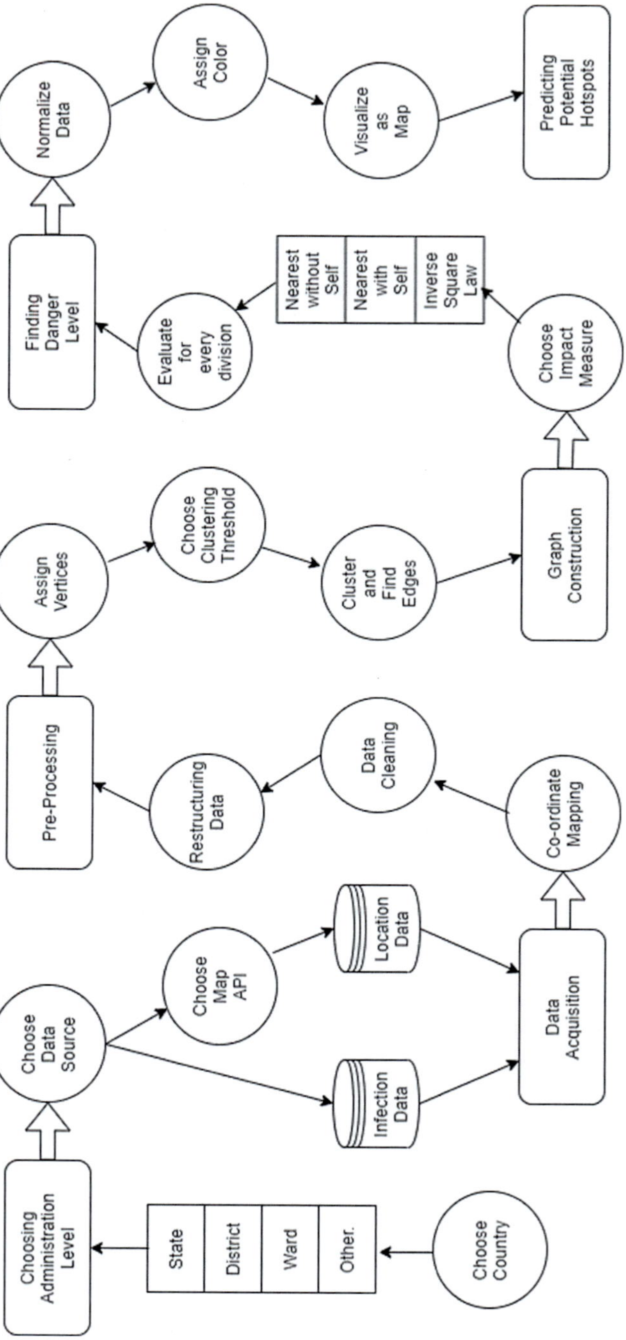

Fig. 1 Methodology overview

SlNo	State	State	District_Key	District	Confirmed	Active	Recovered	Deceased	Migrated_	Delta_Confir	Delta_Acti	Delta_Rec	Delta_Dec
0	UN	State Unassigned	UN_Unassigned	Unassigned	0	0	0	0	0	0	0	0	0
1	AN	Andaman and Nicobar Islands	AN_Nicobars	Nicobars	0	0	0	0	0	0	0	0	0
2	AN	Andaman and Nicobar Islands	AN_North and Middle Andaman	North and Middle Andaman	1	0	1	0	0	0	0	0	0
3	AN	Andaman and Nicobar Islands	AN_South Andaman	South Andaman	51	19	32	0	0	0	0	0	0
0	AP	Andhra Pradesh	AP_Foreign Evacuees	Foreign Evacuees	434	0	434	0	0	0	0	0	0
4	AP	Andhra Pradesh	AP_Anantapur	Anantapur	45300	6020	38913	367	0	753	27	718	8
5	AP	Andhra Pradesh	AP_Chittoor	Chittoor	42561	8979	33109	473	0	927	-35	953	9
6	AP	Andhra Pradesh	AP_East Godavari	East Godavari	66948	14454	52067	427	0	1244	-1339	2578	5
7	AP	Andhra Pradesh	AP_Guntur	Guntur	40135	7597	32119	419	0	703	40	655	8
8	AP	Andhra Pradesh	AP_Krishna	Krishna	18482	2732	15442	308	0	457	87	366	4
9	AP	Andhra Pradesh	AP_Kurnool	Kurnool	48386	5797	42194	395	0	380	-472	848	4
0	AP	Andhra Pradesh	AP_Other State	Other State	2461	0	2461	0	0	0	0	0	0
10	AP	Andhra Pradesh	AP_Prakasam	Prakasam	29074	14109	14645	320	0	1042	496	538	8
11	AP	Andhra Pradesh	AP_S.P.S. Nellore	S.P.S. Nellore	37278	6673	30263	342	0	1299	29	1266	4
12	AP	Andhra Pradesh	AP_Srikakulam	Srikakulam	27984	7065	20656	263	0	818	-28	844	2
13	AP	Andhra Pradesh	AP_Visakhapatnam	Visakhapatnam	40787	6503	33964	320	0	573	-126	694	5
14	AP	Andhra Pradesh	AP_Vizianagaram	Vizianagaram	23911	8876	14855	180	0	593	85	507	1
15	AP	Andhra Pradesh	AP_West Godavari	West Godavari	42903	2832	39718	353	0	1101	18	1078	5
16	AP	Andhra Pradesh	AP_Y.S.R. Kadapa	Y.S.R. Kadapa	31481	8052	23179	250	0	904	27	870	7
17	AR	Arunachal Pradesh	AR_Anjaw	Anjaw	52	10	42	0	0	0	0	0	0
18	AR	Arunachal Pradesh	AR_Changlang	Changlang	525	120	405	0	0	0	0	0	0
19	AR	Arunachal Pradesh	AR_East Kameng	East Kameng	227	34	193	0	0	0	0	0	0
20	AR	Arunachal Pradesh	AR_East Siang	East Siang	320	143	176	1	0	0	0	0	0
21	AR	Arunachal Pradesh	AR_Kamle	Kamle	3	0	3	0	0	0	0	0	0
22	AR	Arunachal Pradesh	AR_Kra Daadi	Kra Daadi	4	3	1	0	0	0	0	0	0
23	AR	Arunachal Pradesh	AR_Kurung Kumey	Kurung Kumey	7	7	0	0	0	0	0	0	0
24	AR	Arunachal Pradesh	AR_Lepa Rada	Lepa Rada	59	31	28	0	0	0	0	0	0
25	AR	Arunachal Pradesh	AR_Lohit	Lohit	151	11	140	0	0	0	0	0	0
26	AR	Arunachal Pradesh	AR_Longding	Longding	45	25	20	0	0	0	0	0	0
27	AR	Arunachal Pradesh	AR_Lower Dibang Valley	Lower Dibang Valley	74	36	38	0	0	0	0	0	0
28	AR	Arunachal Pradesh	AR_Lower Siang	Lower Siang	125	33	92	0	0	0	0	0	0
29	AR	Arunachal Pradesh	AR_Lower Subansiri	Lower Subansiri	60	12	48	0	0	0	0	0	0
30	AR	Arunachal Pradesh	AR_Namsai	Namsai	201	12	188	1	0	0	0	0	0
31	AR	Arunachal Pradesh	AR_Pakke Kessang	Pakke Kessang	28	6	22	0	0	0	0	0	0
32	AR	Arunachal Pradesh	AR_Papum Pare	Papum Pare	1678	413	1263	2	0	0	0	0	0

Fig. 2 Raw infection data

Algorithm 1: Algorithm for finding co-ordinates of every location

Input: Location List
Result: Location with corresponding Co-ordinates

```
READ locationList
APPLY Preprocessing
foreach [district, state, country] in locationList do
    if district != 'Unassigned' or 'Unknown' or 'Other' then
        coordinates = gmapsAPI.geocode(district, state, country)
        APPEND district, coordinates to list L
    end
end
```

6.2 *Location Data*

This is the data of the administrative division location at the required administrative level. This data can be obtained using a Map API, in the form of latitudes and longitudes. The Map API will take the location string as input, and give co-ordinates as output, as shown in Algorithm 1 and Fig. 3. The most popular Map APIs are the Bing Maps API and the Google Maps Geocoding API.

6.2.1 Bing Maps API

There are two ways to find longitudes and latitudes using the Bing Maps API.

First is using the 'Find location by address method'. In this method, a specific URL template is used to obtain longitudes and latitudes. This is done by filling in various parameters in the template URL, like locality, postal code, and street address. To use the API a Bing Maps key is required. Requesting a location using the URL template returns a response containing location data like latitudes and longitudes, location type, and the geographical area that contains the location. The API may be accessed using both HTTP and HTTPS protocols. Information for a location can be obtained by setting either one or more parameters. The output can be obtained either as an XML or JSON response. Both responses will give central latitude and longitude, as well as latitudes and longitudes of the bounding box for that location.

The second is using the 'Find a location by query' method. This method gives longitudes and latitudes, based on a given query. This method is more akin to generally how Map Applications are interacted with. The query can be sent as a structured URL or as an unstructured query string. User Context Parameters may be used for better accuracy. The output is similar to the previous method, although more than one output may be received due to ambiguity. It also likewise supports both HTTP and HTTPS.

State	District	Latitude	Longitude
Andaman and Nicobar Islands	Nicobars	8	93.5
Andaman and Nicobar Islands	North and Middle Andaman	12.6518832	92.8976254
Andaman and Nicobar Islands	South Andaman	11.8311681	92.6586401
Andhra Pradesh	Foreign Evacuees	15.9128998	79.7399875
Andhra Pradesh	Anantapur	14.6818877	77.6005911
Andhra Pradesh	Chittoor	13.217176	79.1003289
Andhra Pradesh	East Godavari	17.32125	82.0407137
Andhra Pradesh	Guntur	16.3066525	80.4365402
Andhra Pradesh	Krishna	16.6099914	80.7214417
Andhra Pradesh	Kurnool	15.8281257	78.0372792
Andhra Pradesh	Prakasam	15.348463	79.560344
Andhra Pradesh	S.P.S. Nellore	14.2580848	79.9192702
Andhra Pradesh	Srikakulam	18.2949165	83.8938018
Andhra Pradesh	Visakhapatnam	17.6868159	83.2184815
Andhra Pradesh	Vizianagaram	18.1066576	83.3955506
Andhra Pradesh	West Godavari	16.9174181	81.3399414
Andhra Pradesh	Y.S.R. Kadapa	14.5767233	78.8382644
Arunachal Pradesh	Anjaw	28.0666845	96.4922423
Arunachal Pradesh	Changlang	27.7421763	96.642433
Arunachal Pradesh	East Kameng	27.4231261	93.0175712
Arunachal Pradesh	East Siang	28.1097051	95.1432068
Arunachal Pradesh	Kamle	27.7999698	94.0750175
Arunachal Pradesh	Kra Daadi	27.726036	93.642161
Arunachal Pradesh	Kurung Kumey	27.7995361	93.5538411
Arunachal Pradesh	Lepa Rada	28.2179994	94.7277528
Arunachal Pradesh	Lohit	27.9039651	96.1739116
Arunachal Pradesh	Longding	26.872816	95.305748
Arunachal Pradesh	Lower Dibang Valley	28.1428728	95.8431018
Arunachal Pradesh	Lower Siang	28.1097051	95.1432068
Arunachal Pradesh	Lower Subansiri	27.6169349	93.8391557
Arunachal Pradesh	Namsai	27.6692115	95.8644018
Arunachal Pradesh	Pakke Kessang	27.145853	93.246261
Arunachal Pradesh	Papum Pare	27.1719019	93.7029002
Arunachal Pradesh	Shi Yomi	28.1693138	94.800147
Arunachal Pradesh	Siang	28.1097051	95.1432068

Fig. 3 Location data

6.2.2 Google Geocoding

It is a part of the Google Maps Platform, a subset of the Google Cloud Platform. This API is similar to the second Bing Maps API method. This API service can be directly accessed using an HTTP request. An API key has to be obtained from the Google Cloud Platform to use the Geocoding API. Google's API also gives output in JSON or XML, as required. Input can be either a string, containing the address or as components with their corresponding values. The output will include both, the central latitude and longitude, and the bounding box latitudes and longitudes. The benefit of using Google Geocoding is the support for up to 5 administrative levels within a country. Districts are at the administrative area level 2 as per the API.

7 Pre-processing

The model requires an administrative division location, with its corresponding active number of cases. Before this can be done, co-ordinates, i.e. latitudes and longitudes for every division, need to be found. The infection data also needs to be cleaned. This includes dropping unnecessary columns and removing rows with undefined divisions. After that, both datasets need to be mapped to give the required restructured data. Since this paper focuses on the district level, the mapping will be done at that level, as shown in Fig. 4.

7.1 Co-ordinate Mapping

This paper explores using the Google Geocoding API for finding latitudes and longitudes of administrative divisions at the required level. The 'googlemaps' package is used, accompanied by the Geocoding API client key. For every district-state pair in the infection data, a query is formed as 'District, State, Country'. For each such query, a corresponding latitude-longitude pair is obtained. All of these obtained location-coordinate pairs are properly stored.

7.2 Data Cleaning

The data obtained from https://covid19india.org/, contains a lot of information, most of which is not required for the study done in this paper. This includes attributes like State Code, Confirmed cases, Recovered cases, etc. The attributes required are State, District, Active cases, and Change in Active cases. Change in Active cases is the increase in the number of active cases from the previous day. This is because a district with a higher number of currently active cases will have a more severe impact on its neighboring districts. The use of each of these parameters is mentioned later in this paper.

State	District	Confirmed	Active	Recovered	Deceased	Delta_Acti	Latitude	Longitude
Andaman and Nicobar Islands	Nicobars	0	0	0	0	0	8	93.5
Andaman and Nicobar Islands	North and Middle Andaman	1	0	1	0	0	12.65188	92.89763
Andaman and Nicobar Islands	South Andaman	51	19	32	0	0	11.83117	92.65864
Andhra Pradesh	Foreign Evacuees	434	0	434	0	0	15.9129	79.73999
Andhra Pradesh	Anantapur	46325	5705	40241	379	0	14.68189	77.60059
Andhra Pradesh	Chittoor	44614	9072	35053	489	0	13.21718	79.10033
Andhra Pradesh	East Godavari	69686	13561	55690	435	0	17.32125	82.04071
Andhra Pradesh	Guntur	41602	7696	33468	438	0	16.30665	80.43654
Andhra Pradesh	Krishna	19064	2726	16020	318	0	16.60999	80.72144
Andhra Pradesh	Kurnool	49216	4770	44045	401	0	15.82813	78.03728
Andhra Pradesh	Prakasam	30950	14570	16043	337	0	15.34846	79.56034
Andhra Pradesh	S.P.S. Nellore	39269	6232	32684	353	0	14.25808	79.91927
Andhra Pradesh	Srikakulam	29048	6911	21867	270	0	18.29492	83.8938
Andhra Pradesh	Visakhapatnam	41618	6165	35123	330	0	17.68682	83.21848
Andhra Pradesh	Vizianagaram	25103	9085	15837	181	0	18.10666	83.39555
Andhra Pradesh	West Godavari	44975	2482	42128	365	0	16.91742	81.33994
Andhra Pradesh	Y.S.R. Kadapa	32729	7794	24671	264	0	14.57672	78.83826
Arunachal Pradesh	Anjaw	77	31	46	0	0	28.06668	96.49224
Arunachal Pradesh	Changlang	547	107	440	0	0	27.74218	96.64243
Arunachal Pradesh	East Kameng	236	35	201	0	0	27.42313	93.01757
Arunachal Pradesh	East Siang	362	170	191	1	0	28.10971	95.14321
Arunachal Pradesh	Kamle	3	0	3	0	0	27.79997	94.07502
Arunachal Pradesh	Kra Daadi	4	3	1	0	0	27.72604	93.64216
Arunachal Pradesh	Kurung Kumey	10	7	3	0	0	27.79954	93.55384
Arunachal Pradesh	Lepa Rada	65	37	28	0	0	28.218	94.72775
Arunachal Pradesh	Lohit	175	40	135	0	0	27.90397	96.17391
Arunachal Pradesh	Longding	71	48	23	0	0	26.87282	95.30575
Arunachal Pradesh	Lower Dibang Valley	91	49	42	0	0	28.14287	95.8431
Arunachal Pradesh	Lower Siang	136	18	118	0	0	28.10971	95.14321
Arunachal Pradesh	Lower Subansiri	73	24	49	0	0	27.61693	93.83916
Arunachal Pradesh	Namsai	207	17	189	1	0	27.66921	95.8644
Arunachal Pradesh	Pakke Kessang	28	2	26	0	0	27.14585	93.24626
Arunachal Pradesh	Papum Pare	1891	490	1398	3	0	27.1719	93.7029
Arunachal Pradesh	Shi Yomi	20	10	10	0	0	28.16931	94.80015
Arunachal Pradesh	Siang	42	37	5	0	0	28.10971	95.14321
Arunachal Pradesh	Tawang	208	71	136	1	0	27.58606	91.85941
Arunachal Pradesh	Tirap	214	84	130	0	0	26.99427	95.54073
Arunachal Pradesh	Upper Dibang Valley	1	0	1	0	0	28.81288	96.1527
Arunachal Pradesh	Upper Siang	81	27	54	0	0	28.75888	95.22667
Arunachal Pradesh	Upper Subansiri	157	32	125	0	0	27.7868	94.33596
Arunachal Pradesh	West Kameng	409	117	289	3	0	27.34276	92.30245
Arunachal Pradesh	West Siang	293	213	80	0	0	28.14725	94.74845
Assam	Baksa	687	235	447	5	0	26.66215	91.35777
Assam	Barpeta	1121	93	1024	4	0	26.33042	91.00405
Assam	Biswanath	467	88	379	0	0	26.72666	93.14785
Assam	Bongaigaon	1303	466	834	3	0	26.50301	90.55357
Assam	Cachar	6111	4585	1498	28	0	24.78213	92.85771
Assam	Charaideo	113	5	105	3	0	26.94131	94.87598
Assam	Chirang	280	59	219	2	0	26.53866	90.69391
Assam	Darrang	1967	225	1730	12	0	26.45229	92.02732
Assam	Dhemaji	612	170	440	2	0	27.48111	94.55728
Assam	Dhubri	1723	504	1215	4	0	26.0207	89.97435
Assam	Dibrugarh	7410	4948	2409	53	0	27.47283	94.91196
Assam	Dima Hasao	237	38	199	0	0	25.3478	93.01757
Assam	Goalpara	658	38	568	1	0	26.16413	90.62515
Assam	Golaghat	2620	954	1656	10	0	26.52385	93.96234
Assam	Hailakandi	1632	509	1116	7	0	24.6811	92.56381
Assam	Hojai	1646	478	1160	8	0	26.00166	92.84774
Assam	Jorhat	5087	3318	1752	17	0	26.75092	94.20367
Assam	Kamrup	4590	2496	2074	20	0	26.31608	91.5984
Assam	Kamrup Metropolitan	34258	19583	14599	76	0	26.07943	91.63722

Fig. 4 Final data

7.3 Restructuring Data

After data has been cleaned, the geolocation data obtained from the Map API is merged with this cleaned data. The final data will contain District, State, Active Cases, Change in Active Cases, and Latitude & Longitude values for every District.

8 Graph Construction

The next step is to construct the graph using the newly restructured data. Each of the divisions/districts shall be considered as a vertex, and few of its connections will be edges. The main aim of creating a graph is to form a high-level view of a country's transport system, or specifically, its road connections. So, the vertices are the districts, and the edges are the shortest paths between neighboring districts, as shown in Fig. 5 and Algorithm 2. Cases are the vertex weights, shortest distances are

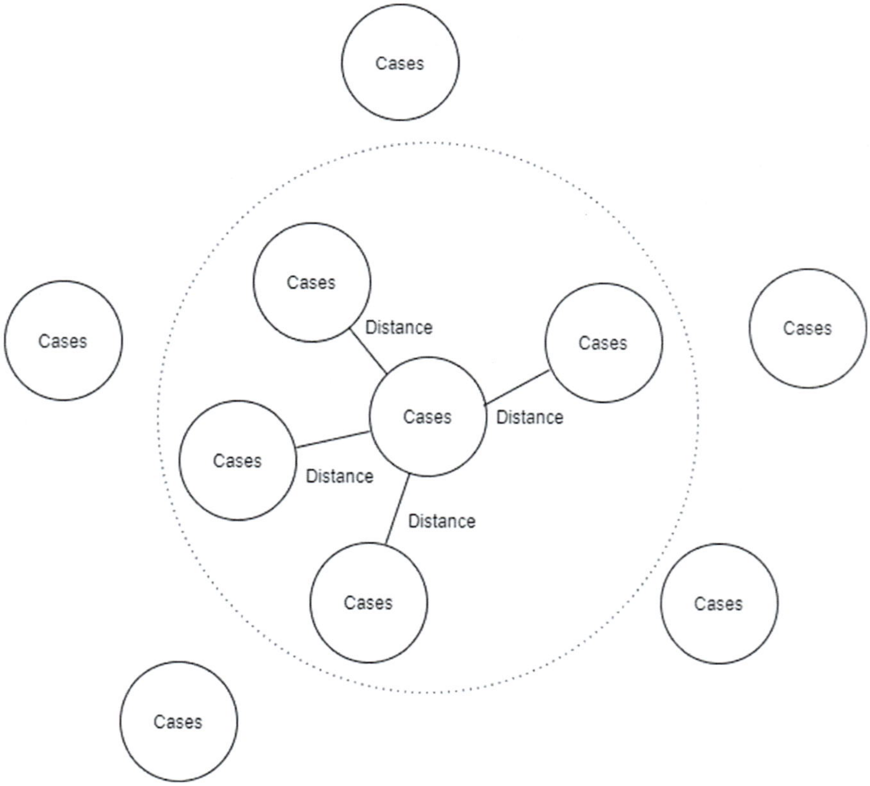

Fig. 5 Sample graph

Algorithm 2: Algorithm for generating Graph

Input: Final Data, Threshold
Result: Connected Graph

```
READ Final Data, Threshold
foreach k in divisions do
    G.addNode(k, cases(k))
end
foreach i in G.nodes do
    foreach j in G.nodes do
        if i != j and i,j not in edge then
            if distance(i,j) <= threshold then
                G.addEdge(i, j, distance(i, j))
            end
        end
    end
end
```

the edge weights, and the dotted line represents the threshold boundary. This entire process is accomplished using 'networkx' package. The final constructed graph is shown in Fig. 6.

8.1 Vertices

For constructing the graph, the district locations are considered as the vertices. The weight for each vertex is the corresponding number of active cases. If any other administrative level is to be considered, the respective division locations are considered as vertices. Accordingly, the vertex weight for each division will be the number of active cases for that division.

8.2 Edges

Edges for the graph are the connections between only some specific pairs of vertices. This is because, districts far away from each other, are difficult to travel to and from, due to air and rail travel restrictions. Also, even if all pairs are considered, the impact from a district very far away, even with a high number of cases, will be negligible. It will only increase the computational complexity. These pairs can be obtained by using some clustering methodology. After that, the weight for an edge can be defined by the distance between its two vertices.

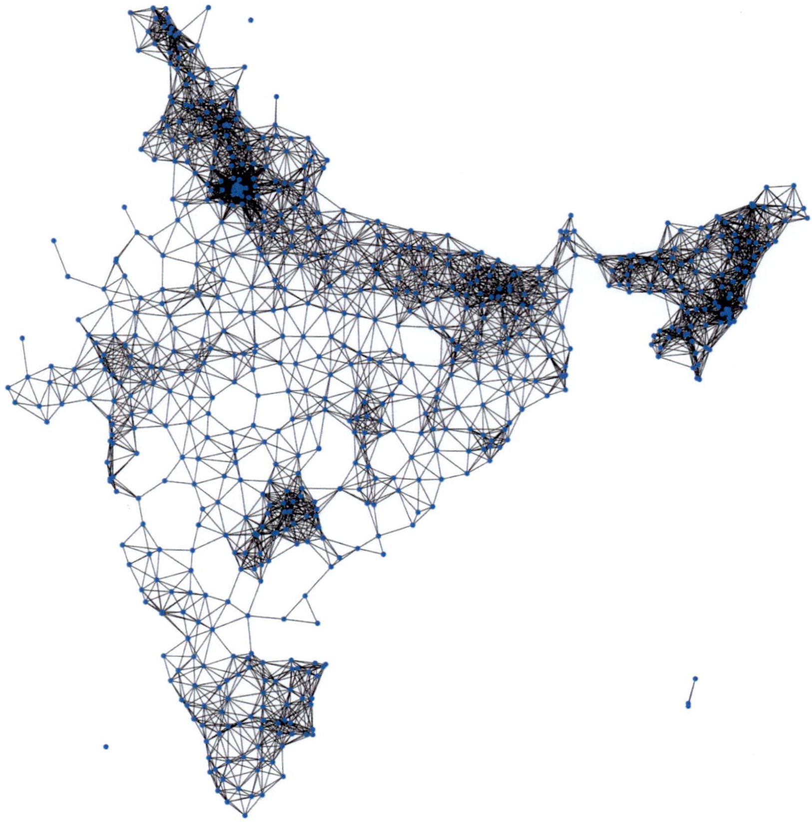

Fig. 6 Constructed graph

8.2.1 Clustering

The simplest and most logical approach to find appropriate pairs is to take inspiration from something like k-means of fuzzy c-means clustering [41]. Basically, for every vertex in the graph, if another vertex is within some threshold range of that vertex, then they are connected. Each vertex is its own cluster center, and cluster centers are not updated. Nearest vertices are identified in a single pass, and a vertex can be in multiple clusters. Every administrative division becomes its own cluster center.

8.2.2 Threshold Value

Two vertices should only be connected based on the above-mentioned threshold value. The threshold value is actually the maximum shortest distance between two districts, so that they may be considered as neighbors. This threshold distance needs to be appropriate and has to be set manually. A reasonably large value is chosen for the threshold, to obtain a balanced number of connections between districts.

8.3 Visualization

Once the edges are defined, and both nodes and edges have their weights assigned, the graph construction is complete. Now, it can be visualized in the graph, that the nodes are the administrative division centers, and the edges are the shortest paths between nearby divisions.

9 Impact Measures and Danger Level

To find the danger level of each individual division, three measures have been defined. Each measure takes into account different factors, so results may be slightly different. These measures give a danger value, that may be color-coded for better visualization. They may be used for more dynamic lockdown restrictions [42].

9.1 Nearest Without Self

In this measure, the impact of all other cluster members, except for the cluster center itself, on the cluster center, is taken into consideration. Meaning that a specific district is affected by other districts near it. Hence, the geodesic distance, and the number of cases from those nearby districts, is taken into consideration. This danger level is defined in Eq. 1 as

$$D_{NWOS}(i) = \frac{\sum_j^N impact(i, j)}{N} \tag{1}$$

here i identifies the division whose danger level needs to be evaluated, j is any division adjacent to division i i.e. $ij \in E$, and N is the number of such pairs, where Eq. 2 defines impact value as

$$impact(i, j) = \frac{V_{scaled}(j)}{E_{scaled}(i, j)} \tag{2}$$

is the impact factor of j division on i, and the scaled vertex weights and edge weights are defined in Eqs. 3 & 4 as

$$V_{scaled}(j) = V(j)/max(V) \tag{3}$$

$$E_{scaled}(j) = E(j)/max(E) \tag{4}$$

In Nearest Without Self, the current region itself is not considered while calculating the danger value. This means that even if a region itself has a high number of cases, it may not have a high danger value if nearby regions have a small number of cases.

9.2 *Nearest with Self*

In this measure, the impact of all cluster members, including the cluster center on itself, is taken into consideration. Meaning that a specific district is affected by other districts near it, and also the cases within that same district. Hence, the geodesic distance, and the number of cases from those nearby districts, and the district itself, is taken into consideration. This danger level is defined in Eq. 5 as

$$D_{NWS}(i) = \frac{\sum_j^N impact(i, j) + impact_{self}(i)}{N + 1} \tag{5}$$

here i identifies the division whose danger level needs to be evaluated, j is any division adjacent to division i i.e. $ij \in E$, and N is the number of such pairs, where $impact(i, j)$ is the same as Eq. 2, where Eq. 6 is

$$impact_{self}(i) = V_{scaled}(i) \times (1 + \Delta V(i)/V(i)) \tag{6}$$

which zero, then self impact will also be zero.

In Nearest With Self, the current region is considered while calculating the danger value. This means that, even if nearby regions have a very small number of cases, the danger value can be high if the current region has a high number of cases.

9.3 *Inverse-Square Law*

In this measure, the impact of all cluster members, including the cluster center on itself, is taken into consideration. It is similar to Coulomb's Law and Newton's Gravitational Law [43]. This danger level is defined in Eq. 7 as

$$D_{ISL}(i) = \frac{\sum_j^N impact(i, j)}{N} \tag{7}$$

here i identifies the division whose danger level needs to be evaluated, j is any division adjacent to division i i.e. $ij \in E$, and N is the number of such pairs, where Eq. 8 is

$$impact_{ISL}(i, j) = \frac{V_{scaled}(i) \times V_{scaled}(j)}{E^2_{scaled}(i, j)} \tag{8}$$

which is the impact factor of j division on i, and the scaled vertex weights and edge weights are same as Eqs. 3 and 4.

All of the above methods are danger value measurement techniques. They define how much impact nearby regions have on each region. Nearest with self, also includes itself in the calculation, whereas nearest without self, does not include itself in the calculation. Inverse Square Law also measures the impact, but here, the impact decreases as the inverse square of the distance, whereas in the other two methods, it decreases linearly.

10 Predicting Potential Hotspots

Using any of the three measures, the danger level value can be easily color-coded. This will allow for ease in visualization, as the various colors will represent different danger levels, in a gradient manner. Three colors are used, which are green, orange, and red. Here, red represents the division that is most severely affected, or about to be severely affected, very soon. This division is either already a hotspot, or will become one in a few days. Orange represents divisions that either are moderately affected or will soon be if correct precautions are not taken. This is the best stage to impose restrictions, to make sure it does not become a red zone i.e. a future disease hotspot. An orange zone may also signify that some of the nearby divisions may already red zones, solidifying the need for major precautions. Finally, green represents a comparatively safe zone, where, as long as there are restrictions on traveling in and out of that zone, lenient rules may be applied. This makes sure that people's lives and their livelihood are not affected.

Since a few highly infected areas may be deemed outliers, a normal distribution is instead used, considering only the second and third quartile. After this, according to danger values, colors are assigned, with the lower danger values getting assigned green, and higher ones as red. The color gradient then helps in easily understanding the spatial situation, and assists in dynamically choosing the rules based on the zone.

In other words, all the regions are mapped to a danger scale with a range of 0 to 1. This range may be divided into any number of clusters. So, the regions themselves get assigned to these clusters. For example, the range may be divided into 3 regions, where 0 to 0.2 is the first region (Green), 0.2 to 0.5 is the second region (Orange) and the remaining, 0.5 to 1 is the third region (Red). Now, every region can be mapped to these clusters based on its danger value.

A gradient may also be used instead of a hard distinction between clusters. Some regions will be more green than other greens, some will be more orange or even red than others. These markers of cluster distinction in the danger range scale, can be adjusted, to make hotspot consideration more lenient or harsh.

11 Experimental Results

Data acquisition, pre-processing, graph construction, and the algorithm implementation, were all done in Python 3, using the Google Collaboratory platform. Experimental evaluation was done using an Intel Xeon 2vCPU @ 2.3 GHz and more than 12 GB of RAM. Infection data was acquired from https://covid19india.org/, while Location Data was obtained using the Google Maps Geocoding API.

With a threshold distance value of 1.3 degrees, a graph was generated (Visualized in Figs. 7, 8, 9 and10). This value was found to be large enough to connect all appropriate nodes while being computationally efficient.

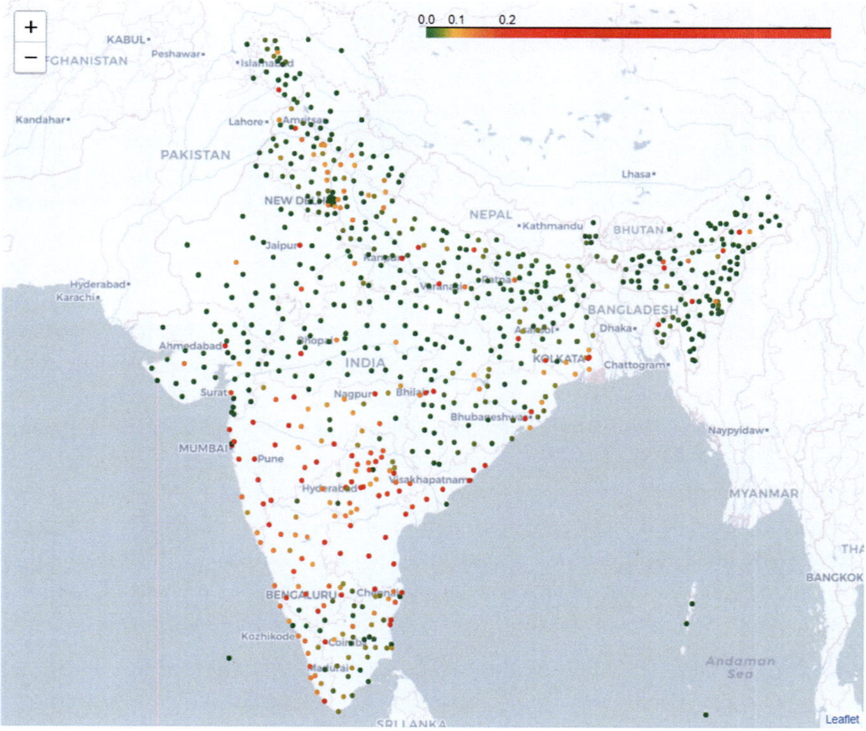

Fig. 7 Original data

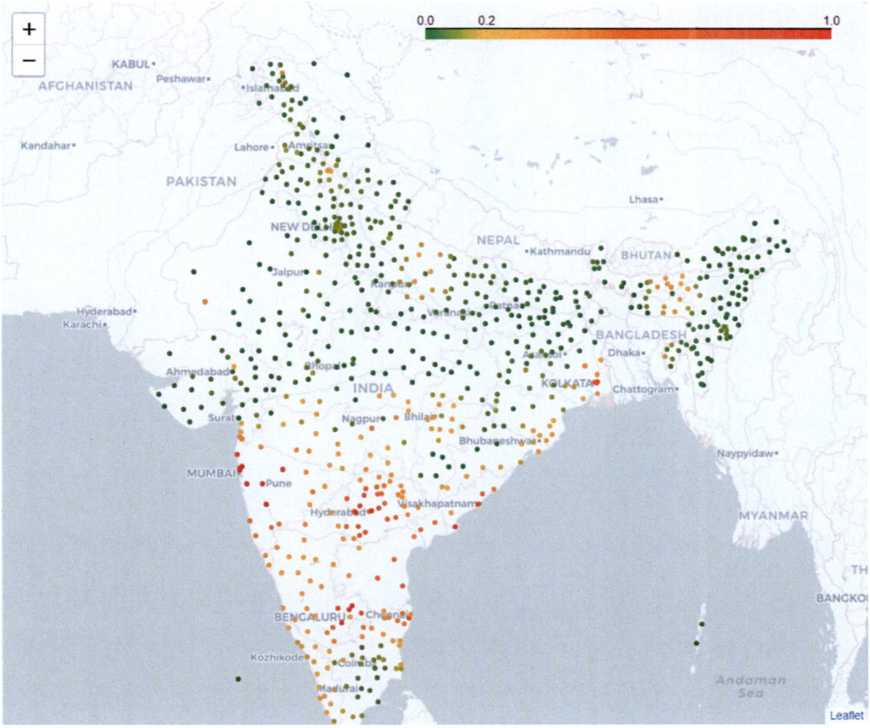

Fig. 8 Nearest without self

To take care of a very huge number of active cases that act as outliers, it was assumed that every node with a danger value of 0.3 and above be equal. The danger value was scaled to a value between 0 and 1. These values were mapped to a color gradient map and the final map visualization of the country was generated. All three measures were used and compared.

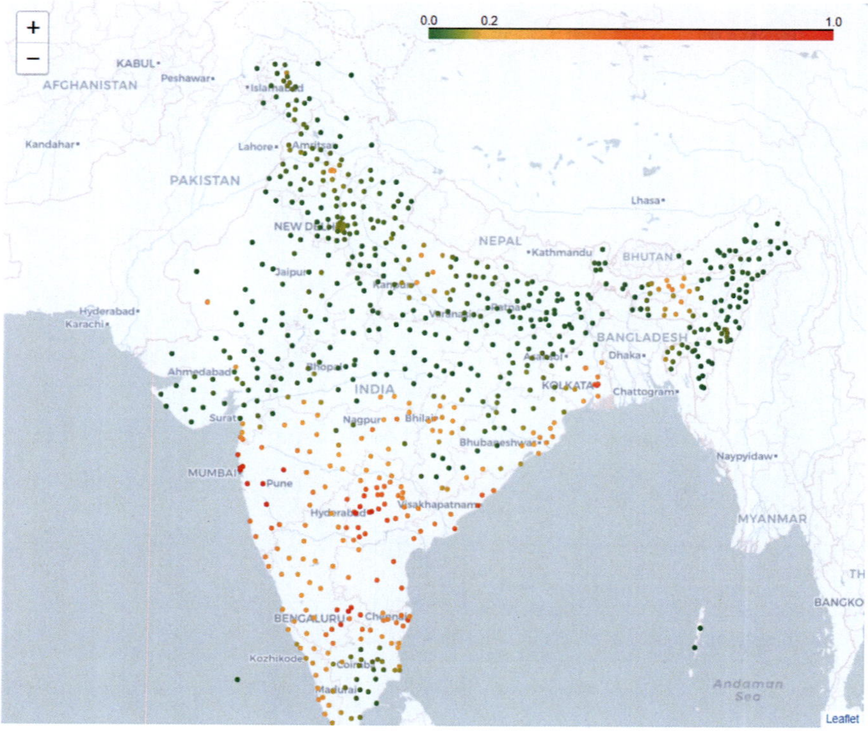

Fig. 9 Nearest with self

The entire system has been converted to a real-time web-app where the computation is scheduled daily and the final visualized map is shown on a dashboard. It is available at https://research.jamlab.in/covid.

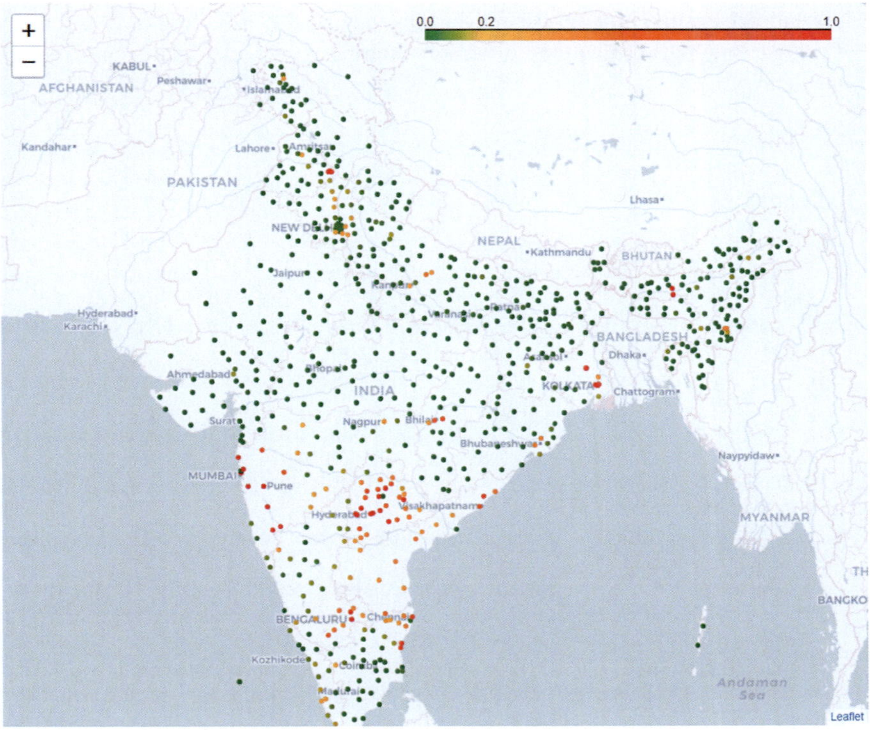

Fig. 10 Inverse square law

12 Conclusion

A system was proposed, to model the spread of COVID-19 cases after lockdown, to define new preventive measures, based on hotspots, using the graph clustering algorithm. The system works on any administrative level, given that data is available at that level. A graph was constructed to model the virus' spread, and danger values were calculated, to assign the danger level of each district. This was visualized using a color map.

Although this system was tested on the COVID-19 India data, it should work on a dataset for any disease, provided all assumptions that are mentioned are met. There is a lot of scope for this methodology, which will be explored in future work.

13 Future Work

A lot of improvements can be made to this methodology. This includes the addition of a temporal dimension, so as prediction can be made for a defined period of time.

A pseudo geospatial time series analysis may be possibly used for this. Also, instead of using a rough geodesic graph, an exact graph-based n the actual road network may also be explored.

References

1. J. Shibo et al., A distinct name is needed for the new coronavirus. Lancet (London, England) **395**(10228), 949 (2020)
2. E. Mahase, Covid-19: WHO declares pandemic because of "alarming levels" of spread, severity, and inaction. BMJ (Clin. Res. ed.) **368**, m1036 (2020)
3. Z. Peng et al., A pneumonia outbreak associated with a new coronavirus of probable bat origin. Nature **579**(7798), 270–273 (2020)
4. Johns Hopkins University. COVID-19 Dashboard by the Center for Systems Science and Engineering (CSSE) at Johns Hopkins University (JHU) (2020)
5. F. Shuo et al., Rational use of face masks in the COVID-19 pandemic. Lancet Respir. Med. **8**(5), 434–436 (2020)
6. M. Lipsitch, D.L. Swerdlow, L. Finelli, Defining the epidemiology of Covid-19–studies needed. New Engl. J. Med. **382**(13), 1194–1196 (2020)
7. S.A. Lauer et al., The incubation period of coronavirus disease 2019 (COVID-19) from publicly reported confirmed cases: estimation and application. Ann. Internal Med. **172**(9), 577–582 (2020)
8. Z. Fei, T. Yu, R. Du, G. Fan, Y. Liu, Z. Liu, J. Xiang et al., Clinical course and risk factors for mortality of adult inpatients with COVID-19 in Wuhan, China: a retrospective cohort study. Lancet **395**(10229), 1054–1062 (2020)
9. S.P. Adhikari et al., Epidemiology, causes, clinical manifestation and diagnosis, prevention and control of coronavirus disease (COVID-19) during the early outbreak period: a scoping review. Infect. Dis. Poverty **9**(1), 1–12 (2020)
10. L. Hien et al., The positive impact of lockdown in Wuhan on containing the COVID-19 outbreak in China. J. Travel Med. **27**(3), taaa037 (2020)
11. F. E. Alvarez, D. Argente, F. Lippi, A simple planning problem for covid-19 lockdown. No. w26981. National Bureau of Economic Research (2020)
12. S.R. Baker, et al., Covid-induced economic uncertainty. No. w26983. National Bureau of Economic Research (2020)
13. Z. Yang et al., Modified SEIR and AI prediction of the epidemics trend of COVID-19 in China under public health interventions. J. Thoracic Dis. **12**(3), 165 (2020)
14. C.O. Stallybrass, *The Principles of Epidemiology and the Process of Infection* (Macmillan Co., New York, 1931)
15. P. Elliot et al., *Spatial Epidemiology: Methods and Applications* (Oxford University: Oxford University Press, Oxford, 2000)
16. R.S. Ostfeld, G.E. Glass, F. Keesing, Spatial epidemiology: an emerging (or re-emerging) discipline. Trends Ecol. Evol. **20**(6), 328–336 (2005)
17. R.S. Kirby, E. Delmelle, J.M. Eberth, Advances in spatial epidemiology and geographic information systems. Ann. Epidemiol **27**(1), 1–9 (2017)
18. L. Beale et al., Methodologic issues and approaches to spatial epidemiology. Environ. Health Perspect. **116**(8), 1105–1110 (2008)
19. D. Kang, H. Choi, J.-H. Kim, J. Choi, Spatial epidemic dynamics of the COVID-19 outbreak in China. Int. J. Infect. Dis. **94**, 96–102 (2020)
20. R. Zheng, Y. Xu, W. Wang, G. Ning, Y. Bi, Spatial transmission of COVID-19 via public and private transportation in China. Travel Med. Infect. Dis. **34**, 101626 (2020)
21. R. Huang, M. Liu, Y. Ding, Spatial-temporal distribution of COVID-19 in China and its prediction: a data-driven modeling analysis. J. Infect. Developing Countries **14**(03), 246–253 (2020)

22. K.L. Cooke, P. Van Den Driessche, Analysis of an SEIRS epidemic model with two delays. J. Math. Biol. **35**(2), 240–260 (1996)
23. S. Tuli et al., Predicting the growth and trend of COVID-19 pandemic using machine learning and cloud computing. Internet Things **11**, 100222 (2020)
24. S. Zhao et al., Preliminary estimation of the basic reproduction number of novel coronavirus (2019-nCoV) in China, from 2019 to 2020: A data-driven analysis in the early phase of the outbreak. Int. J. Infect. Dis. **92**, 214–217 (2020)
25. V. Zarikas et al., Clustering analysis of countries using the COVID-19 cases dataset. Data Brief **31**, 105787 (2020)
26. B.S. Pujari, S.M. Shekatkar, Multi-city modeling of epidemics using spatial networks: Application to 2019-nCov (COVID-19) coronavirus in India. Medrxiv (2020)
27. B. Murugesan et al., Distribution and trend analysis of COVID-19 in India: geospatial approach. J. Geogr. Stud. **4**(1), 1–9 (2020)
28. S. Roy, G.S. Bhunia, P.K. Shit, Spatial prediction of COVID-19 epidemic using ARIMA techniques in India. Model. Earth Syst. Environ. 1–7 (2020)
29. M. Liu, R. Thomadsen, S. Yao, Forecasting the spread of COVID-19 under different reopening strategies. medRxiv (2020)
30. B. Rader et al., Crowding and the shape of COVID-19 epidemics. Nat. Med. **26**, 1–6 (2020)
31. S. Hisada et al., Surveillance of early stage COVID-19 clusters using search query logs and mobile device-based location information. Sci. Rep. **10**(1), 1–8 (2020)
32. S.I. Hay, R.W. Snow, The malaria atlas project: developing global maps of malaria risk. PLoS Med **3**(12), e473 (2006)
33. J.P. Messina et al., A global compendium of human dengue virus occurrence. Sci. Data **1**(1), 1–6 (2014)
34. D.M. Pigott et al., Global database of leishmaniasis occurrence locations, 1960–2012. Sci. Data **1**(1), 1–7 (2014)
35. A. Mylne et al., Comprehensive database of the geographic spread of past human Ebola outbreaks. Sci. Data **1**(1), 1–10 (2014)
36. B. Xu et al., Epidemiological data from the COVID-19 outbreak, real-time case information. Sci. Data **7**(1), 1–6 (2020)
37. E. Dong, D. Hongru, L. Gardner, An interactive web-based dashboard to track COVID-19 in real time. Lancet Infect. Dis. **20**(5), 533–534 (2020)
38. L.L. Wang, K. Lo, Y. Chandrasekhar, R. Reas, J. Yang, D. Eide, K. Funk et al., CORD-19: the Covid-19 open research dataset (2020). arXiv-2004
39. E. Chen, K. Lerman, E. Ferrara, Covid-19: The first public coronavirus twitter dataset. arXiv preprint arXiv:2003.07372 (2020)
40. COVID-19 India Org Data Operations Group, covid19india.org: Coronavirus Outbreak in India, Coronavirus Outbreak in India - covid19india.org (2020). https://covid19india.org/
41. A.K. Jain, Data clustering: 50 years beyond K-means. Patt. Recogn. Lett **31**(8), 651–666 (2010)
42. S. Wang, X. Zhao, Y. Chen, Z. Li, K. Zhang, J. Xia, Negative influence minimizing by blocking nodes in social networks, in Proceedings of the 17th AAAI Conference on Late-Breaking Developments in the Field of Artificial Intelligence (AAAI Press, 2013), pp. 134–136
43. L. Fei, Q. Zhang, Y. Deng, Identifying influential nodes in complex networks based on the inverse-square law. Physica A Stat. Mech. Appl. **512**, 1044–1059 (2018)

COVID-19 Recommendation System of Chest X-Ray Images Using CNN Deep Learning Technique with Optimizers and Activation Functions

Niyati Sinha, Puhup Karjee, Riddhi Agrawal, Ankita Banerjee, and Chittaranjan Pradhan

Abstract Most people who are infected with the COVID-19 virus show respiratory illness. As the virus is newly discovered, the world is not ready with an adequate number of testing kits so as to test each individual in the world. So, a different method that tells if a person may be infected or not will prove to be a big relief and will also help in decreasing the tension and fear among people. This problem can be solved by training machines using deep learning algorithms. The proposed CNN model takes input as chest X-ray images and predicts whether a person needs to go for COVID-19 testing or not. This will decrease the fear of being COVID-19 infected and can be easily implemented even in rural areas; which will also decrease the load of the COVID-19 testing team. At present, the COVID-19 tests are being done clinically through blood tests or nose/throat swab tests which requires around 24 h to give results. The proposed recommendation system takes around 35 min to give the result of a sample and curbs down the chances of virus spread through contact while testing, unlike the presently used methods. The experimental results yielded an accuracy of 97.62% using the chest X-ray scans and requires less computational time.

Keywords Activation function · CNN · COVID-19 · Deep learning · Optimizer · Recommendation system

1 Introduction

The sudden outbreak of COVID-19 virus has impacted many people all over the world. Originally, this virus is a group of some RNA viruses belonging to the Coronaviridae family which affected mammals and birds. But, in late 2019, a specific mutant of this virus infected humans, mainly causing respiratory tract infections.

COVID-19 entered our lives as an uninvited guest exposing health care loopholes and leaving the community in chaos with a global mortality rate of around 3% [1]. Out of 195 countries, there are only a few who have the economic capacity to handle the catastrophe of a situation that the virus has created. Rapid testing and the quarantine

N. Sinha · P. Karjee · R. Agrawal · A. Banerjee · C. Pradhan (✉)
Kalinga Institute of Industrial Technology, Bhubaneswar, India

J. Nayak et al. (eds.), *Understanding COVID-19: The Role of Computational Intelligence*, Studies in Computational Intelligence 963, https://doi.org/10.1007/978-3-030-74761-9_7

of people tested positive is the best way to overcome this crisis. But, unfortunately, many lesser developed countries lack the required funds and health care facilities, and therefore their testing rates drop low. Thus, they are unable to identify COVID-19 positive people and the virus quickly becomes a community outbreak.

As research shows, most of the people who are infected with this virus have abnormalities scattered throughout their lungs which can be identified through the X-ray [1]. Therefore, successive appropriate measures can be taken. The aim of this chapter is to help clinicians decide who would require further COVID-19 testing and hospitalization when the resources are limited. This will create a fair platform for treatment as healthcare resources will be utilized first for the people who need it the most and are in critical stages.

2 Background

This section provides a brief description about the functionality of the convolutional neural network.

2.1 Convolutional Neural Network (CNN)

It is a kind of neural network, which is commonly used in image classification, image recognition, speech recognition etc. [2]. Its ability to detect anomalies in the medical images has proven to reduce time taken to get to the results and improve the decision making process. A classifier, based on Artificial Neural Network (ANN) employing multilayer perceptron with back-propagation algorithm to predict peak events of patients with respiratory diseases primarily asthma, tuberculosis and COPD has been analyzed for drawing this idea [3, 4]. The X-ray images have been scrutinized using the developed CNN model implementing various Keras API features [5]. COVID-Net has been designed for the detection of COVID-19 cases using chest radiography images [6]. CovidGAN has been developed to produce synthetic chest X-ray images [7].

CNN supports spatial information, thereafter deals with data which has direct/indirect reference to specific location/coordinates. They are well suited for image classification. This model is preferred over the other neural network models due to its accuracy. CNN works without any dedicated feature extraction technique. Rather, the system learns to do feature extraction on its own based on back propagation. The key idea involved in CNN is that it takes an image, convolutes it and filters it for generation of invariant features. These features are fed on to the next layer [8]. In the next layer, different filters convolute the features present, to generate more uniform and abstract features. This process will continue till the final features are obtained.

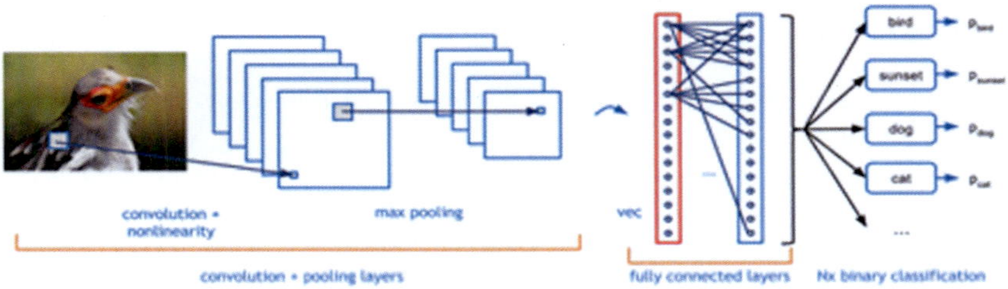

Fig. 1 Layers of CNN [5]

2.1.1 CNN Layers

CNN contains a number of layers which are shown in Fig. 1.

The elementary unit of a convolutional neural network is the convolutional layer which performs an operation named 'convolution' and hence the name. The convolution operation involves the use of a 2D weight array called filter or kernel, which is multiplied with the input by means of a linear operation. The convolution layer stores the relation between the pixels in the image by learning the features using small squares of the input [9]. This is a linear operation in mathematics which multiplies small squares of image matrix with a weighted matrix (known as filter matrix or kernel). The convoluted map obtained is the 'feature map' [2].

Convolution can be performed on 1D signals, 2D signals and further extended to multidimensional signals too. A grayscale image uses 2D convolution. In 2D convolution system, convolution is performed among two signals spanning along mutually perpendicular dimensions [2].

2.1.2 Non-linearity

Rectified Linear Unit (ReLU) introduces non-linearity in the ConvNet. It removes negative values from the ConvNet data, which shows resemblance with real world data.

2.1.3 Stride and Padding

Stride denotes the number of steps movements in each step in convolution. It is one by default. It is the number of pixels that shift over the input matrix. For example, the value of the stride is 1 denotes the filter movement by 1 pixel at a time.

When the filter doesn't fit properly, then fill the image either by zero padding and some valid padding. Padding the picture with zero so that it perfectly fits the input image is known as zero padding. Dropping the image part where filter doesn't fit the image and keeping the remaining valid parts is called as valid padding [10].

2.1.4 Pooling

Large images have way more number of parameters which need to be reduced and is done in pooling layer. This layer works on each one of the feature map to produce a new set pooled feature map which is same in number. Subsampling reduces dimensions of each map while preserving important information. It is also called as spatial pooling. Its types are:

Max Pooling: here, the greatest value in each patch of each feature map is calculated.
Average Pooling: Here, the mean value in each patch of each feature map is calculated.
Sum Pooling: Here, the sum of inputs in each patch of each feature map is calculated.

2.1.5 Hyper Parameters

Before the training stage, the variables determining the network structure are set. These are known as hyper parameters, which contain the followings:

- The layer between input layer and output layer is known as hidden layer. There can be multiple hidden layers.
- Overfitting occurs when many features are taken into consideration and regularization has to be done. Dropout is a regularization technique that avoids overfitting (increases the validation accuracy), thus increasing the generalization power.
- Some non linear activation functions like ReLU, sigmoid and many more are used to allow deep learning models to learn non linear prediction boundaries.
- The learning rate, α, defines the rate at which a network updates its parameters.
- Epochs are the number of times the model gets trained by the whole training data.
- The number of sub samples given to the network is called Batch size. Parameter updation takes place after this.

2.1.6 Fully Connected

The first fully connected layer applies weights to forecast the correct label by taking the inputs from the feature analysis. The function of this layer is to flatten the matrix into vector and feeding it into a fully connected layer like a neural network. The feature map matrix is then converted to a vector by the fully connected output layer. With the fully connected layers, these features are merged to create a model and the activation function is used to classify the output obtained. Here, it performs a binary classification and predicts if the person whose X-ray image has been tested requires COVID-19 testing or not.

2.1.7 Training

Training of CNN is done by back propagation followed by gradient descent.

2.2 *Optimizer Functions*

An optimizer function is a machine learning algorithm [11], which helps in updating the weight parameters used in the algorithm so as to minimize the loss function or the cost function. The aim is to reach the minimum value in the terrain or in other words, the global minima. The optimizer function is guided by the loss function to find the global minima [12].

2.2.1 Stochastic Gradient Descent (SGD)

This technique calculates the gradient descent by considering small batches of the data set instead of taking the entire data set. This helps to reduce the number of computations on a huge data set, thus, reducing the computational burden. This approach can therefore be called as a stochastic approximation of the gradient descent [13]. In Eq. 1 $\partial h/\partial w$ represents the gradient of the loss function with respect to the weights or parameters. It is represented as:

$$w_t = w_{t-1} - \eta\, \partial h/\partial w \tag{1}$$

$$h = \sum_{t=1}^{k} \left(y_{predicted} - y_{actual}\right)^2 \tag{2}$$

where, w stands for the weight/parameter; h is the cost function or loss function as shown in Eq. 2; $y_{\text{predicted}}$ is the estimated value and y_{actual} is the actual value.

2.2.2 Adagrad

Adaptive gradient algorithm or Adagrad estimates the disadvantages of a fixed learning rate. It regularly keeps on updating the learning rate with each iteration, layer, neuron unlike approaches like gradient descent, mini batch stochastic gradient descent etc., thereby eliminating the need of manually fix the learning rate. The learning rate varies according to the Eqs. 3, 4 and 5:

$$w_t = w_{t-1} - \widehat{\eta_t}\, \partial h/\partial w_{t-1} \tag{3}$$

$$\widehat{\eta_t} = \frac{\eta}{\sqrt{\in + \alpha_t}} \tag{4}$$

$$\alpha_t = \sum_{t=1}^{k} \left(\frac{\partial h}{\partial w_t} \right)^2 \tag{5}$$

It can be observed that each term is a positive term. Consequently, w_t value tends to increase with an increased number of iterations, as a result decreasing the adaptive learning rate. This helps to achieve a smooth convergence at the global minima.

2.2.3 Adadelta

The Adadelta optimizer function eradicates the disadvantages of the Adagrad optimizer functions. The amount by which the learning rate increases with each iteration has to be kept in check. This is ensured in this technique by using exponential weighted average in the calculation as given in Eqs. 6 and 7:

$$\widehat{\eta_t} = \frac{\eta}{\sqrt{\in + w_{avgt}}} \tag{6}$$

$$w_{avgt} = w_{avg(t-1)} + (1 - \gamma) \left(\frac{\partial h}{\partial w_t} \right)^2 \tag{7}$$

2.2.4 RMSprop

RMSprop optimizer function is similar to Adadelta function. It tries to resolve Adagrad's drastically diminishing learning rates by using a moving average of the squared gradient [14]. The learning rate is updated by dividing it by the root mean square of the squared gradient as shown in Eqs. 8 and 9:

$$\widehat{\eta_t} = \frac{\eta}{\in + \sqrt{E[g_t^2]}} \tag{8}$$

$$E[g_t^2] = \beta E[g_{t-1}^2] + (1 - \beta) \left(\frac{\partial h}{\partial w_t} \right)^2 \tag{9}$$

2.2.5 Adam-Adaptive Moment Estimation

Adam optimizer is one of the most used and believed to be the most computationally efficient optimizer techniques. As a combination of Adagrad and RMSprop techniques, it includes momentum. Adam uses the exponential moving average [15] of the gradient to update the learning rate.

- Updation of the exponential moving average of the gradient (m_t) [estimate of first moment] and squared gradient (v_t) [estimate of second moment] takes place first.
- Hyper parameters β_1, $\beta_2 \epsilon [0,1)$ determine the exponential decay rates as shown in Eqs. 10 and 11:

$$m_t = \beta_1 m_{t-1} + (1 - \beta_1) g_t^2 \tag{10}$$

$$v_t = \beta_2 v_{t-1} + (1 - \beta_2) g_t^2 \tag{11}$$

Initially, the moving averages are initialized to zero and this can be corrected by using the equations given in Eqs. 12 and 13:

$$\widehat{m_t} = \frac{m_t}{1 - \beta_1^t} \tag{12}$$

$$\widehat{v_t} = \frac{v_t}{1 - \beta_2^t} \tag{13}$$

The parameters are finally updated by Eq. 14.

$$w_{t+1} = w_t - \frac{\eta \widehat{m_t}}{\sqrt{\in + \widehat{v_t}}} \tag{14}$$

2.3 *Activation Functions*

These are used to determine the output of neural networks like yes or no; which can be mapped to the values in between 0 to 1 or −1 to 1. The most commonly used activations functions used are discussed below.

2.3.1 Sigmoid

It is a non-linear mathematical function with s-shaped curve characteristics. It is a bounded, differentiable function defined for all real values and has non-negative slope at all the points. The first derivative of a sigmoidal function is a bell-shaped curve. The range of sigmoid function's output is [0, 1], which can be used as probability measures. As a result, they are quite useful in the last layer of deep learning architectures. Some commonly used sigmoid functions are:

Logistic Function

It is the basic sigmoid function and is represented in Eq. 15:

$$f(x) = \frac{1}{1 + e^{-x}} \tag{15}$$

Hyperbolic Tangent (tanh).

This is a non-linear activation function. It compresses all its inputs in the range $[-1, 1]$. The gradients are not restricted to move in certain directions as this function is zero-centered. It is represented in Eq. 16:

$$f(x) = \frac{e^{x} - e^{-x}}{e^{x} + e^{-x}} \tag{16}$$

2.3.2 Rectified Linear Units (ReLU)

It is a non-linear activation function which is widely used in convolution neural network. The function outputs the input value if the input value is positive and zero otherwise. It is defined in Eq. 17:

$$f(x) = \begin{cases} x, x > 0 \\ 0, x \leq 0 \end{cases} \tag{17}$$

The output obtained from the convolution layer, i.e. the feature map obtained, is given as input (x) to the activation in the CNN [16].

3 Proposed Model

One of the most impelling neural networks to deal with images and videos is CNN. Keras is an API designed for analysis by humans. It makes building deep learning models more convenient for the users by minimizing the number of user actions required for commonly used cases. Here, Keras has been used to build a CNN model that trains on chest X-ray images. Then, the model predicts if the given X-ray images need COVID-19 testing or not.

A collection of data has been taken and pre-processed in terms of rescale, shear range, zoom range and horizontal flip [17]. The image data generator function has been used to produce varying images from a particular image on the basis of the preprocessing parameters. The preprocessed data has been used to train the CNN

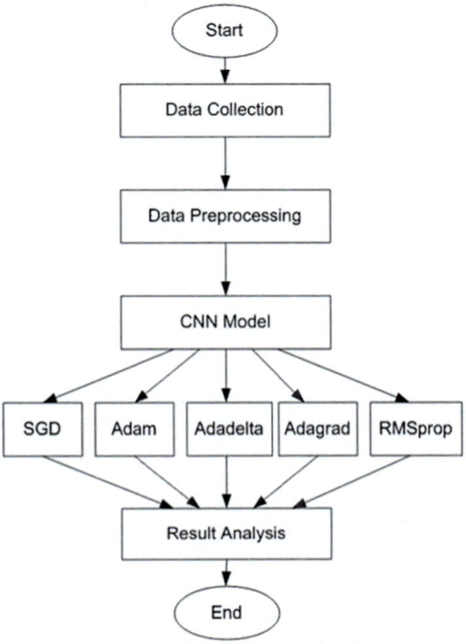

Fig. 2 Overall idea of the proposed model

model. In the CNN architecture, feature extraction and feature classification techniques are implemented. The convolution layer and pooling layer perform feature extraction. The important features are extracted from an anonymous picture (here, X Ray) in this phase. The basic code for implementing CNN has 4 convolutional layers and 1 fully connected layer. The test set has been tested using a various optimizer functions such as SGD, Adam, Adadelta, Adagrad and RMSprop for a particular value of epochs and their results are being compared. The recommendation system performs binary classification and divides the test X-ray image among 'COVID-19 testing required' and 'COVID-19 testing not required' classes. This is shown in Fig. 2.

Figure 3 shows the summary of the model. The proposed model is a sequential model which consists of 4 convolutional layers each followed by an activation layer and a max_pool layer. The MaxPooling2D(pool_size = (2,2)) function is used. The kernel size (2,2) reduces the input image by half. The RELU activation function is used after each convolutional layer and the output obtained is finally passed through the sigmoid activation function to obtain the result in 0 or 1. The rectified linear activation function (ReLU) is used to overcome the vanishing gradient problem. It allows model to perform faster and better. It is also the default activation function used in CNN models. Sigmoid function is used at the last layer because the output of the function lies between 0 and 1. Therefore, it is useful for models where we have to predict the probability which lies between 0 and 1. Flatten function is used to convert the input into a 1-D array. This enables the output obtained from data to act as an input to the next layer. In the model, two dense layers are used, Dense(64)

```
Model: "sequential_1"
_________________________________________________________________
Layer (type)                 Output Shape              Param #
=================================================================
conv2d_1 (Conv2D)            (None, 148, 148, 32)      896
_________________________________________________________________
activation_1 (Activation)    (None, 148, 148, 32)      0
_________________________________________________________________
max_pooling2d_1 (MaxPooling2 (None, 74, 74, 32)        0
_________________________________________________________________
conv2d_2 (Conv2D)            (None, 72, 72, 32)        9248
_________________________________________________________________
activation_2 (Activation)    (None, 72, 72, 32)        0
_________________________________________________________________
max_pooling2d_2 (MaxPooling2 (None, 36, 36, 32)        0
_________________________________________________________________
conv2d_3 (Conv2D)            (None, 34, 34, 48)        13872
_________________________________________________________________
activation_3 (Activation)    (None, 34, 34, 48)        0
_________________________________________________________________
max_pooling2d_3 (MaxPooling2 (None, 17, 17, 48)        0
_________________________________________________________________
conv2d_4 (Conv2D)            (None, 15, 15, 32)        13856
_________________________________________________________________
activation_4 (Activation)    (None, 15, 15, 32)        0
_________________________________________________________________
max_pooling2d_4 (MaxPooling2 (None, 7, 7, 32)          0
_________________________________________________________________
flatten_1 (Flatten)          (None, 1568)              0
_________________________________________________________________
dense_1 (Dense)              (None, 64)                100416
_________________________________________________________________
activation_5 (Activation)    (None, 64)                0
_________________________________________________________________
dense_2 (Dense)              (None, 1)                 65
_________________________________________________________________
activation_6 (Activation)    (None, 1)                 0
=================================================================
Total params: 138,353
Trainable params: 138,353
Non-trainable params: 0
```

Fig. 3 Model summary

and Dense(1), having 64 neurons and 1 neuron respectively, each neuron provides one output to the next layer.

4 Result Analysis

Experiments are being conducted with the following environments:

For different layers, the parameter values are set to 32, 48 and 64. The value of kernel size is set to 3 implying a three pixel window is traversing along the image. The input size of each image data is set to (3, 150, 150). The activation functions used here are ReLU and sigmoid. The pool size is set to 2. This defines a matrix size where value of two pixels is converted to a value.

The data set comprises of 1736 chest X-ray scans, which has been taken from several resources that includes several websites, github repositories [18, 19] and a data set that is available on kaggle for a competition. The data set has been divided into two parts: the training set and test set [20]. The ratio of training data set to the test data set is 78:22.

The loss, accuracy, value loss, value accuracy and result using various optimizers like Adam, RMSprop, SGD and Adadelta with a change in epochs value have been recorded (Tables 1, 2, 3 and 4). Adagrad optimizer has not been considered as it was giving negative results on all test samples and did not stand a place in the comparison.

Table 1 Results with epochs 1–30 for Adam optimizer

Epochs	Loss	Accuracy	Value_Loss	Value_Accuracy	Result
1	0.1366	0.9465	5.67E-02	1.0000	0
2	0.0743	0.9753	3.78E-05	0.9348	1
3	0.0579	0.9805	0.0206	0.9348	1
4	0.0465	0.9842	0.0470	0.9783	0
5	0.0309	0.9888	5.03E-11	1.0000	1
6	0.0334	0.9883	0.0105	0.9348	0
7	0.0240	0.9917	5.92E-05	1.0000	1
8	0.0272	0.9895	7.02E-04	1.0000	0
9	0.0228	0.9919	0.0087	1.0000	1
10	0.0210	0.9932	0.0142	1.0000	1
11	0.0217	0.9920	1.30E-03	1.0000	1
12	0.0181	0.9938	8.36E-05	0.9783	1
13	0.0159	0.9948	1.92E-07	1.0000	1
14	0.0175	0.9953	1.92E-07	1.0000	0
15	0.0163	0.9948	0.0259	1.0000	1
16	0.0123	0.9966	4.61E-04	1.0000	1
17	0.0175	0.9951	8.30E-03	0.9565	1
18	0.0128	0.9959	4.79E-04	1.0000	1
19	0.0172	0.9951	1.59E-02	1.0000	1
20	0.0125	0.9959	4.70E-03	0.9565	1
21	0.0175	0.9953	1.92E-07	1.0000	0
22	0.0085	0.9973	8.47E-05	1.0000	1
23	0.0095	0.9967	1.26E-04	0.9783	1
24	0.0112	0.9969	5.04E-06	1.0000	1
25	0.0125	0.9967	0.0025	1.0000	0
26	0.0013	0.9996	8.99E-08	1.0000	1
27	0.013	0.9948	3.83E-05	1.0000	1
28	0.0092	0.9957	3.18E-05	1.0000	1
29	0.0091	0.9965	5.07E-06	1.0000	0
30	0.0053	0.9985	3.96E-05	1.0000	0

Table 2 Results with epochs 1–30 for SGD optimizer

Epochs	Loss	Accuracy	Value_Loss	Value_Accuracy	Result
1	0.2292	0.9103	0.0398	0.9787	1
2	0.0948	0.9679	0.027	1	1
3	0.0709	0.9741	0.0045	1	1
4	0.0636	0.9789	0.0099	1	1
5	0.0556	0.9786	2.22E-04	1	1
6	0.0468	0.9836	0.023	1	1
7	0.0393	0.9871	0.0107	1	1
8	0.0349	0.9886	0.0033	1	1
9	0.0287	0.9898	6.58E-07	1	1
10	0.0308	0.9882	5.83E-04	1	1
11	0.0271	0.9901	0.0219	1	0
12	0.0224	0.9922	6.05E-04	1	1
13	0.0234	0.9917	0.0269	1	1
14	0.02	0.9931	2.78E-05	1	1
15	0.0186	0.9938	1.42E-05	1	1
16	0.0198	0.9928	2.59E-05	1	1
17	0.0154	0.995	0.0012	1	1
18	0.0195	0.9928	3.56E-04	1	1
19	0.0153	0.9945	3.60E-07	1	1
20	0.0113	0.996	1.77E-08	1	1
21	0.0103	0.9972	6.32E-07	1	1
22	0.0113	0.997	1.51E-05	1	1
23	0.0109	0.9963	0.0015	1	1
24	0.0096	0.9957	4.09E-06	1	1
25	0.008	0.9973	3.24E-05	1	1
26	0.0086	0.9972	2.85E-06	1	1
27	0.0088	0.9975	1.23E-06	1	1
28	0.0071	0.9976	5.00E-05	1	1
29	0.0069	0.9978	1.36E-04	1	1
30	0.0083	0.9975	2.12E-04	1	1

The result 0 represents COVID-19 negative and 1 represents a COVID-19 positive. The test image taken for analysis is a COID-19 positive X-ray.

The accuracy, loss, value-accuracy and value_loss with the epochs variation of 1–30 for Adam optimizer are shown in Figs. 4, 5, 6 and 7 respectively.

The accuracy, loss, value_accuracy and value_loss with the epochs variation of 1–30 for Adadelta optimizers are shown in Figs. 8, 9, 10 and 11 respectively.

Table 3 Results with epochs 1–30 for RMSprop optimizer

Epochs	Loss	Accuracy	Value_Loss	Value_Accuracy	Result
1	0.1854	0.9349	0.0117	0.9348	1
2	0.1005	0.9786	0.0783	1	1
3	0.0974	0.9757	6.78E-04	1	1
4	0.0832	0.9775	1.19E-07	1	1
5	0.081	0.99772	0.1091	0.9783	1
6	0.696	0.9836	7.09E-06	0.9565	1
7	0.0748	0.9831	1.47E-06	1	1
8	0.0566	0.9851	1.19E-07	1	1
9	0.0759	0.9826	1.19E-07	0.9783	1
10	0.071	0.9834	1.19E-07	1	1
11	0.0828	0.9837	1.92E-07	1	1
12	0.0697	0.9879	0.0332	1	1
13	0.4023	0.9689	16.1181	0.5	1
14	0.0659	0.9885	1.19E-07	0.9783	1
15	0.0987	0.9846	1.19E-07	1	1
16	11.4314	0.2913	16.1181	0.5	1
17	0.0488	0.9908	6.78E-05	1	1
18	0.1176	0.9901	1.12E-14	1	1
19	0.0519	0.9907	0.0089	0.9574	1
20	0.1261	0.9867	1.00E-07	0.9783	1
21	6.56E-02	0.9883	2.83E-05	0.9783	1
22	11.5099	0.2846	16.1181	0.5	1
23	0.0838	0.9883	1.19E-07	1	1
24	0.0782	0.9899	1.19E-07	1	1
25	5.6419	0.6474	1.00E-07	0.5	1
26	4.6333	0.7105	15.9424	0.5	0
27	0.0945	0.9896	1.39E-04	1	1
28	0.1242	0.9864	1.00E-07	0.9783	1
29	0.0601	0.9916	1.19E-07	1	1
30	11.5265	0.2855	1.00E-07	0.5	1

The accuracy, loss, value_accuracy and value_loss with the epochs variation of 1–30 for RMSprop optimizers are shown in Figs. 12, 13, 14 and 15 respectively.

The accuracy, loss, value_accuracy and value_loss with the epochs variation of 1–30 for SGD optimizers are shown in Figs. 16, 17, 18 and 19 respectively.

The accuracy and loss with the variation of epochs recorded suggest that the most optimum accuracy and loss if for epoch values 5 and 16. The time taken for testing

Table 4 Results with epochs 1–30 for Adadelta optimizer

Epochs	Loss	Accuracy	Value_Loss	Value_Accuracy	Result
1	0.1386	0.9491	0.0023	1	1
2	0.0622	0.9808	0.2303	0.9348	1
3	0.0458	0.9844	0.0145	0.9348	1
4	0.0378	0.9861	0.0016	1	1
5	0.0287	0.991	9.48E-04	1	1
6	0.031	0.9898	0.0036	1	1
7	0.0258	0.9928	1.93E-04	1	1
8	0.0195	0.9945	0.0037	1	1
9	0.0192	0.9937	1.14E-06	1	1
10	0.0174	0.9948	0.1592	0.9783	1
11	0.0152	0.9954	1.39E-09	1	1
12	0.0171	0.995	3.23E-09	0.96	1
13	0.017	0.9948	6.83E-05	1	1
14	0.0103	0.9981	2.71E-09	0.9783	1
15	0.0104	0.9966	5.28E-07	1	1
16	0.0124	0.9963	8.08E-05	1	1
17	0.0109	0.9965	6.59E-08	1	1
18	0.0088	0.9966	1.69E-06	1	1
19	1.03E-02	0.997	9.84E-06	1	1
20	0.0075	0.9978	4.07E-08	1	1
21	0.0084	0.9982	3.78E-08	1	1
22	0.9969	0.9969	5.57E-08	1	1
23	0.0072	0.9973	0.0013	1	1
24	0.0041	0.9991	7.71E-07	1	1
25	0.0059	0.9984	3.11E-10	1	1
26	0.005	0.9982	2.01E-10	1	1
27	0.9783	0.9988	6.63E-09	1	1
28	0.007	0.9988	4.60E-06	1	1
29	0.0042	0.9984	3.55E-12	1	1
30	0.0064	0.999	1.93E-04	1	1

has also been recorded. Therefore, the analysis has been done with a maximum value of epochs as 30.

The prediction, accuracy and loss have been recorded accordingly. Variation of result with different optimizer functions and fixed epochs of 5 and 16 on proposed recommendation system on COVID-19 samples are shown in Tables 5 and 6. The recommendation system detects a positive result if the X-ray image is similar to the X-ray image of those affected by COVID-19. A negative result obtained from the test

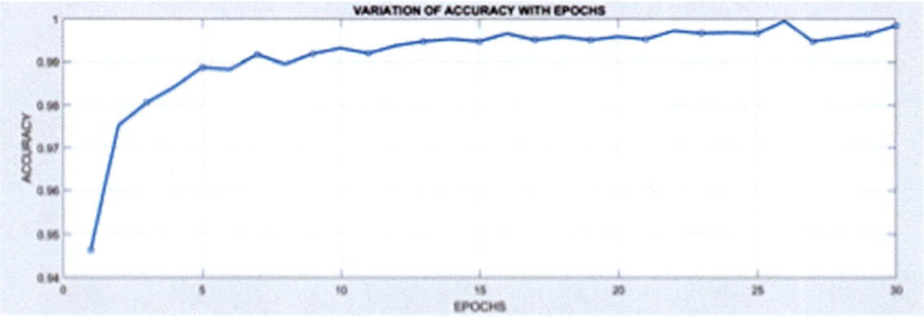

Fig. 4 Accuracy analysis with change in epochs for Adam optimizer

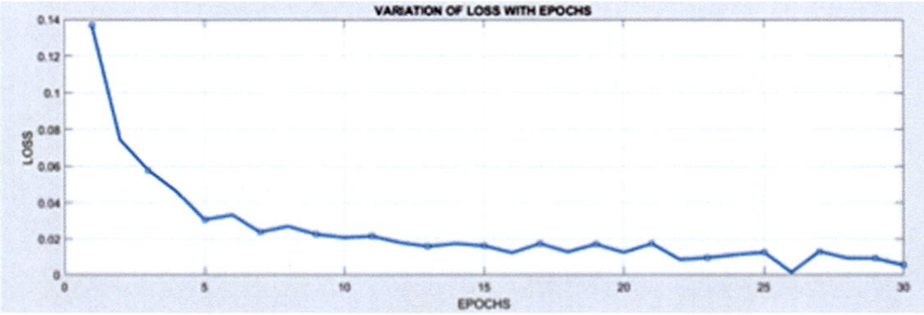

Fig. 5 Loss analysis with change in epochs for Adam optimizer

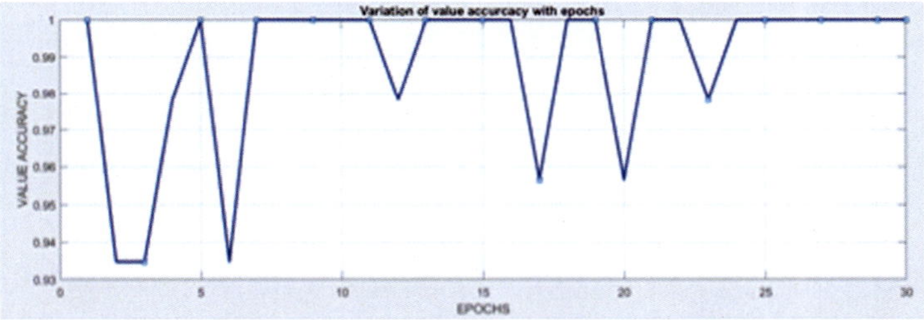

Fig. 6 Value_Accuracy analysis with change in epochs for Adam optimizer

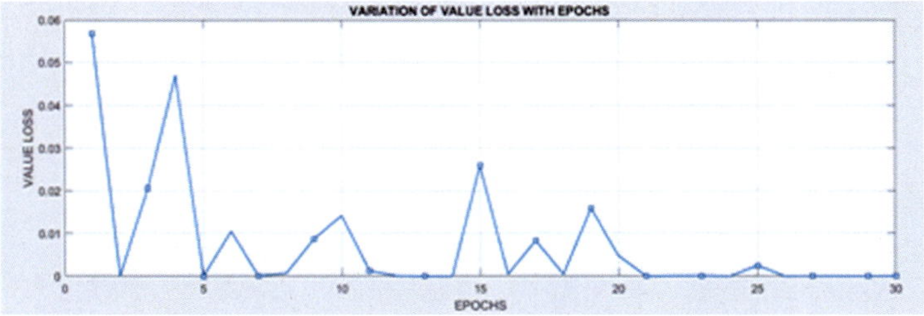

Fig. 7 Value_Loss analysis with change in epochs for Adam optimizer

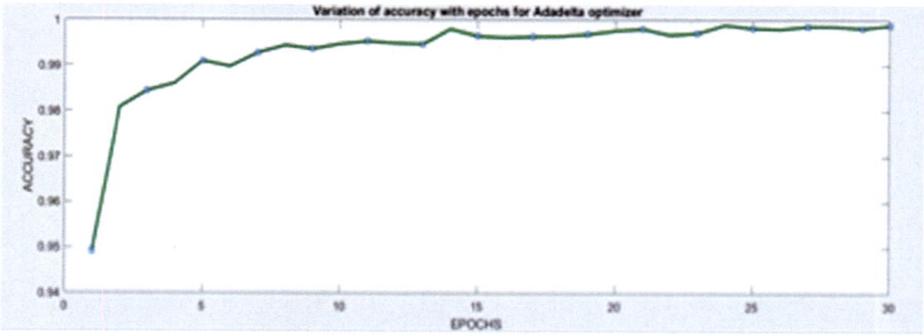

Fig. 8 Accuracy analysis with change in epochs for Adadelta optimizer

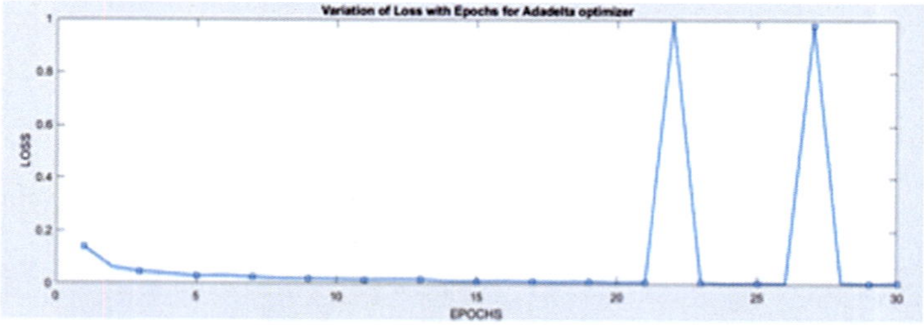

Fig. 9 Loss analysis with change in epochs for Adadelta optimizer

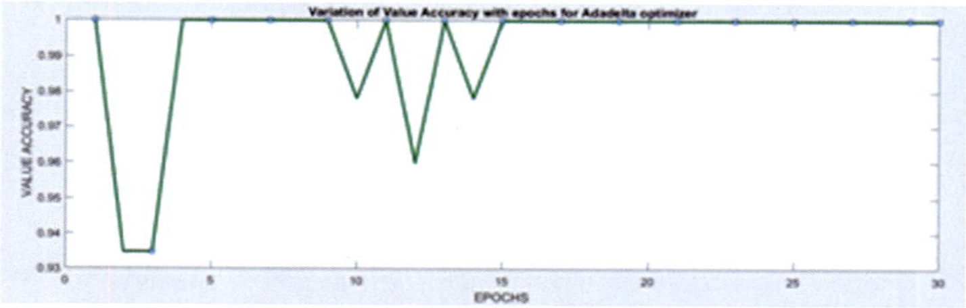

Fig. 10 Value_Accuracy analysis with change in epochs for Adadelta optimizer

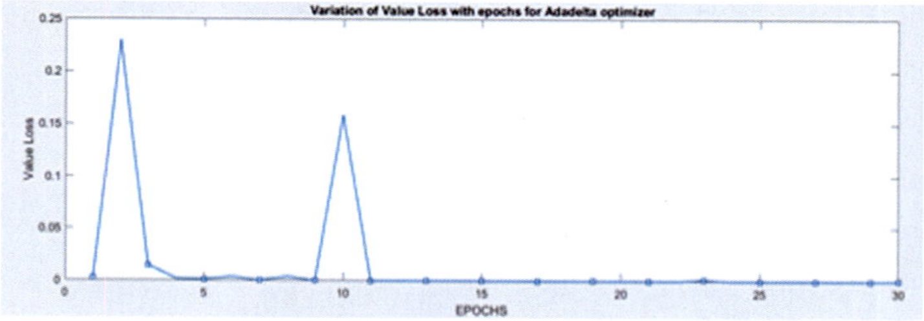

Fig. 11 Value_Loss analysis with change in epochs for Adadelta optimizer

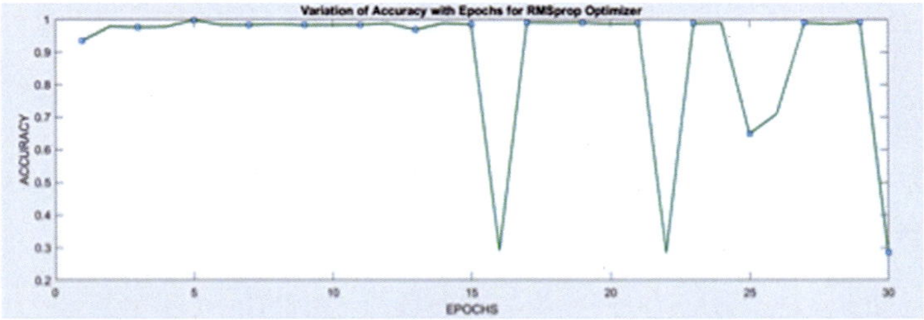

Fig. 12 Accuracy analysis with change in epochs for RMSprop optimizer

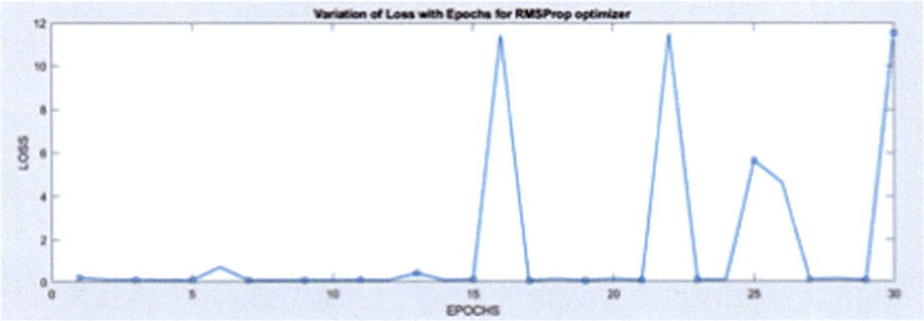

Fig. 13 Loss analysis with change in epochs for RMSprop optimizer

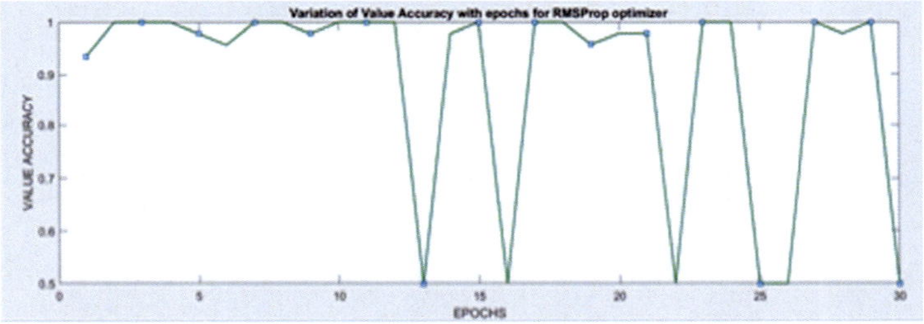

Fig. 14 Value_Accuracy analysis with change in epochs for RMSprop optimizer

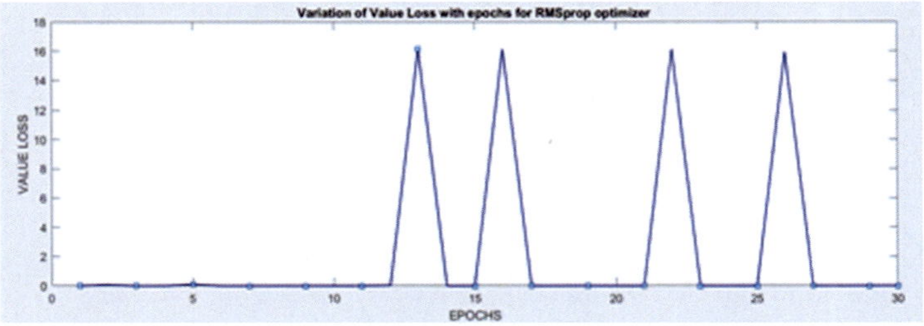

Fig. 15 Value_Loss analysis with change in epochs for RMSprop optimizer

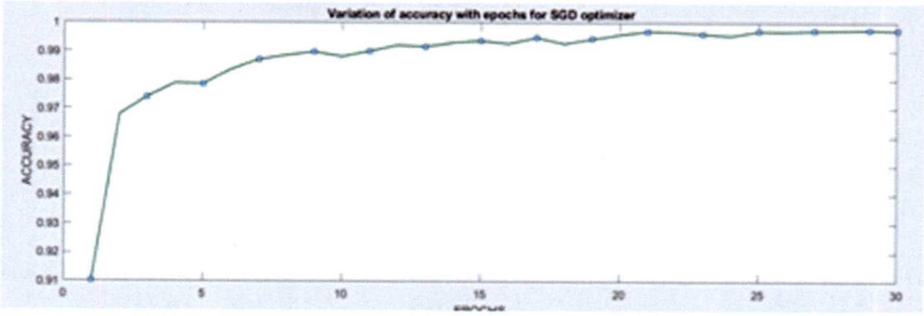

Fig. 16 Accuracy analysis with change in epochs for SGD optimizer

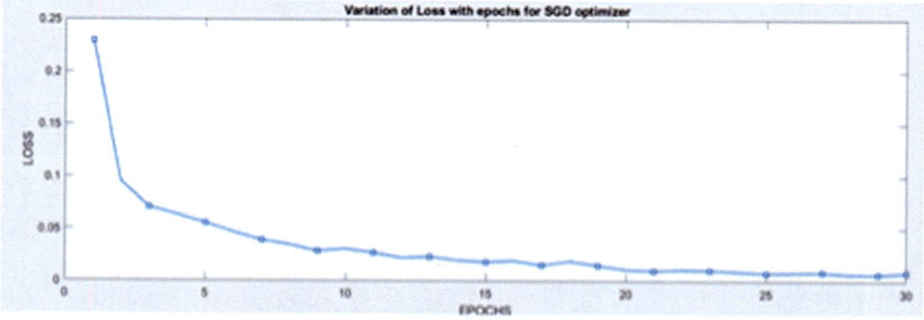

Fig. 17 Loss analysis with change in epochs for SGD optimizer

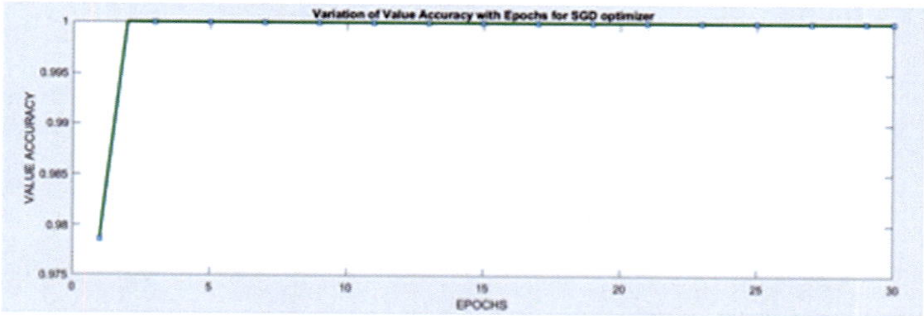

Fig. 18 Value_Accuracy analysis with change in epochs for SGD optimizer

Fig. 19 Value_Loss analysis with change in epochs for SGD optimizer

Table 5 Results with epochs = 5

Optimizer	Loss	Accuracy	Value_Loss	Value_ Accuracy	Result
Adam	0.0309	0.9888	5.03E-11	1.0000	Positive
SGD	0.05	0.9808	1.10E-04	1.0000	Positive
RMSprop	0.063	0.9845	7.61E-07	0.9565	Positive
Adadelta	0.0289	0.9901	3.42E-07	1.0000	Positive

Table 6 Results with epochs = 16

Optimizer	Loss	Accuracy	Value_Loss	Value_ Accuracy	Result
Adam	0.0123	0.9966	4.61E-04	1.0000	Positive
SGD	0.0200	0.9917	1.42E-04	1.0000	Positive
RMSprop	0.0597	0.9883	3.61E-13	1.0000	Positive
Adadelta	0.0164	0.9948	4.19E_07	1.0000	Positive

Table 7 Confusion matrix for epochs 5 – Adam optimizer

	+	–
+	18 (TP)	0 (FP)
-	1 (FN)	23 (TN)

Table 8 Confusion matrix for epochs 5 – SGD optimizer

	+	–
+	16 (TP)	O (FP)
–	3 (FN)	23 (TN)

X-ray suggests that the person is unlikely to be affected by COVID-19 virus. Thus, positive outcome suggests that the X-ray image belongs to the 'COVID-19 testing required' class while the negative outcome suggests that it belongs to 'COVID-19 testing not required' class.

After analyzing all the potted graphs, it was found that Adam and SGD optimizers performed better than other optimizers. Thus, the confusion matrix has been generated for the calculations of epochs 5, using Adam and SGD optimizers as shown in Tables 7 and 8.

TP indicates True Positive, FP indicates False Positive, FN indicates False Negative and TN indicates True Negative.

The test set has been tested and the result recorded in the confusion matrix. Accuracy, precision [11], recall and F1-score of the proposed data model have been calculated using Eqs. 18, 19, 20 and 21 respectively. The values are shown in Tables 9 and 10.

$$Accuracy = TP + TN / TP + FP + FN + TN \quad (18)$$

Table 9 Accuracy, precision, recall and F1-score values – Adam optimizer

Precision	Recall	F1-Score	Accuracy
1.0000	0.9474	0.9730	0.9762

Table 10 Accuracy, precision, recall and F1-score values – SGD optimizer

Precision	Recall	F1-Score	Accuracy
1.0000	0.8421	0.9142	0.9286

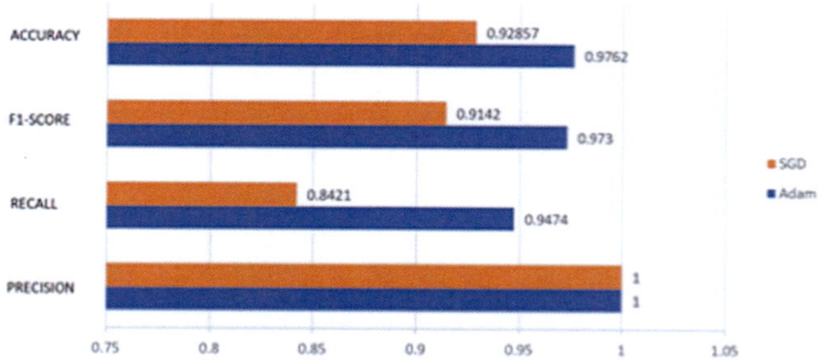

Fig. 20 Comparison of Adam and SGD optimizers

$$Precision = TP/TP + FP \tag{19}$$

$$Recall = TP/TP + FN \tag{20}$$

$$F1Score = 2 * (Recall * Precision/(Recall + Precision) \tag{21}$$

As observed from Fig. 20, Adam optimizer performs better than SGD optimizer.; therefore, it is considered. Accuracy gives the measure of the fraction of correct predictions. An accuracy of 97.62% has been obtained in the model. Precision gives the percentage of relevant results while recall tells the percentage of correctly classified total relevant results which are 1.0000 and 97.74% as obtained from Table 9 respectively. The harmonic mean of the precision and recall [21], which is a measure of the test's accuracy is the F1 score given by 0.9730 in the model.

Further, we have compared the Accuracy, Precision, Recall and F1-Score with two existing work done by L. Wang et al. [6] and A. M. Ud Din Khanday et al. [22]. These are shown in Figs. 21, 22, 23 and 24.

It can be derived that this proposed model is performing better than the existing models.

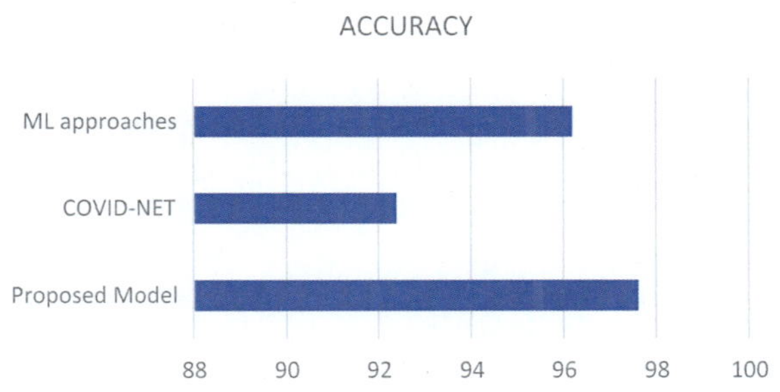

Fig. 21 Accuracy comparison of proposed model with ML approach [22] and COVID-NET [6]

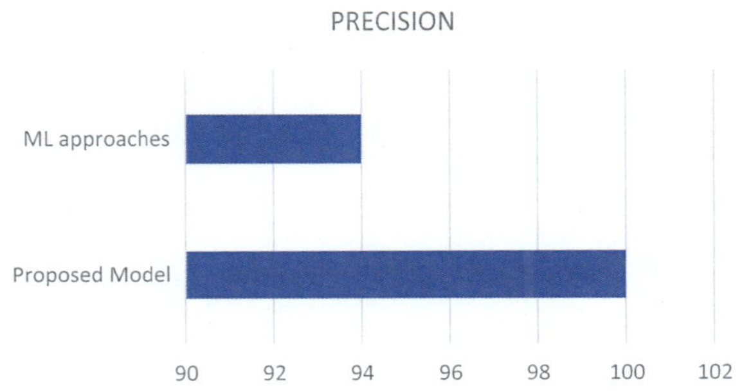

Fig. 22 Precision comparison of proposed model with ML approach [22]

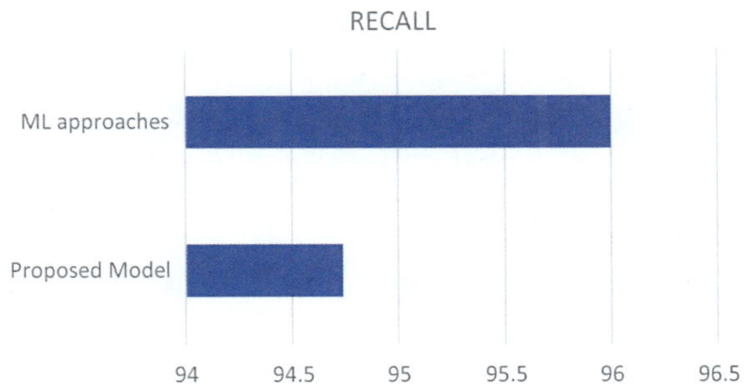

Fig. 23 Recall comparison of proposed model with ML approach [22]

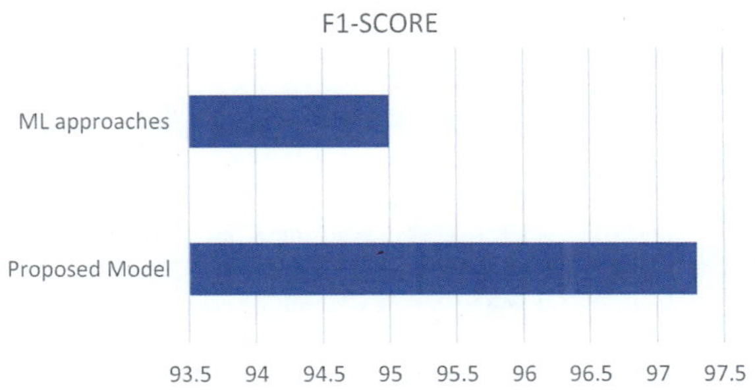

Fig. 24 F1-score comparison of proposed model with ML approach [22]

5 Conclusion

In a country like India with such a huge population, it is difficult to test everyone. So, the primary focus is to implement a model that can be accessed by normal people to self-access their chances of getting the infection and based on the recommendation generated by the model to go for further testing. It will help in reducing the chaos and will help to distinguish between normal flu and COVID-19. The model developed earlier, i.e. the COVID-Net has recorded a test accuracy of 92.4% and other models mentioned in the paper 'Machine Learning based approaches for detecting COVID-19 using Clinical Test Data' have recorded 96.2% test accuracy. The proposed recommendation system, which has been trained on the limited data sources available, has recorded an accuracy of 97.62%.

The future work is directed towards the development of a single web/smart phone operated application so that at an early stage the disease can be detected and the procedure for the proper treatment can be started.

References

1. Mortality Risk of COVID-19. https://ourworldindata.org/mortality-risk-covid, Accessed 24 Aug 2020.
2. Understanding of Convolutional Neural Network (CNN) – Deep Learning. https://medium.com/@RaghavPrabhu/understanding-of-convolutional-neural-network-cnn-deep-learning-99760835f148, Accessed 24 Aug 2020
3. S. Stirenko, Y. Kochura, O. Alienin, O. Rokovyi, Y. Gordienko, P. Gang, W. Zeng, in *Chest X-ray analysis of tuberculosis by deep learning with segmentation and augmentation.* IEEE International Conference on Electronics and Nanotechnology, Ukraine, pp. 422–428 (2018)
4. S. Sathiya, G. Priyanka, S. Jeyanthi, Detection of chronic obstructive pulmonary disease in computer aided diagnosis system with CNN classification. Int. J. Pure Appl. Math. **119**(19), 13815–13821 (2018)
5. S. Asif, Y. Wen-hui, Automatic detection of COVID-19 using X-ray images with deep convolutional neural networks and machine learning. Med. Comput. Sci., 1–7 (2020)
6. L. Wang, Z.Q. Lin, A. Wong, COVID-Net: a tailored deep convolutional neural network design for detection of COVID-19 cases from chest X-ray images. Sci. Rep. **10**, 1–2 (2020)
7. M. Shorfuzzaman, M. Masud, On the detection of COVID-19 from chest X-ray images using CNN-based transfer learning. Comput. Mater. Continua Tech Sci. Press **64**(3), 1359–1381 (2020)
8. N.B. Muppalaneni, in *Handwritten Telugu compound character prediction using convolutional neural network.* IEEE International Conference on Emerging Trends in Information Technology and Engineering, India, pp. 1–4 (2020)
9. R. Kenins, Land cover classification using very high spatial resolution remote sensing data and deep learning. Latvian J. Phys. Tech. Sci. **57**(1–2), 71–77 (2020)
10. A.K. Sahoo, C. Pradhan, H. Das, in *Performance evaluation of different machine learning methods and deep-learning based convolutional neural network for health decision making.* Nature Inspired Computing for Data Science (Springer, Heidelberg, 2020), pp. 201–212
11. A.E. Hassanien, A.B.M. Salem, R. Ramadan, T.H. Kim, in *Advanced machine learning technologies and applications.* International Conference on Advanced Machine Learning Technologies and Applications (Springer, Heidelberg, 2012). ISBN 978-3-642-35326-0.

12. A. Kumar, S. Sarkar, C. Pradhan, in *Malaria disease detection using CNN technique with SGD, RMSprop and ADAM optimizers.* Deep Learning Techniques for Biomedical and Health Informatics (Springer, Heidelberg, 2020), pp. 211–230
13. F. Shang, K. Zhou, H. Liu, J. Cheng, I.W. Tsang, L. Zhang, D. Tao, L. Jiao, VR-SGD: a simple stochastic variance reduction method for machine learning. IEEE Trans. Knowl. Data Eng. **32**(1), 188–202 (2020)
14. M. Yaqub, J. Feng, M.S. Zia, K. Arshid, K. Jia, Z. Ur Rehman, A. Mehmood, State-of-the-art CNN optimizer for brain tumor segmentation in magnetic resonance images. Brain Sci. MDPI **10**(7), 427 (2020)
15. A. Bhattacharjee, S.K. Borgohain, B. Soni, G.K. Verma, X.Z. Gao, in *Machine learning, image processing, network security and data sciences.* MIND (Springer, Hedielberg 2020). ISBN 978-981-15-6315-7
16. S. Dash, B.R. Acharya, M. Mittal, A. Abraham, A. Kelemen, *Deep learning techniques for biomedical and health informatics* (Springer, Heidelberg, 2020). ISBN 978-3-030-33966-1
17. G. Bebis, R. Boyle, B. Parvin, D. Koracin, C. Fowlkes, S. Wang, M.H. Choi, S. Mantler, J. Schulze, D. Acevedo, K. Mueller, M. Papka, in *Advances in visual computing.* International Symposium on Visual Computing, (Springer, Hedidelberg 2012). ISBN 978-3-642-33179-4
18. Hospitals Deploy AI Tools to Detect COVID-19 on Chest Scans. https://spectrum.ieee.org/the-human-os/biomedical/imaging/hospitals-deploy-ai-tools-detect-covid19-chest-scans, Accessed 12 Aug 2020
19. Covid-chestxray-dataset. https://github.com/ieee8023/covid-chestxray-dataset/tree/master/images, Accessed 10 Aug 2020
20. L. Barolli, M. Takizawa, F. Xhafa, T. Enokido, in *Web, artificial intelligence and network applications.* International Conference on Advanced Information Networking and Applications (WAINA), (Springer, Heidelberg, 2019). ISBN 978-3-030-15035-8
21. V.I. Perez-Neuno, A.S. Karaboga, M. Souchet, D.W. Ritchie, GES polypharmacology fingerprints: a novel approach for drug repositioning. J. Chem. Inf. Model. **54**(3), 720–734 (2014)
22. A.M. Ud Din Khanday, S.T. Rabani, Q.R. Khan, N. Rouf, M.M. Ud Din, Machine learning based approaches for detecting COVID-19 using clinical text data. Int. J. Technol. **12**(3), 731–739 (2020)

Application of Deep Learning Techniques for COVID-19 Management

Rajan Gupta, Anshuman Gupta, Manan Bedi, and Saibal K. Pal

Abstract The sudden penetration of coronavirus has taken several countries in a devastating state worldwide. The novelty of the COVID-19 virus and its wide range of symptoms have posed significant difficulties for the governments, authorities, and the medical community in dealing with this global outbreak. Various researchers have been focusing their studies on finding methods for this pandemic's forecasting and diagnosis effectively. Evolutionary advancements in Artificial Intelligence, specifically in Deep Learning, have enabled it to become one of the most promising technologies for solving problems related to the current pandemic and other epidemics in the past. This study presents a Concept Mapping based Research Framework, abbreviated as DL-CoTAUM, to review the usage of Deep Learning Techniques for the management of the COVID-19 pandemic using multiple steps. Through the first step, different concepts around Deep Learning are presented. Multiple deep learning techniques using the basic concepts are studied through the second step. In the next step, various epidemics from the past are learned from the Deep Learning perspective. In the fourth step, multiple use-cases for COVID-19 management are covered up, like trend forecasting, virus and drug discovery, patient diagnosis, clinical analysis, and monitoring for maintaining social distancing & face masks. The last step presents the various evaluation parameters available for analyzing the modeling techniques' performance under Deep Learning. In this review study, it was found that Deep Learning is gaining importance for various applications under COVID-19 management. Techniques like LSTM, CNN, RNN, and GANs were found to be prominently used modeling techniques for growth curve fitting and image analysis. Accuracies of more than 90% have been reported for Deep Learning techniques in COVID-19 use-cases, which is higher than the other baseline methods. This study will be useful for researchers, academicians, and practitioners in Epidemiological Modeling and

R. Gupta (✉)
Centre for Information Technologies & Applied Mathematics, University of Nova Gorica, Nova Gorica, Slovenia

A. Gupta · M. Bedi
Shaheed Sukhdev College of Business Studies, University of Delhi, Delhi, India

S. K. Pal
Defence Research and Development Organization, Delhi, India

J. Nayak et al. (eds.), *Understanding COVID-19: The Role of Computational Intelligence*, Studies in Computational Intelligence 963, https://doi.org/10.1007/978-3-030-74761-9_8

Image Processing. The proposed research framework can be used as a reference point by other researchers for studying the application of a new concept in a different domain. And this research study will also be useful for naïve users in the field of Deep Learning.

Keywords DL-CoTAUM · LSTM · RNN · CNN · GAN

1 Introduction

Francois Chollet defined, "Deep Learning (DL) is a specific subfield of Machine Learning, which is a new take on learning representations from data that emphasizes learning successive layers of increasingly meaningful expressions". The term 'Deep' refers to the concept of consecutive layers of representations 64. Deep Learning is used in a wide range of applications today, almost in every domain. Deep Learning applications may seem impossible to a normal human being, but surprisingly, this concept is making progress globally by solving human problems in almost every possible discipline.

Deep learning is the crucial technology that is driving the idea of autonomous driving and helping in making it a reality. The system feeds on data from millions of datasets for building a model, trained and tested in the perfect environment [1]. Deep Learning can be used to colourize the black and white images automatically. It can use the object and their context to colour the picture. This application uses supervised layers to recreate the image with the addition of colours [1]. One of the most potent applications of Deep Learning is automatic machine translation, where words, phrases, or sentences in one language, are automatically translated into another language, using various techniques under Deep Learning. These concepts leverage the use of stacked networks of large Recurrent Neural Networks (RNN) [1].

Convolutional Neural Network (CNN) can be used to identify the images that have letters or handwritten characters [4]. Drug Discovery, Toxicology, and Predicting Gene ontology annotations (Bioinformatics) use Generative Adversarial Networks (GAN) [57]. Deep learning can be used for the segmentation of customers and also build recommendation systems using Auto-Encoders and Restricted Boltzmann Machines [66]. AlphaGo, which brought a significant breakthrough in the field of Artificial Intelligence in 2016, used Artificial Neural Networks by extensive training, both from humans and computer play, along with various search algorithms like the Monte Carlo tree search algorithm [67].

Deep Learning algorithms work by learning data representations by extracting high level and relatively complex abstractions using a hierarchical learning process. At a given level, complex abstractions are learned from somewhat more straightforward inferences learned at the previous level. The ability of a neural network to analyze and understand an enormous dataset in its native form gives it an advantage in handling big data [2]. Deep Learning techniques are comprised of consecutive layer-based design, with every layer processing a non-linear function on its input, which tries to extract defining features in the data which makes them suitable to work on unprocessed data for finding complicated abstractions automatically [3, 16].

Having seen in-depth & ubiquitous usage of the Deep Learning across various technology applications, it becomes imperative that it is used in the current pandemic condition, i.e., COVID-19 management. COVID-19 has been a weird situation that needs the involvement of advanced technology usage. The present study is an attempt to introduce the concepts of Deep Learning to the naïve user and present the analysis of various DL techniques for COVID-19 under different application areas. This study will help the data scientists, researchers, and practitioners in applying DL techniques for various use-cases of COVID-19 on their customized datasets. The current research work has been divided into seven sections. Section 1 gives the background of the usage of Deep Learning. Section 2 presents the basic DL concepts required for technique understanding. Section 3 offers a description of various DL modeling techniques used prominently in multiple applications. Literature Review and Research methodology are presented in Sect. 4 and 5, respectively. Different DL modeling result, for numerous applications under COVID-19 management, are presented in Sect. 6. This study ends by giving a discussion & conclusion in Sect. 7.

2 Deep Learning Concepts

This section discusses various concepts around Deep Learning, which form the building blocks for understanding different DL modeling techniques.

2.1 Neurons and Neural Network

It is a simple mathematical unit that takes in some finite number of inputs, computes a function from these inputs and outputs it. A neural network with an input layer connected to a single output neuron is called a single layer neural network or a perceptron as shown in Fig. 1 [4]. When several such neurons are stacked up in a layered manner so that the outputs of neurons from one layer are the input to the neurons in another layer, then it is called a multi-layer neural network [4].

2.2 Layers

The layered architecture of a neural network consists of neurons stacked up to form a layer performing a particular operation. A neural network is formed by an input layer that takes in the input features, a group of hidden layers that perform the intermediate computations and transformations, and an output layer which outputs the result as shown in Fig. 2 [4].

2.3 Weights and Bias

Each input to a neuron has some associated value, which determines the contribution of that input to the computed function. These weights are multiplied with the inputs received and the products obtained are added to get a weighted sum of features or inputs. Each node also has a bias term, which is added to this weighted sum, as shown in Eq. 1.1, where w_i is the weight associated with the ith input, b is the bias associated with the neuron, and z is the final weighted sum of the inputs. This information is called the net input to a neuron [6]. The neural network is trained by treating these weights and biases as parameters and approximating them to get an approximate functional mapping, which maps the inputs to the required outputs accurately [6].

$$z = \sum_i w_i x_i + b \tag{1.1}$$

2.4 Activation Functions

A scalar-to-scalar transformation is applied to the net input using a particular function, known as Activation Function, to learn more complex mappings. Some commonly used activation functions are as follows.

(a) Sigmoid function: This function transforms the output values to a smooth range of values between 0 and 1. Equation (1.2) defines a sigmoid function, where z is the weighted sum of inputs obtained from Eq. (1.1). It is a continuous S-shaped function which makes it differentiable over the entire range of real numbers [7].

$$\sigma(z) = \frac{1}{1 + \mathrm{e}^{-z}} \tag{1.2}$$

(b) Tanh function: The hyperbolic tangent function can also be used as the activation function. Equation (1.3) defines a *tanh* function, where z is the weighted sum of inputs obtained from Eq. (1.1). It maps the input x to the interval (−1, 1). Unlike the sigmoid function, it is symmetric about the origin and thus can predict both negative and positive values. It has a steeper gradient than that of the sigmoid function. All these characteristics make it more favourable than sigmoid in most cases [7].

$$\tan(z) = \frac{\mathrm{e}^{z} - \mathrm{e}^{-z}}{\mathrm{e}^{z} - \mathrm{e}^{-z}} \tag{1.3}$$

(c) ReLU function: Rectified Linear Unit or ReLU is used prominently in neural networks. Equation (1.4) defines a ReLU function, where z is the weighted sum of inputs obtained from Eq. (1.1). ReLU is one of the most suitable activation functions, especially for deep neural networks, because of the constant derivative value for all positive inputs, which enables the system to learn faster [4].

$$R(z) = max(0, z) \tag{1.4}$$

2.5 Cost Function

The cost function helps in measuring the error in the predicted output when compared to the actual output. The cost thus obtained is then minimized during training to get a good accuracy over the training dataset [4]. Commonly used Cost functions are Mean Squared Error (for regression), Binary Cross-entropy (for binary classification), and Categorical Cross-entropy (for multi-class classification).

2.6 Gradient Descent

It is the process of updating the weights and bias of the neurons to minimize the loss. In gradient descent, weights are updated such that the cost gets reduced in the order of steepest decline. For each neuron, the update takes place using the Eq. (1.5) where $\overline{W}$ is the vector representing the weights, α is the learning rate, and J is the cost obtained from the current predictions and target values [4].

$$\overline{W} \Leftarrow \overline{W} - \alpha \frac{\partial J(\overline{W}, b)}{\partial \overline{W}} \tag{1.5}$$

2.7 Learning Rate

It specifies the degree by which the weights should be updated during the gradient descent. In Eq. (1.5), α plays the role of the learning rate. The learning rate is selected very carefully. A high learning rate would make parameters to be updated way too much in the direction of gradient descent, and the optimal point would get missed. If a minimal learning rate is chosen, then the neural network might converge to the minima but at a languid pace, and thus the learning process may take a lot of time [4].

2.8 *Batches and Epochs*

Sometimes, the training set is randomly divided into relatively small, equal datasets called Batches. These batches are fed into the neural network one-by-one, and gradient descent is performed for each batch. This activity makes the neural network more generalized as compared to training on the entire training set [4]. When the algorithm performs both forward pass (predictions) and backward pass (gradient descent for all neurons) for the whole of the training set, then it is said to have completed one epoch [8].

2.9 *Forward and Backward-Propagation*

The movement of input features from the input layer to the output layer via the hidden layers and application of various transformations on the input layer-by-layer to get the output is called forward propagation [44]. Once the predicted outcome is obtained, the cost is computed using a cost function. The gradients or derivatives of the cost function concerning the weights and biases in different layers are calculated and are fed back into the network along with the cost [4]. The parameters are then updated layer by layer using Eq. (1.5). This process is called as Backpropagation in neural networks [21].

2.10 *Dropout*

Over-fitting on the training data is one of the major issues while using profound neural networks. In such cases, the Dropout method can be used in which a certain proportion of neurons are randomly dropped from the hidden layer during training. This dropping creates a regularization effect as now the model gets trained on more than one neural network architecture, each having randomly sampled neurons [4].

2.11 *Filters and Pooling*

In Convolutional Neural Networks (CNN), a filter is a weight matrix that is applied to a larger image to transform the image into another form to extract features. These filters are incorporated in a neural network by adding a convolutional layer with one or more such filters. A convolution operation is applied between a filter and various overlapping regions of the image for highlighting specific features [9]. The weights in the filters can be updated using Backpropagation to minimize cost [4]. Pooling layers serve the purpose of reducing the height and width of images (or intermediate transformed images) in CNNs, thus providing a reduction in the number of parameters along with over-fitting avoidance. Max-Pooling is the most popular type, in which the pixel with the maximum value is chosen, and the rest are discarded from a window of fixed size. This process is repeated by sliding the window from top-left to the bottom-right corner of the image [4, 10].

2.12 Padding and Data Augmentation

Padding refers to the addition of extra zeros on the borders of the image to prevent the reduction of output image size after it passes through convolutional layers [4]. In problems where sufficient data is not available, the neural network is not able to generalize well on test data or unseen data, leading to over-fitting. To counter the over-fitting, data augmentation is applied. A new data is derived from the existing dataset by using some boost to obtain a more generalized data distribution. In problems involving images, some of the data augmentation techniques are translation, rotation, scaling, and skewing [4, 11].

3 Deep Learning Techniques

The prominent modeling techniques under Deep Learning are presented in this section.

3.1 Shallow Neural Networks

Shallow Neural Networks are neural networks that have an input layer, a few hidden layers, and an output layer. These consist of only one or two hidden layers [12]. If the hidden layer consists of four nodes, then they perform the required computations as shown in Eqs. (3.1) to (3.4).

$$z_1^{[1]} = w_1^{[1]T} x + b_1^{[1]}, a_1^{[1]} = \sigma(z_1^{[1]}) \tag{3.1}$$

$$z_2^{[1]} = w_2^{[1]T} x + b_2^{[1]}, a_2^{[1]} = \sigma(z_2^{[1]}) \tag{3.2}$$

$$z_3^{[1]} = w_3^{[1]T} x + b_3^{[1]}, a_3^{[1]} = \sigma(z_3^{[1]}) \tag{3.3}$$

$$z_4^{[1]} = w_4^{[1]T} x + b_4^{[1]}, a_4^{[1]} = \sigma(z_4^{[1]}) \tag{3.4}$$

In general, for any layer l with $n^{[l]}$ nodes, if x is the input vector, $w_j^{[l]}$ and $b_j^{[l]}$ are the weights vector and bias, respectively, associated with the jth node of layer l, $z_j^{[l]}$ is the weighted sum of inputs, $a_j^{[l]}$ is the activation output of that node and σ is the activation function, then Eq. 3.5 shows the general output function.

$$z_j^{[l]} = w_j^{[l]T} x + b_j^{[l]}, a_j^{[l]} = \sigma\left(z_j^{[l]}\right), \text{where} 1 \leq j \leq n^{[l]} \tag{3.5}$$

In a network with one hidden layer, the computations can be performed for all the layers using the vectorized equations shown in (3.6) to (3.9), where x is the input vector, $W^{[1]}$ and $W^{[2]}$ are the weight matrix for the two layers, $b^{[1]}$ and $b^{[2]}$ are the bias vector for the two layers, $Z^{[1]}$ and $Z^{[2]}$ are the vectors containing the weighted sum output from all the nodes of layer 1 (hidden layer) and layer 2 (output layer) respectively, $A^{[1]}$ and $A^{[2]}$ are the activation vector of layer 1 and 2, respectively and $\widehat{Y}$ is the output of the network. The activation function output of the first hidden layer $A^{[1]}$ is used to compute $Z^{[2]}$ of the output layer [13]. At the end of the output layer, the predicted value $\widehat{y}$ is obtained.

$$Z^{[1]} = W^{[1]T} x + b^{[1]} \tag{3.6}$$

$$A^{[1]} = \sigma(Z^{[1]}) \tag{3.7}$$

$$Z^{[2]} = W^{[2]T} A^{[1]} + b^{[2]} \tag{3.8}$$

$$\widehat{y} = A^{[2]} = \sigma(Z^{[2]}) \tag{3.9}$$

3.2 *Deep Neural Networks*

An architecture having more than two hidden layers, which are complex in nature, is called a Deep Neural Network. Such networks use sophisticated mathematical modeling for the processing of data, where each layer performs specific types of ordering and sorting in the "Feature Hierarchy" process [14]. Its primary use is to deal with unlabeled and unstructured data. The number of layers the data must pass through in the process of prediction is called the depth of a neural network. Each layer receives different features based on the previous layer's input and more intricate features can be recognized by the nodes present deeper in the network as compared to shallower ones [15]. This architecture enables it to process enormous and high dimensional datasets effectively. The weights are updated after each iteration to reduce cost function as the primary objective [16].

3.3 Radial Basis Function Network

A Radial Basis Function (RBF) network, which is a three-layer feed-forward neural network, predominantly works on the activation function based on radial basis function [19]. The output is a linear combination of function outputs and the neuron parameters [16]. Its use-cases include time-series prediction, function approximation, and systems control. A non-linear RBF is present in the hidden layer and a scalar function of the input vector forms the output. The standard procedure for training includes a selection of the network structure by trial and error. First, the hidden layers are trained using Backpropagation. For each node in the hidden layer, we find receptors and variance. In the second phase, we have to update the weight vectors [18]. Training of RBF Network is faster as compared to a multi-layer perceptron [30].

3.4 Restricted Boltzmann Machines

A double-layered ANN with the ability to generate data is categorized as Restricted Boltzmann Machines (RBM). They can learn probability distribution given a set of inputs. Their usage includes collaborative filtering, reducing dimensions, and topic modeling. An RBM consists of the symmetric bipartite graph in which there is no connection between two units from the same group, but when stacked, gradient descent and Backpropagation can be used for fine-tuning them. The activation is produced through the hidden bias on the forward pass, and the input is reconstructed through visible bias during a backward pass [21].

3.5 Recurrent Neural Network and LSTM

Recurrent Neural Network (RNN) is suitable for sequential data like time-series data such as stock market price, textual data, and audio data. They can process input sequences of variable lengths. A recurrent neuron is similar to a neuron in a feed-forward neural network, except the fact that it has backward connections also. This means that at each time step t along with the input $x_{(t)}$ the recurrent neuron also receives its output from the previous time step $y_{(t-1)}$ as input [21]. There are two sets of weights for every recurrent neuron, w_x for the inputs and w_y for the outputs from the previous time step. Considering the whole layer of recurrent neurons, we would have weight matrices W_x and W_y as the weights for the inputs of the current time step and outputs from the previous time step, respectively. Equation (3.10) represents the output $y_{(t)}$ of a recurrent layer at time step t for a single sample [21], where b is the bias, and σ is the activation function associated with the node.

$$y_{(t)} = \sigma(W_x^T x_{(t)} + W_y^T y_{(t-1)} + b) \tag{3.10}$$

This output of the hidden layer is again passed to the hidden layer for t timestamps and is not given to the next layer until all the timestamps are completed. During Backpropagation, the error is propagated backwards following the unfolded network, as shown in Fig. 3, and the weight updating is done accordingly. This is termed as Back-propagation through time (BPTT) [4]. Long short-term memory or LSTM is a variant of RNN having the capabilities of feedback connections, which can process large datasets as input streams. Gated Recurrent Unit (GRU) is the simplest variant of the LSTM model [52, 64].

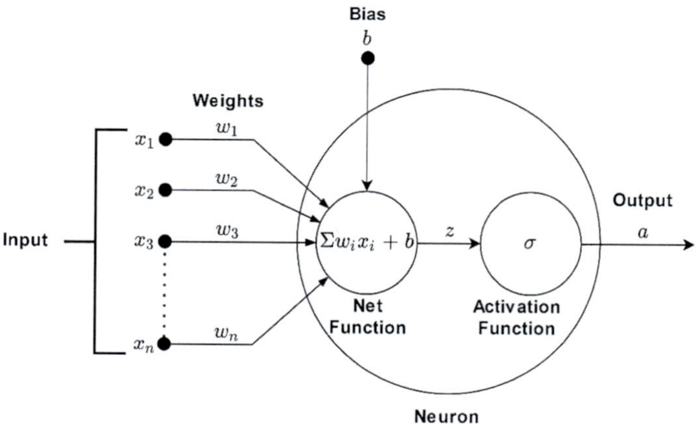

Fig. 1 A simple neuron forming a single layer neural network (Source: Author Created)

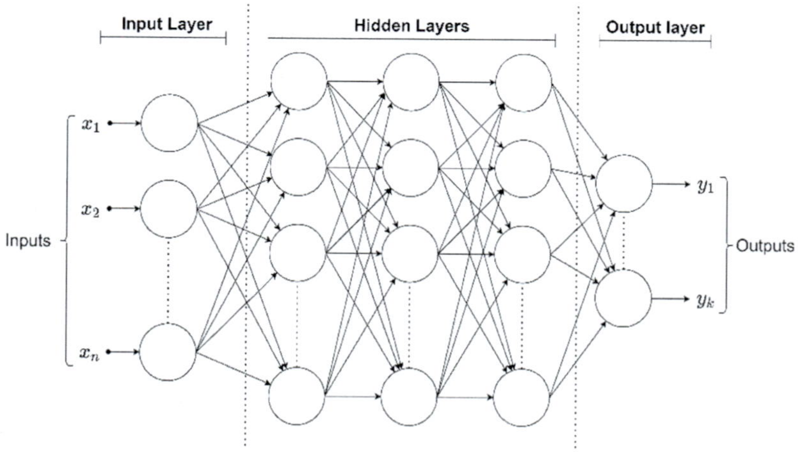

Fig. 2 A multi-layer neural network (Source: Author Created)

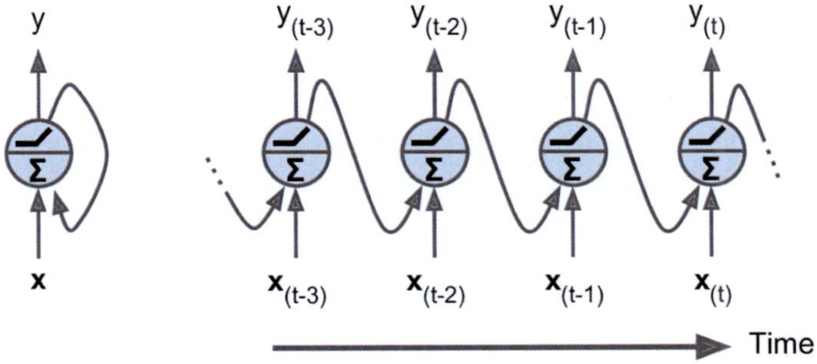

Fig. 3 A recurrent network neuron unrolled through time (Source: [26])

3.6 *Convolutional Neural Networks*

A neural network having the capability of extracting features from complex data is Convolutional Neural Network (CNN). They are highly useful in image recognition systems as they are based on the study of the brain's visual cortex [21]. In CNN, the neurons in each layer form a spatial grid structure. A neuron in a convolutional layer which is not connected to every neuron of the preceding layer, but only to those present in a close region in the preceding layer. This helps in maintaining spatial relationships between relative cells (pixels), which enable the detection of more complex features from simpler ones. Three types of layers are most commonly present in a convolutional network - convolutional layers, pooling layers, and activation layer (ReLU). Additionally, some fully connected layers are also present to produce application-specific outputs [4].

The filters form the backbone of CNN. The convolution operation, when applied to these filters and input images, extract feature maps that highlight the regions that are most similar to the filters. The training process of CNN aims at choosing the best filters to infer the most meaningful features from the images. The first few layers detect simple features like horizontal edges and vertical edges. The deeper convolutional layers highlight more complex features like shapes, patterns, and other complex structures like eyes, nose, and other parts of the face in case of facial recognition [21].

3.7 *Auto-Encoders*

Auto-Encoders are special artificial neural networks that can learn an efficient representation of the input data devoid of any supervision (unsupervised deep learning). They produce input data in some other encoded form called codings, which have lesser dimensions than the original data, thus making them suitable for dimensionality reduction. They can also be used as generative models. They are also capable of being robust feature detectors for unsupervised learning. Autoencoders try to learn an identity function that outputs a copy of the inputs. The architecture of autoencoders and Multi-Layer Perceptron (MLP) are somewhat similar, except for the equality in the number of nodes in both input and output layers. There are two parts of an autoencoder: an encoder that transforms the input into an internal representation and a decoder that can change these internal forms to output. The output generated by the network is called reconstruction as the autoencoder makes an attempt to reconstruct the inputs, and a reconstruction loss is included in the cost, which penalizes the model when the output generated is dissimilar to the information [21].

3.8 Reinforcement Learning and Deep Reinforcement Learning

In Reinforcement Learning, the system learns to engage with an environment with varying levels of complexities to get the best outcome, which will maximize the reward. In this technique, hit-and-trials are done by the system with the motivation of maximizing rewards that it achieves. Reinforcement learning has been very successful in creating intelligent agents for video games, board games, auto-driven cars, and even self-learning robots [67]. Tasks in which evaluation is easy, but the specification is demanding, Reinforcement learning becomes useful [4]. The power of Deep learning in conjunction with reinforcement learning is Deep Reinforcement Learning (DRL). With deep learning, automatic extraction of features from raw sensory inputs is possible, which can be used for reinforcement learning without the use of handcrafted features [26].

3.9 Generative Adversarial Networks

Networks capable of generating realistic samples from a dataset using two adversarial networks are referred to as Generative Adversarial Networks (GANs), which are designed for creating generative models. Two neural networks are simultaneously involved in GANs. One is a generative model that aims to develop artificial samples that are similar to the examples in the original dataset. The second is a discriminative model that aims at discerning whether a given object or sample is synthetic or real. When the latter can correctly predicts an artificial item as fake, it pushes the generative model to update its weights so that the discriminative model has difficulty in distinguishing between real and fake samples. This process is repeated until the discriminative model is unable to determine the synthetic and counterfeit pieces. This technique is instrumental in machine learning for oversampling and data augmentation [4].

4 Literature Review

Deep Learning Techniques have been used prominently by the researchers in the past during different epidemiological modeling and predictions across multiple types of diseases, as discussed below.

4.1 *Influenza*

Influenza is a significant health issue and affects large amounts of the population across the globe [27]. Deep Learning techniques and neural networks are being utilized for forecasting time-series data due to greater computational power. LSTM based models have been used by Venna et al. [28] to estimate initial forecasts following which the error is reduced using an impact factor from weather variables and adjustment factor due to space and time component for incorporating the effect of the spread from nearby regions.

Adhikari et al. [27] proposed EpiDeep for forecasting epidemics. The approach focused on forecasting future cases, the peak intensity of the season, the week of the peak, and onset week using the historical WILI incidence reports. In this approach, the historical surveillance data is fed into the neural network, which learns to map these historical seasons to feature space and then uses these embeddings to find similarities between past seasons and the experimental stage of the current season to make accurate forecasts. The similarity between seasons is judged by leveraging Deep Clustering techniques to learn feature embeddings such that the distance between similar seasons is minimized.

4.2 *Alzheimer's*

Alzheimer's is a medical disease that leads to memory impairment and loss, which, in the future, can cause other symptoms. But, early diagnosis of the disease can help in preventing further brain damage. Alzheimer's disease has seen the application of several DL techniques for early detection [29].

Liu et al. [29] described an automated diagnosis framework having a deep learning architecture with stacked autoencoders followed by softmax layers for Alzheimer's patients' classification from patients with mild cognitive impairment and people without the disease. This approach uses numerical anatomical measurements like volume, shape, and curvature using feature extraction methods, which are then fed to the neural network for learning.

An end-to-end deep learning approach was proposed by Ji et al. [30] for the disease, which uses ConvNets for feature extraction. The method uses techniques like transfer learning and ensemble learning for improving model accuracy. This technique outperforms various traditional machine learning techniques that use hand-engineered features.

4.3 Pneumonia

A large number of the world's population is affected by pneumonia every year, and it takes several lives all across the globe. Chest X-ray has proven to be one of the best available methods for diagnosing pneumonia but requires highly experienced medical practitioners and radiologists for diagnosing the disease correctly. Although, DL advancements have made it possible to develop advanced models that can be utilized for pneumonia detection from chest X-ray images without any human help.

Rajpurkar et al. [31] worked on a Dense Convolutional Network (DenseNet) named CheXNet for diagnosing pneumonia. It is a 121-layer DenseNet trained on the ChestX-ray14 dataset. The initialization of the model's weights was done using a pre-trained model on the ImageNet data. On comparing its performance with practising radiologists, it was found that the deep learning approach was able to classify chest X-ray images into pneumonia patients and regular patients.

Hammoudi et al. [32], in their paper, describe a set of tailored ConvNets for predicting the probabilities of standard cases, viral pneumonia cases, and bacterial cases. These models were built on top of ResNet34, ResNet50, and DenseNet169 with a fully connected layer in the end, including a single output layer as a classifier. It also fetches the hidden layer output of an Inception ResNetV2 architecture with some fine-tuning, which it uses to predict the final categories of different image blocks of an image by making use of a bidirectional LSTM architecture. These tailored deep learning models have displayed good performances with an average accuracy of above 84%. Hence, deep learning architectures have great potential in predicting pneumonia cases amidst this global pandemic.

4.4 Flu Detection Using Social Media

Infectious disease epidemics like influenza and the ongoing pandemic pose a severe threat to the world. Computational Epidemiology can effectively model the growth curve of the disease, but the insufficiency of real-time data is a significant hindrance [33]. Social media has the capability for rapid and elaborate surveillance of disease, but it fails to model the disease well [34]. A semi-supervised Deep Learning Framework can be integrated with the power of the two techniques mentioned above. Some experiments demonstrate that using this technique, useful Spatio-temporal mapping of the disease spread can be obtained [36].

One such approach is SimNest (Social Media Nested Epidemic Simulation), which makes use of self-reported symptoms on social media [37]. Its mechanism is simple - take social media data streams, for example, from twitter, and define a matrix of the posts from users in time *t*. Then, determine a matrix of predefined words related to flu and finally feed it to the Deep Learning Algorithm, which can classify it into infectious or healthy at time *t*. It is an unsupervised model and aims to minimize some application-specific loss function. It aims to detect whether there is some relevance between the post and public health [35]. Whenever any epidemic happens, interventions can reduce the potential impact of the outbreak efficiently. Monitoring the interventions can also be done with the help of a text classifier.

4.5 Other Diseases

Infectious diseases cause massive harm on an immense scale. The trend of the spread of disease is difficult to predict. There is a need to quickly transfer the reports to the authorities who need them to prevent and control the conditions. Several studies have been focusing on predicting the spread of diseases on a real-time basis [38–40]. They are easily accessible and can identify the disease growth trend before the official organizations.

Studies show that there is a considerable impact of weather on the spread of infectious diseases [41, 42]. For instance, a Dengue fever mostly spreads when the temperature is humid. Predicting trends using Time Series Analysis models like Autoregressive Integrated Moving Average (ARIMA) can eliminate the reporting time difference using Big Data, thus minimizing the economic loss. First of all, the data from any social media tool like Twitter can be extracted using a Web Crawler through BeautifulSoup or Selenium library. All the non-clinical search data, along with twitter data, and data on weather-related information, can be inputted into the LSTM and Deep NN models. The dataset can be divided into training, validation, and test data.

4.6 Research Direction

Deep Learning can predict infectious diseases at their onset if it considers the time difference between clinical and non-clinical data, which can be of great significance to control the spread of disease. Thus, Deep Learning can be augmented with surveillance systems of the pandemic, epidemics, or the spread of infectious diseases. Also, there was an absence of literature covering a comparative analysis of various DL modeling techniques under different applications of Deep Learning in managing a particular disease. Thus, it is necessary to conduct this study, which presents a comparative analysis of the multiple DL techniques in various scenarios. This will help public administrators, practitioners, and researchers in selecting appropriate DL modeling techniques for application development related to COVID-19 management in a quick time.

5 Methodology

The novel coronavirus 2019 pandemic has created havoc around the world. It impacted day-to-day working, healthcare, business operations, and the global economy. It becomes crucial to identify positive cases at the earliest to prevent the widespread and to treat affected patients quickly. This study reviewed more than 150 papers related to applications of Deep Learning for the current pandemic, i.e., COVID-19 and earlier pandemics like Influenza, Alzheimer's Disease, and Flu. The search results were limited through the use of AND or OR queries in the search. The papers were searched from primary databases like SCOPUS, SCI, ESCI, and Google Scholar. Preprint archives like SSRN, arXiv, medRxiv, and others, were also used for paper search using their portals directly and through the usage of Collabovid dot org.

The current research work is conducted based on the concept of Conceptual Mapping technique conducted in the field of Epidemiological research [69–71].

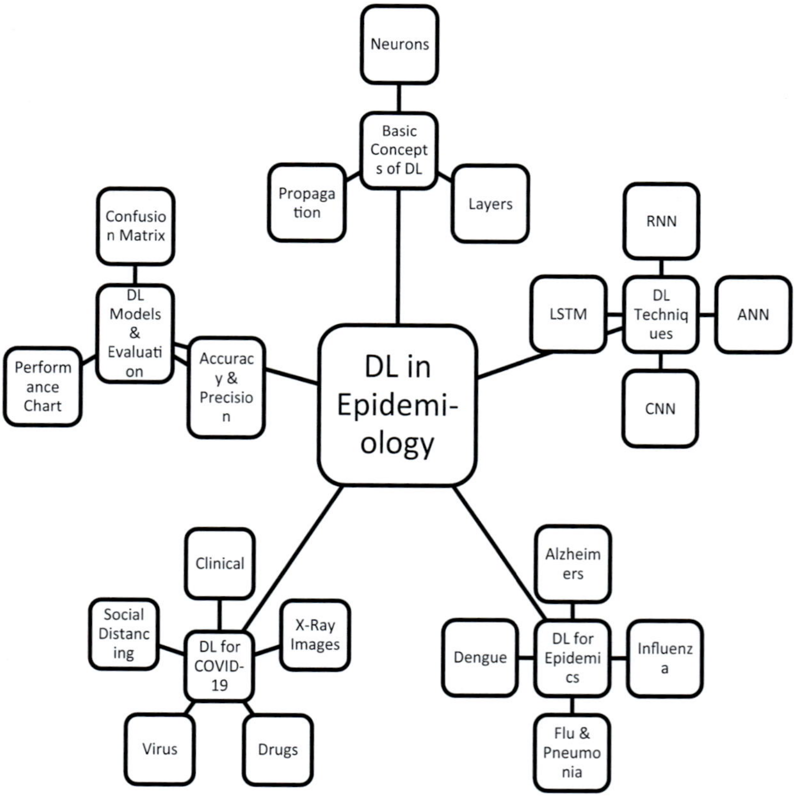

Fig. 4 Conceptual Mapping technique based systematic review of the Deep Learning concepts (Source: Author Created)

Concept Mapping helps in conducting the review based research work and converges various concepts that help in a systematic analysis of the different concepts in the given field. For the current study, Fig. 4 shows the concepts that were mapped around the usage of Deep Learning across the literature. The concept mapping technique has been validate by many researchers in the past for conducting research on newer areas [72, 73] and application of Deep Learning for COVID-19 has been one such emerging area which has been studied using this well established technique.

Based on the concept mapping, the various areas were systematically arranged for the current research work to create a **DL-CoTAUM** framework. The acronyms stands for "**D**eep **L**earning based **Co**ncepts, **T**echniques, **A**pplications, **U**se Cases and **M**odel Evaluation". The five concepts represents five phases in the current research work. Phase 1 covers up the study of basic concepts related to DL, while phase 2 covers up DL techniques prominently used on various pandemics. Phase 3 studies DL's application on multiple diseases, while phase 4 is based on the different use-cases developed under COVID-19 management. Phase 5 considers various evaluation metrics used for analyzing DL models from phase 2 on the applications in phase 3 and phase 4, respectively. The proposed framework, as shown in Fig. 5, is aligned with the Grounded Theory Methods [68] where theoretical concepts of Deep Learning and various techniques are studied first, and then their general applications and COVID-19 related use-cases are reported. The various models studied are evaluated at the end. So the fundamentals of grounded theory validate the proposed research framework, which covers up all major areas.

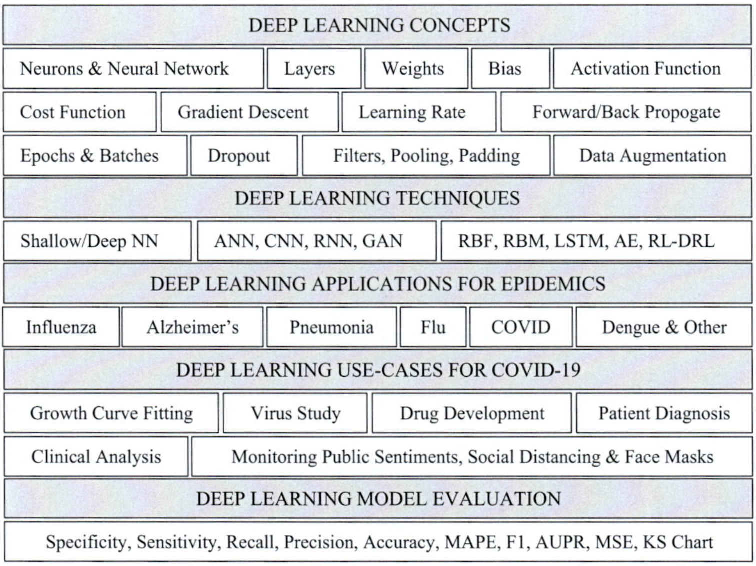

Fig. 5 DL-CoTAUM research framework for the current study (Source: Author Created)

6 Findings

The following are the findings observed from the proposed model.

6.1 Growth Curve Fitting and Trends Forecasting

Forecasting the trends for the number of confirmed cases, deaths, and recovery rate for COVID-19 patients is a crucial task in this pandemic as it can help in planning future actions well in advance. A lot of research has been done to forecast the pandemic trend in different countries through the use of LSTM. Risk of Infection can be predicted based on factors like age, existing medical conditions, number of human interactions using Artificial Neural Network (ANN). According to a study, identifying effective existing drugs by building knowledge graphs and predicting interactions between drugs and viral proteins 45 can be done through DL techniques.

Arora et al. [49] and Shastri et al. [50], in their papers, used different variants of LSTM neural networks, namely, Conv-LSTMs, Stacked-LSTMs, and bi-directional LSTMs models, for forecasting the epidemic in India. Arora et al. carried out state-wise forecasting of the number of new cases per day and also per week in India. The average MAPE values for the three architectures were 5.05, 4.81, and 3.22%, respectively, which shows that bi-directional LSTMs were able to model the situation better. Moreover, on further analysis, the Bidirectional LSTM network gave an error of less than 3% on the daily forecast and 4% to 8% on weekly prediction.

Shastri et al. [50] used the three networks for predicting the number of confirmed and death cases in India and the USA. The summary of the performance on the prediction of patients in India is as shown in Table 1 and Table 2, which show how the convolution variant of LSTM performed better than the others. Chimmula and Zhang [51] used the John Hopkin's dataset and reports from Canadian Health

Table 1 Performance of variants of LSTM on forecasting of confirmed cases in India [50]

Metrics	Models		
	Stacked LSTM	Bi-directional LSTM	Convolutional LSTM
Accuracy	96.00%	96.60%	**97.82%**
MAPE	4.00	3.33	**2.17**

Table 2 Performance of variants of LSTM on forecasting of death cases in India [50]

Metrics	Models		
	Stacked LSTM	Bi-directional LSTM	Convolutional LSTM
Accuracy	93.33%	96.00%	**96.66%**
MAPE	6.66	4.00	**3.33**

Authority for predicting future cases in Canada. LSTM based neural networks were used for the purpose, which achieved an RMSE of 34.83 and an accuracy of 93.4% for short term prediction and an RMSE of 45.70 and accuracy of 92.67% for long term predictions. Thus, LSTMs perform better for short-term forecasting as compared to long-term forecasting.

Dutta and Bandyopadhyay [52] proposed three models – LSTM based RNN, GRU based RNN, and LSTM and GRU combined recurrent neural network – for forecasting positive, negative, death and release cases for the epidemic. On the prediction of the number of deceased, the Accuracy and RMSE of the LSTM model were 59.2% and 4.97; For the GRU model, they were 60.38% and 5.218, and for the LSTM-GRU model, values were 62% and 4.16, respectively. The summary of the results on confirmed cases is shown in Table 3. It can be observed that although GRU performed better with a much lower RMSE, its combination with LSTM drastically improved accuracy, which shows how the amalgamation of more than one type of network can generate more powerful models.

Li et al. [52] proposed Deep learning-enabled approach, which they named ALeRT-COVID, to provide efficient forecasting using transfer learning, RNNs, and attention mechanism techniques. This model also included the lockdown factor in different countries to generate more accurate predictions at various stages of lockdown. The base model for transfer learning was trained on data from countries that were struck by this pandemic quite earlier and thus have accumulated a large amount of data. They also developed some baseline models for comparison – without transfer learning, with transfer learning, with lockdown measure. Table 4 summarizes the performance of the DL variants used in the study in terms of mean MAPE and its standard deviation. It can be concluded that using transfer learning impacts the MAPE tremendously. Also, when the lockdown factor is included along with the attention mechanism, the ALeRT-COVID, thus formed, outperforms baseline models.

Table 3 Comparison of various RNN based models based on their performance on positive cases forecasts [52]

Metric	Models		
	LSTM-RNN	GRU-RNN	LSTM-GRU-RNN
Accuracy	76.60%	76.90%	**87%**
RMSE	53.38	30.95	**30.15**

Table 4 Performance of ALeRT-COVID as compared to other models [52]

Metrics	Baseline models			
	Without transfer learning	With transfer learning	Including lockdown measure	ALeRT-COVID
Mean MAPE	0.100	0.077	0.060	**0.050**
STD	0.087	0.088	0.057	**0.040**

Since countries which have been struck by this pandemic just a while ago do not have sufficient data, transfer learning and attention mechanism can prove to be great tools for accurately predicting the future trends for taking reasonable measures.

6.2 *Study of Virus Characteristics and Drug Development*

One of the reasons why the COVID-19 pandemic became so fatal was the unavailability of effective drugs and vaccines. Thus, the fast development of drugs is necessary so that they can be applied to the patients immediately. Deep learning-based architectures can prove useful in accurately predicting interaction between the target and drug to repurpose the commercially available drugs to work against the COVID-19 virus.

Öztürk et al. [54] proposed a deep learning architecture, called DeepDTA, for estimating Drug-Target Interactions (DTIs). The architecture consisted of multiple convolutional layers and pooling layer for the task and showed promising results on KIBA and DAVIS datasets as compared to the benchmark models. Another Deep Learning approach called MT-DTI was proposed by Shin et al. [56] for the same task. This model used two blocks of neural networks. The first block used SMILES representation of molecules as input and consisted of a Molecular Transformer which contained multiple self-attention and feed-forward layers. The second block was fed with the FASTA sequence of proteins and consisted of convolutional and pooling layers. This model also used transfer learning by initializing the weights from those obtained from a pre-trained model. The comparison of DeepDTA and MT-DTI is as shown in Table 5. It can be observed that although the CNN based DeepDTA showed efficacy, the transformer encoder approach combined with transfer learning improved the results significantly on both the datasets.

Beck et al. [54] used the MT-DTI technique for identifying some commercially available drugs that can be potent against the novel coronavirus. This shows that an in-depth learning approach like this could be useful in predicting drug-target interaction without domain knowledge and thus could speed up the drug-repurposing process for searching suitable medications for COVID-19 patients. Visualization and detection of the novel coronavirus is an essential task for medical purposes. Jamshidi et al. [57] proposed GANs for analyzing the electron microscopy image and the genome sequencing for visualization and detection of the virus. Due to their high

Table 5 Comparison between DeepDTA and MT-DTI [56]

Datasets	Method	CI (std)	MSE	r_m^2(std)	AUPR (std)
Kiba	*DeepDTA*	0.863(0.002)	0.194	0.673(0.009)	0.788(0.004)
	MT-DTI	0.882(0.001)	0.152	0.738(0.006)	0.837(0.003)
Davis	*DeepDTA*	0.878(0.004)	0.261	0.630(0.017)	0.714(0.010)
	MT-DTI	0.887(0.003)	0.245	0.665(0.014)	0.730(0.014)

discriminating power, GANs can tackle these tasks by modeling the latent distribution of the data, which cannot be observed otherwise.

6.3 Patient Diagnosis

COVID-19 has proved to be one of the most complicated illnesses which can emerge in numerous forms showing a wide range of symptoms. Since most of this data, especially the diagnostics data, is unstructured, using Deep learning for gaining insights from these large datasets is a potential approach for finding more about the novel coronavirus [57]. Using scans of the face for detecting symptoms like fever has been conducted through DL application.

Understanding and estimating the extent of cardiac involvement is also a vital issue in handling the COVID-19 virus. As highlighted by Inciardi et al. [57], myocarditis might be a complication in patients with coronavirus infection. Hence, there is a need to recognize the association between COVID-19 and acute myocarditis to monitor the affected patients in a focused manner. Based on this, an LSTM network was proposed by Jamshidi et al. [57], which takes in the features extracted from MRI, cardiac laboratory results, and electrocardiography results as inputs and outputs the estimated cardiac complications caused by the virus. LSTM networks are proposed because of their ability to capture the dynamic temporal behaviour of the data.

Since COVID-19 is a highly infectious virus, it is imperative that modeling of the virus for prediction of their impact be done very accurately without overfitting or under-fitting. Zlatan car et al. used a publicly available dataset containing information on infected, recovered, and deceased patients in 406 locations over 51 days. The dataset was transformed into a regression dataset first and then trained using Multilayer Perceptron (MLP)-ANN [23]. The hyper-parameters were varied using grid search algorithms and evaluated using the Coefficient of Determination (R-Squared). Using the K-Fold algorithm, cross-validation was performed with 5 folds [24]. ReLU activation function was used. This whole combination gave the best R^2 score of 0.94 after performing cross-validation.

Ardakani et al. [48] used Transfer Learning to distinguish between COVID-19 patients from non-COVID-19 patients. Out of AlexNet, VGG-16, VGG-19, SqueezeNet, GoogleNet, MobileNet v2, ResNet-18, ResNet-101, and Xception, ResNet-101 gave the highest accuracy, sensitivity, and AUC score. Thus, various deep learning techniques have a great potential in the diagnostics of COVID-19 and could prove to be useful in increasing the rate at which the virus is detected and also in making the treatment of COVID-19 patients much more effective.

Xiao et al. [59] discussed the transmission of the new virus via the faecal-oral route by providing evidence of gastrointestinal infection of the virus. Inspired by this idea, Jamshidi et al. [57] proposed a GAN network for the prediction of the probability of viral gastrointestinal infection by extracting features from images focusing on target organs like the stomach, rectum, etc. This probability, along with the RT-PCR testing results, can be used for determining and establishing norms for patients.

6.4 Clinical Analysis of Patients Using X-Rays and Scans

According to the current research study of more than 60 papers, it was found that the amalgamation of Deep Learning and radiology imaging can be very helpful for the accurate diagnosis of the disease. This will be of great importance in the places where there is a lack of specialized physicians in remote locations. X-rays & Scans for the lungs' image analysis can be done using a Deep Neural Network-based method nCOVnet [42]. When COVID-19 enters the human respiratory tract, the lungs of a person are affected the most. A fluid gets filled in the lungs, making it inflamed while developing patches called 'Ground-Glass Opacity' [44]. So many features are hard to recognize manually. Training of these images using CNN is difficult, but it will be easier for medical professionals to read the chest X-rays for the patients. It can also be done using transfer learning techniques like VGG-16 and ResNet. Transfer Learning has proved to be a useful technique as per the analysis of many research papers for predicting the number of positive cases, especially in the early stage of the pandemic, when there is not much data available.

Among the many applications of Deep Learning for COVID-19 found through the current review study, the most prominent & common application of DL has been for image processing. A Technology company in China (Deepwin & League of PhD Technology Co. Ltd.) trained and validated 19,291 CT Scans from seven hospitals in China with some criteria 46. Then, clinical doctors recorded the features like Ground-Glass Opacity (GGO), pulmonary consolidation, crazy paving pattern, distribution of lesions, and thoracic lymphadenopathy [47]. There are also a lot of papers available on scanning the X-Ray images for prediction.

Effective, accurate, and efficient screening of patients is a critical factor in dealing with this fast-spreading virus. The most prominent and reliable method used for COVID-19 screening is the RT-PCR testing and is considered a gold standard for its high specificity. But, this approach is slow and laborious and has a complicated manual process which restricts the efficiency of the screening process. Also, the sensitivity of this method is highly variable. Thus, there is a need for faster alternatives that can work alongside RT-PCR and speed up the process. Since the symptoms of infection with this new virus overlap with many other diseases like pneumonia, patients should be identified effectively so that the limited resources are preserved for those affected by the pandemic [60].

Wang et al. [60] proposed a deep learning model, named COVID-Net, which was developed specifically for the detection of SARS-CoV-2 infection from Chest X-ray scans using the COVIDx dataset. The COVID-Net's architecture has been designed in a way such that the sensitivity and COVID-19 positive predictive value are at least 80% so that the number of missed cases is reduced and false-positive cases are limited to avoid overburdening hospitals. Table 6 and Table 7 show that COVID-Net has lower architectural complexity and higher accuracy as compared to VGG-19 and ResNet-50. Moreover, the higher sensitivity and positive predictive value in comparison to other models highlight the fact that COVID-Net has a lesser chance of missing positive cases and also predicting false-positive cases. It was also

Table 6 Comparison of deep CNN architectures in terms of complexity, accuracy and sensitivity [60]

Architecture	Parameters (M)	Accuracy (%)	Sensitivity (%)		
			Normal	Non-COVID19	COVID-19
VGG-19	20.37	83.0	98.0	90.0	58.7
ResNet-50	24.97	90.6	97.0	92.0	83.0
COVID-Net	11.75	93.3	95.0	94.0	91.0

Table 7 Comparison of CNN architectures with respect to sensitivity [60]

Architecture	Positive Predictive Value (%)		
	Normal	**Non-COVID19**	**COVID-19**
VGG-19	98.0	90.0	58.7
ResNet-50	97.0	92.0	83.0
COVID-Net	95.0	94.0	91.0

observed that having more complex architectures may not always be fruitful for the task at hand, and experimenting with selective skip connections and mixing different approaches is an excellent approach to get a more robust model.

An end-to-end deep learning approach was proposed by Wang et al. [61] for the prognosis and diagnosis of COVID-19 using CT images. The system consisted of two neural network architectures, namely, a DenseNet121-FPN and a novel dense-net like COVID-19Net. The former segmented lungs from CT images and the latter was responsible for diagnosis and prognosis. The model achieved an average accuracy of 79.21%, F1-score of 0.7947, and AUC of 0.8749. Also, the predictive analysis of DL features showed that there was a considerable difference in the hospital stay time of low-risk and high-risk patients. Hence, deep learning features have value concerning the prognosis of COVID-19.

DeepPneumonia is a deep learning-based model for detecting pneumonia that can lead to COVID-19 using CT images. Proposed by Song et al. [61], this model consisted of three stages – extraction of central regions of the lungs, followed by obtaining the most relevant details from the images and the predictions at the image level, and lastly, aggregation of image-level predictions to get the patient-level diagnosis. The model proposed by them, DRE-Net, was built on top of a pre-trained ResNet-50 and also used an attention mechanism to give importance to the details in the image. The performance of the model on the prediction of ordinary versus COVID-19 infected is described in Table 8.

Though the two approaches discussed above were trained and tested on different datasets, both of them used deep convolutional models and applied transfer learning on them. Moreover, COVID-19Net was significantly different from DRE-Net as the latter was built on top of ResNet-50, which uses the concept of skip-connections using residual units that make profound neural networks learn better.

Table 8 Performance of the DRE-Net model compared to other prominent deep CNN [61]

Model	Accuracy (%)	F1-Score	AUC
VGG16	90	0.91	0.98
DenseNet	92	0.92	0.97
ResNet	92	0.92	0.99
DRENet	**94**	**0.92**	**0.99**

Table 9 Performance of the HBM classifier with Facebook comments (Source: [64])

Health belief model construct	Performance of GRU-RNN		
	Specificity (%)	Sensitivity (%)	Balanced accuracy (%)
Perceived susceptibility	99.7	79.6	89.6
Perceived severity	98.8	94.3	96.5
Perceived benefit	96.5	90.9	93.7
Perceived barrier	99.0	81.5	90.3

6.5 Monitoring Public Sentiment, Social Distancing, and Mask Usage

In this worldwide pandemic, restricting the spread of the virus by limiting social contact and taking preventive measures has been a significant focus of decision-makers worldwide. Many studies have been focussing on monitoring social distancing and other preventive measures during the outbreak.

Sree [63] proposed a deep learning-based mechanism that alerts the user when other users are within two meters distance from them, thus ensuring social distancing. A system that would automatically process CCTV footage, monitor proper adherence of social distancing practices, and alert the administrator when social distancing norms are violated was also proposed in this paper. An accuracy of 79.3% was achieved using this mechanism, which shows the potential of neural network models for such surveillance tasks and can be improved using more complex architectures and algorithms.

A deep learning approach based on Health Belief Model (HBM) has been proposed by Raamkumar et al. [64], which describes the importance of analyzing social media responses to various activities and initiatives taken by public health authorities. Deep learning-based sentiment analysis of the Facebook comments in response to posts related to COVID-19, social distancing, and other measures taken by the government was done. The text classification model was built using a GRU-based RNN network and achieved balanced accuracy of more than 89%, and specificity of more than 96% for all four HBM constructs, as shown in Table 9 [64].

Table 10 Comparison of using the RetinaFaceMask with MobileNet and Resnet as backbone 25

Model	Face		Mask	
	Precision	Recall	Precision	Recall
RetinaFaceMask + MobileNet	83.0%	95.6%	82.3%	89.1%
RetinaFaceMask + ResNet	91.9%	96.3%	93.4%	94.51%

RetinaFaceMask, proposed by Jiang and Fan [25], is a deep neural network-based system that can detect masks on faces. The model uses an attention mechanism for focusing on essential features and transfer learning for accurate predictions. The author used two base models – MobileNet and ResNet – as backbones for their model and used them separately. The comparison between the two models is summarized in Table 10. The RetinaFaceMask was able to detect faces with and without masks correctly. The model also accurately predicted many tricky images with multiple faces of both categories in the same image and other confusing images.

Thus, deep learning can be applied in numerous ways and multidisciplinary areas for the monitoring, analysis, and management of social distancing and other preventive measures during the COVID-19 pandemic and also for future use-cases.

7 Discussion and Conclusion

This section gives description about the observations, limitations of the study and the conclusions drawn from the proposed system.

7.1 Research Framework Analysis

The research framework developed in this study (shown in Fig. 5) using the concept mapping technique (as shown in Fig. 4) has been implemented through the current study. Section two of this research represents basic concepts of Deep Learning (DL) as discussed in the first phase of the framework. Section 3 represents various techniques of DL as shown in the second phase of the framework. Section four shows the application of Deep Learning in various types of epidemics as evident in the 3rd part of the framework. Section 6 covers various models under different use-cases of DL for COVID-19 and it also presents comparative analysis of different DL modeling techniques on parameters like Accuracy, Recall and Precision, as depicted in last two portions of the research framework. The DL-CoTAUM framework as shown in Fig. 5 can also be used for study of the usage of other emerging technologies in healthcare. For instance, for studying the Computer Vision (CV) usage for COVID-19 management, CV-CoTAUM can be created through systematic literature review using Concept Mapping technique as done in the current study. Basic concepts of CV can be studied first followed by various techniques associated with it. Then its application and use cases can be studied on the epidemic under study. Finally, the comparative analysis of various techniques can be performed from existing literature. Thus, this review framework can be generalized to XX-CoTAUM, where XX represents the technology under study, which would be helpful for researchers to develop and present existing work efficiently.

7.2 Findings Discussion

The use of Deep Learning techniques has been done in numerous ways in the past, and they have a lot of potential in solving many pandemics related problems that can affect the lives of millions. This study discussed how different neural network architectures and deep learning techniques could be used to deal with various aspects of the new coronavirus situation. The current research work analyzed and compared various novel architectures and models that have been proposed for various use-cases.

Based on the findings of this study, one can observe the prominence and effectiveness of LSTM based RNNs and GRU based RNNs in dealing with time-series data. Moreover, when these models are combined together, the forecasting power of the model increases significantly. Different variants of LSTMs like stacked, bidirectional, or convolutional LSTMs can also help to improve the performance of the model. One observable limitation of such LSTM networks is that they can perform well for short term forecasting but not for long term forecasting. However, LSTM and GRU based models have been used earlier for forecasting influenza spread and thus are a clear candidate for accurately predicting positive cases in this global pandemic. The GRU-RNN was successful in the sentiment analysis of social media posts and was able to correctly classify the health belief model constructs with a significantly high accuracy proving that sentiment analysis using GRUs and other sequence-based neural networks can be used for effectively predicting the response of the population on various measures taken by authorities.

Due to their ability to learn long-range relationships, transformer-based models were more successful in predicting drug-target interaction as compared to CNN based model. These transformer-based models can be used for the identification of existing drugs that can be repurposed for fighting the new SARS-CoV-19. Various COVID-19 diagnostic techniques using GANs were also discussed. Due to their high discriminating power, GANs can be used to model situations where the difference is not observable.

CNN-based deep networks have been used prominently for image data and stood up to their reputation by performing remarkably well on radiographs such as chest X-rays and CT-scans for the diagnosis of the COVID-19 virus. Though the new epidemic causing virus has symptoms very close to pneumonia, the deep CNN models with a wide variety of layers were able to distinguish them with decent accuracy. GANs were observed to be able to effectively detect gastrointestinal infection from images of various organs showing their possible usage for radiographs as well. Deep CNN based networks also demonstrated efficacy in facial mask recognition and social distancing surveillance with real-time object detection.

One common technique used by several models discussed in the study was transfer learning. Due to a lack of data sources in the early stages of a pandemic and high data appetite of neural networks, DL based models do not perform well. The transfer learning approaches used in various studies show how it can be used to create better predictive models even with insufficient data.

The attention mechanism was also used in various approaches to focus on specific vital features of input data to increase the performance of data. This reflects that it can be used in combination with major deep learning architectures to enhance their efficacy. The usage of reinforcement learning has been successful in recent years, and still, there are a minimal number of studies that have applied reinforcement learning for dealing with the current outbreak. The application of deep reinforcement learning in this challenging epidemic situation might prove useful and thus is a potential future application of deep learning in this area.

7.3 Limitations of Deep Learning Techniques

Deep Learning is not always the best technique for modeling various problems in epidemiological modeling or pandemic management. It has a certain number of limitations, which restricts its usage.

Overfitting: It refers to the situation in which the neural network can fit the training set entirely or with high accuracy but is unable to generalize well on unseen data. In more formal words, when the trained model has a high variance and a low bias, then it is called overfitting. This situation arises mainly due to a lack of data samples and is a common problem encountered when dealing with neural networks. When a highly complex model with a large number of parameters is trained on a relatively smaller dataset, then the neural network tends to fit the noise present in the data leading to low generalization power. Therefore, carefully designed neural networks are essential to reduce the harmful effects of overfitting. Some of the techniques used to reduce overfitting in neural networks are regularization, early stopping, and ensemble methods like dropouts [4].

Vanishing and Exploding Gradient Problem: Backpropagation in deep neural network systems face an issue concerning the stability of updates. In some neural network architectures, due to the chain-like product computed during the backpropagation step, the updates performed in the layers at shallow depth may either become infinitesimally small, called the vanishing gradients, or exceptionally large, called the exploding gradients. This problem makes the learning process highly unstable in very deep neural networks. ReLU activation, adaptive learning rates, batch normalization and conjugate gradient descent methods are used to overcome this problem [4].

Convergence Problem: In very deep networks, it is difficult to make the learning process reach convergence properly in a reasonable amount of time. In deep networks, there is increased resistance to the training process and thus restrict the smooth flow of gradients through the network. One of the methods to solve the convergence problem is to use residual networks and gating networks [4].

Local Optima: When there are a large number of parameters to learn, a neural network is prone to reach a local optimum and get stuck there when it should instead converge to a global optimum. Since, in most deep learning models, the parameter space is vast, choosing a correct initialization technique, instead of randomizing the parameters, might help in avoiding the problem of local optima [4].

Computational Inefficiency and Resource Requirement: Deep learning algorithms, especially with very deep architectures, are very needy in terms of computational resources and demand very high-end computing systems for fast training. Even with GPUs, more complex models still take a lot of time to train, and thus, computational inefficiency remains to be one of the most pressing issues in the deep learning field. Though the runtime of most neural networks at the prediction phase is significantly low, very deep neural networks may still take some time to output the result, which might pose problems in real-time applications. Thus, deep learning models need to be carefully designed to tackle such time constraints [4].

7.4 Limitations of the Study

Although the study was scientifically carried out in a procedural manner, there are some limitations of this study.

Limited Data Coverage: Due to time constraints, only a limited number of papers (70–80) related to the COVID-19 were incorporated in the study. Thus, some aspects of the topic may have been left out due to limited time and resources.

Theoretical Discussion of the Techniques: This study focused mainly on the theoretical aspects of deep learning and discussed the applications of deep learning techniques in theory. This study did not cover the implementational aspects of the DL techniques discussed above through primary data.

Validation of Techniques: Although this study analysed the performance of several techniques and methodologies, this research relied mainly on the data provided by the existing papers and did not check the validity of methods on more datasets through the self-implementation of the models.

Inclusion of Papers from Preprint Archives: Along with published works, this study also included results from some unpublished papers due to the non-availability of much content outside preprint archives. Nevertheless, proper care has been taken in order to ensure that published sources get more importance as compared to the unpublished ones.

7.5 Future Direction

Re-use of the Proposed Research Framework: The proposed research framework can also be used by other researchers to study the application of Deep Learning or other technologies for existing pandemics and other diseases/ epidemics.

Verification of DL techniques on different datasets: The deep learning techniques mentioned in this study could be implemented and validated on different datasets.

Real-world Implementation: Real-time applications of the aforementioned DL techniques can also be built for testing on real-world medical scenarios which can then be deployed to be used in hospitals and medical facilities for tackling the current and future outbreaks.

7.6 Conclusion

This study shows how deep learning can be applied to tackle COVID-19 related issues. Different deep learning architectures like CNN, RNN, LSTMs, and more complex architectures have been discussed with the aim of explaining their usage in various medical and forecasting applications. The study examined how patient diagnosis, clinical analysis, and growth forecasting could be made using Deep Learning approaches for managing COVID-19 cases. It was also discussed the working and performance of deep learning models as diagnostic and prognostic tools for COVID-19 and other diseases. This in-depth analysis of numerous use-cases of deep learning in multidisciplinary fields highlights the importance and potential of this technology in a global crisis.

This study presented a Concept Mapping Technique based Research Framework to review the usage of Deep Learning Techniques for management of COVID-19 pandemic, which consisted of multiple phases. Through the first phase, different concepts around Deep Learning were presented. Multiple deep learning techniques using the basic concepts were studied through phase two. In the third phase, various epidemics from the past were learned from the Deep Learning point of view. In the fourth phase, multiple use-cases for COVID-19 management were covered up, as discussed in the findings. The fifth phase presented various evaluation parameters available for analyzing the performance of the modeling techniques under Deep Learning. In this review study it was found that Deep Learning is gaining importance for various applications under COVID-19 management. Accuracies of more than 90% have been reported for Deep Learning techniques in COVID-19 use-cases, which is higher than the other baseline methods. This study will be useful for researchers, academicians, and practitioners in the area of Epidemiological Modeling and Image Processing. The proposed research framework can be used as a reference point by other researchers for studying the application of a new concept in a different domain. And this research study will also be useful for naïve users in the field of Deep Learning.

References

1. S. Pouyanfar, S. Sadiq, Y. Yan, H. Tian, Y. Tao, M.P. Reyes, M.-L. Shyu, S.-C. Chen, S.S. Iyengar, A survey on deep learning: algorithms, techniques, and applications. ACM Comput. Surv. (CSUR) **51**(5), 1–36 (2018)
2. M. M. Najafabadi, F. Villanustre, T. M. Khoshgoftaar, N. Seliya, R. Wald, E. Muharemagic, Deep learning applications and challenges in big data analytics. J. Big Data **2**(1), 1 (2015)
3. Y. Bengio, Y. LeCun, Scaling learning algorithms towards AI. Large-scale Kernel Mach **34**(5), 1–41 (2007)
4. C. C. Aggarwal. Neural Networks and Deep Learning. (Springer) (2018). https://doi.org/10.5555/3283445 on 5th August 2020
5. I. Goodfellow, Y. Bengio, A. Courville, Y. Bengio (2016). *Deep learning* (vol. 1, no. 2). MIT Press, Cambridge. https://synapse.koreamed.org/upload/SynapseData/PDFData/1088HIR/hir-22-351.pdf on 15th August 2020
6. P. Sibi, S.A. Jones, P. Siddarth, Analysis of different activation functions using back propagation neural networks. J. Theoretical Appl. Inf. Technol. **47**(3), 1264–1268 (2013)
7. S. Sharma, Activation functions in neural networks. Int. J. Eng. Appl. Sci. Technol. **4**(12), 310–316 (2017)
8. S. Skansi, Modifications and extensions to a feed-forward neural network, in *Introduction to Deep Learning* (Springer, Cham 2018), pp. 107–120
9. C. Zhang, P. Li, G. Sun, Y. Guan, B. Xiao, J. Cong, Optimizing fpga-based accelerator design for deep convolutional neural networks, in *Proceedings of the 2015 ACM/SIGDA international symposium on field-programmable gate arrays*, pp. 161–170. (2015, February)
10. N. Aloysius, M. Geetha, A review on deep convolutional neural networks, in *2017 international conference on communication and signal processing (ICCSP)* (IEEE, 2017, April), pp. 0588–0592
11. C. Shorten, T.M. Khoshgoftaar, A survey on image data augmentation for deep learning. J. Big Data **6**(1), 60 (2019)
12. M. Soltanolkotabi, A. Javanmard, J.D. Lee, Theoretical insights into the optimization landscape of over-parameterized shallow neural networks. IEEE Trans. Inf. Theory **65**(2), 742–769 (2018)
13. D.A. Winkler, T.C. Le, Performance of deep and shallow neural networks, the universal approximation theorem, activity cliffs, and QSAR. Mol. Inf. 36(1–2) (2017)
14. C. Nicholson. A Beginner's Guide to Neural Networks and Deep Learning. Wiki (2019). Accessed from https://skymind.ai/wiki/neural-network on 10th August 2020
15. D.X. Zhou, Universality of deep convolutional neural networks. Appl. Comput. Harmonic Anal. **48**(2), 787–794 (2020)
16. X. Dong, J. Wu, L. Zhou, How deep learning works--The geometry of deep learning. arXiv preprint. arXiv:1710.10784 (2017)
17. D. S. Broomhead, D. Lowe, Radial basis functions, multi-variable functional interpolation and adaptive networks (No. RSRE-MEMO-4148). Royal Signals and Radar Establishment Malvern (United Kingdom). https://apps.dtic.mil/dtic/tr/fulltext/u2/a196234.pdf on 3rd August 2020 (1988)
18. M.J. Er, S. Wu, J. Lu, H.L. Toh, Face recognition with radial basis function (RBF) neural networks. IEEE Trans. Neural Networks **13**(3), 697–710 (2002)
19. R. Zemouri, J.M. Faure, Comparative study between the timed automata and the recurrent radial basis function for discrete event system diagnosis. IFAC Proc. vol. **39**(13), 1455–1460 (2006)
20. A. Alexandridis, H. Sarimveis, H, Control of processes with multiple steady states using MPC and RBF neural networks. in *Computer Aided Chemical Engineering* (vol. 29, Elsevier, 2011), pp. 698–702
21. A. Fischer, C. Igel, An introduction to restricted Boltzmann machines, in *Iberoamerican congress on pattern recognition* (Springer, Berlin, Heidelberg, September 2012), pp. 14–36

22. A. Géron, *Hands-on machine learning with Scikit-Learn, Keras, and TensorFlow: Concepts, tools, and techniques to build intelligent systems* (2019). O'Reilly Media, Canada. https://books.google.co.in/books?hl=en&lr=&id=HHetDwAAQBAJ&oi=fnd&pg=PP1 on 12th August 2020
23. Z. Car, S. Baressi Šegota, N. Anđelić, I. Lorencin, V. Mrzljak, Modeling the spread of COVID-19 infection using a multilayer perceptron. *Computational and Mathematical Methods in Medicine*, Vol. 2020, Article ID 5714714, 10 pages (2020). https://doi.org/10.1155/2020/5714714
24. O. Maimon, L. Rokach,. Introduction to knowledge discovery and data mining, in *Data mining and knowledge discovery handbook* (Springer, Boston, MA), pp. 1–15
25. M. Jiang, X. Fan, RetinaMask: A Face Mask detector. arXiv preprint arXiv:2005.03950 (2020)
26. V. Mnih, K. Kavukcuoglu, D. Silver, A. Graves, I. Antonoglou, D. Wierstra, M. Riedmille, Playing atari with deep reinforcement learning. arXiv preprint arXiv:1312.5602 (2013)
27. B. Adhikari, X. Xu, N. Ramakrishnan, B. A. Prakash, Epideep: Exploiting embeddings for epidemic forecasting, in *Proceedings of the 25th ACM SIGKDD international conference on knowledge discovery & data mining* (pp. 577–586) (July 2019)
28. S.R. Venna, A. Tavanaei, R.N. Gottumukkala, V.V. Raghavan, A.S. Maida, S. Nichols, A novel data-driven model for real-time influenza forecasting. IEEE Access **7**, 7691–7701 (2018)
29. S. Liu, S. Liu, W. Cai, S. Pujol, R. Kikinis, D. Feng (April 2014). Early diagnosis of Alzheimer's disease with deep learning. In *2014 IEEE 11th international symposium on biomedical imaging (ISBI)* (IEEE), pp. 1015–1018
30. H. Ji, Z. Liu, W. Q., Yan, R. Klette, Early diagnosis of Alzheimer's disease using deep learning, in *Proceedings of the 2nd International Conference on Control and Computer Vision* (pp. 87–91) (2019, June)
31. P. Rajpurkar, J. Irvin, K. Zhu, B. Yang, H. Mehta, T. Duan, D. Ding, A. Bagul, C. Langlotz, K. Shpanskaya, M.P. Lungren, Chexnet: Radiologist-level pneumonia detection on chest x-rays with deep learning. arXiv preprint arXiv:1711.05225 (2017)
32. K. Hammoudi, H. Benhabiles, M. Melkemi, F. Dornaika, I. Arganda-Carreras, D. Collard, A. Scherpereel, Deep Learning on Chest X-ray Images to Detect and Evaluate Pneumonia Cases at the Era of COVID-19. arXiv preprint arXiv:2004.03399 (2020)
33. A. Influenza, National and Regional Summary of Select Surveillance Components. Accessed from https://www.izsummitpartners.org/content/uploads/2015/10/CDC-Flu-View-10-10-2015.pdf on 15th August 2020
34. L. Chen, K.T. Hossain, P. Butler, N. Ramakrishnan, B.A. Prakash, Flu gone viral: Syndromic surveillance of flu on Twitter using temporal topic models, in *2014 IEEE international conference on data mining* (IEEE, December 2014), pp. 755–760
35. M.J. Paul, M. Dredze, A model for mining public health topics from Twitter. Health **11**(1), 1–7 (2012)
36. H. Achrekar, A. Gandhe, R. Lazarus, S. H. Yu, B. Liu (2011 April) Predicting flu trends using twitter data, in *2011 IEEE conference on computer communications workshops (INFOCOM WKSHPS)* (pp. 702–707). IEEE.
37. L. Zhao, J. Chen, F. Chen, W. Wang, C. T. Lu, N. Ramakrishnan, Simnest: Social media nested epidemic simulation via online semi-supervised deep learning, in *2015 IEEE international conference on data mining* (2015, November. IEEE), pp. 639–648
38. Y. Zhang, G. Milinovich, Z. Xu, H. Bambrick, K. Mengersen, S. Tong, W. Hu, Monitoring pertussis infections using internet search queries. Sci. Rep. **7**(1), 1–7 (2017)
39. F. Rohart, G. J. Milinovich, S. M. Avril, K. A. Lê Cao, S Tong, W. Hu. Disease surveillance based on Internet-based linear models: an Australian case study of previously unmodeled infection diseases. Sci. Rep. **6**, 38522 (2016)
40. Y. Teng, D, Bi, G. Xie, Y. Jin, Y. Huang, B. Lin, X. An, D. Feng, Y. Tong, Dynamic forecasting of Zika epidemics using Google trends. PloS One **12**(1), e0165085 (2017)
41. Liu, T., Zhang, Y., Lin, H., Lv, X., Xiao, J., W. Zeng, Y. Gu, S. Rutherford, S. Tong, W. Ma, A large temperature fluctuation may trigger an epidemic erythromelalgia outbreak in China. Sci. Rep. 5(1), 1–8 2015)

42. B.H. Noden, M.D. Kent, J.C. Beier, The impact of variations in temperature on early Plasmodium falciparum development in Anopheles stephensi. Parasitology **111**(5), 539–545 (1995)
43. H. Shi, X. Han, N. Jiang, Y. Cao, O. Alwalid, J. Gu, Y. Fan, C. Zheng, Radiological findings from 81 patients with COVID-19 pneumonia in Wuhan, China: a descriptive study. Lancet Infectious Diseases **20**(4), 425–434 (2020). https://doi.org/10.1016/S1473-3099(20)30086-4
44. Ng, M. Y., Lee, E. Y., Yang, J., Yang, F., Li, X., Wang, H., M.M. Lui, C.S.-Y. Lo, B.L., P.-L. Khong, C.K.-M. Hui, K. Yuen, C.K. M. Hui, Imaging profile of the COVID-19 infection: radiologic findings and literature review. Radiol. Cardiothoracic Imaging **2**(1), e200034 (2020)
45. T. Ozturk, M. Talo, E.A.Yildirim, U.B. Baloglu, O. Yildirim, U.R. Acharya, Automated detection of COVID-19 cases using deep neural networks with X-ray images. Computers in Biology and Medicine, 103792 (2020)
46. E.Y. Lee, M.Y. Ng, P.L. Khong, COVID-19 pneumonia: what has CT taught us? Lancet. Infect. Dis **20**(4), 384–385 (2020)
47. L. Deng, D. Yu, Deep learning: methods and applications. Found. Trends Signal Process. **7**(3–4), 197–387 (2014)
48. A.A. Ardakani, A.R. Kanafi, U.R. Acharya, N. Khadem, A. Mohammadi, Application of deep learning technique to manage COVID-19 in routine clinical practice using CT images: results of 10 convolutional neural networks. Comput. Biol. Med. **121**, 103795 (2020)
49. P. Arora, H. Kumar, B.K. Panigrahi, Prediction and analysis of COVID-19 positive cases using deep learning models: a descriptive case study of India. Chaos, Solitons Fractals **139**, 110017 (2020)
50. S. Shastri, K. Singh, S. Kumar, P. Kour, V. Mansotra, Time series forecasting of Covid-19 using deep learning models: India-USA comparative case study. Chaos, Solitons Fractals **140**, 110227 (2020)
51. V.K.R. Chimmula, L. Zhang, Time series forecasting of COVID-19 transmission in Canada using LSTM networks. Chaos, Solitons Fractals **135**, 109864 (2020)
52. S.K. Bandyopadhyay, S. Dutta, Machine learning approach for confirmation of covid-19 cases: Positive, negative, death and release. *medRxiv* (2020). Accessed from on 7th August 2020.
53. Y. Li, W. Jia, J. Wang, J. Guo, Q. Liu, X. Li, G., Xie, F. Wang, ALeRT-COVID: Attentive Lockdown-awaRe Transfer Learning for Predicting COVID-19 Pandemics in Different Countries. medRxiv. Accessed from https://doi.org/10.1101/2020.07.09.20149831 on 8th August 2020 (2020)
54. H. Öztürk, A. Özgür, E. Ozkirimli, DeepDTA: deep drug–target binding affinity prediction. Bioinformatics **34**(17), i821–i829 (2018)
55. B.R. Beck, B. Shin, Y. Choi, S. Park, K. Kang, Predicting commercially available antiviral drugs that may act on the novel coronavirus (SARS-CoV-2) through a drug-target interaction deep learning model. Comput. Struct. Biotechnol. J. **18**, 784–790 (2020). https://doi.org/10.1016/j.csbj.2020.03.025
56. B. Shin, S. Park, K. Kang, J. C. Ho, Self-attention based molecule representation for predicting drug-target interaction. arXiv preprint arXiv:1908.06760 (2019)
57. M. Jamshidi, A. Lalbakhsh, J. Talla, Z. Peroutka, F. Hadjilooei, P. Lalbakhsh, M. Jamshidi, L.L. Spada, M. Mirmozafari, M. Dehghani, A. Sabet, Artificial Intelligence and COVID-19: Deep Learning Approaches for Diagnosis and Treatment. IEEE Access **8**, 109581–109595 (2020)
58. R.M. Inciardi, L. Lupi, G. Zaccone, L. Italia, M. Raffo, D. Tomasoni, D.S Cani, M. Cerini, D. Farina, E. Gavazzi, R. Maroldi, M. Adamo, E. Ammirati, G. Sinagra, C.M. Lombardi, M. Metra, R. Maroldi, Cardiac involvement in a patient with coronavirus disease 2019 (COVID-19). JAMA Cardiology **5**(7), 819-824 (2020). https://doi.org/10.1001/jamacardio.2020.1096
59. F. Xiao, M. Tang, X. Zheng, Y. Liu, X. Li, H. Shan, Evidence for gastrointestinal infection of SARS-CoV-2. Gastroenterology **158**(6), 1831–1833 (2020)
60. L. Wang, A. Wong, COVID-Net: A Tailored Deep Convolutional Neural Network Design for Detection of COVID-19 Cases from Chest X-Ray Images. arXiv preprint arXiv:2003.09871 (2020)

61. Y. Song, S. Zheng, L. Li, X. Zhang, X. Zhang, Z. Huang, J. Chen, R. Wang, H. Zhao, Y. Zha, J. Shen, Y. Chong, Deep learning enables accurate diagnosis of novel coronavirus (COVID-19) with CT images. medRxiv. Accessed from https://doi.org/10.1101/2020.02.23.20026930 on 9th August 2020 (2020)
62. S. Wang, Y. Zha, W. Li, Q. Wu, X. Li, M. Niu, M. Wang, X. Qiu, H. Li, H. Yu, W. Gong, A fully automatic deep learning system for COVID-19 diagnostic and prognostic analysis. Eur. Respiratory J. In press. https://doi.org/10.1183/13993003.00775-2020 on 10th August 2020 (2020)
63. P.K. Sree, Automatic social distancing mechanism for preventing Covid-19 using deep learning. Int. J. Recent Dev. Comput. Technol. Softw. Appl. **4**(1), 6–9 (2020)
64. A.S. Raamkumar, S.G. Tan, H.L. Wee, Use of health belief model-based deep learning classifiers for COVID-19 social media content to examine public perceptions of physical distancing: model development and case study. JMIR Public Health Surveill. **6**(3), e20493 (2020)
65. F. Chollet, *Deep learning with python*, vol. 1. Greenwich, CT: Manning Publications CO. https://faculty.neu.edu.cn/yury/AAI/Textbook/Deep%20Learning%20with%20Python.pdf on 5th August 2020 (2017)
66. Z. Y. Khan, Z. Niu, S. Sandiwarno, R. Prince, Deep learning techniques for rating prediction: a survey of the state-of-the-art. Artif. Intell. Rev., 1–41 (2020). https://doi.org/10.1007/s10462-020-09892-9
67. D. Silver, A. Huang, C. J. Maddison, A. Guez, L. Sifre, G. Van Den Driessche, J. Schrittwieser, I. Antonoglou, V. Panneershelvam, M. Lanctot, S. Dieleman, Mastering the game of Go with deep neural networks and tree search. Nature **529**(7587), 484–489 (2016). https://doi.org/10.1038/nature16961
68. C.S.O. Lo, Enhancing groundedness in realist grounded theory research. Qual. Psychol. **1**(1), 61–76 (2014)
69. T. Hayes, N. Hudek, I.D. Graham, D. Coyle, J.C. Brehaut, When piloting health services interventions, what predicts real world behaviours? a systematic concept mapping review. BMC Med. Res. Methodol. **20**, 1–20 (2020)
70. J.G. Burke, P. O'Campo, G.L. Peak, A.C. Gielen, K.A. McDonnell, W.M. Trochim, An introduction to concept mapping as a participatory public health research method. Qual. Health Res. **15**(10), 1392–1410 (2005)
71. S.C. Conceição, A. Samuel, S.M.Y. Biniecki, Using concept mapping as a tool for conducting research: an analysis of three approaches. Cogent Soc. Sci. **3**(1), 1404753 (2017)
72. N. Péladeau, C. Dagenais, V. Ridde, Concept mapping internal validity: a case of misconceived mapping? Eval. Program Plann. **62**, 56–63 (2017)
73. J.A. Johnsen, D.E. Biegel, R. Shafran, Concept mapping in mental health: uses and adaptations. Eval. Program Plann. **23**(1), 67–75 (2000)

Prediction and Risk Analysis of COVID-19 Susceptibility

An Exploratory Study of Disaster Risk Management Mobile Applications in Pandemic Periods

K Saravanan, K. Ramesh, V. Sri Vigna Hema, and S. Viganesh

Abstract Pandemic is known for primary major disaster in the human life. It is not only affecting the health, but also its consequences leading to the decline of countries socio-economic status for longer period of time. Though the countries have developed better health systems to prevent the pandemic, it is highly necessary to handle and manage the situation without major loss to health and wealth during the period. Computational intelligence technologies can help the citizen and the government in effectively managing the pandemic situations. There are many actors involved in this Disaster Risk Management (DRM) cycle for better coordination and service. This research study focuses on the different mobile applications in need for this emergent timing. The requirements for each of these applications are narrated with future challenges and implementation directions.

Keywords Pandemic applications · COVID 19 · Disaster management · Contact tracker · Mobile E-governance

1 Introduction

Worldwide outbreak of Covid-19 has forced all nations to concentrate on pandemic preparedness, monitoring, planning and management. Most countries have an organizational plan to handle natural disasters like storm, hurricane, flood and earthquakes but lack in pandemic handling system since duration of pandemic is relatively higher

K. Saravanan (✉)
Department of Computer Science and Engineering, Anna University Regional Campus, Tirunelveli, India
e-mail: saravanan.krishnan@auttvl.ac.in

K. Ramesh
Department of Computer Applications, Anna University Regional Campus, Tirunelveli, India

V. S. V. Hema
Department of Information Technology Bannari, Amman Institute of Technology, Erode, India

S. Viganesh
Project Consultant, Tirunelveli Smart City, Tirunelveli, India

J. Nayak et al. (eds.), *Understanding COVID-19: The Role of Computational Intelligence*,
Studies in Computational Intelligence 963, https://doi.org/10.1007/978-3-030-74761-9_9

than natural disaster. Influenza pandemic is an epidemic which affects major population of national or international geographical scope, spread between human and last for many months. Technology enabled management system is a need of the hour to control influenza pandemic which has no vaccine and antibiotic to cure and treat. Currently non-pharmaceutical interventions such as complete lockdown, isolation and quarantine of vulnerable person, maintaining person health and hygiene, avoiding personal contact, social distancing, use of sanitizer, and restricting public gatherings are applied worldwide to control the spread of virus.

Though influenza pandemic has a major impact on socially vulnerable society, it also makes big impact on business and economy because of its long tenure. Like covid-19, another severe pandemic named Spanish flu infected nearly 40% of the world population and claimed 50 million lives worldwide during the year 1918-19. Currently available disaster management tools are not suitable to tackle specific issues such as food supply to isolated patients, orphans and migrant works, providing transport pass to essential service personal and vehicles, linkage of different government service for better coordination, stranded migrant labor tracking, patient tracking, relief material collection from donors, etc.

Government of India initiated social distancing app called "Arogya Setu", a COVID-19 mobile platform, which helps in distancing normal person from infected persons. A computer aided Disaster Risk Management (DRM) apps and tools can be devised to serve the people during lockdown. Major goal of pandemic management system is to have a technology based organizational structure as well as a system to continually assess needs, identify resources, plan the response, and implement the plan to support coordination and monitor the relief process.

This chapter is structured as follows: section one gives the introduction, followed by the different DRM applications in the literature. In section two, the literature survey is discussed. Section three categorizes the different pandemic application requirements in detailed specification. Further, section four listed the Covid 19 mobile apps in different countries currently. Integrated mobile network platform is illustrated in section five. Also, issues in implementing mobile apps during pandemic are discussed in section six.

2 Literature Survey

Mobile applications are important, since users spent 82% of time in Apps whereas 18% in web browser [1]. Mobile Application development for disaster management is developed [2] using existing data sources and ontologies in the semantic web. It has two applications namely Donate-N-Request & WeReport to integrate the Linked data resources in crisis situation. Medical emergency application-CrowdHelp, ensures the realtime patient assessment for medical professionals to give timely & accurate treatment using crowdsourcing of social network platforms [3, 4]. Crisis Informatics [5] deals with usage of IT applications during crisis. There were 57 DRM applications explored [6] for crowdsourcing, collaborating, alerting, collating and notifying activities. Different types of data (raw data, structured/semi-structured/unstructured) will be generated using the crowdsourcing DRM methods [7]. Hence, it is highly required to transform, filter and tagging the data to predict accuracy in the disaster management cycle.

Another work [8] explores the community outreach mobile applications, which are freely available in India. Mobile Apps specifically developed to use before, after or during a disaster, are known DRM applications. Although Apps were available for crowdsourcing, they lack with issues such as installation problem, OS update, data not relevant to users, loading problem, security issues, etc. PetaJakarta.org project [9] gathers geospatial data using crowdsourcing techniques to map information about flooding and water for Jakarta residents in real time. An IoT application using real-time thermal 3D imaging [10] is developed for human detection in effective way. IoT based applications and big data analytics [11–14] also helps in disaster management, since modern sensors placed in high alert and sensitive places can predict the disaster earlier. Smart cities can use image analytics to analyze and handle the disaster management operations in predictable way [15]. Computational intelligence based AI applications to handle Disaster management are discussed in detail [16].

Mobile applications [17] are crucial to tackle and control the pandemic like COVID 19 by large dissemination of data across the country using data channel. People tracking can be achieved by GPS based mobile Apps. Symptom Tracker mobile application is developed [18] to identify the herald symptoms, clinical outcomes, and geographical hot spots for the COVID 19. The correlation between mobile data & COVID 19 and how effectively mobile applications helps to manage and control social distancing is analyzed in detail [19]. Online classroom apps are also getting momentum during pandemic to engage the students in active self-learning [20, 21]. These Apps have the features such as video conferencing, chat box, online quiz, and discussion forum, collaborate with peers, etc.

Online doctor's collaboration and consultation to treat the COVID-19 patients is implemented effectively in Ruijin Hospital located in China [22–24]. Knowledge sharing through mobile Apps among medical professionals enables faster treatment. Health-check apps can able to curb the disease spread by ubiquitous mobile data sharing to the concerned authorities [25]. People can be warned and isolated by proper awareness through mobile Apps. The pandemic symptom is identified based on the self-testing results in the mobile phone application and trigger further actions based on the doctor report [26]. Mobile App is developed to analyze the cough sound and predict the status of the pandemic using Artificial Intelligence techniques [27, 28]. Medicine clinicians used Virtual Visit iPad application [29] to manage the clinical workflows in COVID-19 pandemic period. Industry 4.0 applications [30] plays major role in combat against the pandemic and hugely employs mobile sensor applications.

3 Mobile Applications

This section describes the working mechanism and limitations of mobile applications used for tackling COVID-19 pandemic situation.

3.1 Medical Store APPs

Amidst the enduring worldwide lockdown to contain COVID-19 pandemic, several countries have issued an official notification that the medical retailers will be indorsed to supply an indispensable medicines/drugs to the consumers at their doorway. Moreover, the sales of a medicine must be based on the prescription's receipt either by physically or via e-mail. In order to receive the consumer's prescription detail via e-mail, the licensee shall succumb authorization with the licensing authority. Correspondingly, the prescription shall be distributed only when it is accessible to the licensee within the period of seven days in critical cases. Furthermore, the licensee must uphold the records of all transactions as well as the return e-mail for the bill or cash memo.

There are several existing apps for online medicine, where the eminent apps in Android and IoS platform comprise of Netmeds, PharmEasy, 1 mg, and Practo. The actors involved in the online medical store management system are super administrator, system user, vendor, delivery person and consumers. The functionalities of the actors:

- **Super administrator:** Access full application, and manage medical stores and drugs.
- **System user:** Manage drug stocks, company, and retails.
- **Vendor:** Generate bills, maintain stocks, and manage transaction records.
- **Delivery person:** Collect medicines from shop keepers, deliver medicines, and manage delivery records.
- **Consumer:** Search drugs, place order, and make payment.

However, certain rules are necessitated to follow while placing an order for indispensable medicine. There should be minimal cart values, delivery charges, and general practitioner's prescription for high-risk medicines. The major benefits of using these apps are doorstep delivery to consumers, order from anywhere at any time, usage of discounts, maintain social distancing, and avoid epidemic. The demerits of online medicine apps are misusing of drugs, and delivery time delay. These limitations can be upgraded by increasing number of pharmacies by regional-wise.

3.2 *Collection of Relief Materials Food*

Despite the fact, almost a billion of people around the world are starving for food. Meanwhile, one-third of the globe's foods are landfilled. United States and China, top-most manufacturers of the production are facing the problems in wastage of food and outpouring of carbon. In order to alleviate the disruption, various organizations as well as individuals around the globe have established numerous applications, which help to connect cultivators & mediators. Moreover, the world's most familiar apps to collect and prevent food waste are Feedie, Flashfood, Food Cowboy, No Food Waste, Food for All, GoMkt, Goodr, etc. The major actors and their functionalities in these apps (Fig. 1) are:

- **Food donors:** supply either relief materials food or demand the app to deliver the donation.
- **System Administrator:** manage full application, arranges efficient communication among food donors and food receivers, and authenticates and enters hunger spots in their database.
- **Volunteer or Collector:** fast collection and delivery of food.
- **Food receivers:** give request for food.

These apps have smart mechanisms to reduce tons of food waste and help to eliminate lakhs of hunger all over the world. During this worldwide lockdown of COVID-19, numerous non-profitable people are getting beneficial via these smart apps.

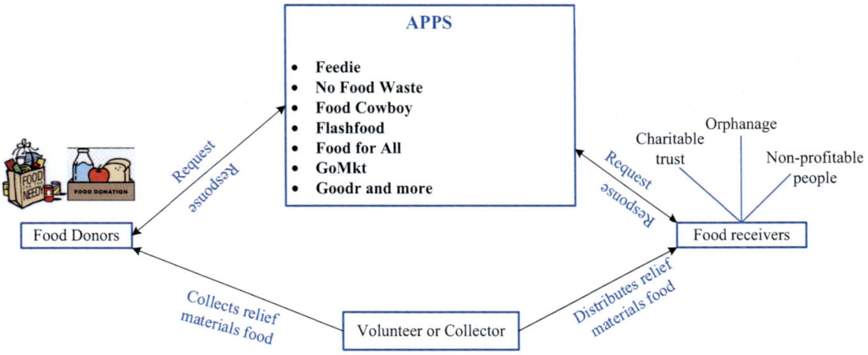

Fig. 1 Relief materials supply

3.3 Online Vocational Training Courses and Support to Unorganized Sector Workers

Pandemic devastates the life of the unorganized sector workers such as weavers, handloom workers, fishermen and fisherwomen, toddy tappers, leather workers, plantation laborers, beedi workers, etc. Due to lockdown for longer period, they lost their jobs and have to survive with their family. Hence, it is required to implement the system to identify these workers and provide them the government subsidies in effective manner without any middleman. It can be achieved through collecting their bank accounts from their respective association bodies and remit the government fund directly. They might feel insecure about the job in post-pandemic, hence the system should also give online counseling (phone/IVRS/video/audio/text messages) at periodical intervals and get to know their life status. Counselors can have the questionnaire asking for basic need criteria of these workers from time to time.

The COVID19 has shaken the entire world and have affected almost every single individual in the world not only by infection, but also trouble them economically to the core. The need of the hour is to prevent us from the pandemic and also equip us with better skills which make us well prepared physically and mentally to combat any future pandemics. Also help the other sector people particularly the needy skilled laborers working in the unorganized sectors. Hence, computationally intelligent solutions are designed in such a way that online vocation training courses & support to unorganized sector workers are addressed with proper mapping. In order to give online vocation training courses, platform can be created by government with online payment integration option, so that more number of registrations becomes feasible. Skill development ministry can take the initiatives on issue of online certificates. The website should basically comprise of 4 parts: schools, colleges, working professionals & Fitness, Arts, Other skills.

Let us discuss in detail about all of these sections. User should register in the website to update profile and must express the area of interest among these four areas.

- **Schools –** Upon selecting the school, student is directed to respective class ranging from class 1 to class 12. Each and every class must contain their syllabus, previous year question papers and the online teaching facility, wherein any teacher can upload their video and get the rating from the students across any schools. The final evaluation of the modules has to be done under the gamification model, student is given an option to unlock various stages upon completing each section and if the individual section score is good he/she will get a bonus point that would be a value addition to the final evaluation. Hence, each student can unlock the subsequent sections only upon successful completion of each and every section. Finally, upon successful completion student is given online certificate and would be rated as per the score and published in the Atal Tinkering Labs rating as a certified candidate with scoring of highest marks in each subject.
- **Colleges –** Same as above, similar method can be used here as well for Arts/Engineering/any other studies, for the case of live project in final year the

problem statement can be obtained from the registered user area in unorganized sector and people around it and try resolving it as per their expertise from the selective backgrounds. This project must have 3 phases and each phase will get unlocked only upon proper completion and approved by the right mentor assigned to each student. The 3 phases would be Introduction & Problem Approach, Ideation & Execution, and Implementation & Maintenance. The tasks will be assigned by the mentor and for each and every task proper research and study has to be done for successful finish of the live project. Students get the evaluation and certification that can be considered for scoring in semester.

- **Official Purpose –** This section is basically for the working professionals, wherein they can upgrade themselves with better knowledge on new subjects and concepts such that, the concerned skills will help for better promotion in the office. The training will be done on different subjects with all subject matter expertise and again for evaluation gamification model is followed assigned with one mentor for each and every individual to unlock and get through to reach the final stage. In this section an additional option of clarification of doubts upon working from home will be given such that the individual can get the doubts clarified to the assigned mentor using this website on the difficulties he/she faces during the work.
- **Fitness –** In this section, yoga, dance, arts, liberal studies subject matter expertise will be available to teach and mentor. The final evaluation for certification is done is a niche way that the concerned study should be taken as a video and must be uploaded in the user social networks, depending on the reach and the post heat map gained the user will get the points and certified.

By doing the above, skills were enhanced under these four categories to assist the unorganized sector workers as per the expertise certification obtained from the vocational training website as a competency mapping. e.g., If a college student is certified in website development, he/she will be given a problem statement from the tailor of same region to market the manufactured mask and sell it through e-commerce, regarding the same the college student will be assigned with a mentor to give hand holding support and execute the work, in a way extending help to unorganized sector worker and showcases student skill. 5% of collected fees from all the courses in online vocational training website can be given to the unorganized sector people for their initial production or skill activity.

3.4 Hospital Admissions and Cluster Geo-Fencing

The COVID19, most dangerous pandemic not only affects the physical ability, but also disturbs the emotional stability. It affects the citizen's daily life, by prevent him to do his routine activity. In order to address this issue, two basic areas must be focused to help the citizens for immediate hospital admission, then cluster geo-fencing the complete affected area. Basic mobile application has to be designed in such a way that it must be user friendly application and the uploaded feature in play store must execute in such a way that upon downloading the application it must seek the details of the user: Name, Address, Age, Medical treatments taken (Diabetics, BP, Cancer, etc. or any medicines taken for any problem), Mail id, Local Guardian along with their mobile number, Registered mobile number of users etc.

Once the user downloads the mobile application, user must login with the correct password and the application must have two features, (i) Symptoms selfie, and (ii) Quarantined area Selfie. Any user whom they feel like normal symptoms of COVID19 such as cold, cough & headache, then they can immediately take a selfie under the symptoms selfie and send it to the nearby corporation office filling up the basic questions posted in the application - how long they have been suffering from cold, cough, fever and some other relevant questions pertaining to COVID19. Similarly, the same features can be given in the quarantined area selfie as well where these citizens have already been identified by the concern corporation and have been isolated for quite some days. Once the data reaches the concern corporation data base they have a track of record on the collected data and the region, zone from where they obtained these information. Meanwhile in the application upon seeking the medical treatments taken by the user during registration and the selfie photo information can be mapped using Machine Learning Algorithm and can be designed for an alert of corona into 3 colors namely Red, Orange, Green, wherein green is not corona symptoms, and red being mostly corona symptoms and orange as the intermediate stage. Depending on the color fetched for that particular individual an appointment to the nearby hospital will be fetched directly through the mobile application, indicating the nearby hospital server and the nearby hospital can send the ambulance immediately to help in evacuation of the general patient/quarantined patient as per the obtained color.

In this way when such an application is done, it becomes easier for the concerned zonal office and the corporation to geofence the complete zone and cluster as corona identified area and can update in social media or to the television to avoid such areas for travelling and alerting the people in such areas to be in safe zone with utmost care. e.g., If the corporation gets continuous selfies from specific cluster that particular cluster becomes easy to isolate and alert people to avoid further spreading of COVID19 and take necessary actions, as executed in the Greater Chennai Corporation, wherein they have executed a similar kind of mobile application to combat COVID19.

In this way with the help of Mobile Application design, COVID19 problem can be addressed by seeking assistance and giving clarity in infection and admissions to nearby hospital with a proper cluster Geo-fencing in order to avoid further spread.

3.5 Volunteers Registration and Assigning Work

One of the most devastating and unpredictable things that occurs to the countries in the world is a natural disaster. Natural and man-induced disasters can wreak chaos on a huge population and substantially long salvage process regardless of economic or political stand-up of a country. In order to help physically as well as ardently re-erect affected communities, many people and agencies as volunteers would often call for such critical situations. Due to the crisis of COVID-19, numerous volunteers are desirable in several categorizations. The volunteer registration and assigning works for disaster risk management offers a variety number of useful apps includes: Community Emergency Response Teams (CERT), American Red Cross, World Vision India, Medical Reserve Corps, UN Youth Volunteer, Giving Way, and so on.

There are four major actors involved in these criteria, (1) volunteer, (2) agencies, (3) agency manager, and (4) system administrator. Primarily, the volunteers, who can help sustainable development either in their own county or another country around the world. Moreover, the works of volunteer are wide ranging, where they may work in primary healthcare sector, disaster management sector, food and materials supplier, medical provider, or many other sectors/areas. The volunteers may work directly in an agency or be allocated to one of its partner administrations, or the government. In each agency, there will be an agency manager, who is responsible for recruiting and assigning a volunteer for serious crisis and communicate over affected areas.

The recruited volunteers must be highly capable for world development, adaptive to several situations, motivational, enthusiastic, and asset. Meanwhile, they must undertake either 'short-term' assignments or 'long-term' assignments. Short-term assignments are less than six months of period within the country, whereas, long-term assignments are national and international assignments between the period of 6 months and 24 months. Before and after assigning works to the volunteers, the agencies would provide proper training and orientation. Depend upon the assignment allocated, the educational background as well as experienced are also required for the volunteers. Lastly, the main task of the system administrator is to manage the full application so as to accomplish both the agencies and the volunteers.

On the other hand, with the advent of novel tools and apps, one can also become a virtual volunteer in this global impact of COVID-19. Virtual volunteering can take several practices: volunteer with student, offer web development skills, deal marketing, or perhaps labor in worldwide health. The remarkable benefits of this virtual volunteering are: the volunteers can uphold maximum flexibility, and agencies may save money on conveyance. Henceforth, these are the ins and outs to make use of virtual volunteering that could be right for current state of affairs.

3.6 Home Treatment and Alarm System for Routine

The ongoing COVID19 pandemic has made every individual to lockdown himself without any stringent action taken to do so. When the disease spreads in large scale it becomes difficult for everyone and even the corporation officials to manage a separate isolation ward. So, in that case home treatment would be the ideal solution and treating the patient in their home will be an intelligent decision. In order to do so, when an individual gets positive corona, the immediate action that has to be done is of isolating and quarantine the immediate blood relatives. In doing so the affected patient must be kept in an isolated room having CCTV camera enabled with face recognition software, wherein which it helps in knowing the temperature of the affected individual in different time zones at regular intervals. This will help us in tracking the continuous health monitoring of patient to some extent in terms of fever related symptoms that would help in reduction of frequent attendant of nurse/doctor near patient. On top of it upon continuous monitoring, one can actually understand the needs and requirement of the patient. Also, a watch like tracker can be tied in the hands of the affected patient that helps in getting the heart beat status, blood pressure level, etc., which helps in better understanding the patient.

So, these 2 are the essential parameters that would be required to a great extent in order to keep a track of record on patient health status and any distortion or change in body happens through the watch kind of device an alarm alert can be sent to the nearby doctor and immediate evacuation and necessary steps can be taken immediately. An additional feature of the thought process sensor in built in the watch will help in understanding the emotional stability of the affected patient, which is crucial for speedy recovery and also to give them the better service staying in the home isolation ward. Doctors and medical team can have a monitoring in the thought process and emotional stability of the patients for their better well-being and speedy recovery. If any counselling required or any psychological treatment to be given, a speaker with external control can be fixed inside the isolation room and can be done without reaching close to the patient and shielding the counsellor or doctor.

3.7 *Food Supply to Isolated Patients*

The COVID-19 affected patients are generally isolated from the remaining public, in order to avoid further spread and make others shielded from the affected patients. So these patients will be taken to the isolated ward for continuous monitoring and observation, when it is done a regular supply of food for their speedy recovery must be ensured without spreading the virus to other fellow people. In order to meet these issue 3 basic things can be done with the help of ICT (Information Communication Technology) application and they are:

- Line follower Robot
- Voice Automation Device Installation in Isolation ward
- Community Kitchen

3.7.1 Line Follower Robot

This becomes cost effective and easier for the immediate implementation wherein a black colour sticker has to be stick from isolation ward to exit door, wherein the ordered food and delivery can be done through the line follower robot that helps in social distancing of human delivering it to the affected individual.

3.7.2 Voice Automation Device Installed in Isolation Ward

A voice automation device like Alexa or Google home has to be installed in the isolation ward, such that it helps the patient in better assistance to understand what they need and through the received voice, the order gets placed to nearby kitchen whom they have been designed exclusively to give the required food for infected people and what has to be given to them. The algorithm in the device can be designed in such a way that COVID-19 affected people can only order the doctor prescribed foods and other than that no other food can be ordered through this device. When the food gets processed and is ready to dispatch, it must be handed over to the line follower robot and can be watched until it delivers with the inbuilt camera in the robot.

3.7.3 Community Kitchen

The community kitchen has to be designed and camped in proximity to the isolation ward in less than half a kilometre to the ward, such that the menu received from the particular bed in the isolation ward can be dispatched at the right time. The patient upon ordering from the isolation ward through voice automation device gets recorded as placed order in the community kitchen server and the order processing can be done in the kitchen within the stipulated time and can ensure hot delivery of the food.

In this way the food delivery to the COVID-19 affected patients can be delivered at regular intervals without any delay and make them help better for a speedy recovery.

3.8 Essential Service Transport Pass and Personal Pass Service Request

The most essential parameter during the pandemic situation is of the pass utilization in order to meet the regular requirements. In order to avoid demand of goods, inflation situation, proper flow of goods & materials has to be there for in order to lead our routine life, to address this issue and also the pass avail for personal reasons a website can be created having issues related only about the pass request, wherein which the website can seek all the details and the reason to generate e-pass and finally land to the approval of concern city corporation commissioner. By doing so, the city monitoring body can have a track of record on people in avoiding the unnecessary movements and the spread control through these unnecessary movements of people.

In order to address this as described above a web based solution/ mobile application classified into 2 types of service transport pass comprising of essential commodities movement, agriculture items, grocery items, medical & related services, emergency vehicles, Government vehicles will only be generated the e-pass upon getting

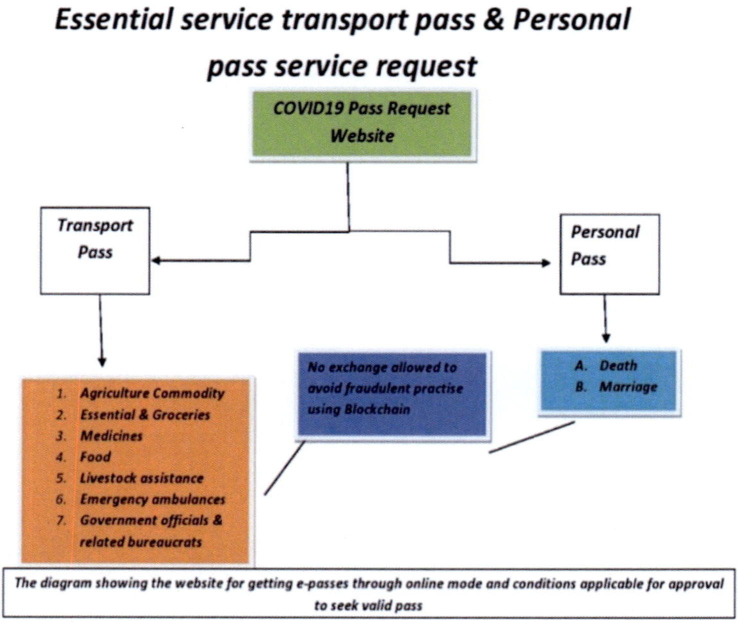

Fig. 2 Essentials services app workflow

relevant information from the end-user side and proper verification done from corporation side by giving an acknowledgment approval. In order to avail personal pass only two situations are accepted one is the death of immediate blood relative and the other is marriage purpose of immediate blood relative, only these two cases are accepted for personal pass. So, when a user login to this website and reveals the details for getting the pass, only these situation and conditions are accepted apart from this another scenario are considered as not valid and the designed algorithm goes for a immediate rejection in pass. In case of fraudulent activities, where the exchange of passes is done from personal reason to transport access, to avoid such a situation, Blockchain technology is being implemented in order to avoid the tampering of given information by the user such that only this person is eligible for availing this pass as per the records and can prevent the fraudulent activities to a great extent. The proposed system is as shown in Fig. 2 as follows:

4 COVID-19 Pandemic Contact Tracker Applications

A novel COVID-19 pandemic is resulting in high mortality rates and it requires prevention measures to discontinue the chain of transmission. In order to overcome this issue, several contact tracker mechanisms should be implemented rather than manual contact tracing. As the viral transmission is too fast, it could be controlled by most featured contact tracking apps that are more efficient and faster at scale. These apps build a memory of proximity interactions and further alerts the positive cases contacts, which can accomplish pandemic control once used by sufficient individuals.

However, the whole obstacle of COVID-19 is still in research, the computational problem here is about half of all contagions occur before the symptoms appear. Therefore, every single individual needs to act quickly to reduce contagions by finding all close proximity people with an affected individual. Mathematical procedures of the epidemic depict that fast contact tracking integrated with a testing program of the contagion might be able to deferral and stop the pandemic completely. Moreover, this current quarantine around the world could be turns out normal once the digital contact tracking apps would help rather than manual process.

The usage and functionalities of several COVID-19 apps in different countries are as follows:

4.1 TraceTogether App, Country: Singapore

TraceTogether App has high privacy preservation policy implemented for its use by not reveal the information to any system. solely throughout the registration alone, personal data is gathered by the App. once this App is connected with probably infected patients, information is valid while not poignant the privacy. Also, App provides the flexibleness to the users to opt from the system at any time. Consequently, their contact knowledge is deleted from the MOH, that means any log entries they seem in will now not be matched with them. To trace users the Ministry of Health (MOH) problems time finite anonymous temporary IDs square measure issued to patients by Ministry of Health (MOH) department to trace & establish to any or all third parties. Whenever 2 App users square measure connected, the devices begin to exchange their temporary IDs and store them within the history log. This log chronicles each user by assembling the last twenty-one days history and is kept solely on the user's device. If the App user sadly tests positive, then MOH fetches the contact log from the user history. It's sent to the MOH during which keep temporary IDs square measure they matched with the contact data. This app could be a community-driven contact tracing app that broadcasts Bluetooth signals for determinant whether or not the users had shut communication with a recognized COVID-19 case.

4.2 Smittestopp App, Country: Norway

This app can provide automatic notification if you happen to fulfil the infected person (who is additionally the users of this app) within the recent past. This notification is shipped via SMS in addition as within the app, which supplies pointers and procedures on the way to defend the users. National health register referred to as Communicable Diseases Notification System stores the coronavirus infection patient's details.

App follows the statutory obligations within the Health Register Act and MSIS rules. This data is coupled with mobile numbers hold on within the public contact & reservation information exploitation the birth details. The app will find the shut contact individuals with associate affected person. A text message with recommendation notification would be sent if they need been in shut interaction with a diagnosed COVID-19 case.

4.3 StopKorona App, Country: North Macedonia

The app uses Bluetooth technology to speak with different app users and exchanges encrypted, anonymized knowledge regarding the space of all close devices, at a distance relevant to the spreading of the infection, for a restricted amount of the past fourteen days. just in case Associate in Nursing app user tests positive for COVID-19, the Ministry of Health can request that user to divulge heart's contents to the Ministry an inventory of his/her app knowledge, where-by authorities would then find phone numbers and inform the persons that they need been involved with someone positive to COVID-19, while absolutely protective the privacy. The most goals of this app are: (a) to produce a quicker identification of shut contact person with Associate in Nursing affected one. Moreover, it offers a quicker response to the tending consultants regarding those persons, and (b) breakdown the chain of COVID-19 transmission.

4.4 HaMagen App, Country: Israel

HaMagen is Associate in the Nursing app that's supported by the Ministry of Health. It will tell the United Nations agency are within the presence of any-one United Nations agency has been diagnosed with coronavirus. This app cross verifies the GPS history of itinerant with the historical geographic information of patients. This app helps to spot the contact between detected patients. Mean-while, it conjointly provides the knowledge United Nations agency came in touch within the fourteen days before the affected person. It offers alerts so as to avoid spreading the virus and safeguard the close individual.

4.5 CoronaApp, Country: Colombia

This app helps in reporting symptoms and those of family members and receives recommendations according to the state of health after making the self-diagnosis. It obtains the mobility status through a QR code with some exceptions provided by the Colombian Government and access official information on Coronavirus Cases in

Colombia. It aids to discover the affected areas as well as to trace the nearby people with a positive case of COVID-19.

4.6 Gerak Malaysia App, Country: Malaysia

Gerak Malaysia could be a tracing app that permits police and Ministry of Health to track and analyze user's movement and register for permission to permit state border crossing.

4.7 MySejahtera App, Country: Malaysia

MySejahtera is Associate in Nursing app build by National council and Ministry of Health to retrieve data concerning the updated information and statistics of the pandemic.

4.8 The Corona DataSpende, Country: Germany

This application detects the signs and symptoms of the vital illness. It gathers information by using wearable smart devices from volunteers.

4.9 NHS COVID-19 Tracker, Country: UK

This app is mainly used for reporting the COVID-19 symptoms by the users to the National Health Service (NHS) officials, which evade spreading of pandemic.

4.10 eRouska (eFacemask), Country: Czech Republic

The app helps to trace all the users (whom they come into communicate with) via a Bluetooth device. It will send alert notification to health officials upon identifications of positive cases during an epidemiologic investigation.

4.11 Aarogya Setu Coronavirus Tracker App, Country: India

The key objective behind the app is to apprise users if they are available in reality with any COVID-19 positive person at any given purpose, and whether or not their area unit probabilities to induce infected. It conjointly provides tips to shield from coronavirus. Finally, this app offers info associated with the virus, facilitate centres, Self-Assessments Tests and additional. This pursuit app employs the smartphone's options of GPS moreover as Bluetooth for tracing the COVID-19 contagion. it's expedient for recognizing the infected person (who is at intervals six feet nearer) by scanning via information that contains the famous cases across India.

4.12 Unmaze App, Country: India - Kerala

The application is mapped with the portable variety of the user. It picks up the geo-location of the user and therefore the details of the devices that came within the locality of the device (5 – ten meters). Just in case someone is known to be infected with corona, the administrator can enter the small print on the server. The answer can after acquiring information from those devices and establish all the locations they visited and therefore the different devices they came involved with. Automatically, a notification is shipped to users requesting them for self-quarantine. After, the devices are often monitored to ensure that they're not violating the quarantine. A separate alert is shipped to the administrator informing him of the areas which require to be dis-infected.

4.13 Sahyog App by Survey of India, Country: India

SAHYOG can collate knowledge collected at the state-level and geo-tag it therefore the Survey of Bharat will analyze this data because the information grows. the target is to assist the govt. improve its response system.

4.14 CG Covid-19 ePass, Country: India - Chhattisgarh

Government of Chhattisgarh has launched this app to issue State-wide and Intra-district e-Pass for conveyance movement throughout the lock-down amount to modify the seamless transportation of essential commodities like Food grains, Vegetables, Fruits, Milk, Fuel etc. at intervals the State of Chhattisgarh. The someone will get Associate in Nursing e-Pass by submitting Photograph, Id Proof (Aadhaar Card) and

Business proof. Someone is going to be ready to choose the travel date, time and destination.

4.15 Quarantine Watch App, Country: India - Karnataka

Karnataka's revenue department has developed 'Quarantine Watch' that's obligatory for individuals in-home quarantine UN agency use it to send across their details, as well as a selfie each hour except between ten p.m and seven a.m.

4.16 SMC App, Country: India - Gujarat

COVID-19 huntsman to trace World Health Organization people that folks that those who have abroad or interstate travel history and persons who have are available in direct contact with positive COVID-19 individual. when putting in the app, the user is needed to declare his or her current location and travel history. Then each hour the user is needed to push a button known as "Send Location" to update their whereabouts. If the user is found to own withdrawn of the house or the desired space, he or she is shipped to institutional quarantine.

4.17 T-COVID 19 App, Country: India

T-COVID 19 an associate degree all-in-one app that provides official information, preventive tips and a mess of alternative services within the best interest of voters. All the data associated with government orders, media bulletins, government announcements, list of essential services, is additionally enclosed within the application, besides one-click choice to directly connect with 24 $\times$ 7 emergency helpline. Further, the app permits voters to require approved self-assessment tests to grasp his/her medical conditions and avoid supererogatory panic. The appliance additionally has associate degree integrated telemedicine module of 'Call Health' that enables the patient to remotely book associate degree appointment with a medical skilled. Additionally, the appliance additionally includes info regarding approved labs and check centres, isolation wards and quarantine centres. The critical analysis and usage of these apps is tabulated here in Table 1.

Table 1 Apps for different usage in pandemic periods

S. no	Name of app	Usage
1	TraceTogether	Community-driven contact tracing
2	Smittestopp	Bluetooth-based contact tracing
3	StopKorona	Bluetooth-based contact tracing
4	HaMagen	GPS based notification system alert the containment location history
5	CoronaApp	Bluetooth-based contact tracing
6	Gerak malaysia	Interstate travelers Movement control and contact tracing
7	MySejahtera	Preventive health risk assessment application
8	The Corona DataSpende	Smartwatch app to monitor the vital signs of Covid 19 and pandemic deceases
9	NHS COVID-19 tracker	Trace, Alert and symptoms verification
10	eRouska (eFacemask)	Bluetooth-based electronic facemask for Smart quarantine
11	Aarogya Setu	Bluetooth and GPS based self symptoms awareness and contact tracing app
12	Unmaze	Awareness, Alerts and Information dissemination app
13	Sahyog App	Users can collect, store and share the ground verified GIS data
14	CG Covid-19 ePass	Interstate travelers Movement control and contact tracing
15	Quarantine Watch	GPS based quarantine monitoring using selfie app
16	SMC	Contact tracing and Information dissemination app for city municipal corporation
17	T-COVID 19	Self assessment, contact tracing and Information app

4.18 Usage Statistics

A study by researchers at Oxford University suggests that the epidemic can be brought to an end if 56% of population of United Kingdom download and use a contact-tracing app. But the download and usage rate of the mobile app is only around 20% in an average. Germany has a download rate around 21%, Italy has 14%, and in France just 3%. The highest download rate is in Iceland and Singapore with 40%. Download rates is the vital measure which helps to increase the rate of trace of person who had contact with virus infected person. If 20% of the population uses the app the chance of a person coming in contact with another app user is only 4%. The rate increases by four fold to 16% if the download rate is 40%. Another report from Oxford researchers suggest that if 15% of the population of a country use the app with human or manual assisted system could reduce infections or spread by 4–12% and casualty or death by 2–15%.

Qatar leads with the maximum penetration rate of 91% and countries like Singapore, Australia and Norway have a better penetration because of its smaller population. India's Aarogya Setu is the world's most downloaded contact-tracing app with 100 million users but it is ranked 15th as it is sparse when compared to its 1.38 billion population. Detail of contact tracing app with its number of downloads and its penetration percentage is presented in Table 2.

5 Integrated Mobile Platform Application Framework

There are various actors involved in the pandemic mobile apps. They can be categorized into essential workers and other users. Essentials workers involved are – doctors, nurses, police, government officials, bankers, etc. Patients, employees, delivery persons are the other actors involved. All these actors use the mobile apps either to access or to generate the data. Integrated mobile network framework (Fig. 3) is necessary to consolidate the data generated by each app and summarize it to the government officials. The initial level e-governance is managed by the city municipal corporation authorities, and then followed by state level and national level. The greater data insights can be derived from the integrated mobile network, since they share the collected data from each citizen to this centralized command center network.

6 Challenges in Implementation of Mobile Apps

Though the Apps are highly necessary to control and prevent pandemic diseases, there are some issues in implementation of these apps to general public.

6.1 Network Connectivity

All the mobile Apps need to be installed and regularly updated, which requires mobile data consumption. During pandemic, Internet Service Providers (ISP) faces difficulties in managing the network and providing seamless services to the millions of consumers. It is also observed that mobile data consumption rises rapidly in the lockdown period. Mobile Apps in the remote locations such as rural areas may not receive the proper network connectivity. Hence, it is necessary that mobile apps should be designed with thin layers, so that it consumes less data. Extensible Markup Language (XML) based web services are the best options.

6.2 People Awareness

The literacy rate of the people plays major role in mobile services usage. Language translation in regional languages is necessary to reach all the states of the country.

Table 2 Usage statistics of apps (as of 30 October 2020) [31]

S. no	Countries	App name	Users (million)	Penetration (% of population)
1.	Qatar	Ehteraz	25,31,620	91.00%
2.	Singapore	TraceTogether	27,00,000	47.88%
3.	Finland	Koronavilkku	25,00,000	45.31%
4.	Iceland	Rakning C-19	1,40,000	38.45%
5.	Australia	COVIDSafe	71,60,909	28.64%
6.	UK	NHS COVID-19 App	1,90,00,000	28.51%
7.	Gibraltar	Beat Covid Gibraltar	9,000	26.69%
8.	Norway	Smittestopp	14,27,000	26.58%
9.	Israel	HaMagen	20,00,000	22.51%
10.	Germany	Corona-Warn-App	1,80,00,000	21.68%
11.	Turkey	Hayat Eve Sığar	1,41,86,000	17.30%
12.	Italy	Immuni	97,69,449	16.19%
13.	Canada	COVID Alert	53,14,026	14.03%
14.	New Zealand	NZ COVID Tracer	5,88,800	12.10%
15.	India	Aarogya Setu	16,30,00,000	12.05%
16.	Belgium	Coronalert	9,70,000	8.37%
17.	Austria	Stopp Corona	6,00,000	6.77%
18.	Japan	COCOA	77,00,000	6.09%
19.	Switzerland	SwissCovid	5,00,000	5.83%
20.	Thailand	MorChana	3,55,000	5.11%
21.	Iran	AC19	40,00,000	4.89%
22.	France	TousAntiCovid	24,00,000	3.58%
23.	Fiji	CareFiji	27,000	3.06%
24.	Czech	eRouska	2,77,000	2.60%
25.	Philippines	StaySafe	20,00,000	1.87%
26.	Indonesia	PeduliLindungi	46,00,000	1.72%
27.	Bulgaria	ViruSafe	55,000	0.79%
28.	Cyprus	CovTracer	9,000	0.77%
29.	Vietnam	BlueZone	3,82,160	0.40%
30.	Malaysia	MyTrace	1,00,000	0.32%
31.	Tunisia	E7mi	23,140	0.20%
32.	Poland	ProteGO	41665	0.11%
33.	Hungary	VirusRadar	10,000	0.10%
34.	Mexico	CovidRadar	50,000	0.04%

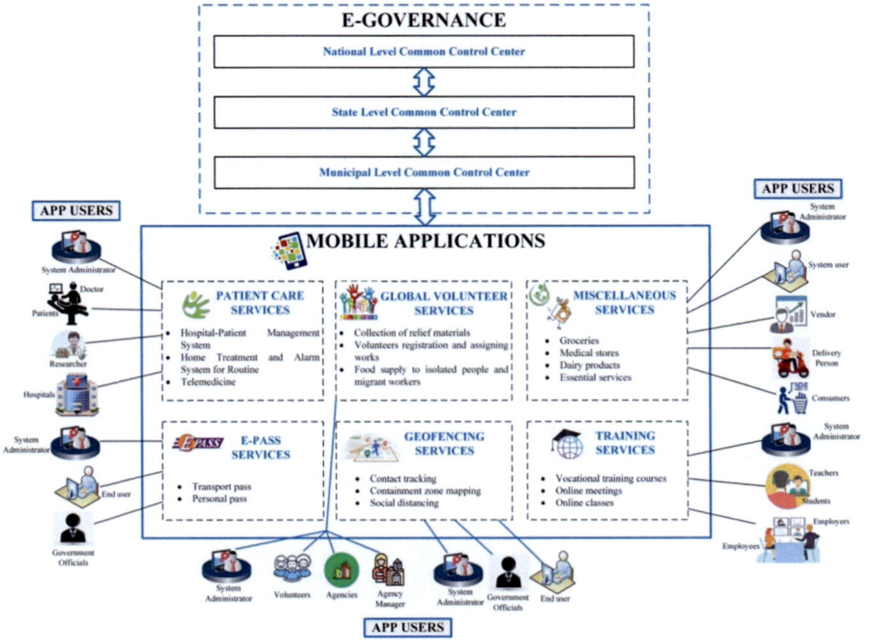

Fig. 3 Integrated mobile app framework architecture

People might have installed the app, but not aware of effectively usage of it. Proper user manual with simple steps can be placed in the mobile help section.

6.3 *Security and Privacy Issues*

The health care and personal data collected by the mobile apps are vulnerable to cyber attack. These sensitive personal details might be hacked by the potential attackers. Even the cloud computing offers best security practices, still placing personal health care data in data center lacks trustworthiness. Since the users are forced to grant access to several services in the mobile, hackers can make potential harm to the mobile device and can access the consumer data which affect the privacy of individual. Mobile apps need powerful security layer adding to the mobile security options.

7 Conclusion

This chapter explores the detailed study of the mobile application requirements in the pandemic. Different mobile apps in the past, current are narrated with its specification.

Also, an integrated mobile network platform is presented with all the actors involved in the COVID pandemic management. This paper gives a statistical insight in usage and its percentage of impact in tackling COVID-19. It also elaborates the challenges faced by various counties while using mobile apps and gives an elucidate outline for application developers.

References

1. S Gupta, For mobile devices, think apps, not ads. Harvard Bus. Rev. **91**(3), March 2013
2. F. Shih, et al. Democratizing mobile app development for disaster management. In: Joint proceedings of the workshop on AI problems and approaches for intelligent environments and workshop on semantic cities (2013)
3. L. I. Besaleva, A. C. Weaver, Applications of social networks and crowdsourcing for disaster management improvement, in *2013 International Conference on Social Computing*. IEEE, (2013)
4. C. Martin-Shields, The Technologist's Dilemma: Ethical Challenges of Using Crowdsourcing Technology in Conflict and Disaster-Affected Regions. Geo. J. Int'l Aff. **14**, 157 (2013)
5. C. Hagar, Crisis informatics In: Encyclopedia of Information Science and Technology, 3rd edn. ed by M. Koshrow-Pour. Hershey. (2015), pp. 1350–1359
6. M. L. Tan, et al. Mobile applications in crisis informatics literature: a systematic review. Int. J. Disaster Risk Reduction **24**, 297–311 (2017)
7. M. Poblet, E. García-Cuesta, P. Casanovas, Crowdsourcing roles, methods and tools for data-intensive disaster management. Inf. Syst. Front. **20**(6), 1363–1379 (2018)
8. V. Sukhwani, R. Shaw, Operationalizing crowdsourcing through mobile applications for disaster management in India. Progress Disaster Sci. **5**, (2020)
9. R. I. Ogie, et al. Crowdsourced social media data for disaster management: lessons from the PetaJakarta. org project. Comput. Environ. Urban Syst. **73**, 108–117 ((2019))
10. A. V. Arunraj, et al., An IoT application in disaster management using real-time thermal imaging system, in *AIP Conference Proceedings*, vol. 2222. No. 1. (AIP Publishing LLC, 2020)
11. A. Adeel, et al., A survey on the role of wireless sensor networks and IoT in disaster management, in *Geological disaster monitoring based on sensor networks*. (Springer, Singapore, 2019), pp. 57–66
12. K. Ali, et al., Internet of things (IoT) considerations, requirements, and architectures for disaster management system, in *Performability in internet of things*. (Springer, Cham, 2019), pp. 111–125
13. S. A. Shah, et al. The rising role of big data analytics and IoT in disaster management: recent advances, taxonomy and prospects. IEEE Access **7**, 54595–54614 (2019)
14. A. Sinha, et al. Impact of internet of things (IoT) in disaster management: a task-technology fit perspective. Ann. Oper. Res. **283**(1–2), 759–794 (2019)
15. N. Chaudhuri, I. Bose Application of image analytics for dis-aster response in smart cities, in *Proceedings of the 52nd Hawaii International Conference on System Sciences* (2019)
16. W. Sun, P. Bocchini, B. D. Davison, Applications of artificial intelligence for disaster management. Natural Hazards (2020), 1–59
17. N. Oliver, et al., Mobile phone data for informing public health actions across the COVID-19 pandemic life cycle (2020): eabc0764
18. D. A. Drew, et al. Rapid implementation of mobile technology for real-time epidemiology of COVID-19. Science (2020)
19. N. Oliver, et al., Mobile phone data and COVID-19: Missing an opportunity?. arXiv preprint arXiv:2003.12347 (2020)

20. G. Basilaia, D. Kvavadze, Transition to online education in schools during a SARS-CoV-2 coronavirus (COVID-19) pandemic in Geor-gia. Pedagogical Res. **5**(4), 1–9 (2020)
21. I. A. M. S. Widiastuti, I. B. N. Mantra, H. Sukoco, Mobile internet-based learning to cultivate students'speaking skill during coronavirus pandemic. Int. J. Appl. Sci. Sustainable Dev. (IJASSD) **2**(1), 6–10 (2020)
22. K. J. Danjuma, et al., Mobile application for Ebola virus disease diagnosis (EbolaDiag), in *Censorship, Surveillance, and Privacy: Concepts, Methodologies, Tools, and Applications.* (IGI Global, 2019), pp. 419–432
23. J.H. Wright, R. Caudill, Remote treatment delivery in response to the COVID-19 pandemic. Psychother. Psychosom. **89**(3), 1 (2020)
24. A. Yan, Z. Yi, D. A. Mirchandani. How hospitals in main-land China responded to the outbreak of COVID-19 using IT-enabled ser-vices: an analysis of hospital news webpages. Journal of the American Medical Informatics Association (2020)
25. M. Ienca, E. Vayena, On the responsible use of digital data to tackle the COVID-19 pandemic. Nature Med. **26**(4), 463–464 (2020)
26. P. A. McCullough, et al. Urgent need for individual mobile phone and institutional reporting of at home, hospitalized, and intensive care unit cases of SARS-CoV-2 (COVID-19) infection. Rev. Cardiovascular Med. **21**(1), 1–7 (2020)
27. K. Ganasegeran, S. A. Abdulrahman, Artificial intelligence applications in tracking health behaviors during disease epidemics, in *Human Behaviour Analysis Using Intelligent Systems.* (Springer, Cham), 2020, pp. 141–155
28. M. Z. Iqbal, Md F. I. Faiz, Active surveillance for COVID-19 through artificial intelligence using real-time speech-recognition mobile application, in *2020 IEEE International Conference on Consumer Electronics-Taiwan (ICCE-Taiwan).* IEEE (2020)
29. K. A. Wittbold, et al. iPad deployment for virtual evaluation in the emergency department during the COVID-19 pandemic. The American Journal of Emergency Medicine (2020)
30. M. Javaid, et al. Industry 4.0 technologies and their applications in fighting COVID-19 pandemic. Diabetes & Metabolic Syndrome: Clinical Research & Reviews (2020)
31. https://www.health-house.be/en/corona/mit-technology-review-covid-tracing-tracker/ Accessed on November 25, 2020

Potential of Deep Learning Algorithms in Mitigating the Spread of COVID-19

Oluwafemi A. Sarumi, Ons Aouedi, and Lawan J. Muhammad

Abstract COVID-19 pandemic has become endemic and has plunged the global community into a perilous situation pervaded with an economic recession, loss of jobs, and the death of thousands of people. It spreads exponentially around the world, affects 213 countries and territories as well as two international conveyances. Yet, the pandemic has neither clinically proven drugs nor vaccines. Therefore, it is now evident that non-medical approaches such as deep learning, data mining, expert system, software agents, and other artificial intelligence techniques are urgently needed to combat the pandemic, provide alternative solutions to alleviate the huge burden on the limited health care systems available around the world and curtail the future outbreak of the COVID-19 pandemic. Specifically, deep learning (DL) techniques evolved from machine learning (ML) concepts over a period of time and have been amply embraced in many real-life applications because of its unique nature and features for solving problems. Moreover, it is a powerful method of data exploration, and more importantly, has outperformed human efforts in several areas such as computer vision and health-related applications. Therefore, DL can be employed for combating and mitigating the proliferation of COVID-19 virus among humans. This chapter introduces the concept of deep learning and its potentials for combating the current spread COVID-19 pandemic and mitigating future outbreaks, discussed ongoing efforts of deep learning as one of the non-clinical approaches to alleviate the spread and curtail the further outbreak COVID-19 pandemic as well as the challenges of deep learning in combating COVID-19 pandemic and future directions.

Keywords COVID-19 · Deep learning · Computational intelligence · Machine learning · Pandemic

O. A. Sarumi (✉)
The Federal University of Technology, Akure, Nigeria
e-mail: oasarumi@futa.edu.ng

O. Aouedi
University of Nantes, Nantes, France
e-mail: ons.aouedi@ls2n.fr

L. J. Muhammad
Federal University Kashere, Gombe State, Nigeria
e-mail: lawan.jibril@fukashere.edu.ng

J. Nayak et al. (eds.), *Understanding COVID-19: The Role of Computational Intelligence*, Studies in Computational Intelligence 963, https://doi.org/10.1007/978-3-030-74761-9_10

1 Introduction

Severe Acute Respiratory Syndrome Coronavirus (SARS-Cov-2) which is popularly known as or called COVID-19 is a recent mutation of coronavirus newly discovered in humans. COVID-19 emerged in late 2019 from Hubei Province called Wuhan [1, 2]. It was widely believed that COVID-19 originated from bats, and subsequently spread to humans through intermediate hosts [2, 3]. Fever, shortness of breath, and cough are the major symptoms of COVID-19 which have a resemblance to the major symptoms of flu. COVID-19 had been declared by the World Health Organization (WHO) as a pandemic, but it is now evidently becoming endemic because there is neither an approved vaccine nor cure for the virus yet [1]. COVID-19 virus is transmitted among humans through close contact with an infected patient that produces respiratory droplets while sneezing or coughing [4]. Severe COVID-19 cases may lead to other serious medical conditions such as kidney failure, acute respiratory diseases, and many other health complications. Besides, people with adverse medical situations such as diabetes, heart diseases, kidney diseases, and pneumonia are more likely to have severe health challenges when infected with SARS-Cov-2 [1]. Therefore, early detection and prediction of COVID-19 infection are essential in reducing the spread of the virus and curtailing the infection rate among people [3, 4]. Also, It is now evident that there is a global shortage of testing kits and laboratories for COVID-19 across the globe needed for screening and testing the increasing suspected cases of COVID-19 infections for appropriate quarantine and treatment to curtail the further spread of the disease, especially in developing nations. However, non-clinical approaches such as deep learning, data mining, and ML methods have shown great potentials for timely detection, precise prognosis, and diagnosis of the COVID-19 infections [5, 6]. The effectiveness of theses non-clinical techniques have been demonstrated by many researchers for detection, prediction, diagnosis, and treatment of many health-related challenges such as malaria diagnosis [7], cancer prediction [8] heart failure risk prediction [9], and chronic kidney disease diagnosis [10].

Deep learning (DL) is used to develop intelligent systems and models that learn from datasets with various degrees of abstractions using multiple processing layers [11]. Sequel to the ability of DL for developing many high computational tasks, DL algorithms have been applied by researchers to extract from COVID-19 dataset specific graphical features and generate clinical diagnosis ahead or before pathogenic lab tests to save critical time for the containment of the disease [12]. In the current epoch of COVID-19 pandemic, researchers can further explore the potentials of DL for combating and mitigating the communal growth of the virus. In this context, this chapter presents two algorithms from the background of supervised deep learning for predicting future case of COVID-19. In brief, the contributions of this chapter are stated as follows.

- Firstly, the concept of Deep Learning and its potentials for combating the current spread of COVID-19 virus and mitigating future outbreak were highlighted.
- Secondly, two DL algorithms, Long Short-Term Memory (LSTM) and Convolutional Neural Network (CNN) were employed to develop a mechanism to

predict future cases of COVID-19 using a real-life dataset (China and Canada as case studies). Also, some other ongoing efforts on the use of DL as one of the non-clinical approaches in alleviating the effects and curtailing the spread of the COVID-19 pandemic were discussed in this study.
- Finally, some of the challenges of using Deep Learning in combating the effects of this pandemic were presented and future research directions on the use of DL for mitigating the challenges of COVID-19 were discussed.

The rest of this paper is ordered as follows: Sect. 2 presents the Deep Learning algorithms and their features. The process of feeding the LSTM and CNN model for predicting the future cases is described in Sect. 3. Evaluation and results are presented in Sect. 4.Discussions and comparative analysis were given in Sect. 5. DL challenges was presented in Sect. 6 and conclusions were given in Sect. 7.

2 Deep Learning

In this section, the essential background on DL and a succinct overview of established DL models as well as key advantages that gave them high recommendations and usage among many researchers were discussed DL is a novel research direction in machine learning (ML) and hence a branch of Artificial Intelligence (AI). AI is a subset of computer science with focus on developing intelligent systems from human knowledge, reasoning, self-learning, and problem-solving for fast and enhanced operations. [13]. Whereas, machine learning is a branch of AI that is used to develop a computational model in which a machine learns from experience [14]. Then as the technology evolved, the amount and the complexity of data increased, in turn, making the DL paradigm popular in 2010 as shown in Fig. 1. It had in recent years set existing new trends in ML and promising to reshape the future of AI because of its unique nature and features for solving complex problems.

In contrast to conventional ML models (e.g. SVM, Random Forest) that rely on features extracted manually by an expert, DL algorithms can learn hidden features through multiple non-linear processing units (i.e., neurons) in deep architecture. It consists of many layers of presentation, derived through a nonlinear composition of each transforms that is a representation of one level right raw data input into a representation at a slightly higher and more abstract level [15] as presented in Fig. 2. Therefore, the key advantage of DL is its ability to extract relevant features from raw data [15].

Theoretically, DL is based on the foundation of classical artificial neural network (ANN) but unlike ANN, it uses many hidden neurons and added two hidden layers that make DL design has an advantage over ANN architecture [16]. Also, like traditional machine learning models, DL has been used as supervised learning, unsupervised learning, semi-supervised learning, and reinforcement learning model as discussed below:

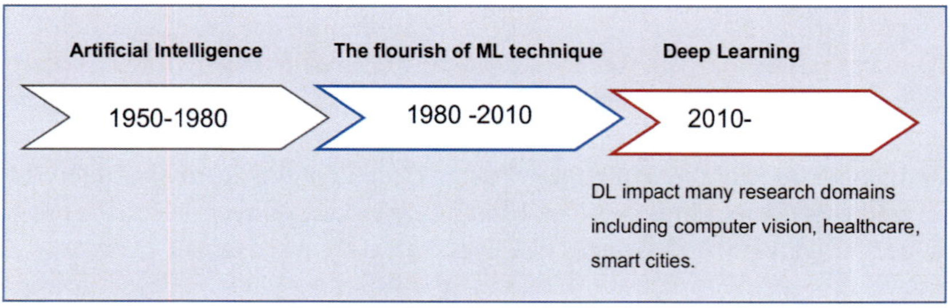

Fig. 1 DL evolution from AI and ML paradigms

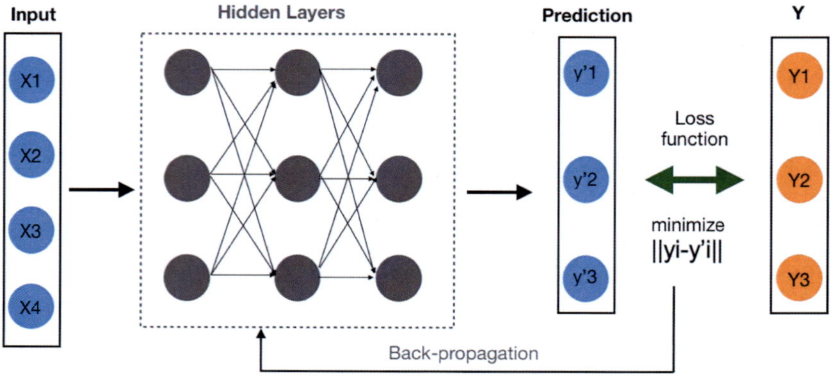

Fig. 2 The training process of a DL model

- In supervised learning, the DL model use labeled data in which there is a set of inputs and the corresponding output. There are several surpervised DL model such as Convolutional Neural Network (CNN), Recurrent Neural Network (RNN), Long Short Term Memory (LSTM), and Gate Recurrent Units (GRU).
- In unsupervised learning, DL models use unlabeled data in which there is no set of inputs and the corresponding output. The model uncovers hidden structures or relationships within the source data by learning their internal representation or key features of the input data. AutoEncoders (AE) is one of the most used for unsupervised DL.
- In semi-supervised deep learning, DL techniques use partially labeled data in which not all the set of inputs has a corresponding output in the dataset. DL models such as Generation Adversarial Network (GAN), GRU and LSTM, AutoEocnder with NN layer are used for semi-supervised learning tasks.

- Reinforcement deep learning, is the combination of the deep neural network and reinforcement learning (e.g., Q-learning) which is firstly used Google deep-mind on some gaming applications using the deep Q network for Q value-action function approximation [17].

2.1 DL Algorithms

In this section, background information and conceptual architecture were provided on common DL algorithms; which are CNN, LSTM, RNN, and AE.

- **2.1.1 Convolutional Neural Network:** CNN is configured to process multi-dimensional data like time series and image using feature extraction methods and computation of weights during its training phase [17, 18]. It was proposed by Fukushima in 1988, but was not popular due to computer hardware limitations of training the network [19]. Then, in 1990 got the much-needed attention when the gradient-based learning algorithm was applied to it for handwritten classification problem [20]. Today, CNN is among the most widely used algorithms in computer vision and it gets its name from some of its hidden layers which include the convolutional layer, pooling layer, fully connected layers, and normalization layer [21]. Each layer receives its output from the immediate previous layer as its input and passes it as output to the next layer [18]. Moreover, it derived its name through the use of a convolution operator in solving intricate operation. Therefore, the primary advantage of CNN is that it provides automatic feature extraction through its layers (i.e. convolutional layer, pooling layer) [22] as shown in Fig. 3. Convolutional contains a digital filter to accomplish the convolutional action on the input data while polling layers are used as a layer for dimensionality reduction and also used to decide the threshold. Instead of using a normal activation function, CNN uses convolution and pooling functions as its activation functions for introducing non-linearity [22].
- **2.1.2 Recurrent Neural Network:** RNN is a DL algorithm that is designed for analyzing a stream of data through hidden units and is naturally suited for time series and other sequential data processing [18, 19]. It provides more interdependencies for the input data and the available parameters are determined by the output produces at a given time (Fig. 4). Moreover, RNN usually keeps two kinds of inputs which include the present input and most recent input to generate new data for the output [23]. However, due to the vanishing gradient problem, several variations of RNNs have emerged which include LSTM and Gated Recurrent Unit (GRU) to address the vanishing gradient problem [19]. LSTM and GRU have capacity of sustaining long term interrelations unlike RNN and nonlinear dynamics for time series input dataset.

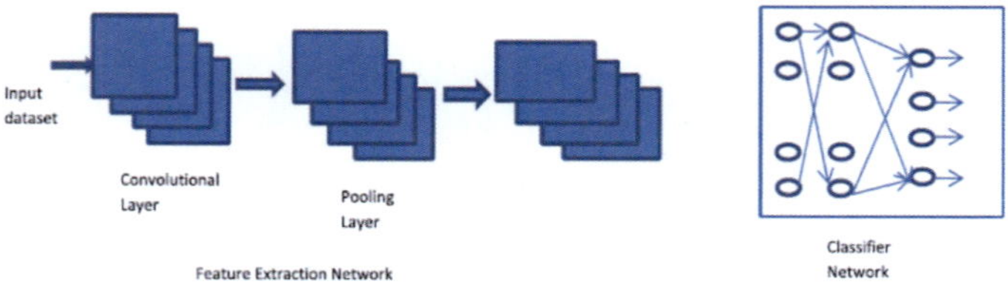

Fig. 3 Overall architecture of CNN *Source* [22]

Fig. 4 Architecture of RNN *Source* [17]

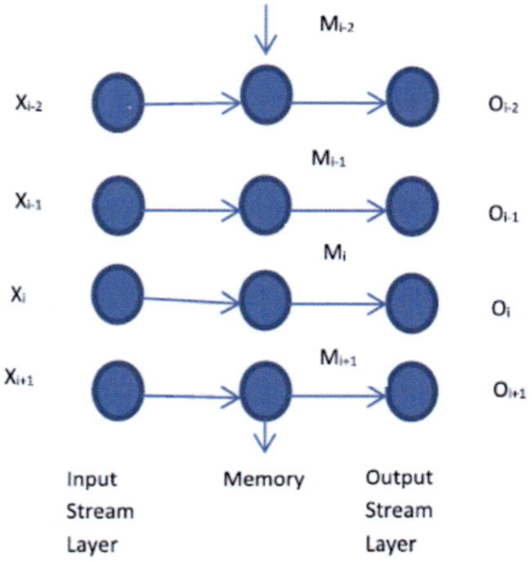

- **2.1.3 Long Short Term Memory:** LSTM is a variant of RNN apt to learn from experiences commonly used for ranking process and prediction of time series data especially cases with lengthy-time lags of undetermined size between cardinal events. Also, is a variant of RNN designed to handle the limitation of classical RNN architectures. LSTM as shown in Fig. 5 and standard RNN has a similar control flow–data is processed by passing on information as it propagates forward. The key differences of LSTM from other RNN are the mechanisms within the LSTM's cells which is made up of LSTM blocks as a complement to the classical RNN network units. An LSTM typology of a "smart" network unit with the capacity to remember records for a stochastic period. LSTM networks are configured to learn long-term dependencies with strong capacities to overcome the previously inherent problems of RNNs. Also, LSTM is well suited for time-series prediction of cases with long and short term dependencies. An additional advantage of LSTM is its ability to map process between input and output sequences using context-sensitive information.
- **2.1.4 AutoEncoder:** Autoencoder is a DL algorithm that used a data-driven learning technique to extract features in an unsupervised manner [15]. It consists of encoder and decoder layers. As presented in Fig. 6, the encoder layer extracts relevant attributes from the input row data while the decoder layers to construct this input.

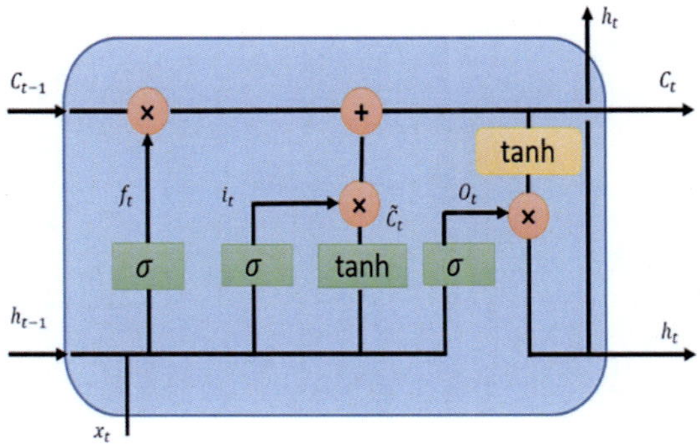

Fig. 5 Architecture of LSTM *Source* [41]

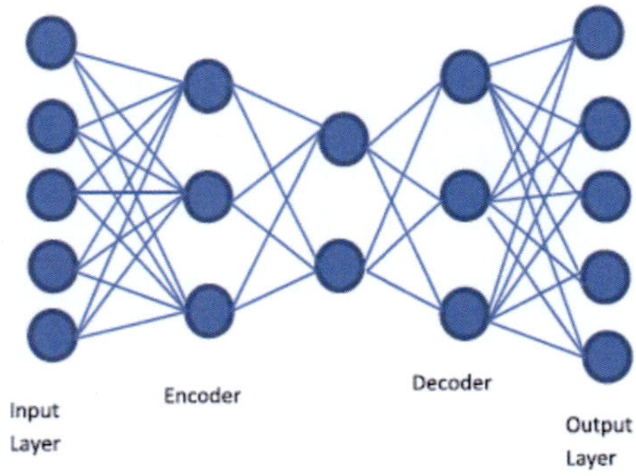

Fig. 6 Architecture of the autoencoder model *Source* [17]

To show the potential of DL algorithms in mitigating the spread of COVID-19, predictive models for COVID-19 future cases of the Republic of China and Canada were developed in this study using CNN and LSTM deep learning algorithms. CNN has attracted a lot of attention among researchers due to its efficiency in object recognition [24, 25] and application in time series classification problems [26, 27]. Also, in many prediction cases, LSTM generates better outcomes than other recurrent model structures [28, 29].

3 Methodology

In this section, a description of the dataset, preprocessing methods, and our CNN and LSTM DL models for forecasting the probable future cases and trends of COVID-19 infection cases–using China and Canada as studies was discussed. An overview of the architecture of our proposed deep learning model is presented in Fig. 7.

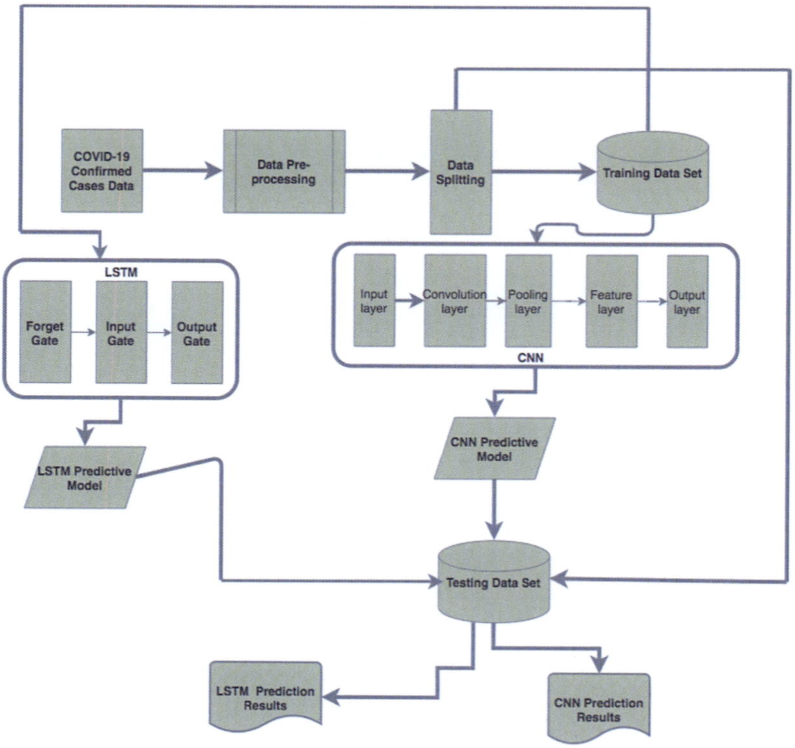

Fig. 7 An overview of our deep learning architecture

CNN and LSTM were employed in this study because they have been successfully used for time series prediction, handwriting recognition and natural language processing among others. [30, 31]. In CNN, the convolutional layers capture very well the temporal connection of the input data [32]. Moreover, LSTM was developed to overcome the vanishing and exploiting gradient problems peculiar to the regular RNNs.

3.1 Description of the Dataset and the Pre-processing Methods

The daily confirmed cases of COVID-19 were downloaded from the online repository for COVID-19 Time Series Data "https://data.world/shad/covid-19-time-series-data" [47]. The monthly confirmed cases for all the 33 states in China and the 14 provinces in Canada from January 2020 to October 2020 was computed and used for training the DL algorithms in this study. Equation 1 describes the total dataset (D) containing a set of records or instances of the confirmed monthly cases in all the states in China and Canada. Also, a feature scaling of the input data was carried out to normalize every column of the dataset to a uniform scale as shown in Eq. 2.

$$D = \left\{ \sum_{i=1}^{n} X_i S_j \right\} \forall S = (S_1, S_2, S_3, ..., S_j) \quad (1)$$

where: X is the daily number of confirmed cases
S represent each state in China or province in Canada
n is the total days in a month
j represent the total number of states in China or provinces in Canada.

$$V' = \frac{V - V_{min}}{V_{max} - V_{min}} \quad (2)$$

where, V represent an attribute value of individual record in D, V' represent the normalized form of the attribute value, V_{min} equals the minimum value of V, and V_{max} gives the maximum value of V.

3.2 *Convolutional Neural Networks (CNN)*

At the input layer of our CNN model, the neuron number, weights, and bias for the classification process using a random number was generated. Also, the learning rate, and select sigmoid function (f) for the activation function was activated as shown in Eq. 3.

$$f(x) = sigmoid(x) = \frac{1}{1 + e^{-x}} \quad (3)$$

At the convolution layer, a set of filters were used to perform convolution operations on the COVID-19 time-series data from the input layer. Equation 4 describes the output of the convolution layer output.

$$C_r(n) = f(\sum_{i=1}^{l} \sum_{j=1}^{k} t(i + s(n-1), j) w_r(i, j) + b(r)) \quad (4)$$

where $\boldsymbol{l}$ represents the convolution filter length, $\boldsymbol{k}$ is the variate number of the COVID-19 time series data. Therefore, the size of the convolution filter is computed as $\boldsymbol{l}$ x $\boldsymbol{k}$, and the convolution stride is given as $\boldsymbol{s}$. The input COVID-19 time series data is represented as $\boldsymbol{t}$, C_r denotes the nth component of the feature map, while w_r and b_r represent the convolution weights and bias respectively.

$$P_r(n) = y(C_r((n-1)l + 1), C_r((n-1)l + 2), ..., C_r(nl)) \quad (5)$$

At the pooling layer, the feature map is halved into identical segment lengths. This helps in reducing the convolution sample output bands, thus speeding up the convolution process. The result of the pooling layer can be expressed as shown in Eq. 5. Such that the pooling strategy is activated using the function y which is a max pooling operation.

At the feature layer, the results of the convolution and pooling operations are represented as a series of feature maps. The product of the feature layer is fed as input into the output layer. The result of the output layer can be represented as shown in Eq. 6.

$$O(j) = f(\sum_{i=1}^{M} Z(i)w_f(i, j) + b_f(j)), j = 1, 2, ..., n \tag{6}$$

where the outputs from the feature layer is represented by **z**, the output layer bias denoted as b_f and w_f represents the weights connections between the feature and the output layers.

3.3 Long-Short Time Memory (LSTM) Algorithm

At the forget gate, irrelevant information is discarded from the COVID-19 normalized time series data using the forget gate function f(t) as shown in Eq. 7.

$$f^{(t)} = \sigma\left(W^f[h^{t-1}, x^t] + b^f\right) \tag{7}$$

where x^t denotes the input vector at current time stamp, h^{t-1} describes the previous cell output, σ represent the recurrent transfer function, W^f denotes the weight of the input and recurrent connections, and b^f the gate bias value.

$$I^{(t)} = \tanh\left(W^i[h^{t-1}, x^t] + b^i\right) \tag{8}$$

The input gate is used to update the cell contents as given in Eq. 8, and oversee the input activations discharge into the memory cell with tanh as the activation function. The output gate as given in Eq. 9 is use to control the output flow by filtering the cell activations and determine what flows into the rest of the neural network.

$$O^{(t)} = \sigma\left(W^o[h^{t-1}, x^t] + b^o\right) \tag{9}$$

3.4 Performance Evaluation Metrics

The performance evaluation of our proposed deep learning models–CNN and LSTM for predicting the growth rate of COVID-19 was evaluated based on standard performance metrics for deep learning algorithms which are Mean Square Error, Accuracy, Root Mean Squared Error, and Mean Absolute Error.

1. **Mean Square Error (MSE):** This is used to compute the mean of the squared difference between the predicted parameter and the observed parameter as shown in Eq. 10 The smaller the MSE values of a model, the better the performance of the model.

$$MSE = \sum_{i=1}^{n} x_i - \bar{x}_i \tag{10}$$

   where x_i denotes the actual values, $\bar{x}_i$ represents the predicted discharge values and n is the overall sum of COVID-19 confirmed cases.
2. **Accuracy (AC):** This provide the overall general performance of the model. It is estimated by calculating the percentage ratio of the sum of difference between the actual and the predicted values to the total number of actual value as shown in Eq. 11.

$$AC = \frac{T_{diff}}{T_{actual}} * 100 \tag{11}$$

3. **Root Mean Squared Error (RMSE):** Denotes the standard deviation of the difference between the predicted COVID-19 cases and the actual confirmed cases of COVID-19 as shown in Eq. 12.

$$RMSE = \sqrt{\frac{1}{n}\sum_{i=1}^{n}(x_i - \bar{x}_i)^2} \tag{12}$$

4. **Mean Absolute Error (MAE):** This is a measure of the closeness of the predicted COVID-19 cases to the actual confirmed cases as shown in Eq. 13.

$$MAE = \frac{1}{n}\sum_{i=1}^{n} |x_i - \bar{x}_i| \tag{13}$$

4 Experimental Set up and Results

Our algorithms–LSTM and CNN were executed using Python programming language in Windows Operating System environment and implemented on HP Elitebook with Intel(R) Core i5 CPU at 2.6GHz, 8GB RAM, 64-bits OS, and 500GB hard storage. COVID-19 confirmed cases in the 33 states in China and 14 provinces in Canada from January 2020 till September 2020 was used as the training dataset, and the dataset of the confirmed cases for the month of October 2020 was used as the testing dataset which is equivalent to a data splitting ratio of 90% training data and 10% testing data.

Table 1 Comparison of CNN and LSTM predictions with the confirmed cases (China)

Duration	Actual Cases	CNN Predictions	LSTM Predictions
21-09-20	90381	90176	87696
22-09-20	90399	90195	87709
23-09-20	90409	90216	87722
24-09-20	90424	90235	87734
25-09-20	90441	90255	87745
26-09-20	90456	90274	87756
27-09-20	90483	90290	87766
28-09-20	90505	90310	87776
29-09-20	90528	90328	87786
30-09-20	90545	90346	87797
01-10-20	90567	90364	87808
02-10-20	90584	90385	87819
03-10-20	90604	90405	87830
04-10-20	90629	90422	87840
05-10-20	90652	90443	87851
06-10-20	90667	90462	87863
07-10-20	90687	90480	87874
08-10-20	90728	90497	87885
09-10-20	90751	90520	87897
10-10-20	90778	90542	87910

A comparison of the results of LSTM and CNN predictions with the actual confirmed cases from 21-September, 2020 to 10-October, 2020 in China and Canada are shown in Tables 1 and 2 respectively. It is evident from Tables 1 and 2 that predictions of LSTM and CNN for the COVID-19 cases for this period in China and Canada were close to the actual number of confirmed cases–with CNN given more accurate predictions for the two countries. Also, Figs. 8 and 9 respectively show that CNN prediction is more accurate than LSTM prediction for the two countries– with the daily predictions of CNN closer to the valid values of the daily confirmed cases than the values predicted by LSTM over the period used for this study. Table 3, shows a further evaluation of the performance of our deep learning models using standard performance metrics. Table 3, shows that CNN has higher accuracy than LSTM for both China and Canada. Besides, the MSE value for CNN–which is computed as the mean of the squared difference between the values predicted CNN and the actual confirmed cases is far smaller than the LSTM MSE value-thus this also corroborates that CNN has a better performance than LSTM. Besides, the MAE which was used to compute the closeness of the actual confirmed cases to the predicted COVID-19 cases by our deep learning models–CNN and LSTM show that the values predicted by CNN were closer to the actual number of confirmed cases more than the values predicted by the LSTM. Thus, all the evaluation performance metrics used to evaluate the performance of the 2 deep learning models–CNN and LSTM in predicting future cases of COVID-19 show that CNN performed better than LSTM.

Table 2 Comparison of CNN and LSTM predictions with the confirmed cases (Canada)

Duration	Actual Cases	CNN Predictions	LSTM Predictions
21-09-20	147583	140777	133171
22-09-20	148855	141681	133784
23-09-20	149939	142628	134435
24-09-20	151087	143604	135113
25-09-20	152717	144550	135809
26-09-20	153770	145584	136534
27-09-20	155428	146680	137271
28-09-20	157219	147763	138032
29-09-20	158991	148940	138827
30-09-20	161107	150180	139659
01-10-20	162930	151549	140540
02-10-20	165054	152972	141464
03-10-20	166360	154441	142433
04-10-20	168501	155891	143415
05-10-20	171384	157401	144419
06-10-20	173756	159095	145475
07-10-20	175380	160850	146580
08-10-20	177730	162557	147700
09-10-20	180625	164341	148840
10-10-20	182767	166266	150016

Table 3 Performance Evaluation of CNN and LSTM predictions for China and Canada

Performance metrics	CNN (CH)	LSTM (CA)	CNN (CA)	LSTM
Accuracy	99.56	97.27	97.03	92.74
MSE	2.73E−06	0.00075	0.00115	0.00503
RMSE	0.00523	0.02733	0.03393	0.07092
MAE	0.00432	0.02709	0.02226	0.05446

Figures 10 and 11 show respectively the CNN and LSTM forecast from 11-October, 2020 to 31-October, 2020 of the probable cases of COVID-19 infection in both China and Canada respectively. CNN forecast for this period for China shows a continuous flattened curve in the growth of new COVID-19 infections in the 33 states of China, which follows the trend of the actual number of confirmed cases for July 2020. LSTM forecast for the probable cases in China for the period under consideration indicates a downward progression as shown in Fig. 11 thus, it does not follow the actual trends of the confirmed cases in July 2020. In the case of Canada, both CNN and LSTM predicts a sharp drop in the cases of COVID-19 which is more deviating from the actual cases. This is as a result of the cold start problem [46] of

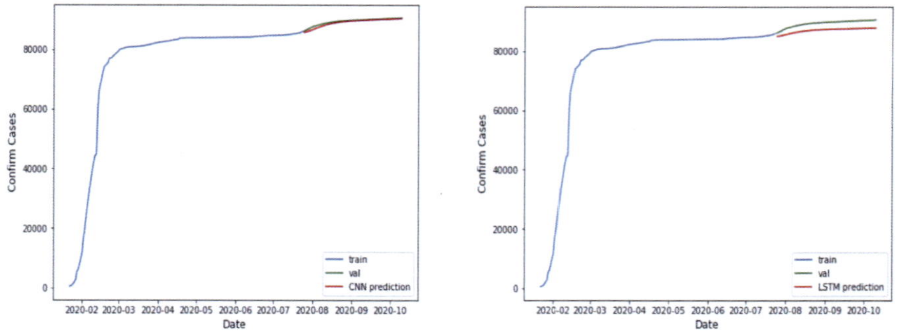

(a) CNN Prediction (b) LSTM Prediction

Fig. 8 Predictions using China COVID-19 confirmed cases

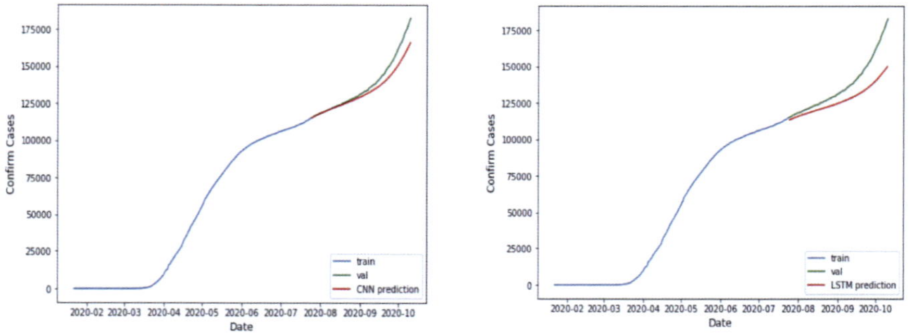

(a) CNN Prediction (b) LSTM Prediction

Fig. 9 Predictions using Canada COVID-19 confirmed cases

CNN and LSTM stimulated by the staggering numbers of confirmed cases during the first wave of COVID-19 infection in Canada. Notwithstanding, CNN predictions are more accurate and closer to the actual number of infections cases for this period.

Furthermore, Fig. 9 shows that Canada experienced a mild wave of COVID-19 infection at the beginning stages of COVID-19 pandemic, while China had a severe wave of infection and a number of cases at the initial stage as shown in Fig. 8. As a point when the infection rate in China starts to flatten, Canada, started experiencing a second wave which is more severe than the first wave in Canada.

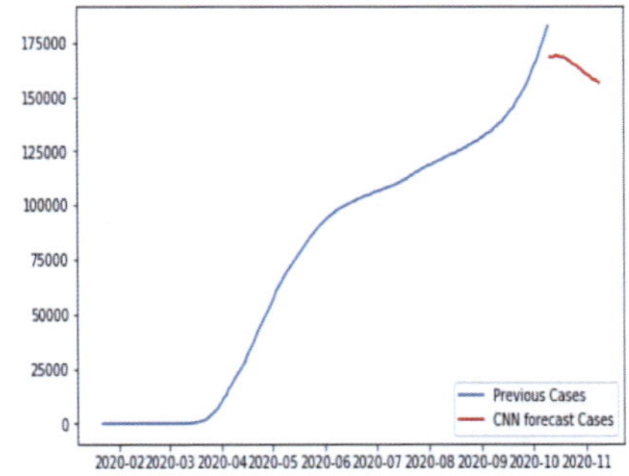

(a) CNN Forecast

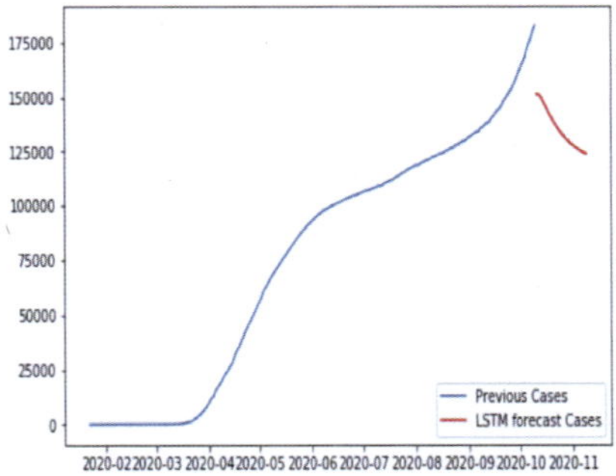

(b) LSTM Forecast

Fig. 10 Forecast using Canada COVID-19 confirmed cases

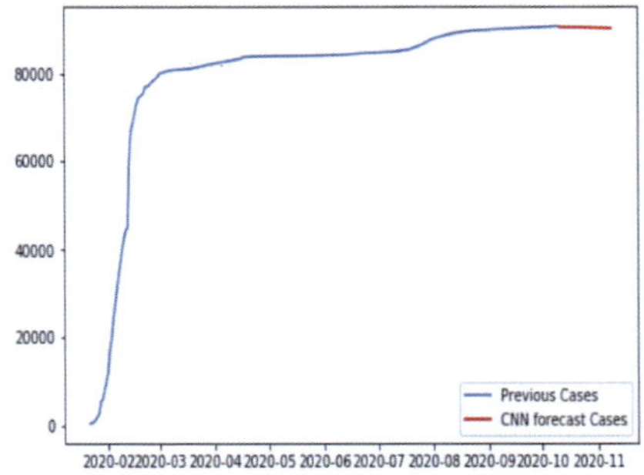

(a) CNN Prediction

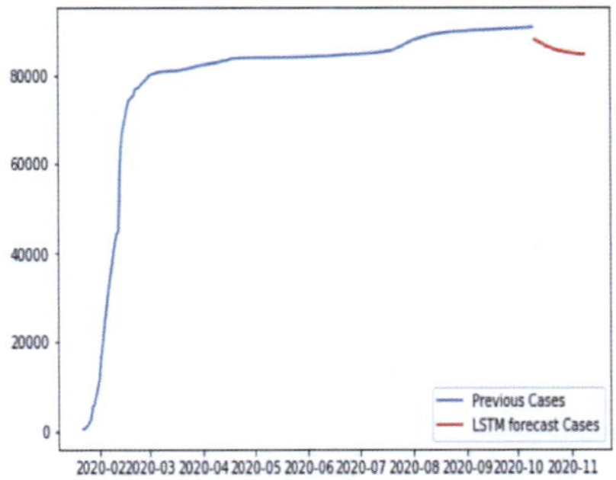

(b) LSTM Prediction

Fig. 11 Forecast using China COVID-19 confirmed cases

5 Discussions

Recent advances in DL algorithms have enabled them to be used in healthcare [33], smart environment [34], wireless network [35]. Our findings in this research further accentuate the strength of CNN in predicting future probable cases of COVID-19 infections. Some other researchers have proposed different algorithms and methods that show the potential of DL algorithms in addressing different forms of challenges posed by the COVID-19 pandemic. In [13], a DL approach based color visualization was used to generate a predictive model from x-ray and CT scan images for the detection of COVID-19 infected cases. Similarly, [38] engaged the use of four different DL algorithms to predict COVID-19 infection using chest x-ray images. Also, [36, 43] developed a DL model for diagnosing COVID-19 infection form X-ray images. In [37] predictive models were developed using LSTM, GRU, and Bi-LSTM for forecasting confirmed cases, recoveries, and death cases of COVID-19 infected people. Also, [38] proposed the use of LSTM, autoregressive integrated moving average (ARIMA), and prophet algorithm (PA) algorithms for predicting future recovery

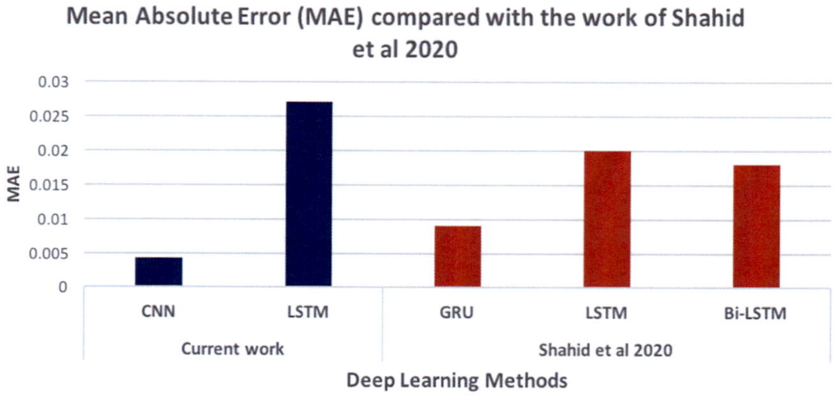

Fig. 12 Deep learning methods compared

cases in Australia. Besides, [44] presented a novel susceptible-exposed-infectious-recovered (SEIR) dynamic model modified which is an improvement on the classic SEIR model to develop a diagnostic system for predicting the transmission trend of the COVID-19 virus. Figure 12 presents a comparative analysis of the results of our proposed DL models with the DL models used in the work of [37]. The analysis shows that our CNN model has the lowest MAE, thus given the highest prediction accuracy, followed by GRU used in [37]. Also, the accuracy of the LSTM and BI-LSTM models presented in the work of [37] is better than the result of our proposed LSTM model.

6 Deep Learning Challenges

To ascertain the accuracy and the effectiveness of these DL algorithms [45], many factors should be taken into account during the development of the DL algorithms. In this regard, some of the research challenges and issues that should be addressed when developing models using DL algorithms were presented as follows:

1. **DL architecture settings** The DL architecture used for learning has an important influence on its final performance. However, There is no clear understanding of which architectures should perform better than others. Also, DL models have several hyper-parameters and no mathematical theorems help to tune them. Therefore, it is difficult task to find the optimal architecture (i.e. number of layer and neurons) and the appropriate hyper-parameters values (e.g. learning rate, dropout).
2. **Highly unbalanced data:** Besides the DL architecture, the amount and type of training data an algorithms needs are very important things. If the data is unbalanced can degrade the performance of DL models such that decision tends

to favor the larger class and misclassifies the smaller class samples and treated them as noise. However, in many situations, classes with lower instances are the most useful ones. In the case of the COVID-19 pandemic, the number of people that have tested positive belongs to the minority class compared to the global population. Therefore, it is more expensive to give a wrong classification to the minority class than classifying the majority class wrongly [39].

3. **Lack of Large Dataset:** Note that the model training performance is highly dependent on the dataset which is a crucial component in DL that demonstrates more potential capacity and be more accurate with a large amount of data. However, due to the privacy and the lack of data availability, it is very difficult to find appropriate public datasets to be used for training phase. This issue can limit the learning ability of the DL models. Therefore, the major impediment in implementing DL models is data–as they need big data to achieve more accuracy. In this context, it is suggested that research-based COVID-19 data should be publicly available to encourage the proposition of robust models.
4. **Preprocessing:** Most DL models need a preprocessing step to yield good accuracy which depends on the data scaling capacity. Therefore, preparing the data in a suitable representation to be fed is an important task. For instance, scaling or normalizing the input data into a specific range at the pixel level or transforming them into a standard representation can improve CNNs (Conventional neural networks) accuracy, which one of the most used image processing model [40].
5. **Selection of DL model:** The quality of classification/regression is depending on the DL model. Also, there here is no best DL model since each model is trained on its own environment (i.e. data). Some models are suitable to extract robust feature and other for classification. Moreover, some models work well with labeled data and others are better when our data is unlabeled. Therefore, the choice of the model is a crucial task.
6. **Model over-fitting:** Over-fitting is a dilemma problem for DL models. Since they consist of many parameters, hence there is a great probability to be over-fit. That is why the researchers innovate and proposed several techniques to avoid the over-fitting problem such as dropout [41] and denoising code [42].
7. **Data quality:** Building a model does not mean accurate results. If the data used is not of high quality, deep learning models tend not to generalize well such as the number of features and instances. For example, the mismatch between the numbers of features and instances sizes lead to reduced computational accuracy and easily trigger overfitting and hence less generalization ability.
8. **Overconfidence in applying deep learning algorithms:** Besides the data quality, the overconfidence in using DL algorithms stimulates a big problem in the machine learning research communities. The use of DL is depending on the nature of our problem and not by default. Therefore, understanding the nature of the problem and our data are of paramount importance in the choice between DL or traditional machine learning models.

7 Conclusion

The alarming rate at which the COVID-19 virus is spreading globally and the growing number of death cases is a major concern. The pandemic has adversely affected every sector of life–education, agriculture, tourism, and the global economy. Recently, many countries have relaxed lockdown measures and people are gradually returning to their normal life–facing the reality that we may have to live with the virus for a while before the world has a tested and confirmed medical treatment for the virus. Therefore, it is imperative that government agencies start planning for an effective way of living with the virus–such that we can gradually return to our normal livelihood with non or minimal negative consequences of the COVID-19 virus in our communities. A system that can predict future cases of potential infection in an environment will be useful in helping the government make adequate plans–such as increasing the capacities of the isolation centers, recruiting more medical personnel, and proper evaluation of the effectiveness of the control measures in curtailing the spread of the virus in a community. Thus, helping the government to take proactive steps and make informed decisions that can alleviate the negative impacts of the COVID-19 infections. In this research, we have shown the potential of deep learning algorithms in predicting future cases of COVID-19 infection in a community–using China and Canada as a case studies. In the future, the algorithm would be implemented for more countries with additional features such as–probable death and recovery cases.

References

1. S. Sharma, Drawing insights from COVID-19-infected patients using CT scan images and machine learning techniques: a study on 200 patients. Environ. Sci. Pollut. Res. Int., 1-9 (2020)
2. L.J. Muhammad, M.M. Islam, S.S. Usman et al., *Predictive Data Mining Models for Novel Coronavirus (COVID-19) Infected Patients* (Recovery", Springer Nature Computer Science, 2020)
3. L. J. Muhammad, S. S. Usman, Power of artificial intelligence to diagnose and prevent further COVID-19 outbreak: a short communication (2020). arXiv:2004.12463
4. S. K. Saxena , S. Kumar, V. K Maurya , R. Sharma, H. R. Dandu et al., Current Insight into the Novel Coronavirus Disease 2019 (COVID-19). Coronavirus Disease 2019 (COVID-19). 2020, pp. 1–8
5. M. B. Jamshidi et al., Artificial intelligence and COVID-19: deep learning approaches for diagnosis and treatment, in *IEEE Access*, https://doi.org/10.1109/ACCESS.2020.3001973
6. B.A. Ojokoh, O.A. Sarumi, K.V. Salako, A.J. Gabriel, A.E Taiwo, O.V. Johnson, I.P. Adegun, O.S. Babalola, O.S. "Modelling and Predicting the Spread of COVID-19: A Continental Analysis". 2020 Elsevier Data Science for COVID-19
7. A. Oguntimilehin, O. Adetunmbi, I. Osho, Towards achieving optimal performance using stacked generalization algorithm: a case study of clinical diagnosis of malaria fever. Int. Arab J. Inf. Technol. **16**, 1074–1081 (2019)
8. O.A. Sarumi, C.K. Leung, Exploiting anti-monotonic constraints in mining palindromic motifs from big genomic data, in *2019 IEEE Int. Conf. Big Data (Big Data)* (2019)

9. O.W. Samuel, G.M. Asogbon, A.K. Sangaiah, P. Fang, G. Li, An integrated decision support system based on ANN and Fuzzy AHP for heart failure risk prediction. Expert Syst. Appl. **68**(2017), 163–172 (2017)
10. O.A. Jongbo, A.O. Adetunmbi, R.B. Ogunrinde, B. Badeji-ajisafe, Development of an Ensemble Approach to Chronic Kidney Disease Diagnosis, Sci. African (2020)
11. Y. LeCun, Y. Bengio, G. Hinton, Deep learning. Nature **521**, 436–444 (2015)
12. J. Weston, F. Ratle, H. Mobahi, R. Collobert, Deep Learning via Semi-supervised Embedding, in *Neural Networks: Tricks of the Trade* eds. by G. Montavon, G.B. Orr, K. R. Müller. Lecture Notes in Computer Science, 2020, vol. 7700
13. H. Panwar, P.K. Gupta, M.K. Siddiqui, R. Morales-Menendez, P. Bhardwaj, V. Singh. A deep learning and grad-CAM based color visualization approach for fast detection of COVID-19 cases using chest X-ray and CT-Scan images, 2020
14. L.J. Muhammad, E.A. Algehyne, S.S. Usman, Predictive supervised machine learning models for diabetes mellitus. SN Comput. Sci. **1**, 240 (2020)
15. Y. LeCun, Y. Bengio, G. Hinton, Deep learning. Nature **521**(7553), 436 (2015)
16. L. Deng, A tutorial survey of architectures, algorithms, and applications for deep learning. APSIPA Trans. Signal. Info. Process **3**, 1–29 (2014)
17. V. Mnih, K. Kavukcuoglu, D. Silver, A.A. Rusu, J. Veness, M.G. Bellemare, A. Graves, M. Riedmiller, A.K. Fidjeland, G. Ostrovski et al., Human-level control through deep reinforcement learning. Nature **518**(7540), 529–533 (2015)
18. T. M. Navamani, Efficient deep learning approaches for health informatics, in *Deep Learning and Parallel Computing Environment for Bioengineering Systems*, 2019, pp. 123–137
19. M.Z. Alom, T.M. Taha, C. Yakopcic, S. Westberg et al., A state-of-the-art survey on deep learning theory and architectures. Electronics **8**, 292 (2019)
20. Y. Bengio, Learning deep architectures for AI. Found. Trends Mach. Learn. **2**, 1–127 (2009)
21. E. Choi, A. Schuetz, W. F. Stewart, J. Sun Using recurrent neural network models for early detection o Heart failure onset. J. Am. Med Inform. Assoc. **24**, 361–370 (2016)
22. M.D. Zeiler, R. Fergus, Visualizing and understanding convolutional networks. arXiv 2013, arXiv:1311.2901
23. H. W. Ian, F. Eibe, A. H. Mark, J. P. Christopher, Deep learning, Practical Machine Learning Tools and Techniques, 2017, pp. 417–466
24. P. Swietojanski, A. Ghoshal, S. Renals. Convolutional neural networks for distant speech recognition. IEEE Signal Process
25. S. Hayat, S. Kun, Z. Tengtao, Y. Yu, T. Tu, Y. Du, A deep learning framework using convolutional neural network for multi-class object recognition, in *2018 IEEE 3rd International Conference on Image, Vision and Computing (ICIVC)*, Chongqing, 2018, pp. 194–198
26. Y. Zheng, Q. Liu, E. Chen, et al. Time series classification using multi-channels deep convolutional neural networks, in *Proc. of the 15th International Conference on Web-Age Information Management*, 2014, pp. 298–310
27. B. Zhao, H. Lu, S. Chen, J. Liu, D. Wu, Convolutional neural networks for time series classification. J. Syst. Eng. Electron. **28**(1), 162–169 (2017). https://doi.org/10.21629/JSEE.2017.01.18
28. X. Ma, Z. Tao, Y. Wang, H. Yu, Y. Wang, Longshort-termmemory neural network for traffic speed prediction using remote microwave sensor data. Transp. Res. Part C **54**, pp. 187–197 (2015)
29. Y. Liu, Z. Su, H. Li and Y. Zhang, An LSTM based classification method for time series trend forecasting, in *2019 14th IEEE Conference on Industrial Electronics and Applications (ICIEA)* (Xi'an, China, 2019), pp. 402-406, https://doi.org/10.1109/ICIEA.2019.8833725
30. L. Marcus, et al. A novel approach to on-line handwriting recognition based on bidirectional long short-term memory networks, in *Proc. 9th Int. Conf. on Document Analysis and Recognition*. vol. 1 (2007)
31. Y. Kim, Convolutional neural networks for sentence classification, in *Proceedings of the 2014 Conference on Empirical Methods in Natural Language Processing (EMNLP 2014)*, 2014, p. 17461751

32. D. Tran, L. Bourdev, R. Fergus, L. Torresani, and M. Paluri, earning spatiotemporal features with 3D convolutional networks, in 2015 IEEE International Conference on Computer Vision (ICCV), December 2015, pp. 4489–4497
33. O. Aouedi, M. A. Bach Tobji, A. Abraham, An ensemble of deep auto-encoders for healthcare monitoring, in *Hybrid Intelligent Systems*, pp. 96–105 (Springer International Publishing, 2020)
34. O. Aouedi, M. A. Bach Tobji, A. Abraham, Internet of things and ambient intelligence for mobile health monitoring: a review of a decade of research. Int. J. Comput. Inf. Syst. Ind. Manag. Appl. **10**, 261–270 (2018)
35. Q. Mao, F. Hu, and Q. Hao, Deep learning for intelligent wireless networks: a comprehensive survey IEEE Commun. Surveys Tuts. **20**(4), 2595–2621 (2018). 4th Quart
36. F. Ucar, D. Korkmaz. COVIDiagnosis-Net: Deep Bayes-SqueezeNet based diagnosis of the coronavirus disease 2019 (COVID-19) from X-ray images. Med Hypotheses, 2020
37. F. Shahid, A. Zameer, M. Muneeb, *Predictions for COVID-19 with deep learning models of LSTM* (GRU and Bi-LSTM, Chaos Solitons Fractals, 2020)
38. M. Alazab, A. Awajan, A. Mesleh, A. Abraham, V. Jatana, S. Alhyari, COVID-19 prediction and detection using deep learning. Int. J. Comput. Inf. Syst. Ind. Manage. Appl. **12**, 168–181 (2020)
39. S.M.A. Elrahman, A. Abraham, A review of class imbalance problem. J. Network Innov. Comput. **1**, 332–340 (2013)
40. A. Krizhevsky, I. Sutskever, and G. E. Hinton, ImageNet classification with deep convolutional neural networks, in *Proc. Adv. Neural Inf. Process. Syst.*, 2012, pp. 1097–1105
41. N. Srivastava, G.E. Hinton, A. Krizhevsky, I. Sutskever, R. Salakhutdinov, Dropout: a simple way to prevent neural networks fromoverfitting. J. Mach. Learn. Res. **15**(1), 1929–1958 (2014)
42. P. Vincent, H. Larochelle, I. Lajoie et al., Stacked denoising autoen-coders: learning useful representations in a deep network with a local denoising criterion. J. Mach. Learn. Res. **11**, 3371–3408 (2010)
43. S. Minaee, R. Kafieh, M Sonka, S. Yazdani, G. J. SoufiDeep-covid: predicting covid-19 from chest x-ray images using deep transfer learning. Med. Image Anal. (65), 101794 (2020). https://doi.org/10.1016/j.media.2020.101794
44. J. Xie, M. Wang, R. Liu, Deep learning-based COVID-19 diagnosis and trend predictions. In: *Intelligent Systems and Methods to Combat Covid-19* ed. by Joshi A., Dey N., Santosh K. (SpringerBriefs in Applied Sciences and Technology. Springer, Singapore). https://doi.org/10.1007/978-981-15-6572-4_7
45. M. Z. Alom, T. M. Taha, C Yakopcic, S Westberg et al. A state-of-the-art survey on deep learning theory and architectures. Electronics (8): 292. 2019
46. J. Wei, J. He, K. Chen, Y. Zhou and Z. Tang, Collaborative filtering and deep learning based hybrid recommendation for cold start problem, in *IEEE 14th Intl Conf on Dependable, Autonomic and Secure Computing, 14th Intl Conf on Pervasive Intelligence and Computing, 2016*, pp. 874-877, https://doi.org/10.1109/DASC-PICom-DataCom-CyberSciTec.2016.149
47. S. Reynolds, COVID-19 Time Series Data. https://data.world/shad/covid-19-time-series-data. Accessed 10 Oct 2020

Predicting Antiviral Drugs for COVID-19 Treatment Using Artificial Intelligence Based Approach

Ayan Banerjee and Soumen Kumar Pati

Abstract Today's world is seriously affected and medical infrastructure is challenged due to the novel COVID-19 epidemics. On 15th May 2020, Naturemedicine published a detailed survey report which concluded that drugs like aminoquinolines, chloroquine, hydroxychloroquine targeted the virus replication cycle by keeping the virus out of host cells and helping people to overcome the disease. But, the exponential incidence rate cannot be stopped and some limitations also detected to treat the patients using these drugs. In these circumstances, suitable medicines are immediately needed that can also help in vaccine production. In this chapter, a novel AI based architecture has been proposed to obtain influencial drugs for the COVID-19 treatment. Here, The COVID-19 (MT050943 and MT012098) data are taken from the NCBI database, submitted by the Government of India to build tests and find drugs. A detailed analysis of these sequences has been performed to find the non-coding percentage, gene density, and abundances in functional gene categories using Virus Amplicon Sequencing Assembly Pipeline (V-ASAP). This analysis finds the unique regions in the virus genome as a target region. This knowledge is feed into the AI model. Firstly, the informative genes are selected from the target region. Then the Advanced Matched Molecular Pair (AMMP) analysis is performed to study the possible local region of a drug and its impact. At last, a Generative Adversarial Convolution Neural Network (GACNN) is trained with the information gain and performs a gradient-based optimization to predict the drug to target for the validate genome. The result of the proposed methodology obtains higher accuracy than the other existing methods and identified effective drugs (like, Dexamethasone, Remdesivir etc.) for the study of the COVID-19 vaccine.

A. Banerjee
Department of Computer Science and Engineering, Jalpaiguri Goverment Engineering College, Jalpaiguri 735102, WB, India
e-mail: ab2141@cse.jgec.ac.in

S. K. Pati (✉)
Department of Bioinformatics, Maulana Abul Kalam Azad University of Technology, Haringhata, WB, India

J. Nayak et al. (eds.), *Understanding COVID-19: The Role of Computational Intelligence*, Studies in Computational Intelligence 963, https://doi.org/10.1007/978-3-030-74761-9_11

Keywords COVID-19 · Virus amplicon sequencing assembly pipeline · Advanced matched molecular pair · Generative adversarial convolution neural network · Drug repurposing · Genome sequencing

1 Introduction

In today's world, there are nearly 1.2 million people already affected and 54 billion people died till November, 2019 due to this novel COVID 19 epidemics. Due to the lack of effective treatment facilities, various strategies (including drug repurposing) have been applied but the incidence rate is still increasing exponentially all over the world. Recently, all the researchers conclude that hydroxychloroquine targets the virus replication cycle by keeping the virus out of the host cells and helps people to overcome the pandemic situation. The World Health Organization (WHO) also appreciates this experiments but on 17th June, 2020, they announced to stop the Solidarity Trial of hydroxychloroquine for COVID 19 treatment. In these circumstances, proper medicines are immediately needed that can also help in vaccine production. But for drug repurposing, a brief knowledge of the virus genome has been needed. According to Ray et al. [1] and Beck et al. [2], coronavirus strain, and Middle East Respiratory Syndrome Coronavirus (MERS-CoV), and Severe Acute Respiratory Syndrome Coronavirus (SARS-CoV-2), are enveloped, contagious, highly virulent single-stranded RNA virus spread acute respiratory disease all over the world. Till June 2020, the researchers find four genera about this novel coronavirus namely α-coronavirus, β-coronavirus, δ-coronavirus and γ-coronavirus and they mainly spread infection in mammals, birds, and humans. Not only, the strain MERS-CoV and SARS-CoV-2 both caused a high mortality rate in humans (34.4% for MERS-CoV and 9.6% for SARS-CoV-2). So, a quick solution has to be find to get rid of the grasp of this novel coronavirus otherwise once the commonality transmission started the situation became an impossible one to control.

1.1 Literature Review

On 15th March 2020, the World Health Organization (WHO) declared COVID-19 as a pandemic until then several computational and research techniques have been proposed by renowned researchers to reuse the existing drugs. Among them, Zeng et al. [3] find a notable solution to discover therapeutics for COVID-19 and repurpose open data about the disease using a deep learning model. Not only that, Alimadadi et al. [4] also contribute to this fight using artificial intelligence and machine learning techniques. Besides that, Gysi et al. [5] identified the drug repurposing opportunities using the network medicine framework. Wang et al. [6] boosted this research by constructing a literature knowledge graph and publishing a report of the drug repurposing. Mohanty et al. [7] also contributed to this fight by showing the applications of drug repurposing with the help og artificial intelligence in COVID-19. Along with this, now there is a toolkit DeepPurpose proposed by Huang et al. [8] for drug repurposing. But this era is not ending here. Bullock et al. [9] mapped the landscape applications of artificial intelligence to continue this fight. Ioannidis et al. [10] used the graph neural networks to repurpose the drugs via shot link prediction. Li et al. [11] provide the network bioinformatics analysis to fight against the COVID-19 pandemic. Among this Zhavoronkov et al. [12] provides a notable contribution to find potential noncovalent protease inhibitors using a generative deep learning model. A contribution also comes from Ke et al. [13] to fight against the coronavirus using drug repurposing techniques. Avchaciov et al. [14] also use artificial intelligence techniques to investigate new drugs in this scenario. Han et al. [15] accelerate this research by proposing a drug mechanism model applied with large scale gene expression profiles. Along with this, Naudé et al. [16] shows the constraints, limitations, and pitfalls of artificial intelligence to resist the COVID-19 paandemic. Besides that, Liu et al. [17] performed the ontological analysis of anti-coronavirus drugs and showed the implication of drug repurposing to fight against COVID-19 pandemic. Later for treatment purposes, Heiser et al. [18] proposed an artificial intelligence enabled phonemic analysis of infected human cells to identify potential treatments against the pandemic. Patankar et al. [19] also makes a notable contribution to computational drug discovery to inhibit the RNA dependent RNA polymerase. In between, Baker et al. [20] repurpose the quaternary ammonium compound as a potential treatment for COVID-19 but recently, WHO stops the trial of quaternary ammonium compounds. Besides that, Skelton et al. [21] predicts the drug repurposing for the COVID-19 using probabilistic networks and crowd sourced curation. Last but not least, Sanders et al. [22] obtains a detailed review of the pharmacologic treatment to fight against this pandemic. But this research does not end here. A group of researchers across the world are engaged to fight against this novel coronavirus pandemic. But according to the findings, these are the most relevant literature for this chapter to date. All the existing methodology discussed above are validated using precision, sensivity, specificity, accuracy, f1-score, cohen's kappa. The comparative study is shown is the result section.

1.2 *Objective and Proposed Outcome*

All the state-of-the-art approaches discussed above are basically worked upon the protein data which is not so fundamental as the genome data is still not available for all. But from April 2020, WHO makes the genome data available for all. Not only that, several review articles also show the pros and cons of the existing artificial intelligence techniques and state the reason for less accurate drug repurposing against this pandemic. To overcome the mentioned drawbacks this chapter proposed a novel AI based methodology for identification of the potential drugs and their target regions for the treatment of the COVID-19 from the knowledge of genome sequencing. Two notable sequences of genomes (MT050943 and MT012098) have been taken from the NCBI database, submitted by the Government of India, to build tests and find drugs or vaccines. Then, a detailed analysis of these sequences has been performed to find the noncoding percentage, gene density, and abundances in functional gene categories using Virus Amplicon Sequencing Assembly Pipeline (V-ASAP). This analysis finds out the unique regions in the virus genome as a target region for drug development. This knowledge is fed into the proposed AI model where first the feature extraction [23] of the region is performed through a biological interpretation based on dimension reduction technique. Then, it checks the validity of the region by our proposed validation algorithm. Along with this, the Advanced Matched Molecular Pair (MMP) analysis is performed to study the possible localized change of a drug and its impacts on the selected region. In the Later stage, a Generative adversarial convolution neural network (GACNN) is trained with the information gain and performs a gradient-based optimization to predict the drug target for the validated genome. In the result section, a liberal comparison this method with the existing methodology like AlphaFold [24], DeepChem [25], MOSES [26], Biopandas [27], ReLeaSe [28], HTMD [29], Gnina [30], DeepNeuralNet-QASR [31], NNScore [32], SIEVE Score [33], Hit Dexter [34], etc. has been performed and concluded that proposed methodology achieves higher accuracy in drug development from the rest. At the end of this chapter, it obtains with the idea that drugs like Dexamethasone, Remdesivir, Nafamostat, Famotidine, Camostat, Umifenovir, Nitazoxanide, Ivermectin, Tocilizumab, Sarilumab, Bevacizumab, Fluvoxamine, and the mixture of Nafamostat and Camostat, Tocilizumab and sarilumab, and Dexamethasone and Remdesivir should be considered as a better treatment for in the development study for the COVID-19 vaccine.

2 Methodology

In this section, the chapter describes the Artificial Intelligence based approach to predict the most suitable chemical components for COVID-19 using the knowledge of genome sequencing. This proposed methodology has three defined layers. After passing through all the steps, the most suitable drugs can be obtained for the experimental target genome with higher accuracy. So, it can increase computational power in terms of accuracy as well as computational time. The Fig. 1 is showing a brief idea of the overall methodology of this chapter.

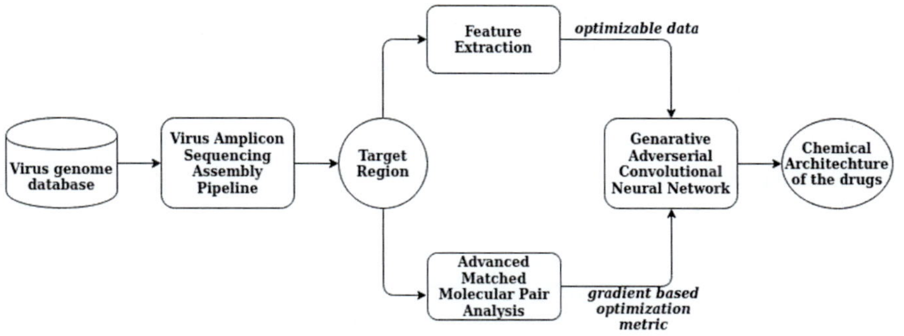

Fig. 1 Overall flow sequence of the proposed method

2.1 *Virus Amplicon Sequencing Assembly Pipeline*

It is a data development stage to perform the genome sequencing. Though, this chapter is using the synthetic data from the NCBI database (MT050943 and MT012098) but with the prior knowledge of data collection, the error factor will be increased in the drug prediction. Figure 2 gives a brief idea about this V-ASAP process.

As described by Maurier et al. [24], it is to convert the extracted RNA to cDNA through hexamers primer and reverse transcriptase enzyme. These cDNAs are amplified by PCR with Precision Master Mix. The obtained multiple primer pairs are then divided into several groups according to the experimental samples (here, 12 groups). Then, the amplified products are purified through NucleoMag NGS and the purified amplicons are quantified by Qubit dsDNA HS. Lastly, these amplicons are pooled together after normalize to the same equimolar concentratio to collect one pool per sample. This pooled DNA formed sequencing libraries with NxSeq AmpFREE Low DNA Library Kit having a 6 bases index quantified by the KAPA library and checked by microfluidic electrophoresis. Then the pools are merged to form the Reads. These Reads are then filtered, sorted by amplicon, trimmed by primer sequences, and formed clusters through CD-HIT for each amplicon. Lastly, these Reads clusters are assembled with CAP-3 to obtain the target region.

2.2 *Feature Extraction*

The target regions obtain through V-ASAP are in unstructured, discrete form and not suitable to feed in any AI algorithm. The suitable biological interpretation based novel component analysis technique has been used to convert it. Figure 3 provides a brief idea of this step.

Here, the discrete target region $(t_1, t_2, \ldots, t_n)$ is obtained through V-ASAP and clustered on the basis of their ontological similarity. Firstly, a mapping is performed between the genomes in $(t_1, t_2, \ldots, t_n)$ to the virus genome ontological library. As

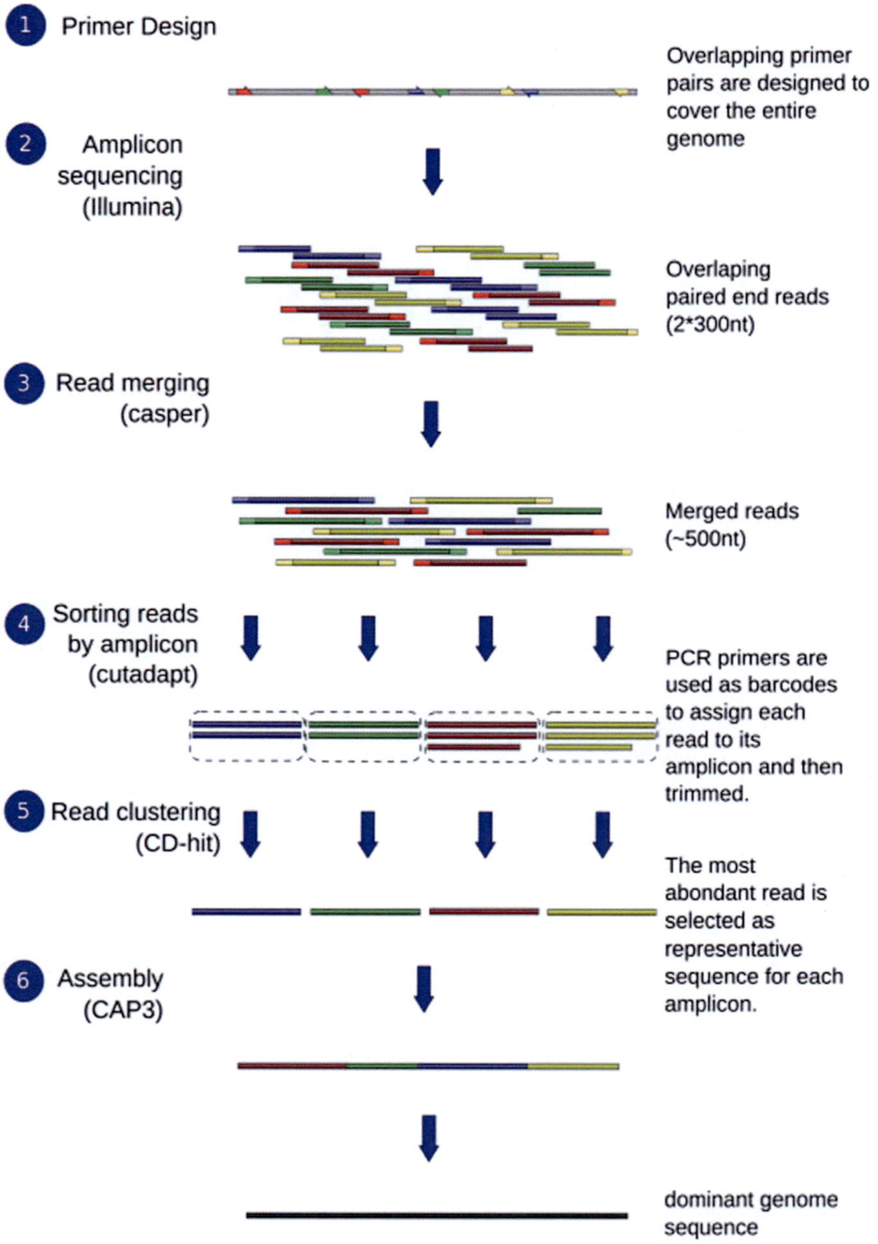

Fig. 2 Virus Amplicon Genome Sequencing

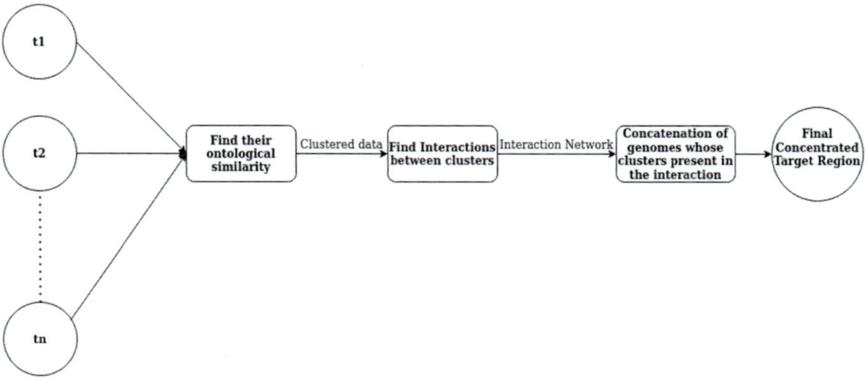

Fig. 3 Flow diagram for feature extraction

a result, the genes are clustered according to their biological process with targeted cellular components. This mapping is defined in the Eq. (1).

$$\theta_{ij} = \theta_j - \alpha * \frac{1}{mn} * \sum_{i=1}^{m} \sum_{j=1}^{n} (h_\theta(\theta_{ij}) - \theta_j) * \theta_{ij} \tag{1}$$

Where, θ_{ij} refers to the jth genome of the ith target region, α is the similarity index, m, n are the no. of the target region, and number of the genomes belong to that target region, respectively. Here, h_θ is the hypothesis of the ontological similarity which is extracted the functional group from their respective genome class.

Still, this clustered data is in discretized form. Now, the genome interaction knowledge is needed to form connections between them. To obtain this, protein-protein interaction from these clustered genes has been used (describes in Algorithm 1). It obtains an adjacency matrix where 1/0 represents the presence/absence of connections. From the adjacency matrix, if the connection between two clusters exist, then these clusters are collected and concatenate using Eq. (2).

$$P_i || P_j = P_i * b^{l(P_j)} + P_j \tag{2}$$

Where, P_i and P_j are the present interactions in the network and b is the base of the concatenation. Compute the length of the clustered genome sequence using Eq. (3).

$$l(x) = \lfloor log_b x \rfloor + 1 \tag{3}$$

Compute the length of the clustered genome sequence. After completing this step, the final concentrated target region feasible to feed any AI technique.

Algorithm 1: *Finding Protein-Protein Interactions from clustered data*

```
Input: A data matrix d[i][j] having genome clusters information.
Output: An adjacency matrix represents the presence/absence of
interactions of genome clusters.
begin
    For each i in n:    //n is the number of genes
        For each j in m:     // m is the number of samples
            W[i][j] = d[i][j]*w[i][j];
        End for
    End For
    For each i in n:
        For each j in (i+1, m):
            link.list  = W[i,j-1];
    // Find the link i.e. interactions of genes
            if(W[i][j-1]   W[i][j]) then
                link.list  = link.list  W[i,j];
            End if
        End for
    End for
    For each k in link.list:
        if(link.list[k]  link.list[k+1]) then
            Delete link.list[k]; // Remove the chance of duplicacy.
        End if
    End for
    Obtain the undirected graph from the link.list.
    Find the adjacency matrix of the graph.
    Obtain the graph data frame with only the presence interactions.
    Obtain the network from the presence interactions.
end.
```

2.3 Advanced Matched Molecular Pair (AMMP) Analysis

The AMMP analysis searches the location of changes of a single drug candidate on the basis of molecular properties and bioactivity of the Protein-Protein Interaction Network. It is a combined method of the supervised and unsupervised technique. The supervised method generates the chemical transformation and unsupervised methods find all possible pairs between a set of chemical compounds based on protein structure. The Fig. 4 Shows the steps involved in AMMP analysis.

Definition and Explanation AMMP Analysis: Some important definition needed in this section is as follows.

Definition 1 (MMP): It is a process to predict the bioactivity of the amino acid structure with substitution through property optimization and determine the chemical environment for drug discovery related to the input disease database.

Definition 2 (Chemical Environment Matching): It is matched the corresponding Chemical Environment based on similar core and addictive property. For that purpose, it breaks the amino acid structure until getting any known molecular organic structure

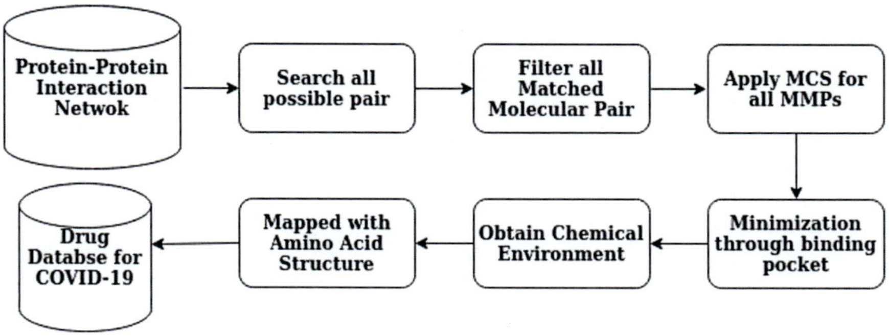

Fig. 4 Advanced matched molecular Pair(AMMP) analysis

(i.e. Aryl group, Alkyl group, etc.). Then it's checking their core and addictive/non-addictive property. If they have a similar core and addictive property, then established a match between the corresponding chemical environment is named on the core.

Example 1: This is described in the Fig. 5, where an amino acid compound is divided into its subgroup until any known chemical structure is found. Here, all the subgroups have the same core (benzene) and they have addictive properties. So, they can be matched through AMMP and the chemical environment is obtained. And, all the drugs are included into drug database whose core is benzene.

Definition 3 (Maximum Common Substructure(MCS)): It is a metric to search chemical similarity and predict its activity. The MCS of two chemical compounds is the largest substructure that appears in both.

AMMP Analysis: The potential of the AMMP algorithm (Algorithm 2) all molecules of the amino acid structure of the target region have been extracted that is calculated their affinity using Eq. (4). If it gets higher affinity, then it is proceeded to the next step where it calculates the structure activity landscape index (SALI) using Eq. (5).

$$A_{s_1,s_2} = |log_1 0(a_{m_2}) - log_1 0(a_{m_1})| \tag{4}$$

Where, A_{s_1,s_2} is the overall affinity index between s_1 and s_2 structure, and $a_{m_i} = K_i/K_d \pm val$ is the affinity of the a_{m_i} molecule K_i and K_d is the Kappa index of the molecule and whole data respectively.

$$SALI_{i,j} = \frac{|Ac_i - Ac_j|}{|1 - sim(Ac_i, Ac_j)|} \tag{5}$$

Where, Ac_i and Ac_j are the activity of the ith and j^{th} molecule, respectively. The protein-protein interaction network of the target region obtained in Subsect. 2.2 and $sim(Ac_i, Ac_j)$ is the similarity coefficient between them.

The AMMP analysis algorithm based on the above facts is defined as follows.

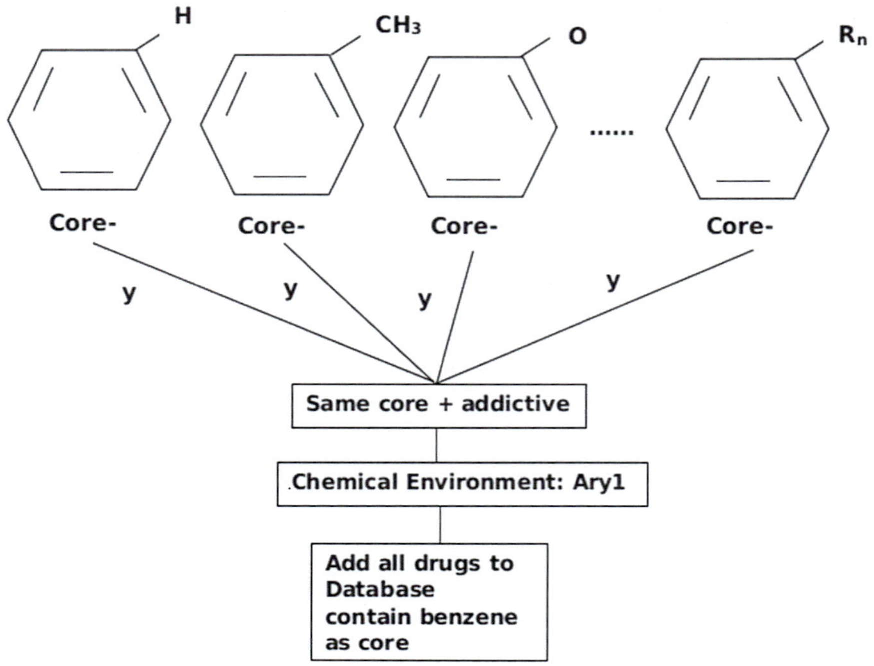

Fig. 5 Example of AMMP analysis

Algorithm 2: *Advanced Molecular Matched Pair Analysis.*

```
Input: Amino acid structure of the target region obtained in Subsection 2.2.
Output: Drug database.
begin
Function subgroup(Amino acid structure(AAS)):
    If AAS -> atomic then:
        Return AAS;
    Else:
        Divide the AAS into a smaller subgroup;
    End if.
End.
Function AMMP(AAS):
Struct= subgroup(AAS);
For i in Struct:
    For j:i+1 to Struct:
        If A_{struct_i,struct_j} -> 1then:
            For i in molecule:
                For j:i+1 to molecule:
                    If SALI_{i,j} -> 1 then:
                        Find the core of the molecule.
                        Find all drugs having the same core.
                        Add them to the drug database.
                    End if.
                End for.
            End for.
        End if.
```

```
    End for.
End for.
End.
end.
```

After this algorithm, core work has been created for drug discovery. This information is now fed into the Generative Adversarial Convolutional Neural Network (GACNN) (describe in Subsect. 2.4) to predict the antiviral drugs suitable to treat the COVID-19 disease.

2.4 *Generative Adversarial Convolutional Neural Networks*

It has two different parts for two different models one is called a generative model and another is the discriminator model. The generative model is created the data suitable for the prediction which is fed into the discriminator model and makes the prediction. A brief overview of the architecture is obtained in Fig. 6.

Generative Model: The input features from previous sections are transformed into feature space to calculate their potential to feed into the next layer using the query function (defined in Eq. (6)) and key function (defined in Eq. (7)).

$$q(x) = W_q \times x \tag{6}$$

$$k(x) = W_k \times x \tag{7}$$

Where, x is the input features and W_f and W_k is the weighted connectivity of the previous node and current node, respectively. From this knowledge, the data are synthesized for the next node using Eq. (8).

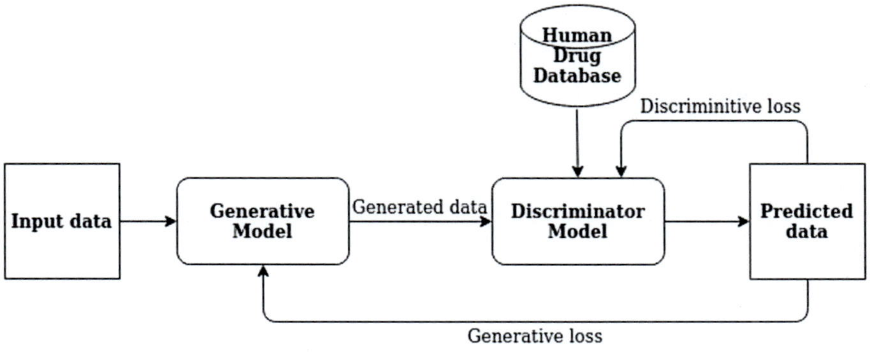

Fig. 6 The overall architecture of the proposed AI technique

$$B_{j,i} = \frac{e^{s_{ij}}}{\sum_{i=1}^{n} e^{s_{ij}}} \tag{8}$$

Where, the synthesized data is defined in Eq. (9).

$$s_{ij} = q(x_i)^T \times k(x_j) \tag{9}$$

The synthesized data is now fed into the next layer where a hypothesis is applied to this input data using Eqs. (10) and (11). And the output data is obtained using Eq. (12).

$$h(x) = W_h \times v_i \tag{10}$$

$$v_i = (W_v \times x_i)^T \tag{11}$$

Where, h(x) and v represent the hypothesis and value, respectively.

$$O_j = v_i \times (\sum_{i=1}^{n} B_{j,i} * h(x_i)) \tag{12}$$

Now, this output is feed into the discriminative layer and it is scaled by learning γ to obtain final output as defined in Eq. (13).

$$y_i = \gamma \times O_j + x_j \tag{13}$$

After all the information are gathered to train the generative one, the model is trained in an alternating fashion to optimize the adversarial loss using Eq. (14).

$$L_g = min(e^{h(x_i)} - e^{h(y_i)}) \tag{14}$$

A brief idea of this generative model is shown in Fig. 7 and the data generates in this stage is ready to feed into the discriminator level to obtain the final prediction.

The Discriminative Model: Here, a four-layer one dimensional CNN has been used for the drug prediction. The Fig. 8 shows the proposed architecture for this prediction consisting of four convolutions, four pooling, and two fully connected layers. The convolutional layers are convoluted with stride 1 having 16 filters and Kernel size 3×1. In Conv-1, the input vector is 229×1 is obtained and output obtained through ReLu activation with element-wise non-linear activation. This output feature map works as an input in the pool-1. Here, the pooling size is 2×1 with stride 2. It extracts the maximum element of the feature map and obtains a 114×1 feature map as output. This feature goes as an input in Conv-2 with kernel size 3×1 and stride 1 gives the 112×1 feature map. Similarly, this output goes through as an input in pool-2 which gives 56×1 vector as an output. Conv-3 receives this as input and produces an output vector of 54×1 after convolution. The pool-3 works on it

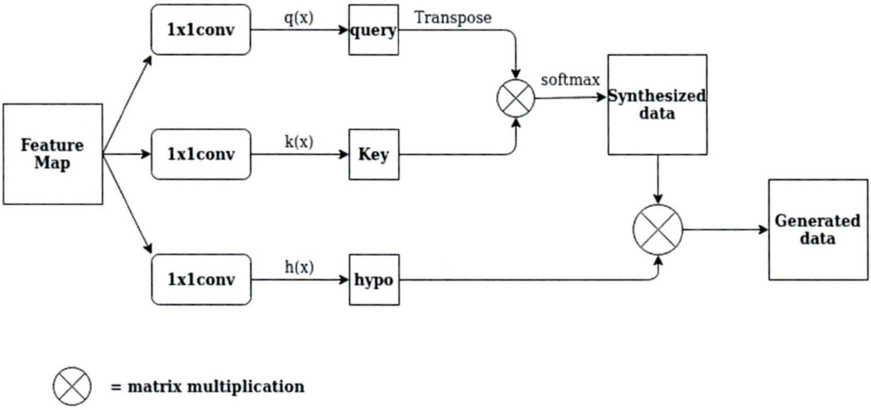

Fig. 7 The generative model

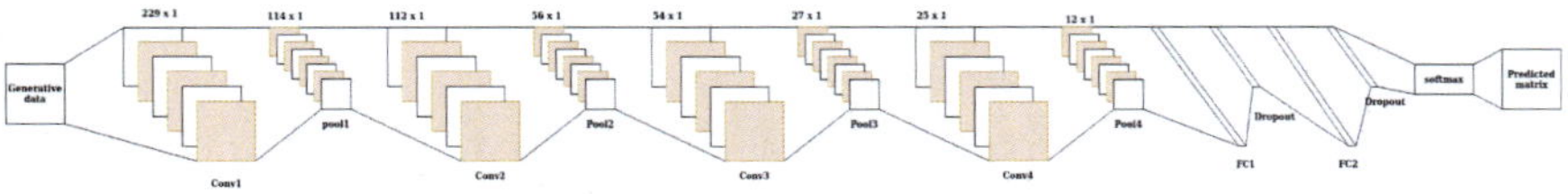

Fig. 8 The architecture of the convolutional neural network

and produces an output of 27 × 1. Later, this output is fed into the Conv-4 which produces an output of size 25 × 1. This vector is fed into the pool-4 which reduces the size is 12 × 1. Here, one thing should be notified that all the Conv layer has kernel size 3 × 1 and the pooling layer has 2 × 1 kernels but they all work in stride 1. The 12 × 1 vector is obtained in the pool-4 now goes through a fully connected layer. In this architecture, there are two fully connected layers FC1 and FC2 which are connected with two dropout layers and self-activated. Here, the weight and bias of all the layers are initialized with random values having $\mu = 0$ and $\sigma = 1$. These weights and bias are updated accordingly, the deviation from actual to the predicted probability distribution. Lastly, the softmax cross-entropy loss has been adopted to obtain the loss function and optimize the network through Adam optimizer and trained it with a learning rate of 0.00001.

After performing all the steps listed above, this experiment is concluded with most significant drugs to resist the novel COVID-19 pandemic. In the next section, all the results are obtained corresponding to the methodology, discuss their importance, and make a liberal comparison with a similar strategy to prove its efficiency.

3 Result Analysis and Performance Evaluation

This experiment is performed on two publicly available benchmark complete genome data those are taken from the NCBI database submitted by the Government of India (MT050943 [35] and MT012098 [36]). These genome data contain all significant genomes are collected from Riboviria, Orthornavirae, Pisuviricota, Pisoniviricetes, Nidovirales, Cornidovirineae, Coronaviridae, Orthocoronavirinae, Betacoronavirus, Sarbecovirus which are linearly inseparable. All the features of these dataset are listed in Table 1.

3.1 *Performance Evaluation*

Virus Amplicon Sequencing Assembly Pipeline: A cluster size distribution study has been performed here to validate the dominant variation for each amplicon, where it defines the number of the identical reads. Figures 9 and 10 are shown the distribution for MT050493 and MT012098, respectively.

The above distribution concludes that the number of genomes can be varied between the clustered amplicons and it is obtained discrete genome clusters. But, it is maintained the dominant sequencing systematically between the clusters. Not only that, this variability problem is shown by the non-overlapping region and not suitable for further analysis. For this reason, all these clusters are connected through the knowledge of the protein-protein interaction to obtain a single target region.

Feature Extraction: The cluster of genome data is obtained through V-ASAP in discrete form and not suitable for further analysis. This step prepares the data are obtained according to Subsect. 2.2 in a form to perform the AMMP analysis and feed the data into the AI model. Actually, it obtains an interaction network from

Table 1 Dataset summary

Data	Biological process	Cellular components	Assembly method
MT050493	29851	orf3a protein orf6 protein orf7a protein orf8 protein orf10 protein envelope protein orf1ab polyprotein membrane glycoprotein surface glycoprotein nucleocapsid phosphoprotein	CLC Genomics v. 11.1
MT012098	29854		

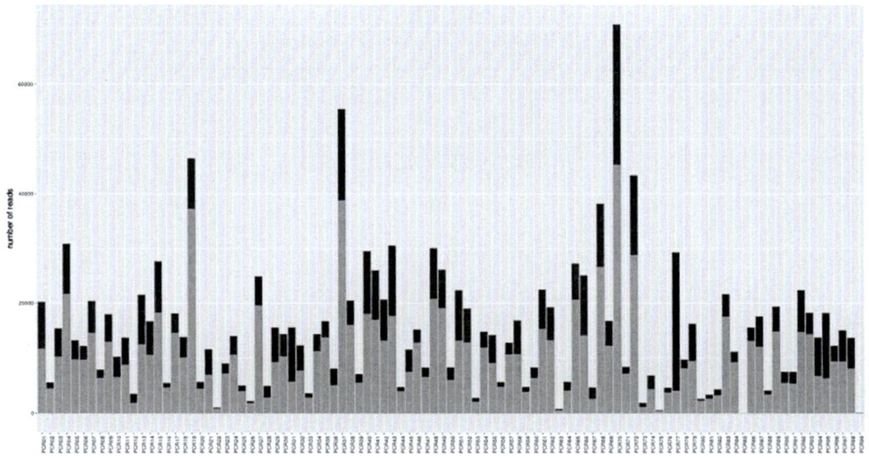

Fig. 9 Distribution of cluster of the amplicon for MT050493 sample

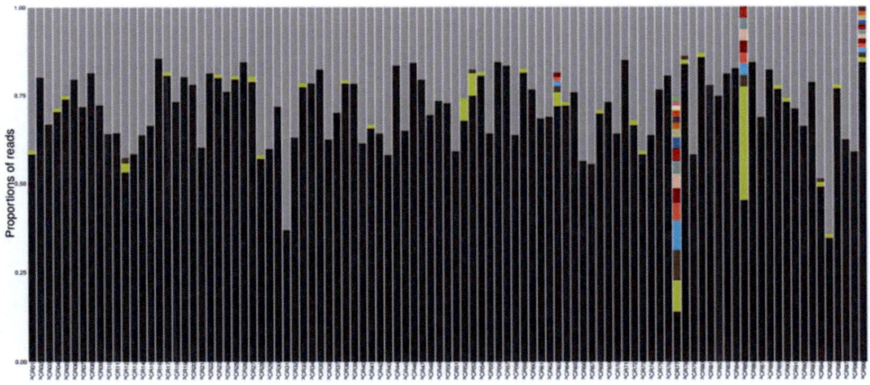

Fig. 10 Distribution of cluster of the amplicon for MT012098 respectively

all the genome clusters by analyzing the characteristics of their respective protein groups. Figures 11 and 12 show the results of the PPI network of MT050493 and MT012098, respectively. The target region is obtained in this section are used for the AMMP analysis in the next step for drug prediction.

Advanced Molecular Matched Pair Analysis: The main objective of this section is to obtain the chemical environment suitable for the drug prediction and then find the core similarity and addictive property between them and listed all drugs having the same core to obtain a final drug database suitable to feed into the AI module. Figures 13 and 14 show the AMMP analysis results. Figure 13 shows the classification of 97 drugs having four similar core (red, purple, sky, and blue dot region) based on the correlation probability score. Figure 14 shows the listing of drugs into the drug database based on bioactivity and molecular reactivity. As the drug database

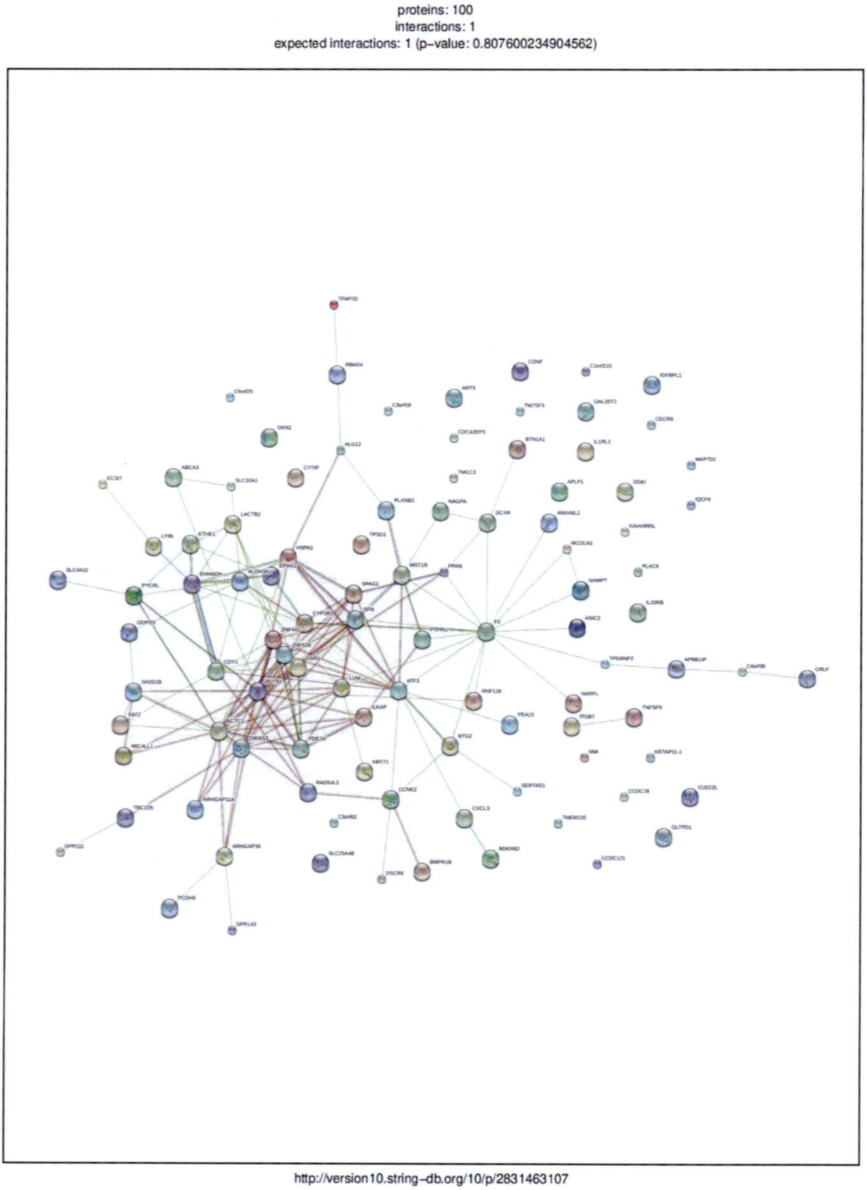

Fig. 11 Target region for drug prediction of MT050493 dataset

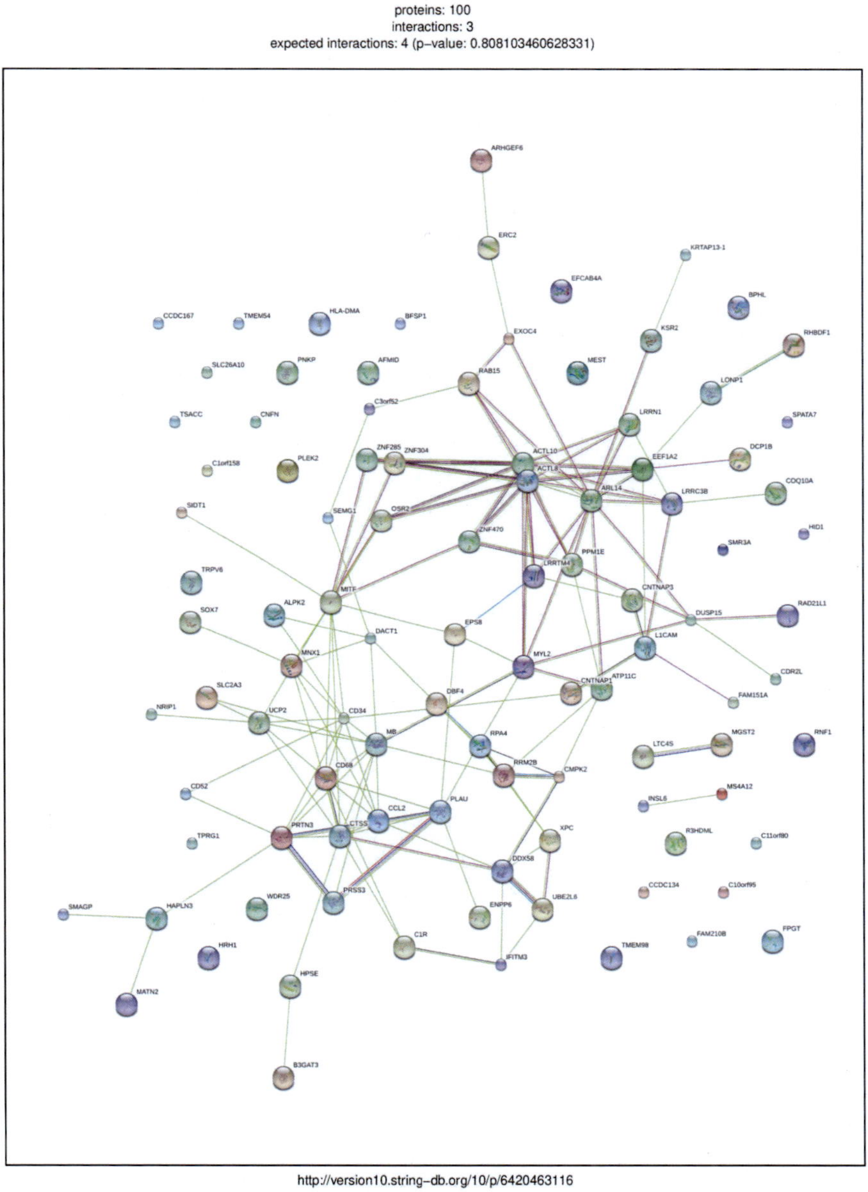

Fig. 12 Target region for drug prediction of MT012098 dataset

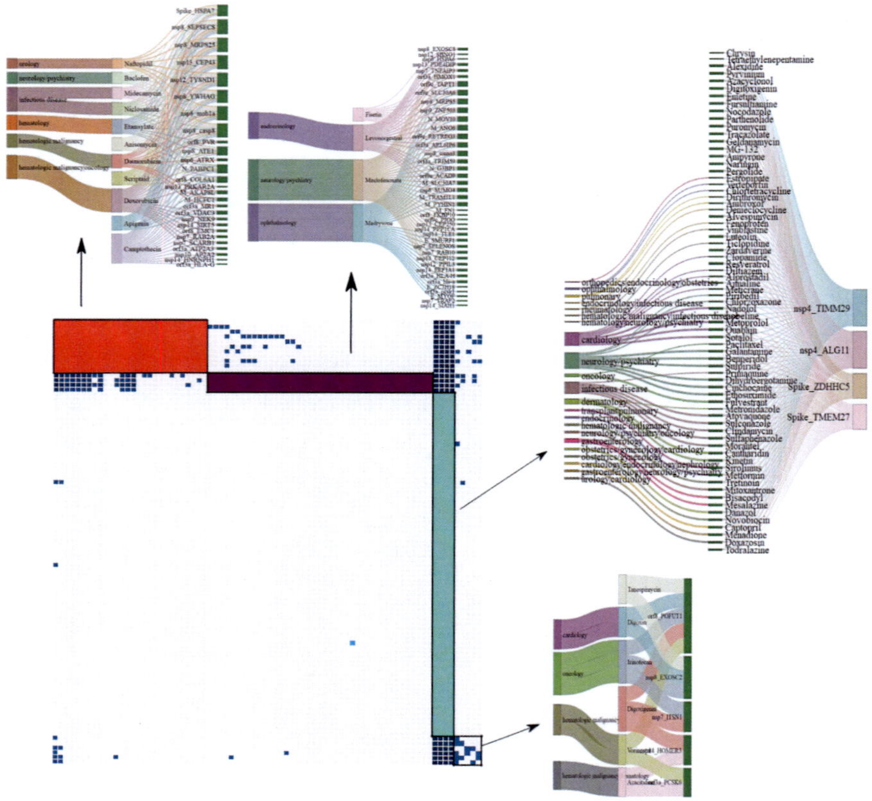

Fig. 13 AMMP analysis of MT050493: correlation viewpoint

is obtained in this step and target region obtained in the previous step (according to Subsect. 2.3), let's feed both the data into GACNN for further prediction.

3.2 Generative Adversarial Convolution Neural Network

For the drug-genome interaction prediction, the GACNN has been exploited where the target region is given as one input in the form of adjacency matrix, (1/0) represents the presence and absence of the clustered genome interaction and the drug database are given as other input. A gradient-based optimization has been performed where the interaction coefficient is obtained as a gradient to reduce the side effects of the drugs. The performance analysis of both the dataset is obtained in Table 2. The description of the predicted drugs, their chemical property, and significance are listed in Table 3. Figure 15 shows the ROC curve for both the dataset. From the Table 2 and Fig. 15, it can be concluded that the used GACNN is 98% accurate for the attain

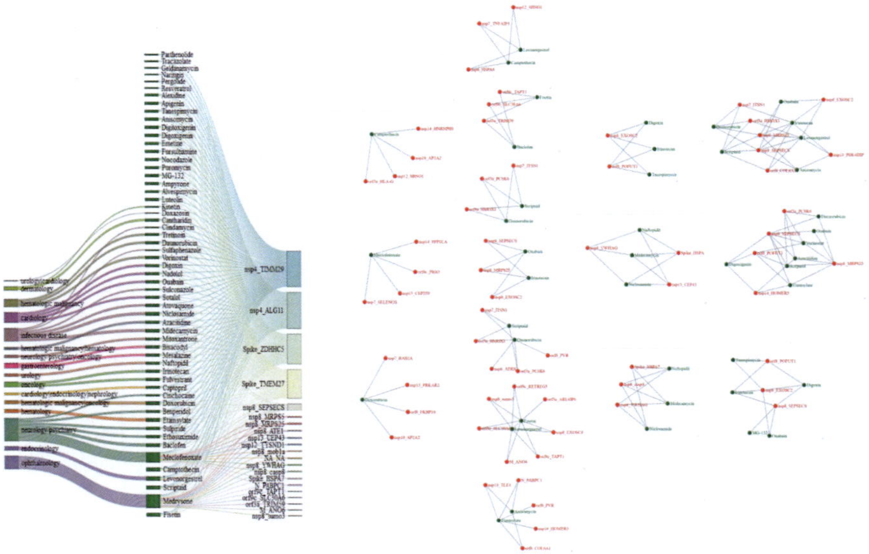

Fig. 14 AMMP analysis of the MT012098 dataset: interaction viewpoint

Table 2 Performance evaluation of GACNN

Statistical parameters	Dataset	
	MT050493	MT012098
Precision	0.9394	0.9412
Sensitivity	0.9211	0.9233
Specificity	0.9413	0.9422
Accuracy	0.9837	0.9872
f1-score	0.9635	0.9622
Cohens-kappa	0.9213	0.9122

target regions. The method is predicted that Dexamethasone, Remdesivir, Camostat, Nafamostat, Famotidine, Umifenovir, Ivermectin, Nitazoxanide, Tocilizumab, sarilumab, Fluvoxamine, Bevacizumab, Baclofen, Clopamide, Digoxin, Ouabain, Tanespimycin, Camptothecin, Alvespimycin, Emetine and the mixture of Nafamostat and Camostat, Tocilizumab and Sarilumab, and Dexamethasone and Remdesivir are obtained as most suitable drugs to treat the COVID-19 diseases. Though, it is predicted 20 drugs out of 97 drugs to fight against the disease which are listed in Table 3 with highest percentage of accuracy.

From Table 2, Fig. 15, and Table 3, it can be concluded that the proposed methodology satisfied the objectives of this chapter. But, to obtain the generosity of this approach a detailed comparative study has been made in Subsect. 3.3.

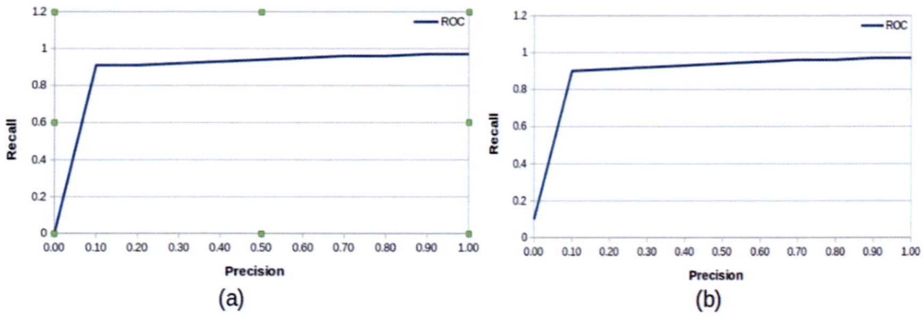

Fig. 15 ROC curve: (a) MT050493 and (b) MT012098 dataset

Table 3 Predicted drug description

Drugs predicted	Chemical components	Usage	% of accuracy
Dexamethasone	$C_{22}H_{29}FO_5$	Inhibits the enzyme phospholipase A2 and blocks the synthesis of the inflammatory mediators	96.93
Remdesivir	C27H35N6O8P	Inhibits viral RNA polymerases	96.37
Nafamostat	C19H17N5O2	Block the entry of SARS-CoV-2 into cells	95.67
Camostat	C20H22N4O5	Serine protease inhibitors	95.45
Famotidine	C8H15N7O2S3	Papain-like protease that is encoded by the SARS-CoV-2 genome	95.22
Umifenovir	C22H25BrN2O3S	Binding the viral lipid membrane and affecting cellular trafficking of the virus	94.77
Nitazoxanide	C12H9N3O5S	Blocking maturation of the viral nucleocapsid	94.32
Ivermectin	C47H72O14(H2B1b)	Destabilizing cell-transport proteins used to enter the nucleus	94.11
Tocilizumab	C6428H9976N1720O2018S42	Block different cytokines	93.78
Sarilumab	C6388H9918N1718O1998S44	Monoclonal-antibody antagonists of the IL-6 receptor	93.44
Bevacizumab	C6638H10160N1720O2108S44	Reduce vascular permeability	93.08
Fluvoxamine	C15H21F3N2O2	Selective serotonin reuptake inhibitors	92.92
Baclofen	C10H12ClNO2	Benzodiazepine receptor agonist	92.73
Clopamide	C14H20ClN3O3S	Cotransporter inhibitor	92.68
Digoxin	C41H64O14	ATPase inhibitor	92.31
Ouabain	C29H44O12	ATPase inhibitor	92.01
Tanespimycin	C31H43N3O8	HSP inhibitor	91.67
Camptothecin	C20H16N2O4	Topoisomerase inhibitor	91.30
Alvespimycin	C32H48N4O8	HSP inhibitor	90.92
Emetine	C29H40N2O4	Protein synthesis inhibitor	90.51

Table 4 Comparative study

Methods	MT050493						**MT012098**					
	pre	*sens*	*spec*	*acc*	*f1*	*CK*	*pre*	*sens*	*spec*	*acc*	*f1*	*CK*
AlphaFold [23]	0.8322	0.8482	0.8324	0.8412	0.8262	0.8116	0.8312	0.8348	0.8415	0.8211	0.8166	0.8121
DeepChem [24]	0.8421	0.8583	0.8425	0.8313	0.8363	0.8217	0.8413	0.8449	0.8516	0.8312	0.8267	0.8112
MOSES [25]	0.8521	0.8693	0.8623	0.8514	0.8564	0.8418	0.8614	0.8411	0.8717	0.8513	0.8466	0.8313
Biopandas [26]	0.8721	0.8713	0.8824	0.8715	0.8665	0.8519	0.8715	0.8612	0.8818	0.8614	0.8568	0.8614
ReLeaSe [27]	0.8922	0.8800	0.8996	0.8896	0.8778	0.8611	0.8817	0.8894	0.8899	0.8815	0.8879	0.8715
HTMD [28]	0.8234	0.8211	0.8412	0.8332	0.8413	0.8213	0.8312	0.8213	0.8411	0.8124	0.8221	0.8491
Gnina[[29]	0.8312	0.8382	0.8321	0.8411	0.8261	0.8117	0.8312	0.8342	0.8416	0.8211	0.8165	0.8121
DeepNeuralNet-QASR [30]	0.8411	0.8483	0.8422	0.8312	0.8366	0.8211	0.8413	0.8444	0.8518	0.8312	0.8266	0.8111
NNScore [31]	0.8511	0.8593	0.8622	0.8512	0.8566	0.8411	0.8613	0.8412	0.8718	0.8512	0.8466	0.8311
SIEVEScore [32]	0.8711	0.8613	0.8822	0.8712	0.8666	0.8511	0.8710	0.8612	0.8812	0.8612	0.8566	0.8611
Hit Dexter [33]	0.8912	0.8900	0.8997	0.8892	0.8777	0.8612	0.8812	0.8891	0.8897	0.8812	0.8877	0.8712
GACNN	0.9394	0.9211	0.9413	0.9837	0.9635	0.9213	0.9412	0.9233	0.9422	0.9872	0.9622	0.9122

3.3 Comparative Study

From the performance evaluation step, it is now an established fact that the proposed methodology has the potential to fight against the novel COVID-19 pandemic by predicting suitable antiviral drugs. But as mentioned to make it generic and check its novelty a liberal comparison is required in this section with the existing AI based drug discovery techniques. The same dataset and the same experimental setup have been used to obtain this comparative study. This comparison is based on the six well accepted statistical parameters (i.e. precision (pre), sensitivity (sens), specificity (spec), accuracy (acc), f1-score (f1), and Cohens-kappa (CK)). The result is listed in the Table 4.

From the Table 4, it can be concluded that the proposed methodology is generic and provide a better result than the existing methodology. Not only that, but it also takes less time only (22.88 min) for training where other methodology takes several hours even days for training whereas GACNN needs only a few data for training which are selected from original dataset. Also, it has a quick response time in drug predicting (1.023 s). It can be said that this is a novel methodology has a huge potential to fight against the novel COVID-19 pandemic by considering all the above facts. Figures 16 and 17 show the comparison with respect to time and response, respectively.

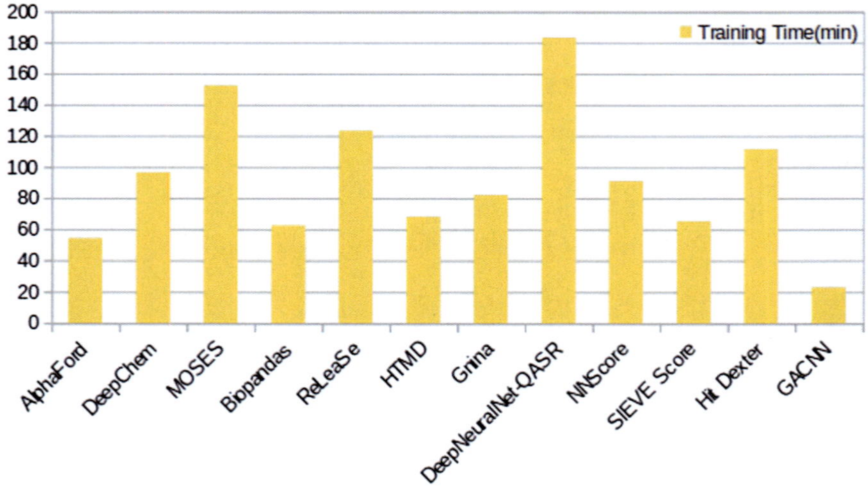

Fig. 16 A comparative study based on training time

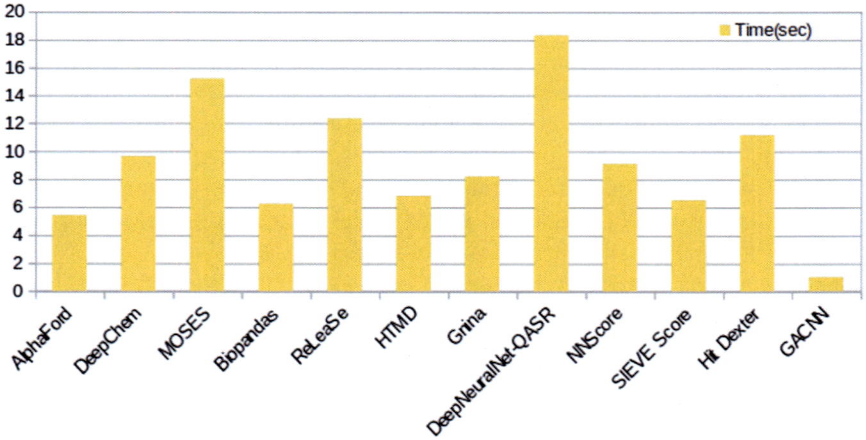

Fig. 17 Comparative Study based on the prediction response time

4 Discussion

A novel structure has been proposed in this chapter to find the most suitable antiviral drugs to fight against the novel COVID-19 pandemic. This virus is now spread worldwide and already 10 million people died due to this acute respiratory disease. Several researchers are involved to find a vaccine against this virus but till now it has very low success rate.

This chapter tried to predict antiviral drugs through artificial intelligence techniques on the basis of the knowledge of genome sequencing. The genome datasets are sequenced through V-ASAP to obtain discrete genome clusters. Then from these clusters, a feature extraction has been performed and from these features, a protein-protein interaction network has been obtained to create a unique target region. An AMMP analysis has been performed to obtain the corresponding chemical environments and obtain a drug database based on core similarity. At last, the target region and drug database is feed into a GACNN to predict the drugs to treat the disease formed by COVID-19. This model has 98% accuracy and very less training time (22.88 min) and quick response time (1.023 s) and performs better than existing approaches.

But some section needs to be updated in the future. Till now it can predict the drug and mixture of two or three drugs to fight against the disease. But, it unable to predict what is the chemical composition (in mg/L) of the mixture is needed for the perfect treatment.

References

1. S. Ray, S. Lall, A. Mukhopadhyay, S. Bandyopadhyay, A. Schönhuth, Predicting potential drug targets and repurposable drugs for COVID-19 via a deep generative model for graphs. arXiv preprint arXiv:2007.02338 (2020)
2. B.R. Beck, B. Shin, Y. Choi, S. Park, K. Kang, Predicting commercially available antiviral drugs that may act on the novel coronavirus (SARS-CoV-2) through a drug-target interaction deep learning model. Comput. Struct. Biotechnol. J. **18**, 784–790 (2020)
3. X. Zeng, X. Song, T. Ma, X. Pan, Y. Zhou, Y. Hou, Z. Zhang, K. Li, G. Karypis, F. Cheng, repurpose open data to discover therapeutics for COVID-19 using deep learning. J. Proteome Res. **19**(11), 4624–4636 (2020)
4. A. Alimadadi, S. Aryal, I. Manandhar, P.B. Munroe, B. Joe, X. Cheng, Artificial intelligence and machine learning to fight COVID-19. Physiol. Genomics **52**, 200–202 (2020)
5. D.M. Gysi, Í.D. Valle, M. Zitnik, A. Ameli, X. Gan, O. Varol, H. Sanchez, R.M. Baron, D. Ghiassian, J. Loscalzo, A.L. Barabási, Network medicine framework for identifying drug repurposing opportunities for COVID-19. arXiv preprint arXiv:2004.07229 (2020)
6. Q. Wang, M. Li, X. Wang, N. Parulian, G. Han, J. Ma, J. Tu, Y. Lin, H. Zhang, W. Liu, A. Chauhan, COVID-19 Literature Knowledge Graph Construction and Drug Repurposing Report Generation. arXiv preprint arXiv:2007.00576 (2020)
7. S. Mohanty, M.H.A. Rashid, M. Mridul, C. Mohanty, S. Swayamsiddha, Application of Artificial Intelligence in COVID-19 drug repurposing. Diabetes & Metabolic Syndrome: Clinical Research & Reviews (2020)
8. K. Huang, T. Fu, C. Xiao, L. Glass, J. Sun, DeepPurpose: a Deep Learning Based Drug Repurposing Toolkit. arXiv preprint arXiv:2004.08919 (2020)
9. J. Bullock, K.H. Pham, C.S.N. Lam, M. Luengo-Oroz, Mapping the landscape of artificial intelligence applications against COVID-19. arXiv preprint arXiv:2003.11336 (2020)
10. V.N. Ioannidis, D. Zheng, G. Karypis, Few-shot link prediction via graph neural networks for Covid-19 drug-repurposing. arXiv preprint arXiv:2007.10261. 20 July 2020
11. X. Li, J. Yu, Z. Zhang, J. Ren, A.E. Peluffo, W. Zhang, Y. Zhao, K. Yan, D. Cohen, W. Wang, Network bioinformatics analysis provides insight into drug repurposing for COVID-2019 (2020)
12. A. Zhavoronkov, B. Zagribelnyy, A. Zhebrak, V. Aladinskiy, V. Terentiev, Q. Vanhaelen, D.S. Bezrukov, D. Polykovskiy, R. Shayakhmetov, A. Filimonov, M. Bishop, Potential non-covalent SARS-CoV-2 3C-like protease inhibitors designed using generative deep learning approaches and reviewed by human medicinal chemists in virtual reality (2020)

13. Y.Y. Ke, T.T. Peng, T.K. Yeh, W.Z. Huang, S.E. Chang, S.H. Wu, H.C. Hung, T.A. Hsu, S.J. Lee, J.S. Song, W.H. Lin, Artificial intelligence approach fighting COVID-19 with repurposing drugs. Biomed. J. **43**(4), 355–362 (2020)
14. K. Avchaciov, O. Burmistrova, P.O. Fedichev, AI for the repurposing of approved or investigational drugs against COVID-19. Research Gate 10 (2020)
15. L. Han, G.C. Shan, H.Y. Wang, S.Q. Gao, W.X. Zhou, Accelerating drug repurposing for COVID-19 via modeling drug mechanism of action with large scale gene-expression profiles. arXiv preprint arXiv:2005.07567. May 15 2020
16. W. Naudé, Artificial intelligence vs COVID-19: limitations, constraints and pitfalls. Ai Soc. **35**(3), 761–765 (2020)
17. Y. Liu, W.K. Chan, Z. Wang, J. Hur, J. Xie, H. Yu, Y. He, Ontological and bioinformatic analysis of anti-coronavirus drugs and their Implication for drug repurposing against COVID-19 (2020)
18. K. Heiser, P.F. McLean, C.T. Davis, B. Fogelson, H.B. Gordon, P. Jacobson, B.L. Hurst, B.J. Miller, R.W. Alfa, B.A. Earnshaw, M.L. Victors, Identification of potential treatments for COVID-19 through artificial intelligence-enabled phenomic analysis of human cells infected with SARS-CoV-2. bioRxiv (2020)
19. S. Patankar, Deep learning-based computational drug discovery to inhibit the RNA Dependent RNA Polymerase: application to SARS-CoV and COVID-19 (2020)
20. N. Baker, A.J. Williams, A. Tropsha, S. Ekins, Repurposing quaternary ammonium compounds as potential treatments for COVID-19. Pharm. Res. **37**, 1–4 (2020)
21. D.J. Skelton, A. Alsobhe, E. Anastasi, C. Atallah, J. Bird, B. Brown, D. Didon, P. Gater, K. James, D.D. Lennon Jr, J. McLaughlin, Drug repurposing prediction for COVID-19 using probabilistic networks and crowdsourced curation. arXiv preprint arXiv:2005.11088 (2020)
22. J.M. Sanders, M.L. Monogue, T.Z. Jodlowski, J.B. Cutrell, Pharmacologic treatments for coronavirus disease 2019 (COVID-19): a review. Jama **323**(18), 1824–1836 (2020)
23. A. Banerjee, S.K. Pati, M.K. Gupta, dimension reduction of microarray data using gene ontology and correlation filter. In: Computational Intelligence in Pattern Recognition (Springer, Singapore, 2020), pp. 303–313
24. M. AlQuraishi, AlphaFold at CASP13. Bioinformatics **35**(22), 4862–4865 (2019)
25. B. Ramsundar, K. Leswing, Molecular machine learning with DeepChem. In: Proceedings of the Abstracts of Papers of the American Chemical Society: Amer Chemical Soc (2018)
26. E.L. MosesKolko, J.C. Price, M.E. Thase, C.C. Meltzer, D.J. Kupfer, C.A. Mathis, W.D. Bogers, S.R. Berman, P.R. Houck, T.N. Schneider, W.C. Drevets, Measurement of 5HT1A receptor binding in depressed adults before and after antidepressant drug treatment using positron emission tomography and [11C] WAY100635. Synapse **61**(7), 523–530 (2007)
27. S. Raschka, BioPandas: Working with molecular structures in pandas DataFrames. J. Open Source Softw. **2**(14), 279 (2017)
28. C.C. Lin, A.T. Metters, Hydrogels in controlled release formulations: network design and mathematical modeling. Adv. Drug Delivery Rev. **58**(12–13), 1379–1408 (2006)
29. S. Doerr, M.J. Harvey, F. Noé, G. De Fabritiis, HTMD: high-throughput molecular dynamics for molecular discovery. J. Chem. Theor. Comput. **12**(4), 1845–1852 (2016)
30. D. Koes, GNINA: Deep learning for molecular docking. In: Abstracts of papers of the American chemical society (Vol. 256). 1155 16TH ST, NW, Washington, DC 20036 USA: Amer Chemical Soc (August 2018)
31. V. Rausch, A. Hansen, E. Solowjow, C. Liu, E. Kreuzer, J.K. Hedrick, Learning a deep neural net policy for end-to-end control of autonomous vehicles. In: 2017 American Control Conference (ACC), pp. 4914–4919. IEEE (May 2017)
32. J.D. Durrant, J.A. McCammon, NNScore 2.0: a neural-network receptor–ligand scoring function. J. Chem. Inf. Model. **51**(11), 2897–2903 (2011)
33. B. Houshmand, N. Janbakhsh, F. Khalilian, M.R.T. Ardakani, Efficacy of conventional laser irradiation versus a new method for gingival depigmentation (Sieve method): a clinical trial. J. Lasers Med. Sci. **8**(2), 88 (2017)
34. C. Stork, Y. Chen, M. Sicho, J. Kirchmair, Hit Dexter 2.0: machine-learning models for the prediction of frequent hitters. J. Chem. Inf. Model. **59**(3), 1030–1043 (2019)

35. V. Potdar, M. Choudhary, P.D. Yadav, A. Shete-Aich, Maximum Containment Laboratory, National Institute of Virology, Pashan, Pune, Maharashtra 411021, India. www.ncbi.nlm.nih.gov/nuccore/MT050493
36. P.D. Yadav, V. Potdar, P. Abraham, Maximum Containment Laboratory, National Institute of Virology, Pashan, Pune, Maharashtra 411021, India. www.ncbi.nlm.nih.gov/nuccore/MT012098

Geographic Spread and Control of 2019-nCoV in the Absence of Vaccine

Himadri Shekhar Mondal, Motahar Reza, and Prasanta Sahoo

Abstract Presently the world is facing an extremely tough time due to the prevalence of the Novel Coronavirus, 2019-nCoV or COVID-19, which has been declared a pandemic by WHO. The virus usually transmits via droplets of saliva or discharge from the nose when an infected person coughs or sneezes. Since there is no vaccine to prevent the disease, social distancing and proper quarantine of infected persons are needed. To include and quantify the spatial effect of the pandemic regarding the geotemporal development of 2019-nCoV, a mathematical model of partial differential equations is essential. In this chapter a diffusion model has been developed by dividing the total population of constant size into four classes: susceptible population, infected population, quarantined population and recovered population. Here the disease transmission factor for both infected and quarantined population into the susceptible population are in more general form. Looking for a travelling wave solution this model can give the wave speed at which the disease 2019-nCoV spread. Additionally it can be predicted whether the total population in forward time will become susceptible or not in the absence of vaccine.

Keywords Epidemic model · Diffusion model · Reproductive rate · Re-parametrisation · Phase plane analysis · Wave speed

1 Introduction

Viruses existed on Earth since the evolution of first living cells. They are the most numerous biological entity on Earth and can live in almost every ecosystem. Viruses can infect all life forms including human beings. Therefore viral infections always had terrible impact over human life and welfare. The recent pandemic throughout the globe is due to the novel coronavirus 2019, commonly designated as COVID-19 [29]. Though human coronaviruses are not new but COVID-19, a new strain, is dangerous

H. S. Mondal · P. Sahoo
Department of Mathematics, Midnapore College (Autonomous), Midnapore 721101, WB, India

M. Reza (✉)
Department of Mathematics, GITAM Deemed to be University, Hyderabad 502329, India
e-mail: mreza@gitam.edu

J. Nayak et al. (eds.), *Understanding COVID-19: The Role of Computational Intelligence*,
Studies in Computational Intelligence 963, https://doi.org/10.1007/978-3-030-74761-9_12

in the seance that it is exceptionally infectious [4]. China country office of WHO, for the first time on December 31st, 2019, come to know about pneumonia cases with causes not known being identified which was found in Wuhan city, Hubei city in China. Up to January 3rd, 2020, a total of 44 cases of pneumonia with unknown cases were reported. Subsequently on January 7th, 2020, it was identified that the pathogenic agent behind the cases of pneumonia was coronavirus of a new strain. On January 13th, 2020, the first imported case of novel coronavirus (2019-nCoV) was reported by the Ministry of Public Health, Thailand [11, 17, 28]. With the passage of time the whole world have been hit by the rapid spread of COVID-19. As reported by WHO [29] a total of 21294845 confirmed cases of COVID-19 positive and total of 761779 deaths have taken place by 16th August, 2020. To predict the disease pattern for COVID-19, the reported data for COVID-19 infection is essential. In this chapter the reported data of COVID-19 infection for 217 countries from all over the world is considered, among which the following Table 1 shows a list of top 12 countries with respect to transmission of COVID-19 from 21st January, 2020 to 16th August, 2020.

In Table (1): (i) 'Cumulative COVID-19 infection cases' interpret the total number of infected individuals from 21st January, 2020 to 16th August 2020, due to COVID-19. (ii) 'New cases in last 24 h of 16th August, 2020' represent the number of new COVID-19 patient who are newly infected in last 24 h of 16th August 2020. (iii) 'Cumulative deaths due to COVID-19' interpret the total number of individuals who are died from the beginning of COVID-19 to 16th August 2020, due to COVID-19. (iv) 'New deaths in last 24 h of 16th August, 2020' represent the number of new COVID-19 patient who are dies out in last 24 h of 16th August 2020.

Table 1 Laboratory- confirmed COVID-19 cases and deaths of 12 countries in the World topping in the reported list: data as of 16th August, 2020 [29].

Country name	Cumulative COVID-19 infection cases	New cases in last 24 h of 16th August, 2020	Cumulative deaths due to COVID-19	New deaths in last 24 h of 16th August, 2020
United States of America	5258565	55359	167201	1206
Brazil	3275520	50644	106523	1060
India	2589682	63490	49980	944
Russian Federation	922853	4969	15685	68
South Africa	583653	4513	11677	121
Peru	516296	8300	25856	208
Mexico	511369	5618	55908	615
Colombia	445111	11306	14492	347
Chile	383902	1791	10395	55
Spain	342813	0	28617	0
Iran	341070	2245	19492	161
The United Kingdom	316371	0	41358	1

A high mortality rate is associated with acute respiratory distress syndrome in patients with COVID-19 pneumonia and it is a major cause of deaths in SARS-COV 2 infected patients. It has also been reported that Myocardial injury has strong correlation with severe outcome for patients who are without that much cardiovascular problem but elevated tropopin level. Surveys have indicated that persons with co-morbidity such as hypertension and cardiovascular disease, obesity, type-II diabetics as well as type-I diabetics are at risk of severe COVID-19, [26]. Severe acute respiratory syndrome in COVID-19 infection might be more worsening factor for people with diabetes as it can have direct negative effects on β-cell function resulting in diabetic ketoacidosis.

From the beginning of global pandemic named as COVID-19 human population of all over the world has started to face an extremely tough time. Scientist from all over the world are trying to predict the future of the disease and that how to control it in the absence of vaccine.

The present know how of the actions and menifestations of the novel coronaviruses SARS-CoV, MERS-CoV and COVID-19 in the senario of deadly COVID-19, and pointout their differences and similarities are established by [13]. [18] provided an updated information on COVID-19. According to them the spread of SARS-CoV-2 is not restricted on particular age group. Any individual of human population can be effected by this virus [35]. A kit for microscopic slide named as "Virus Goes Viral" is designed by [8]. This can explore some aspects of animal virus interactions with cells. [31] used Nested RT-PCR of RNA-dependent RNA polymerase (RdRp) cistron from rectal swabs for CoV spotted and the breed of CoV was affirmed by ordering. [1] predicted the covid-19 positive cases by using the deep learning approach. They analysed a descriptive case study by using Deep learning (RNN). [27] developed a classical SEIR-type epidemic model on Covid-19 pandemic in India by introducing the contact tracing and other interventions such as social distancing, isolation, quarantine, lockdown etc. They demonstrated that COVID-19 is heavily depends on the enforcement of quarantine, isolation and precautionary measures. [10] analysed a dynamical population model on the epidemic spreading of Covid-19 in China, Germany and Italy. [34] employed a segmented Poisson model to examine the new cases data of covid-19 outbreaks in western countries, to predict the turning point of the attack rate. They also analysed the duration period and the percentage of attack rate of total population of these countries. [12] has developed a paper based on autophagy and analysed the system with the help of statistical analysis. A dynamics of contagion model of COVID 2019 is represented by [22]. In that model they studied the local stability of the disease free equilibrium by calculating the canonical reproduction number. [7] has developed a BHRP network model for the simulation of potential transmission for COVID-19. In this paper they estimate the canonical Reproduction Number $\mathcal{R}_0$ and studied the stability analysis of the system with $\mathcal{R}_0$. [19] have analysed the dynamical behaviour of the COVID in terms of the canonical reproduction number. Further, a short-term trend of COVID-19 for highly affected states in India like Maharashtra, Delhi, and Tamil Nadu is predicted by them. An additional mechanism on Susceptible-Infectious-Quarantined- Recovered (SIQR) model is analysed by [6]. The collected data at Rio de Janeiro, they implemented the policies of social isolation [2]. A control-theoretical epidemic model to predict the variation of disease in many state in India where infection rate is high and new identification method was used for investigate [25]. The mathematical models on the population dynamics for COVID in Nigeria are discussed by [16, 23] to examine the impact of various non-pharmaceutical controls. [30] estimated the different tend of the epidemic COVID-19 and their control strategies.

Any natural phenomena can be described by mathematical model. Corona virus usually transmits via salivary droplets or nasal discharge from coughs or sneezes of an infected person. As a result there is a high risk of spread of COVID-19 through contact. For the daily life survival, people have to move in their surroundings. Then the virus can spread out as a result of the movement of infected individuals. When such a movement in regularity for a particular population, then a diffusion process can be thought of. Thus to incorporate and express the measurements of spatial effects. Thus to include and quantify spatial effects of COVID-19, a diffusion model is essential to describe the spread of COVID-19 [3, 5].

In this chapter a more general diffusion model is considered. This model can analyse the stability of the pandemic COVID-19 and can predict the wave speed of COVID-19 at which the disease is spread.

2 The Model

Let Ω be an open subset of $\mathbb{R}^n$. For $(x, t) \in \Omega \times (0, \infty)$, let $S(x, t)$ is the susceptible population, $I(x, t)$ is the infected population, $Q(x, t)$ is the quarantine population, $R(x, t)$ is the recovered population. Then the model for the geographic spread of COVID-19 is

$$\frac{\partial S}{\partial t} = D_S \frac{\partial^2 S}{\partial x^2} - \frac{\alpha_1 I^{p_1}}{1 + \beta_1 I^{q_1}} S - \frac{\alpha_2 Q^{p_2}}{1 + \beta_2 Q^{q_2}} S \tag{1}$$

$$\frac{\partial I}{\partial t} = D_I \frac{\partial^2 I}{\partial x^2} + \frac{\alpha_1 I^{p_1}}{1 + \beta_1 I^{q_1}} S + \frac{\alpha_3 Q^{p_3}}{1 + \beta_3 Q^{q_3}} I - \gamma_1 I \tag{2}$$

$$\frac{\partial Q}{\partial t} = D_Q \frac{\partial^2 E}{\partial x^2} + \frac{\alpha_2 Q^{p_2}}{1 + \beta_2 Q^{q_2}} S - \frac{\alpha_3 Q^{p_3}}{1 + \beta_3 Q^{q_3}} I - \gamma_2 Q \tag{3}$$

$$\frac{\partial R}{\partial t} = D_R \frac{\partial^2 R}{\partial x^2} + \gamma_1 I + \gamma_2 Q \tag{4}$$

$\frac{\partial S}{\partial \eta} = \frac{\partial I}{\partial \eta} = \frac{\partial Q}{\partial \eta} = \frac{\partial R}{\partial \eta} = 0, x \in \partial\Omega, t > 0$
$S(x, 0) = S_0(x), I(x, 0) = I_0(x), Q(x, 0) = Q_0(x), R(x, 0) = R_0(x), x \in \overline{\Omega}$.
where D_S, D_I, D_Q and D_R are the diffusion coefficients for susceptible, infected, quarantined and recovered population respectively; γ_1 and γ_2 are respectively the

rate at which I and Q release their individuals from their respective classes. For non-linear transmission rate [14], the rate of infectiousness due to I class in susceptible population is measured by $\alpha_1 I^{p_1} S$ and the reduced infectiousness occur from behavioural change as the number of infected individuals increases is given by $\frac{1}{(1+\beta_1 I^{q_1})}$, with $1 \leq p_i \leq q_i + 1$, $i = 1, 2, 3$. The other terms are defined in similar way. Since the terms for the disease transmission are Lipschitz continuous function of $S(x,t)$, $I(x,t)$ and $Q(x,t)$ in open first quadrant. It is possible to extend their definition in the entire first quadrant by assigning its value as zero whenever $S = 0$ or $I = 0$ or $Q = 0$. Also for the positive number of infected and quarantined population we must have

$$\int_\Omega I(x,0)dx > 0 \text{ and } \int_\Omega Q(x,0)dx > 0$$

with $S(x,0) \geq 0$, $I(x,0) \geq 0$, $Q(x,0) \geq 0$ and $R(x,0) \geq 0$ in Ω.

Then the maximum principle for semi-linear parabolic system [9, 24] ensure the existence of positive and bounded solution for the system (1)–(4) in $\overline{\Omega} \times [0, \infty)$. Define

$$N = \int_\Omega [S(x,0) + I(x,0) + Q(x,0) + R(x,0)]\,dx > 0$$

as the total population at time $t = 0$ in Ω. Now from the system (1)–(4) we have

$$\frac{\partial}{\partial t} \int_\Omega [S(x,0) + I(x,0) + Q(x,0) + R(x,0)]\,dx$$

$$= \int_\Omega \frac{\partial^2}{\partial x^2} \left[D_S S(x,t) + D_I I(x,t) + D_Q Q(x,t) + D_R R(x,t)\right] dx = 0, \forall\, t > 0.$$

This shows that the total population in Ω is always constant and is given by $N = \int_\Omega [S(x,t) + I(x,t) + Q(x,t) + R(x,t)]\,dx, \forall\, t > 0.$

3 The Basic Reproduction Number $\mathcal{R}_0$

Let $U = (U_S, U_I, U_Q, U_R) = (S, I, Q, R)$, $d = (D_S, D_I, D_Q, D_R)$, $f(U,x) = (f_S(U,x), f_I(U,x), f_Q(U,x), f_R(U,x))$ and

$f_S(U,x) = -\frac{\alpha_1 I^{p_1}}{1+\beta_1 I^{q_1}} S - \frac{\alpha_2 Q^{p_2}}{1+\beta_2 Q^{q_2}} S,$

$f_I(U,x) = \frac{\alpha_1 I^{p_1}}{1+\beta_1 I^{q_1}} S + \frac{\alpha_3 Q^{p_3}}{1+\beta_3 Q^{q_3}} I - \gamma_1 I,$

$f_Q(U,x) = \frac{\alpha_2 Q^{p_2}}{1+\beta_2 Q^{q_2}} S - \frac{\alpha_3 Q^{p_3}}{1+\beta_3 Q^{q_3}} I - \gamma_2 Q$ and

$f_R(U,x) = \gamma_1 I + \gamma_2 Q.$

Then the system (1)–(4) can be represented as

$$\frac{\partial U_i}{\partial t} = \nabla.(D_i \nabla U_i) + f_i(U,x), t > 0, x \in \Omega, \forall\, i \in \{S, I, Q, R\}$$

That is,

$$\begin{bmatrix} \frac{\partial S}{\partial t} \\ \frac{\partial I}{\partial t} \\ \frac{\partial Q}{\partial t} \\ \frac{\partial R}{\partial t} \end{bmatrix} = \begin{bmatrix} \nabla.(D_S \nabla U_S) \\ \nabla.(D_I \nabla U_I) \\ \nabla.(D_Q \nabla U_Q) \\ \nabla.(D_R \nabla U_R) \end{bmatrix} + \begin{bmatrix} f_S(U, x) \\ f_I(U, x) \\ f_Q(U, x) \\ f_R(U, x) \end{bmatrix}$$

with the boundary condition $\frac{\partial U_i}{\partial \eta} = (0, 0, 0, 0), t > 0, x \in \partial\Omega, \forall i \in \{S, I, Q, R\}$.

Let $f(U, x) = \mathcal{F}(U, x) - \mathcal{V}(U, x)$ and $\mathcal{V}(U, x) = \mathcal{V}^-(U, x) - \mathcal{V}^+(U, x)$, where the ith component of

$$\mathcal{F}(U, x) = \big(\mathcal{F}_S(U, x), \mathcal{F}_I(U, x), \mathcal{F}_Q(U, x), \mathcal{F}_R(U, x)\big)$$
$$= \left(0, \frac{\alpha_1 I^{p_1}}{1+\beta_1 I^{q_1}} S + \frac{\alpha_3 Q^{p_3}}{1+\beta_3 Q^{q_3}} I, \frac{\alpha_2 Q^{p_2}}{1+\beta_2 Q^{q_2}} S, 0\right)$$

represents the rate at which a new infected individual enter in the ith compartment,
$\mathcal{V} = \big(\mathcal{V}_S(U, x), \mathcal{V}_I(U, x), \mathcal{V}_Q(U, x), \mathcal{V}_R(U, x)\big)$
$= \left(\frac{\alpha_1 I^{p_1}}{1+\beta_1 I^{q_1}} S + \frac{\alpha_2 Q^{p_2}}{1+\beta_2 Q^{q_2}} S, \gamma_1 I, \frac{\alpha_3 Q^{p_3}}{1+\beta_3 Q^{q_3}} I + \gamma_2 Q, -\gamma_1 I - \gamma_2 Q\right)$,

the ith component of $\mathcal{V}^-(U, x)$ represents the removal rate of the individuals from the ith compartment and the ith component of $\mathcal{V}^+(U, x)$ represents the rate at which an individual transferred in the ith compartment by means of birth, migration etc. for all $i \in \{S, I, Q, R\}$. Then the above system becomes

$\frac{\partial U_i}{\partial t} = \nabla.(D_i \nabla U_i) + \mathcal{F}_i(U, x) - \big(\mathcal{V}_i^-(U, x) - \mathcal{V}_i^+(U, x)\big), t > 0, x \in \Omega, \forall i \in \{S, I, Q, R\}$
$\frac{\partial U_i}{\partial \eta} = (0, 0, 0, 0), t > 0, x \in \partial\Omega, \forall i \in \{S, I, Q, R\}$.

In this model total population can be separated into two compartments. One is the uninfected compartment which contain S and R classes and the other is the infected compartment which contain I and Q classes. Then the disease free state for the total population can be defined as $U_0 = (\frac{N}{|\Omega|}, 0, 0, 0)$, where $|\Omega|$ represents the volume of Ω. Define $U_{Disease\ free} = (S, R)$ and $U_{Infected} = (I, Q)$ for disease free compartment and Infected compartment respectively; $d_{Disease\ free} = (D_S, D_R)$ and $d_{Infected} = (D_I, D_Q)$ as diffusion coefficients for disease free compartment and Infected compartment respectively. Also
$f_{Disease\ free} = (f_S, f_R) = \left(-\frac{\alpha_1 I^{p_1}}{1+\beta_1 I^{q_1}} S - \frac{\alpha_2 Q^{p_2}}{1+\beta_2 Q^{q_2}} S, \quad \gamma_1 I + \gamma_2 Q\right)$ and

$$f_{Infected} = (f_I, f_Q) = \left(\frac{\alpha_1 I^{p_1}}{1+\beta_1 I^{q_1}} S + \frac{\alpha_3 Q^{p_3}}{1+\beta_3 Q^{q_3}} I - \gamma_1 I, \quad \frac{\alpha_2 Q^{p_2}}{1+\beta_2 Q^{q_2}} S - \frac{\alpha_3 Q^{p_3}}{1+\beta_3 Q^{q_3}} I - \gamma_2 Q\right)$$

Let $\nabla.(d_{Disease\ free} \nabla U) = (\nabla.(D_S \nabla S), \nabla.(D_R \nabla R))^T = (D_S \frac{\partial^2 S}{\partial x^2}, D_R \frac{\partial^2 R}{\partial x^2})^T$, here T stands for transpose. Also we assume that the diffusion coefficients are constant.

Then the linear diffusion system for the uninfected compartment in a neighbourhood of the disease free equilibrium point $(\frac{N}{|\Omega|}, 0, 0, 0)$ is given by

$\frac{\partial U_{Disease\ free}}{\partial t} = \nabla.(d_{Disease\ free} \nabla U_{Disease\ free}) + M_0 U_{Disease\ free}, t > 0, x \in \Omega$
$\frac{\partial U_i}{\partial \eta} = (0, 0), t > 0, x \in \partial\Omega,$
for all $i \in \{S, R\}$ and $M_0 = \begin{bmatrix} \frac{\partial f_S}{\partial S}(U_0, x) & \frac{\partial f_S}{\partial R}(U_0, x) \\ \frac{\partial f_R}{\partial S}(U_0, x) & \frac{\partial f_R}{\partial R}(U_0, x) \end{bmatrix} = \begin{bmatrix} 0 & 0 \\ 0 & 0 \end{bmatrix}.$
Then M_0 is cooperative on $\overline{\Omega}$, i.e. all of its off diagonal elements are non-negative. From the properties of $\mathcal{F}(U, x)$ and $\mathcal{V}(U, x)$ it is obvious that

$$D_U \mathcal{F}(U_0, x) = \begin{bmatrix} F(x) & 0 \\ 0 & 0 \end{bmatrix} \quad D_U \mathcal{V}(U_0, x) = \begin{bmatrix} V(x) & 0 \\ J(x) & -M_0(x) \end{bmatrix}.$$

where $F(x) = \begin{bmatrix} \frac{\partial \mathcal{F}_I}{\partial I}(U_0, x) & \frac{\partial \mathcal{F}_I}{\partial Q}(U_0, x) \\ \frac{\partial \mathcal{F}_Q}{\partial I}(U_0, x) & \frac{\partial \mathcal{F}_Q}{\partial Q}(U_0, x) \end{bmatrix} = \begin{bmatrix} \alpha_1 \frac{N}{\Omega} & \alpha_3 \\ 0 & \alpha_2 \frac{N}{\Omega} \end{bmatrix}$
which is non-negative and

$$V(x) = \begin{bmatrix} \frac{\partial \mathcal{V}_I}{\partial I}(U_0, x) & \frac{\partial \mathcal{V}_I}{\partial Q}(U_0, x) \\ \frac{\partial \mathcal{V}_Q}{\partial I}(U_0, x) & \frac{\partial \mathcal{V}_Q}{\partial Q}(U_0, x) \end{bmatrix} = \begin{bmatrix} \gamma_1 & 0 \\ 0 & \gamma_2 \end{bmatrix}$$

Together with $J(x)$ is a 2×2 matrix.

Let $X := C\left(\mathbb{R}^2, \overline{\Omega}\right)$ and $X_+ := C\left(\mathbb{R}^2_+, \overline{\Omega}\right)$. Then (X, X_+) is an ordered Banach space. Also consider $T(t)$ as a solution semi group on the Banach space X, associated with the linear diffusion system

$\frac{\partial U_i}{\partial t} = \nabla.(D_i \nabla U_i) - V U_i, t > 0, x \in \Omega$
$\frac{\partial U_i}{\partial \eta} = (0, 0), t > 0, x \in \partial\Omega$, for all $i \in \{I, Q\}$
Then $-V$ is cooperative on $\overline{\Omega}$.

For the prime reproduction number of the system (1)–(4), assume that all the variables for state vector are in a small neighbourhood of the disease free equilibrium point U_0. Let $\varphi : \Omega \to X$ represents the initial density distribution of the COVID-19 infection. Then the distribution of infection produced by an infective individual into the total population is given by $T(t)\varphi(x)$. Therefore the new infection is distributed by the function $F(x)T(t)\varphi(x)$ (these distribution are not the probability distribution). Thus the total new infection is distributed by the formula

$$\int_0^\infty F(x)T(x)\varphi(x)dt$$

Let

$$\mathcal{L}(\varphi)(x) = \int_0^\infty F(x)T(t)\varphi(x)dt = F(x) \int_0^\infty T(t)\varphi dt$$

Then $\mathcal{L}$ is a positive operator, which is continuous and maps the distribution of initial infection φ to the total infected population during the infection period. By the concept of next generation operator, it is obvious that the basic reproduction number for the system (1)–(4) is the spectral radius of the operator $\mathcal{L}$, i.e.,

$$\mathcal{R}_0 = r(\mathcal{L})$$

Theorem 1. *For positive diffusion coefficients considered in the system (1)–(4) if $F(x)$ and $V(x)$ are independent of $x \in \overline{\Omega}$ then $\mathcal{R}_0 = r(FV^{-1})$.*

Proof. The proof this theorem is done in the paper [32].

Now $FV^{-1} = \begin{bmatrix} \frac{\alpha_1 N}{\gamma_1 \Omega} & \frac{\alpha_3}{\gamma_2} \\ 0 & \frac{\alpha_2 N}{\gamma_2 \Omega} \end{bmatrix}$

Then the spectral radius of FV^{-1} is given by $r(FV^{-1}) = \frac{\alpha_1\gamma_2+\alpha_2\gamma_1}{\gamma_1\gamma_2}\frac{N}{\Omega}$.
Therefore $\mathcal{R}_0 = r(\mathcal{L}) = \frac{\alpha_1\gamma_2+\alpha_2\gamma_1}{\gamma_1\gamma_2}\frac{N}{\Omega}$.

4 Re-parametrisation

For every mathematical model it is essential to estimate the values of parameters used in the model. In most cases there are some parameters which are very hard to estimate. For diffusion model one such parameter is the diffusion coefficient. A re-parametrisation of these models can solve the difficulty in such a way that the particular parameter can scaled away. For the above system (1)–(4) a re-parametrisation of all dependent variables by initial susceptible population S_0, the time variable by t_0 and the space variable by x_0 is given by $S = S_0 S^*$, $I = S_0 I^*$, $Q = S_0 Q^*$, $R = S_0 R^*$, $t = t_0 t^*$ and $x = x_0 x^*$. Then the system (1)–(4) becomes

$$\frac{S_0}{t_0}\frac{\partial S^*}{\partial t^*} = D_S \frac{S_0}{x_0^2}\frac{\partial^2 S^*}{\partial x^{*2}} - \frac{\alpha_1 S_0^{p_1} I^{*p_1}}{1+\beta_1 S_0^{q_1} I^{*q_1}} S_0 S^* - \frac{\alpha_2 S_0^{p_2} Q^{*p_2}}{1+\beta_2 S_0^{q_2} Q^{*q_2}} S_0 S^*$$

$$\frac{S_0}{t_0}\frac{\partial I^*}{\partial t^*} = D_I \frac{S_0}{x_0^2}\frac{\partial^2 I^*}{\partial x^{*2}} + \frac{\alpha_1 S_0^{p_1} I^{*p_1}}{1+\beta_1 S_0^{q_1} I^{*q_1}} S_0 S^* + \frac{\alpha_3 S_0^{p_3} Q^{*p_3}}{1+\beta_3 S_0^{q_3} Q^{*q_3}} S_0 I^* - \gamma_1 S_0 I^*$$

$$\frac{S_0}{t_0}\frac{\partial Q^*}{\partial t^*} = D_Q \frac{S_0}{x_0^2}\frac{\partial^2 Q^*}{\partial x^{*2}} + \frac{\alpha_2 S_0^{p_2} Q^{*p_2}}{1+\beta_2 S_0^{q_2} Q^{*q_2}} S_0 S^* - \frac{\alpha_3 S_0^{p_3} Q^{*p_3}}{1+\beta_3 S_0^{q_3} Q^{*q_3}} S_0 I^* - \gamma_2 S_0 Q^*$$

$$\frac{S_0}{t_0}\frac{\partial R^*}{\partial t^*} = D_R \frac{S_0}{x_0^2}\frac{\partial^2 R^*}{\partial x^{*2}} + \gamma_1 S_0 I^* + \gamma_2 S_0 Q^*$$

That is,

$$\frac{\partial S^*}{\partial t^*} = D_S \frac{t_0}{x_0^2}\frac{\partial^2 S^*}{\partial x^{*2}} - \frac{\alpha_1 S_0^{p_1} I^{*p_1}}{1+\beta_1 S_0^{q_1} I^{*q_1}} t_0 S^* - \frac{\alpha_2 S_0^{p_2} Q^{*p_2}}{1+\beta_2 S_0^{q_2} Q^{*q_2}} t_0 S^*$$

$$\frac{\partial I^*}{\partial t^*} = D_I \frac{t_0}{x_0^2}\frac{\partial^2 I^*}{\partial x^{*2}} + \frac{\alpha_1 S_0^{p_1} I^{*p_1}}{1+\beta_1 S_0^{q_1} I^{*q_1}} t_0 S^* + \frac{\alpha_3 S_0^{p_3} Q^{*p_3}}{1+\beta_3 S_0^{q_3} Q^{*q_3}} t_0 I^* - \gamma_1 t_0 I^*$$

$$\frac{\partial Q^*}{\partial t^*} = D_Q \frac{t_0}{x_0^2}\frac{\partial^2 Q^*}{\partial x^{*2}} + \frac{\alpha_2 S_0^{p_2} Q^{*p_2}}{1+\beta_2 S_0^{q_2} Q^{*q_2}} t_0 S^* - \frac{\alpha_3 S_0^{p_3} Q^{*p_3}}{1+\beta_3 S_0^{q_3} Q^{*q_3}} t_0 I^* - \gamma_2 t_0 Q^*$$

$$\frac{\partial R^*}{\partial t^*} = D_R \frac{t_0}{x_0^2}\frac{\partial^2 R^*}{\partial x^{*2}} + \gamma_1 t_0 I^* + \gamma_2 t_0 Q^*$$

Where the initial conditions for the scaled system are given by $S^*(x, 0) = 1$, $I^*(x, 0) = \frac{I_0}{S_0}$, $Q^*(x, 0) = \frac{Q_0}{S_0}$, $R^*(x, 0) = \frac{R_0}{S_0} = 0$ and I_0, Q_0, R_0 represents the initial population of I class, Q class, R class respectively.

Letting $\gamma_2 t_0 = 1$, $D_S \frac{t_0}{x_0^2} = 1$, $D_I \frac{t_0}{x_0^2} = 1$, $D_Q \frac{t_0}{x_0^2} = 1$, $D_R \frac{t_0}{x_0^2} = 1$ and $\beta_1 S_0^{q_1} = 1$ we obtain $t_0 = \frac{1}{\gamma_2}$ and $x_0 = \sqrt{\frac{D}{3\gamma_2}}$, where $D = D_S + D_I + D_Q + D_R$. Then the scaled system becomes

$$\frac{\partial S^*}{\partial t^*} = \frac{\partial^2 S^*}{\partial x^{*2}} - \frac{\alpha_1\left(\frac{1}{\beta_1}\right)^{\frac{p_1}{q_1}} I^{*p_1}}{\gamma_2(1+I^{*q_1})} S^* - \frac{\alpha_2\left(\frac{1}{\beta_1}\right)^{\frac{p_2}{q_1}} Q^{*p_2}}{\gamma_2\left(1+\beta_2\left(\frac{1}{\beta_1}\right)^{\frac{q_2}{q_1}} Q^{*q_2}\right)} S^* \quad (1.1)$$

$$\frac{\partial I^*}{\partial t^*} = \frac{\partial^2 I^*}{\partial x^{*2}} + \frac{\alpha_1\left(\frac{1}{\beta_1}\right)^{\frac{p_1}{q_1}} I^{*p_1}}{\gamma_2(1+I^{*q_1})} S^* + \frac{\alpha_3\left(\frac{1}{\beta_1}\right)^{\frac{p_3}{q_1}} Q^{*p_3}}{\gamma_2\left(1+\beta_3\left(\frac{1}{\beta_1}\right)^{\frac{q_3}{q_1}} Q^{*q_3}\right)} I^* - \frac{\gamma_1}{\gamma_2} I^* \quad (1.2)$$

$$\frac{\partial Q^*}{\partial t^*} = \frac{\partial^2 Q^*}{\partial x^{*2}} + \frac{\alpha_2\left(\frac{1}{\beta_1}\right)^{\frac{p_2}{q_1}} Q^{*p_2}}{\gamma_2\left(1+\beta_2\left(\frac{1}{\beta_1}\right)^{\frac{q_2}{q_1}} Q^{*q_2}\right)} S^* - \frac{\alpha_3\left(\frac{1}{\beta_1}\right)^{\frac{p_3}{q_1}} Q^{*p_3}}{\gamma_2\left(1+\beta_3\left(\frac{1}{\beta_1}\right)^{\frac{q_3}{q_1}} Q^{*q_3}\right)} I^* - Q^* \quad (1.3)$$

$$\frac{\partial R^*}{\partial t^*} = \frac{\partial^2 R^*}{\partial x^{*2}} + \frac{\gamma_1}{\gamma_2} I^* + Q^* \quad (1.4)$$

5 Method of Solution

To investigate the local spread and control [33] of infection among uniform susceptible population [15] for the system (1)–(4), we need to determine the conditions of existence of such travelling wave and speed of its propagation.

We look for travelling wave solutions [21], consider the model as one dimensional problem and

$S^*(x^*, t^*) = S^*(z)$, $I^*(x^*, t^*) = I^*(z)$, $Q^*(x^*, t^*) = Q^*(z)$, $R^*(x^*, t^*) = R^*(z)$, $z = x^* - ct^*$

where the wave speed c has to determine. Along the direction of positive x^*-axis a travelling wave of constant shape can be obtained by the above settings. Substitution of the above consideration transform the system (1)–(4) to the system as

$$-c\frac{dS^*}{dz} = -\frac{\alpha_1\left(\frac{1}{\beta_1}\right)^{\frac{p_1}{q_1}} I^{*p_1}}{\gamma_2\left(1+I^{*q_1}\right)} S^* - \frac{\alpha_2\left(\frac{1}{\beta_1}\right)^{\frac{p_2}{q_1}} Q^{*p_2}}{\gamma_2\left(1+\beta_2\left(\frac{1}{\beta_1}\right)^{\frac{q_2}{q_1}} Q^{*q_2}\right)} S^*$$

$$-c\frac{dI^*}{dz} = \frac{\alpha_1\left(\frac{1}{\beta_1}\right)^{\frac{p_1}{q_1}} I^{*p_1}}{\gamma_2\left(1+I^{*q_1}\right)} S^* + \frac{\alpha_3\left(\frac{1}{\beta_1}\right)^{\frac{p_3}{q_1}} Q^{*p_3}}{\gamma_2\left(1+\beta_3\left(\frac{1}{\beta_1}\right)^{\frac{q_3}{q_1}} Q^{*q_3}\right)} I^* - \frac{\gamma_1}{\gamma_2} I^*$$

$$-c\frac{dQ^*}{dz} = \frac{\alpha_2\left(\frac{1}{\beta_1}\right)^{\frac{p_2}{q_1}} Q^{*p_2}}{\gamma_2\left(1+\beta_2\left(\frac{1}{\beta_1}\right)^{\frac{q_2}{q_1}} Q^{*q_2}\right)} S^* - \frac{\alpha_3\left(\frac{1}{\beta_1}\right)^{\frac{p_3}{q_1}} Q^{*p_3}}{\gamma_2\left(1+\beta_3\left(\frac{1}{\beta_1}\right)^{\frac{q_3}{q_1}} Q^{*q_3}\right)} I^* - Q^*$$

$$-c\frac{dR^*}{dz} = \frac{\gamma_1}{\gamma_2} I^* + Q^*$$

Which can be represented as

$$S^{*''} + cS^{*'} - \frac{\alpha_1\left(\frac{1}{\beta_1}\right)^{\frac{p_1}{q_1}} I^{*p_1}}{\gamma_2\left(1+I^{*q_1}\right)} S^* - \frac{\alpha_2\left(\frac{1}{\beta_1}\right)^{\frac{p_2}{q_1}} Q^{*p_2}}{\gamma_2\left(1+\beta_2\left(\frac{1}{\beta_1}\right)^{\frac{q_2}{q_1}} Q^{*q_2}\right)} S^* = 0 \tag{5}$$

$$I^{*''} + cI^{*'} + \frac{\alpha_1\left(\frac{1}{\beta_1}\right)^{\frac{p_1}{q_1}} I^{*p_1}}{\gamma_2\left(1+I^{*q_1}\right)} S^* + \frac{\alpha_3\left(\frac{1}{\beta_1}\right)^{\frac{p_3}{q_1}} Q^{*p_3}}{\gamma_2\left(1+\beta_3\left(\frac{1}{\beta_1}\right)^{\frac{q_3}{q_1}} Q^{*q_3}\right)} I^* - \frac{\gamma_1}{\gamma_2} I^* = 0 \tag{6}$$

$$Q^{*''} + cQ^{*'} + \frac{\alpha_2\left(\frac{1}{\beta_1}\right)^{\frac{p_2}{q_1}} Q^{*p_2}}{\gamma_2\left(1+\beta_2\left(\frac{1}{\beta_1}\right)^{\frac{q_2}{q_1}} Q^{*q_2}\right)} S^* - \frac{\alpha_3\left(\frac{1}{\beta_1}\right)^{\frac{p_3}{q_1}} Q^{*p_3}}{\gamma_2\left(1+\beta_3\left(\frac{1}{\beta_1}\right)^{\frac{q_3}{q_1}} Q^{*q_3}\right)} I^* - Q^* = 0 \tag{7}$$

$$R^{*''} + cR^{*'} + \frac{\gamma_1}{\gamma_2} I^* + Q^* = 0 \tag{8}$$

where prime represents differentiation w.r.t. z. The above system consists of finding range of parameters considered above, for which there exists a solution with positive wave speed c and non-negative S^*, I^*, Q^* such that

$$I^*(-\infty) = I^*(\infty) = 0,\ Q^*(-\infty) = Q^*(\infty) = 0,\ 0 \le S^*(-\infty) < S^*(\infty) = 1$$

The conditions on I^* and Q^* imply a pulse wave of infective population which propagates into the uninfected population. As time goes, the system (5)–(8) changes with $S^* \to 1$, $Q^* \to I^*$ and $I^* \to 0$ and this gives

$$S^{*''} + cS^{*'} - \frac{\alpha_2 \left(\frac{1}{\beta_1}\right)^{\frac{p_2}{q_1}} I^{*p_2}}{\gamma_2 \left(1 + \beta_2 \left(\frac{1}{\beta_1}\right)^{\frac{q_2}{q_1}} I^{*q_2}\right)} S^* \approx 0 \tag{9}$$

$$I^{*''} + cI^{*'} + \frac{\alpha_1 \left(\frac{1}{\beta_1}\right)^{\frac{p_1}{q_1}} I^{*p_1}}{\gamma_2 (1 + I^{*q_1})} + \frac{\alpha_3 \left(\frac{1}{\beta_1}\right)^{\frac{p_3}{q_1}} I^{*p_3+1}}{\gamma_2 \left(1 + \beta_3 \left(\frac{1}{\beta_1}\right)^{\frac{q_3}{q_1}} I^{*q_3}\right)} - \gamma_1 I^* \approx 0 \tag{10}$$

$$Q^{*''} + cQ^{*'} + \frac{\alpha_2 \left(\frac{1}{\beta_1}\right)^{\frac{p_2}{q_1}} Q^{*p_2}}{\gamma_2 \left(1 + \beta_2 \left(\frac{1}{\beta_1}\right)^{\frac{q_2}{q_1}} Q^{*q_2}\right)} - Q^* \approx 0 \tag{11}$$

$$R^{*''} + cR^{*'} \approx 0 \tag{12}$$

Now linearisation of equations (10) and (11) with $p_1 = p_2 = p_3 = 1$ gives

$$I^{*''} + cI^{*'} + \left(\frac{\alpha_1}{\gamma_2}\left(\frac{1}{\beta_1}\right)^{\frac{1}{q_1}} - \gamma_1\right) I^* \approx 0 \tag{13}$$

$$Q^{*''} + cQ^{*'} + \left(\frac{\alpha_2}{\gamma_2}\left(\frac{1}{\beta_1}\right)^{\frac{1}{q_1}} - 1\right) Q^* \approx 0 \tag{14}$$

5.1 *Phase Plane Analysis*

Equation (13) represents a solution for a typical wave front where I^* is at one steady state for $z \to -\infty$ and for $z \to \infty$ it is at the other. In this case the problem is to determine the value(s) of c, for which I^* is a non-negative solution of the equation (13) and satisfy

$$\lim_{z \to -\infty} I^*(z) = 1, \ \lim_{z \to \infty} I^*(z) = 0.$$

In (I^*, X) phase plane

$$I^{*'} = X, X' = -cX - \left(\frac{\alpha_1}{\gamma_2}\left(\frac{1}{\beta_1}\right)^{\frac{1}{q_1}} - \gamma_1\right) I^*$$

Then the phase plane trajectories are represented by the solution of

$$\frac{dX}{dI^*} = \frac{-cX - \left(\frac{\alpha_1}{\gamma_2}\left(\frac{1}{\beta_1}\right)^{\frac{1}{q_1}} - \gamma_1\right) I^*}{X}$$

Then the system $(X,\ I^*)$ has $(0,\ 0)$ as the only singular point. Thus the matrix for the system $(X,\ I^*)$ corresponding to the singular point $(0,\ 0)$ is given by

$$A = \begin{pmatrix} -c & -\left(\frac{\alpha_1}{\gamma_2}\left(\frac{1}{\beta_1}\right)^{\frac{1}{q_1}} - \gamma_1\right) \\ 1 & 0 \end{pmatrix}$$

whose eigenvalues are

$$\frac{1}{2}\left[-c \pm \sqrt{c^2 - 4\left(\frac{\alpha_1}{\gamma_2}\left(\frac{1}{\beta_1}\right)^{\frac{1}{q_1}} - \gamma_1\right)}\right] \implies$$
$$\begin{cases} stable \quad node \ c^2 > 4\left(\frac{\alpha_1}{\gamma_2}\left(\frac{1}{\beta_1}\right)^{\frac{1}{q_1}} - \gamma_1\right) \\ stable \quad spiral \ c^2 < 4\left(\frac{\alpha_1}{\gamma_2}\left(\frac{1}{\beta_1}\right)^{\frac{1}{q_1}} - \gamma_1\right) \end{cases}$$

The singular point $(0,\ 0)$ is a stable node if $c \geq c_{min} = 2\sqrt{\frac{\alpha_1}{\gamma_2}\left(\frac{1}{\beta_1}\right)^{\frac{1}{q_1}} - \gamma_1}$ and a degenerate node if $c = c_{min}$. If $c^2 < 4\left(\frac{\alpha_1}{\gamma_2}\left(\frac{1}{\beta_1}\right)^{\frac{1}{q_1}} - \gamma_1\right)$ then $(0,\ 0)$ is a stable spiral; i.e., I^* oscillates in the neighbourhood of the origin.

There does not exists any wave solution for $\frac{\alpha_1}{\gamma_2}\left(\frac{1}{\beta_1}\right)^{\frac{1}{q_1}} < \gamma_1$. This is the necessary condition for the spread of epidemic wave for I^* class.

For $c \leq 2\sqrt{\frac{\alpha_1}{\gamma_2}\left(\frac{1}{\beta_1}\right)^{\frac{1}{q_1}} - \gamma_1}$, There exists a typical travelling wave solution. In this case $I^* < 0$ for some z and they are physically unrealistic, because in that case I^* spirals around the origin. In this seance $I^* \to 0$ at the leading edge with decreasing oscillation around $I^* = 0$.

Equation (14) represents a solution for a typical wave front where I^* is at one steady state for $z \to -\infty$ and for $z \to \infty$ it is at the other. In this case the problem is to determine the value(s) of c, for which I^* is a non-negative solution of the equation (14) and satisfy

$$\lim_{z\to-\infty} I^*(z) = 1, \ \lim_{z\to\infty} I^*(z) = 0.$$

In (Q^*, Y) phase plane

$$Q^{*'} = Y, Y' = -cY - \left(\frac{\alpha_2}{\gamma_2}\left(\frac{1}{\beta_1}\right)^{\frac{1}{q_1}} - 1\right) Q^*$$

Then the phase plane trajectories are represented by the solution of

$$\frac{dY}{dQ^*} = \frac{-cY - \left(\frac{\alpha_2}{\gamma_2}\left(\frac{1}{\beta_1}\right)^{\frac{1}{q_1}} - 1\right) Q^*}{Y}$$

Then the system $(Y, \ Q^*)$ has $(0, \ 0)$ as the only singular point. Thus the matrix for the system $(Y, \ Q^*)$ corresponding to the singular point $(0, \ 0)$ is given by

$$A = \begin{pmatrix} -c & -\left(\frac{\alpha_2}{\gamma_2}\left(\frac{1}{\beta_1}\right)^{\frac{1}{q_1}} - 1\right) \\ 1 & 0 \end{pmatrix}$$

whose eigenvalues are

$$\frac{1}{2}\left[-c \pm \sqrt{c^2 - 4\left(\frac{\alpha_2}{\gamma_2}\left(\frac{1}{\beta_1}\right)^{\frac{1}{q_1}} - 1\right)}\right] \implies$$

$$\begin{cases} stable \ \ node \ \ c^2 > 4\left(\frac{\alpha_2}{\gamma_2}\left(\frac{1}{\beta_1}\right)^{\frac{1}{q_1}} - 1\right) \\ stable \ \ spiral \ c^2 < 4\left(\frac{\alpha_2}{\gamma_2}\left(\frac{1}{\beta_1}\right)^{\frac{1}{q_1}} - 1\right) \end{cases}$$

The singular point $(0, \ 0)$ is a stable node if $c \geq c_{min} = 2\sqrt{\frac{\alpha_2}{\gamma_2}\left(\frac{1}{\beta_1}\right)^{\frac{1}{q_1}} - 1}$ and a degenerate node if $c = c_{min}$. If $c^2 < 4\left(\frac{\alpha_2}{\gamma_2}\left(\frac{1}{\beta_1}\right)^{\frac{1}{q_1}} - 1\right)$ then $(0, \ 0)$ is a stable spiral; i.e., Q^* oscillates in the neighbourhood of the origin.

There does not exists any wave solution for $\frac{\alpha_2}{\gamma_2}\left(\frac{1}{\beta_1}\right)^{\frac{1}{q_1}} < 1$. This is the necessary condition for the spread of epidemic wave for Q^* class.

For $c \leq 2\sqrt{\frac{\alpha_2}{\gamma_2}\left(\frac{1}{\beta_1}\right)^{\frac{1}{q_1}} - 1}$, There exists a typical travelling wave solution. In this case $Q^* < 0$ for some z and they are physically unrealistic, because in that case Q^* spirals around the origin. In this seance $Q^* \to 0$ at the leading edge with decreasing oscillation around $Q^* = 0$.

5.2 *Analysis of Analytic Solution*

Then the solution of (13) and (14) are given by

$$I^*(z) \propto exp[\frac{(-c \pm \{c^2 - 4(\frac{\alpha_1}{\gamma_2}\left(\frac{1}{\beta_1}\right)^{\frac{1}{q_1}} - \gamma_1)\}^{\frac{1}{2}})z}{2}] \tag{15}$$

and

$$Q^*(z) \propto exp[\frac{(-c \pm \{c^2 - 4(\frac{\alpha_2}{\gamma_2}\left(\frac{1}{\beta_1}\right)^{\frac{1}{q_1}} - 1)\}^{\frac{1}{2}})z}{2}] \tag{16}$$

As $Q^*(z) \to I^*(z)$ and $I^*(z) \to 0$ with $Q^*(z) > 0$ and $I^*(z) > 0$, these solutions can not oscillate about $Q^* = 0$ and $I^* = 0$; otherwise $Q^*(z) < 0$ and $I^*(z) < 0$ for some z. Then the travelling wave speed c which can derived from (15) and (16) must satisfy

$$c \geq c_{min} = 2 \quad max\left\{\sqrt{\frac{\alpha_1}{\gamma_2}\left(\frac{1}{\beta_1}\right)^{\frac{1}{q_1}} - \gamma_1}, \sqrt{\frac{\alpha_2}{\gamma_2}\left(\frac{1}{\beta_1}\right)^{\frac{1}{q_1}} - 1}\right\} \tag{17}$$

we expect such travelling waves derived from fully non-linear system of equations will evolve into a travelling waveform with the minimum wave speed given by the equation (19), except in exceptional conditions. The wave velocity for COVID-19, V say, in dimensional term is then given as

$$V = \left(\sqrt{\gamma_2 D}\right) c,$$

where c is given by the relation (17).

The travelling wave solution for susceptible population S cannot have a local maximum, since $S' = 0$ there and the equation for S, (9) shows that $S^{*''} = \frac{\alpha_2\left(\frac{1}{\beta_1}\right)^{\frac{p_2}{q_1}} I^{*p_2}}{\gamma_2\left(1+\beta_2\left(\frac{1}{\beta_1}\right)^{\frac{q_2}{q_1}} I^{*q_2}\right)} S^* > 0$, which indicate a local minimum. So $S^*(z)$ is a monotone increasing function of z. Linearisation of the equation for S^* as $z \to \infty$, where $p_2 = 1$ and $S = 1 - s$, with s small gives

$$s'' + cs' - \alpha_2 I^* = 0$$

with which $I(z)$ from (15), it is obvious that

$$S^*(z) \sim 1 - O(exp[\frac{(-c \pm \{c^2 - 4(\gamma_1 - 1)\}^{\frac{1}{2}})z}{2}])$$

and so, as $z \to \infty$, $S^*(z) \to 1$ exponentially.

6 Experimental Set up

In order to verify the validity of a mathematical model, the experimental set up is essential. In this chapter the infected population and the migrant population are responsible for the transmission of the disease and β_1 is a sensitive parameter for the proposed model. Thus an experimental set up is done for the infected as well as the migrant population with the variation of β_1.

For the experimental set up the system (9), (12), (13) and (14) with $p_1 = p_2 = p_3 = q_1 = q_2 = q_3 = 1$, can be represented as

$$\gamma_2 \left(\beta_1 + \beta_2 I^*\right) \left(S^{*''} + cS^{*'}\right) - \alpha_2 S^* I^* \approx 0 \tag{18}$$

$$\beta_1\gamma_2 \left(1 + I^*\right) \left(\beta_1 + \beta_3 I^*\right) \left(I^{*''} + cI^{*'}\right) + \left(\alpha_1\beta_1 + \beta_1^2\gamma_1\gamma_2\right) I^* + \left(\alpha_1\beta_3 + \alpha_3\beta_1 + \beta_1^2\gamma_1\gamma_2 + \beta_1\beta_3\gamma_1\gamma_2\right) I^{*^2} + \left(\alpha_3\beta_1 + \beta_1\beta_3\gamma_1\gamma_2\right) I^{*^3} \approx 0 \tag{19}$$

$$\gamma_2 \left(\beta_1 + \beta_2 Q^*\right) \left(Q^{*''} + cQ^{*'}\right) + \left(\alpha_2 - \beta_1\gamma_2\right) Q^* - \beta_2\gamma_2 Q^{*^2} \approx 0 \tag{20}$$

$$R^{*''} + cR^{*'} \approx 0 \tag{21}$$

with the boundary conditions
$I^*(-\infty) = I^*(\infty) = 0$, $Q^*(-\infty) = Q^*(\infty) = 0$, $0 \le S^*(-\infty) < S^*(\infty) = 1$.

The mathematical model for this problem has been solved numerically by Matlab function ode45()which is based on the shooting technique with Runge-Kutta 4th order method. 500-1000 iterations are considered to correct accuracy up to 10^{-4}.

The physical parameters used in this mathematical model are estimated using the Statistical technique from [29] for 217 countries from all over the world and the value of parameters are given by
$\alpha_1 = 0.3, \alpha_2 = 0.55, \alpha_3 = 0.6$
$\beta_1 = 0.73, \beta_2 = 0.75, \beta_3 = 0.8$
$\gamma_1 = 0.5, \gamma_2 = 0.7, c = 1$.

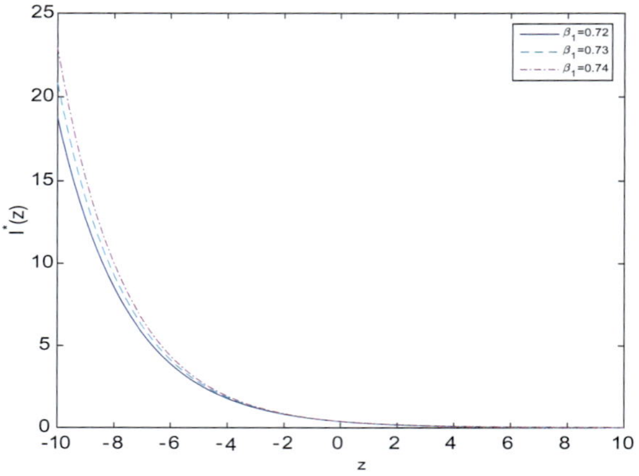

Fig. 1 Variation of the Infected population with z when $\alpha_1 = 0.3; \alpha_2 = 0.55; \alpha_3 = 0.6;, \beta_1 = 0.73, \beta_2 = 0.75; \beta_3 = 0.8; \gamma_1 = 0.5; \gamma_2 = 0.7; c = 1$

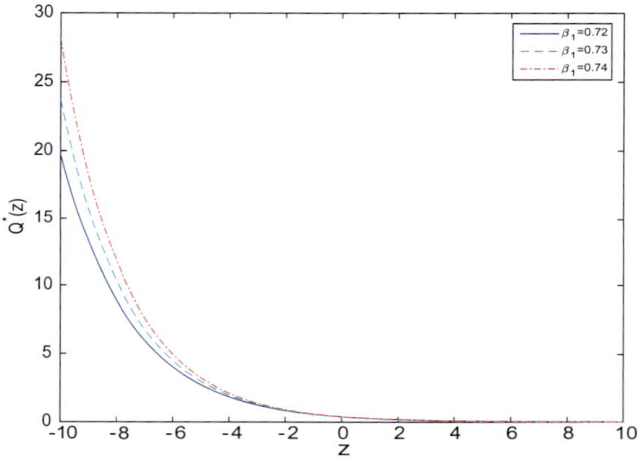

Fig. 2 Variation of the Quarantined population with z when $\alpha_1 = 0.3; \alpha_2 = 0.55; \alpha_3 = 0.6;, \beta_1 = 0.73; \beta_2 = 0.75; \beta_3 = 0.8; \gamma_1 = 0.5; \gamma_2 = 0.7; c = 1$

Since total population is constant, so from this graphs tells infection rate and recovery rate is balance as per asymptotic nature.

By solving the linearized system (18)–(21) with boundary conditions $I^*(-\infty) = 1$, $I^*(\infty) = 0$, $0 \leq S^*(-\infty) < S^*(\infty) = 1$, by numerical method, we get the solution for I^*, Fig. 1, which tends to 0 as z approaches to ∞; solution for I^*, Fig. 2, which tends to 0 as z approaches to ∞. Which guarantee us that the number of susceptible population will increase in forward time and after certain time all population becomes susceptible.

7 Result Analysis and Discussion

For the proposed model of COVID-19, investigation of both analytical solution as well as experimental set up are done. There are so many methods to solve a diffusion mathematical model for an infectious disease. Among all of them, the travelling wave solution method for a diffusion model is one of the best to justify the spatial effects of the infection and to calculate some typical but important result like the wave velocity of the infection, which is considered in this chapter. From the analysis of analytical solution it is obvious that the total population will becomes susceptible after certain time. Again the experimental set up ensure that, due to proper quarantine of infected as well as migrant population, the rate of infection into the total population will decrease in forward time. As a result total population will become susceptible in forward time. This also agree with the analytical solution. As both the infected and quarantined populations are responsible for the transmission of COVID-19, the wave velocity of COVID-19 due to both the quarantined as well as the infected populations are obtained and finally they represent the wave velocity of COVID-19 into the total population by the relation (17).

To validate our results, for specific location, the $S(t)$ (susceptible population), $I(t)$ (Infected population), $Q(t)$ (Quarantine population) and $R(t)$ (recovered population), are considered as function of time (t) only. The coupled linearized form of the diffusion equations (1.1) to (1.4) are solved numerically by Matlab function ode45() as the coupled initial value problems using Runge-Kutta 4th order method with correct accuracy up to 10^{-4}.

A number of simulations have been done to verify and compare the results (see the Fig. 3 and Table 2) with existence literature work [20], analytical results. Our results are consistent and it can be verified that 14 d quarantine is valid from the asymptotic results. It can be explained that in proper quarantine process will help to control the infected population and spread of COVID-19.

Table 2 Time series solutions

Time(t)	S (t)	I (t)	Q(t)	R (t)
0	0.8	0.2	0.2	0
0.01084298	0.79712323	0.19997592	0.19918775	0.0037131
1.03850115	0.59559617	0.17590595	0.12083956	0.30765832
2.94726064	0.44220197	0.09297624	0.03734018	0.62748161
5.38376996	0.38765927	0.03091142	0.0080987	0.77333062
8.23671657	0.37254254	0.00768807	0.00141177	0.81835761
10.16948687	0.36981639	0.0029501	0.00043707	0.82679644
14.21933826	0.36844006	0.00039346	0.00003785	0.83112863
16.31934595	0.36831133	0.00013819	0.00001063	0.83153985
16.77684449	0.36829746	0.00011011	0.00000808	0.83158435
18.73434304	0.36826403	0.00004155	0.00000248	0.83169194
20	0.36825474	0.00002216	0.00000116	0.83172193

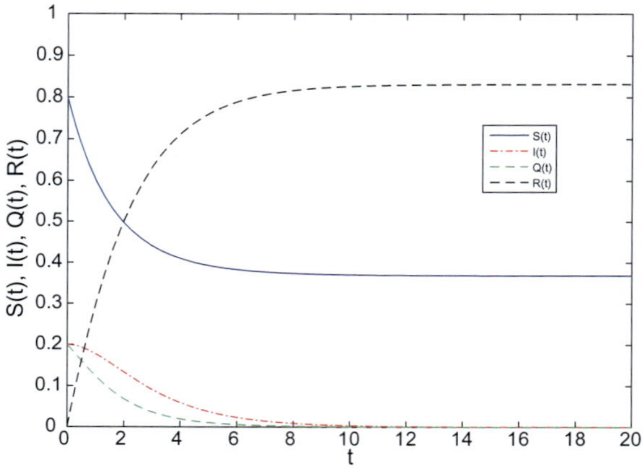

Fig. 3 Variation of the $S(t)$, $I(t)$, $Q(t)$ and $R(t)$ with time (t) $\alpha_1 = 0.3; \alpha_2 = 0.55; \alpha_3 = 0.6; \beta_1 = 0.73; \beta_2 = 0.75; \beta_3 = 0.8; \gamma_1 = 0.5; \gamma_2 = 0.7$

8 Conclusion

In the current scenario, 2019-nCoV is the third major infection due to corona viruses throughout the world. So understanding of COVID-19 is essential, specially in the absence of vaccine. Mathematical model is one such tool to understand it. In this article the investigation on the necessity of quarantine of infected as well as the migrant population proves that, maintenance of social distancing and proper quarantine of infected as well as migrant population can reduce the spread of COVID-19. This model is valid for all kind of infectious disease which can transmit through contact.

References

1. P. Arora, H. Kumar, B.K. Panigrahi, Prediction and analysis of COVID-19 positive cases using deep learning models: A descriptive case study of India. Chaos Solitons Fractals **139**, 110017 (2020). https://doi.org/10.1016/j.chaos.2020.110017
2. R.M. Anderson, B. Anderson, R.M. May, *Infectious Diseases of Humans: Dynamics and Control* (Oxford University Press, Oxford, 1991). https://doi.org/10.1002/hep.1840150131
3. L.J.S. Allen, B.M. Bolker, Y. Lou, A.L. Nevai, Asymptotic profiles of the steady states for an SIS epidemic reaction-diffusion model. Discrete Contin. Dyn. Syst. **21**, 1–20 (2008). https://doi.org/10.3934/dcds2008.21.1
4. A. Brandenburg, Quadratic growth during the 2019 novel coronavirus epidemic. arXiv preprint: http://arxiv.org/abs/2002.03638
5. E. Bonabeau, L. Toubiana, A. Flahault, The geographical spread of influenza. Proc. Roy. Soc. Lond. B **265**, 2421–2425 (1998). https://doi.org/10.1098/rspb.1998.0593d
6. N. Crokidakis, COVID-19 spreading in Rio de Janeiro, Brazil: do the policies of social isolation really work? Chaos Solitons Fractals **136**, 109930 (2020). https://doi.org/10.1016/j.chaos.2020.109930
7. T. Chen, J. Rui, Q. Wang et al., A mathematical model for simulating the phase-based transmissibility of a novel coronavirus. Infect. Dis. Poverty **9**, 24 (2020). https://doi.org/10.1186/s40249-020-00640-3

8. G.A.P. de Souza, V.F. Queiroz, M.T. Lima et al., Virus goes viral: an educational kit for virology classes. Virol. J. **17**, 13 (2020). https://doi.org/10.1186/s12985-020-1291-9
9. L.C. Evans, *Partial Differential Equations*, 2nd edn. (American Mathematical Society, Providence, 2010). Electronic ISBN 978-1-4704-1144-2
10. D. Fanelli, F. Piazza, Analysis and forecast of COVID-19 spreading in China, Italy and France. Chaos, Solitons Fractals **134**, 109761 (2020). https://doi.org/10.1016/j.chaos.2020.109761
11. W. Feng, Y.H. Quan, Trend and forecasting of the COVID-19 outbreak in China. J. Infect. **80**, 469–496 (2020)
12. N.C. Gassen, D. Niemeyer, D. Muth et al., SKP2 attenuates autophagy through Beclin1-ubiquitination and its inhibition reduces MERS-Coronavirus infection. Nat. Commun. **10**, 5770 (2019). https://doi.org/10.1038/s41467-019-13659-4
13. A. Gulati, C. Pomeranz et al., A comprehensive review of manifestations of novel coronaviruses in the context of deadly COVID-19 global pandemic. Am. J. Med. Sci. **360**(1), 5–34 (2020)
14. H.W. Hethcote, P. van den Driessche, Some epidemiological models with nonlinear incidence. J. Math. Biol. **29**, 271–287 (1991). https://doi.org/10.1007/BF00160539
15. A. Kallen, P. Arcuri, J.D. Murray, A simple model for the spatial spread and control of rabies. J. Theor. Biol. **116**, 377–393 (1985)
16. S. Lahmiri, S. Bekiros, The impact of COVID-19 pandemic upon stability and sequential irregularity of equity and cryptocurrency markets. Chaos, Solitons Fractals **138**, 109936 (2020). https://doi.org/10.1016/j.chaos.2020.109936d
17. Q. Li, X. Guan, P. Wu et al., Early transmission dynamics in Wuhan, China, of novel coronavirus-infected pneumonia. N. Engl. J. Med. **382**(13), 1199–1207 (2020). https://doi.org/10.1056/NEJMoa2001316
18. C.-C. Lai, Y.H. Liu et al., Asymptomatic carrier state, acute respiratory disease, and pneumonia due to severe acute respiratory syndrome coronavirus 2 (SARS-CoV-2). J. Microbiol. Immunol. Infect. **53**, 404–412 (2020)
19. M. Mandal, S. Jana, S.K. Nandi, A. Khatua, S. Adak, T.K. Kar, A model based study on the dynamics of COVID-19: prediction and control. Chaos Solitons Fractals **136**, 109889 (2020). https://doi.org/10.1016/j.chaos.2020.109889
20. B.K. Mishra et al., COVID-19 created chaos across the globe: three novel quarantine epidemic models. Chaos Solitons Fractals **138**, 109928 (2020). https://doi.org/10.1016/j.chaos.2020.109928
21. J.D. Murray, *Mathematical Biology*, 2nd edn. (Springer, Berlin, 1993). ISBN 978-3-662-08542-4
22. F. Ndairou, I. Area, J.J. Nieto, Mathematical modeling of COVID-19 transmission dynamics with a case study of Wuhan. Chaos Solitons Fractals **135**, 109846 (2020). https://doi.org/10.1016/j.chaos.2020.109846
23. D. Okuonghae, A. Omame, Analysis of a mathematical model for COVID-19 population dynamics in Lagos, Nigeria. Chaos Solitons Fractals **139**, 110032 (2020). https://doi.org/10.1016/j.chaos.2020.110032
24. Y. Pinchover, J. Rubinson, *An Introduction to Partial Differential Equations* (Cambridge University Press, New York, 2005)
25. D. Rafiq, S.A. Suhail, M.A. Bazaz, Evaluation and prediction of COVID-19 in India: a case study of worst hit states. Chaos Solitons Fractals **139**, 110014 (2020). https://doi.org/10.1016/j.chaos.2020.110014
26. D.C. Rotzinger, C. Beigelman-Aubry, C. von Garnier, S.D. Qanadli, Pulmonary embolism in patients with COVID-19: time to change the paradigm of computed tomography. Thromb. Res. **190**, 58–59 (2020). https://doi.org/10.1016/j.thromres.2020.04.011
27. K. Sarkar, S. Khajanchi, J.J. Nieto, Modeling and forecasting the COVID-19 pandemic in India. Chaos Solitons Fractals **139**, 110049 (2020). https://doi.org/10.1016/j.chaos.2020.110049
28. S. Tian, N. Hu, J. Lou, K. Chen, X. Kang, Z. Xiang, H. Chen et al., Characteristic of COVID-19 infection in Beijing. J. Infect. **80**, 401–406 (2020)
29. World Health Organization, Global Research on Coronavirus Disease (COVID-19) (2020)

30. H. Wang, Z. Wang, Y. Dong et al., Phase-adjusted estimation of the number of Coronavirus Disease 2019 cases in Wuhan. China. Cell Discov **6**, 10 (2020). https://doi.org/10.1038/s41421-020-0148-0
31. S. Wacharapluesadee, P. Duengkae, A. Chaiyes et al., Longitudinal study of age-specific pattern of coronavirus infection in Lyle's flying fox (Pteropus lylei) in Thailand. Virol. J. **15**, 38 (2018). https://doi.org/10.1186/s12985-018-0950-6
32. W. Wang, X.Q. Zhao, Basic reproduction numbers for reaction-diffusion epidemic models. SIAM J. Appl. Dyn. Syst. **11**(4), 1652–1673 (2012)
33. S. Ying, F. Li, X. Geng, Z. Li, X. Du, H. Chen, et al.: Spread and control of COVID-19 in China and their associations with population movement, public health emergency measures, and medical resources (medRxiv, 2020), https://doi.org/10.1101/2020.02.24.20027623
34. X. Zhang, R. Ma, L. Wang, Predicting turning point, duration and attack rate of COVID-19 outbreaks in major Western countries. Chaos Solitons Fractals **135**, 109829 (2020). https://doi.org/10.1016/j.chaos.2020.109829
35. L. Zhong, L. Mu, J. Li, J. Wang, Z. Yin, D. Liu, Early prediction of the 2019 novel coronavirus outbreak in the mainland China based on simple mathematical model. IEEE Access **8**, 51761–51769 (2020). https://doi.org/10.1109/ACCESS.2020.2979599

Machine Learning Based Anxiety Prediction of General Public from Tweets During COVID-19

Shubham Tribedi, Anuraag Biswas, Swarup Kr Ghosh, and Anupam Ghosh

Abstract In this crisis of COVID19, everyone is staying in touch with the world through social media. This has led to social media becoming a significant source of new information for many people and unfortunately this phenomenon has given birth to a lot of misinformation, chaos and fear in people's minds. This fear is often due to the inadequate and wrong information. Therefore, there is a important need to understand this crisis. Patterns need to be established between popular tweets and its effect on the public's sentiments, especially their fear. So, tweets of three different countries namely United States of America, Federative Republic of Brazil and Republic of India. Sentiment analysis reveals that fear of this unknown and mysterious nature of the coronavirus is dominant among the public. Predominant analysis of tweets within past two months will be done and then a model will be built to predict future reaction of the general public based on the crisis level in the country. Machine Learning algorithms such as 'Logistic Regression (LR)', 'Multinomial Naïve Bayes' and 'Support Vector Machine (SVM)'are used for classification purpose preceded by the pre-processing steps of raw data from each country. 90% of accuracy has been achieved from sentiment classification result. Insights to the fear, sentiments have also been provided. Tweets with negative sentiment and emotion indicates the cases for the pandemic outbreak.

Keywords Fear · Sentiments · Negative emotion · Twitter · Machine learning · Public awareness

1 Introduction

Sentiment Analysis is the classification, understanding, interpretation, definition of all sorts of emotions which one uses in his or her text. In one way it can be said that this is an analysis of text technique which allows one to understand the sentiment

S. Tribedi · A. Biswas · S. K. Ghosh (✉)
Sister Nivedita University, Kolkata 700156, India
e-mail: swarup.kg@snuniv.ac.in

A. Ghosh
Netaji Subhash Engineering College, Kolkata 700152, India

J. Nayak et al. (eds.), *Understanding COVID-19: The Role of Computational Intelligence*, Studies in Computational Intelligence 963, https://doi.org/10.1007/978-3-030-74761-9_13

of the user towards a particular thought or expression [1, 4]. It is used by many to understand the feelings of the target people and deal with them. As it is known that in this type of analysis, data is needed, in this case, when all are in a lockdown situation due to the COVID-19 pandemic no one can go out to collect data. So, the primary source of data is social media with the most accurate data available to us via Twitter [2]. The other websites such as Facebook, Instagram and particularly WhatsApp has settled in the trend of sharing fake news in a great scale and hence can't be really relied upon. The events which are now occurring in a day-to-day basis is shared on social media and hence it has become the primary way of people expressing their thoughts and sentiments in the virtual world which in a way is good because one gets to know the mind sets of different people from all over the globe and not really confined to people in a particular place or region. Coronavirus also known as COVID-19 started in Wuhan, China has been one of the most discussed topics on the social media. It has become a hotspot of every person's thoughts and it has changed the way people look at their lives. According to the World Health Organisation (WHO) more than 20 million people over the world have been in contact with this virus and the number is still rising. More than 157 thousand people and counting have died because of this virus and hence it can be understood that the scale of this thing is huge and hence it has become he hotspot of discussion on the social media. Form this data it can be seen that this is one of the biggest biological outbreaks that the world has witnessed in the last two decades. Here, the fear sentiment of people via their tweets are taken into consideration and the expression which is visible through the same [3, 5]. The same data has been taken from twitter using python which allows easy extraction of data using modules such as GetOldTweets3. This model was used over the Twitter's Official API because the official one has the bother limitation of time constraints and tweets which are older than a week can't be extracted. There are some other tools as well but they are paid hence this module named GetOldTweets3 was the best feasible for us. Keywords such as "COVID-19", "coronavirus", "Wuhan Virus", "Wuhan", "China" and some other similar words are used to collect the data and extract the tweets and then sentiment analysis is performed on the same for which the same programming language Python is chosen. In the ultimate summary of all the things mentioned above sentiment analysis methodology is applied to the data collected and a detailed experiment along with explanation is conducted [2, 4, 5]. The discussion is initiated with the exploratory Word Cloud provided in the Fig. 1.

Tweets from three different countries namely USA, Brazil and India have been collected. The whole process can be separated into two different categories: first part contains the pre-processing steps, since the work is done on an unstructured data the results of the data analytics directly associate with this step. After cleaning all the tweets and removing duplicates, preliminary analysis of fear sentiment for all three countries was performed. Second part describes the data analytics methods which include three different and well-known machine learning algorithms namely 'Logistic Regression (LR)', 'Multinomial Naïve Bayes' and 'Support Vector Machine (SVM)'. The three algorithms are used to analyse the tweets and predict their sentiments, individually for the given countries and then compared their accuracies among the three models for the three countries.

Fig. 1 The word clouds created with all tweets for, **a** Brazil, **b** India, **c** USA

The remaining part is organised as follows: Sect. 2 contains related work in the literature. The methodology with data processing and machine learning algorithms for sentiment classification have been discussed in Sect. 3. The experimental evaluation and sentiment classification results with different graphs have been illustrated in Sect. 4, The conclusion with discussion and some areas for future development are given in Sect. 5.

2 Related Work

A large number of researchers have worked and are currently working on this and getting some key features in the sentiments of people regarding this virus. Kaur and

Sharma [1–4] analysed the sentiments related to COVID-19 and the different opinions people had related to this outbreak. They used the twitter API for the extraction of tweets and then analysed the emotions ad positive, negative or neutral with the help of machine learning tools. They used NLTK library for pre-processing of the data and text blob dataset for analysing the tweets. Prabhakar Kaila et al. [23] also did some work that collected raw data that was suitable and worthy to be used in this experiment. The Latent Dirichlet Allocation was applied in the document taken from the datasets and that led to discovery of huge information regarding this COVID-19 like fear and positive sentiments such as trust. Medford et al. [6], prepared a list of different types of hashtags related to this pandemic for a two-week time period from 14th Jan to 28th Jan, 2020. The tweets were extracted using the API and the same is stored in text format. After getting the keywords sentiment analysis was carried on the data to identify the emotions (anger, disgust, fear, joy, happiness or sadness). The tweets were finally analysed and the identification was done using an Unsupervised Machine Learning method.

C.K. Pastor [7] did the sentiment analysis of Filipinos about the effect of the extreme effects of quarantine of the community caused by COVID-19. He also did the analysis of the effects of this pandemic in common life. This all was done on the basis of tweets done by the users. The method used by him was the natural language processing and the opinions were treated as data for the analysis. Rajput, et al. [8] investigated the tweets in addition to the data collection and the investigation was done on the basis of two main aspects, the occurrence of words and the sentiment analysis on the same. His work also stated that the number of tweets in February and March were much more spiked in accordance to the pandemic which showed a trend in this aspect. Samuel et al. [21] showed in their research tweet sentiment classification in the stock market while Skoric [22] performed meta-analysis on public opinion forecasts with social media data.

Like this many others also worked in the same field and by combining all the works done till date it can be understood that how individuals of different countries reacted to this outbreak. There are some rules and regulations that is needed to be followed during this like pre-processing and removing any tweets which are irrelevant during this process. However, it was found that people are mostly positive in their emotions and are hoping to get back to normal life soon. However, some states in Europe think that the situation can't be trusted because it has affected large number of people worldwide and the population scale is very high. Though this is not what every sector is thinking and by doing further analysis something serious about this pandemic was also found [2, 4–8].

Chen et al. [29] also showed us an approach to improve the sentiment analysis at sentence level using sentence type classification. They found that sentences which separate having targets of different options boost the outcome of the sentence-level sentiment analysis. Mironczuk et al. [30] devised a multi-view framework for text classification. The frame work was 2 layers viz- Multi-layered and information fusion layer. They also proposed a method to assess various combinations of the information fusion layer using heuristic ranking. The experimental results indicated information fusion layer may significantly increase the classification quality when a meta-classifier works with a features projection method. Grob-Klubmann et al. [31] presented novel results on the relationship between sentiment indicators mined from Twitter messages and global financial market developments. Their results showed that past Twitter expert sentiment signals are competitive when compared to signals combining past stock index. Chen et al. [32] presented their works on citizen satisfaction on government responses during this pandemic and their results show that people pay more attention on the results of battle against COVID-19 by their own government (number of confirmed cases and deaths per million population) rather than what policies they initiate. Chandrasekaran et al. [33] identified dominant sentiments and changing trends in tweets about the COVID-19 pandemic which can help governments, health care agencies, and other bodies create appropriate responses to prevent and control the spread of the pandemic. Mansoor et al. [34] used deep learning methods like LSTM and ANN to analyse sentiments and find worst hit countries during this pandemic.

3 Methodology and Dataset

In this context, three different types of machine learning methods namely 'Logistic Regression (LR)', 'Multinomial Naïve Bayes' and 'Support Vector Machine (SVM)'s which have been individually applied to the three countries namely United States of America, Brazil and India have been used. First, 100 thousand tweets for USA, 100 thousand tweets for India are 35 thousand tweets for Brazil were gathered. Then the tweets were pre-processed i.e. cleaned them to use them with the above-mentioned statistical methods by removing hashtags, URLs, emojis, duplicate tweets etc. Thereafter, three dictionaries were created namely 'fear dictionary', 'fearlessness dictionary', and 'neutral dictionary' on the basis of which data has been analysed. These dictionaries contain unigrams which are further used for identifying fear level in each individual tweets. Bigrams and trigrams were also designed using the aforesaid three types of unigram dictionaries. Before using these statistical methods, a preliminary check was done to see the fear levels of the users. After that the methods mentioned above has been used and then compared for accuracy in the results section. After every analysis, graphs were created to summarise the study and predictions.

3.1 Dataset

In this the module GetOldTweets3 was used over Twitters official API because the former has the ability to get tweets within a time span specified by the user. Twitters official API has the limitation that it can only extract the data of 1 week from the current date in which the extraction is done. As three particular countries viz: United States of America (USA), India and Brazil [1] were targeted.

Table 1 The module used for the above is GetOldTweets3

Time span	Country	Number of tweets	Max tweets available
01-04-2020 to 30-06-2020	USA	1,00,000	>1,00,000
01-04-2020 to 30-06-2020	India	1,00,000	>1,00,000
01-04-2020 to 30-06-2020	Brazil	32,268	<33,000

Table 2 Number of tweets after filtering

Country	Number of tweets
USA	85,888
Brazil	27,191
India	74,052

These countries were selected because these almost analyse all the sections of the world from their own perspectives. Keeping these three countries in hand approximately 232,000 tweets were extracted [10, 11]. These tweets are not from Coronavirus as a whole but only these 3 countries in particular. The time span considered for the extraction of the data is 01-04-2020 to 30-06-2020 from twitter. This module has inbuild function which prevents data redundancy. The data which was collected has been stored in CSV format. The detailed statistics of the data collected is given in Table 1.

Table 1 describes the tweets obtains from the three different countries namely USA, Brazil and India. First column shows the time span of the tweets i.e., three months which are given in Fig. 1. Table 6 represents the data split in terms of Training and Testing considering the 80 and 20% split.

3.1.1 Data Collection Procedure

In this segment the collection of data is discussed in detail. As previously mentioned, that the social media platform chosen for this data collection was Twitter and the module used was GetOldTweets3. The entire procedure in divided into many steps and each of them is discussed below [2, 9].

Firstly, the module GetOldTweets3 is imported along with the time and csv package. Then the csv in which the data is to be stored is created or opened if already present using the open function and write mode. The Writing function is then invoked and connected to the csv file in which the data is to be stored. The time required in this extraction is also calculated in this approach. The criteria of the searching and extraction of tweets in then set using the methods and the time span for which the data is needed is also specified with the maximum number of Tweets set at 100 thousand [12]. The criteria then merged with the extraction is started. The

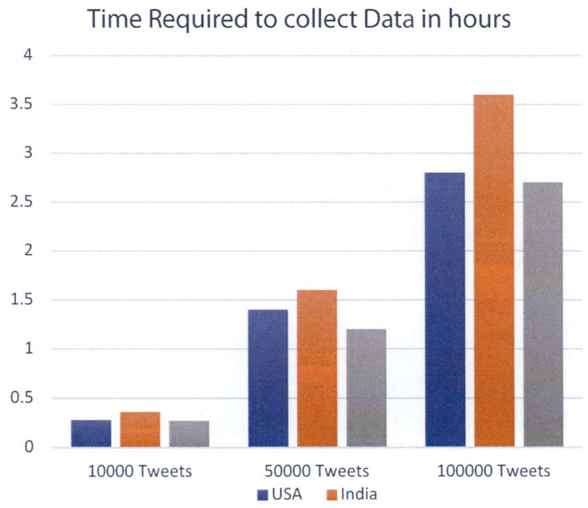

Fig. 2 Graphical representation of the time required to extract the tweets with respect to the three different countries

time is also printed for reference and getting an idea of how much time would be required for getting the specified number of tweets. After this process in completed then the data is written in the.csv file and is ready to be viewed. Following this whole procedure, the Number of Tweets and the time elapsed is also printed. Finally, the method is closed and the program is terminated. To apply this procedure, Python 3 and Ubuntu Operating System which is a distribution of Linux was used. Figure 2 represents the statistical measurement of tweets data in a given time intervals from three countries [7].

More data could not be collected as whenever higher number of tweets extraction was done either the program timed out or the data could not be written to the csv files. The workflow of this extraction has been illustrated in Fig. 3.

The Data collected was mostly constructed from alphabetical characters. Now a day's people are also expressing their feelings through emojis but emojis were not taken into consideration. The data collected for different countries were kept in three different csv files [13].

3.1.2 Data Collection Implementation

All this data was collected with just few keywords like "COVID-USA", "COVID-INDIA", "COVID-BRAZIL", "COVID-19 USA", "COVID-19 INDIA", "COVID-19 BRAZIL". This also helped in gaining a huge number of tweets from the specific region and a large number of tweets were retrieved using these. Finally, after all the tweets were collected, and pre-processed after the cleaning of the tweets the same was sent to the next step in which many different types of Algorithms were used and the results are explained by doing different experiments [8].

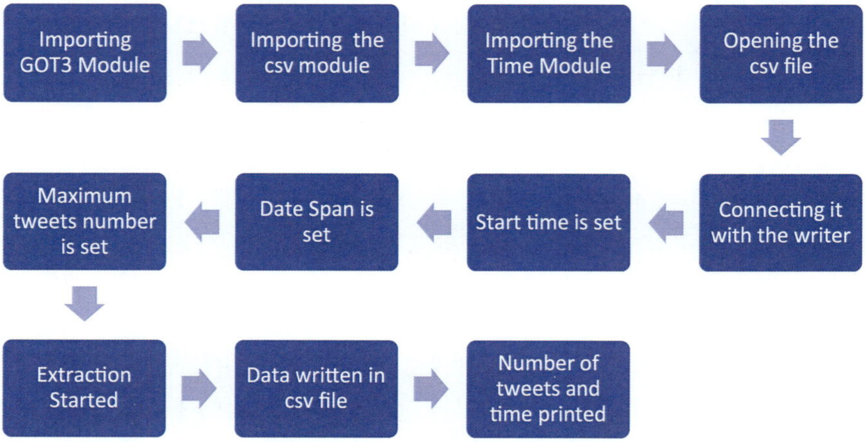

Fig. 3 Block diagram of the proposed model

3.2 *Sentiment Analysis Procedure*

For sentiment analysis [9] and classification of tweets into three different categories namely Fear, Neutral and Fearless, three well knows machine learning models namely 'Logistic Regression (LR)', 'Multinomial Naïve Bayes' and 'Support Vector Machine (SVM)' were used. Combining their usage with word list algorithms and python language was conducted after analysis.

3.2.1 Data Summary

All the tweets collected within the three months were stored in a.csv file. Below are the snapshots of some of the tweets after they have been loaded into a pandas data frame. The datasets extracted from twitters has been demonstrated in Fig. 4.

3.2.2 Text Pre-processing

The following are the steps for text pre-processing.

Step 1: Removing URL.

Any URL contained in the tweets were removed:

Example: *#Brazil "This is a project of great importance for the life of the rural producer because it allows both treatment and* prevention*..." said nurse Maria F. #CooperativaDosCostas* https://bit.ly/3ieIR2T*#WeAreCLAC #FairtradeTogether #Fairtrade #SmallProducersBigHeroes #COVID19.*

	Tweets
0	#Brazil "This is a project of great importance...
1	Illegal miners had invaded #Yanomami indigenou...
2	#boletim #piripiri #covid19 #coronavirus em Pi...
3	Brazil coronavirus death toll reaches 58,314 #...
4	As the death toll of #COVID19 climbs in Brazil...
...	...
35253	Asi con #EEUU EL ASESINO UNIVERSAL #Argentina ...
35254	Please, I'm Brazilian, President Bolsonaro is ...
35255	Pega a visão! #covid19 #coronavirus #covid #co...
35256	If Brazil takes a serious economic blow becaus...
35257	Quando acabar a quarentena todo mundo vai tira...

35258 rows × 1 columns

(a)

	Tweets
0	To my Asians partners are you listening #covid...
1	#COVAXIN is India's first #indigenous vaccine ...
2	Hyderabad based pharmaceutical firm #BharatBio...
3	So tired of "America failed::covid" narrative...
4	COVID-19 LIVE | India's 1st vaccine candidate ...
...	...
99995	@goibibo Unable to reschedule my Air India fli...
99996	If you want a change, you have to take a decis...
99997	K'taka to continue corona tests on dead patien...
99998	मार्च: #कोरोना अप्रैल: घर पे रहोना मई: #आत्मनि...
99999	India me...dunya me sabse zyada Covid Cases aa...

100000 rows × 1 columns

(b)

	Tweets
0	Trump ignores Covid-19 risk in renewed attack ...
1	. The Coronavirus numbers are worse now - Japa...
2	"After 3 months of recovery I can report that ...
3	Ya veremos cómo avanza lo de esta nueva cepa o...
4	#WTF? #Virus is off today?" THOUSANDS rally in...
...	...
99995	Is that an upward #Covid_19 turn in the #USA??...
99996	They are falling behind on DNA testing (Covid-...
99997	Usa - Cina: aumenta la tensione. Le news -) ht...
99998	Usa - Cina: aumenta la tensione. Le news -) ht...
99999	Nach Austrittsdrohung der #USA - einigen sich ...

100000 rows × 1 columns

(c)

Fig. 4 Snap shot of tweets data for different countries: **a** Brazil, **b** India, and **c** USA

After removal of URL:

#Brazil "This is a project of great importance for the life of the rural producer because it allows both treatment and prevention…" said nurse Maria F. #CooperativaDosCostas.

Step 2: Removing Hashtags.

Any hashtags contained in the tweets were removed:

Example: #Brazil "This is a project of great importance for the life of the rural producer because it allows both treatment and prevention…" said nurse Maria F. #CooperativaDosCostas.

After removal of hashtags:

"This is a project of great importance for the life of the rural producer because it allows both treatment and prevention…" said nurse Maria F.

Step 3: Removing stop words.

Any stop words contained in the tweets were removed. Stop words present in the NLTK library was used for this purpose.

Examples: "a," "and," "but," "how," "or," and "what."

Step 4: Removing emojis.

Any emoji present in the tweets were removed.

Step 5: Duplicate tweets removal.

Several tweets were found to be duplicates. To prevent bias in the analysis all duplicate tweets were removed. Count of tweets after removal of duplicates. Table 2 shows the count of tweets obtains from the three different countries after they have been processed and clean for analysis and all the duplicates/retweets have been removed.

*S*tep 6: Tokenisation.

An URL contained in the tweets were removes:

Example: All the tweets were tokenized for TF- IDF vectorisation:

Example:

death,toll,covid19,climbs,brazil,“,”,many,fear,virus,disproportionately,impact, country’s,indigenous,communities,“,”,including,inflicting,irreparable,damage, tribal,knowledge,history,“,”,culture,natural,practices,.,read,:

Step 7: Stemming and Lemmatisation.

Stemming and lemmatisation was done alternatively to see which gives better results. Lemmatisation was finally used for analysis.

3.2.3 Creating a Dictionary of Fear Related Words

Before starting the analysis, a unigram dictionary was created with words related to fear sentiment which can be further used to classis the tweets [14, 25]. The words using for the analysis are provided in Table 3. Table 3 contains all the words present in the fear words dictionary, on the basis of which the tweets from the three countries have been analysed.

3.2.4 Creating a Dictionary of Fearlessness Related Words

Before creating a fear related words dictionary, a unigram dictionary was created with words related to fearless/bravery/faith sentiment which can be further used to classis the tweets into positive category [17]. Table 4 contains all the words present in the fearless words dictionary, on the basis of which the tweets from the three countries have been analysed. Table 5 represents the scoring pattern used for our analysis.

Table 3 Some words in ‘fear dictionary’

Shout	Tremble	Cringe	Bolt	Recoil	Cower	shrink
Hyperventilate	Choke	Gulp	Scream	Scare	Grovel	Anxious
Shriek	Panic	Nervous	Angst	Concern	Sweat	Misery
Uncertain	Unease	Trouble	Suffer	Distress	Shiver	Shake

Table 4 Few words in ‘fearlessness dictionary’

Delight	Happy	Faith	Calm	Peace	Content	Cheer
Joy	Brave	Please	Lively	Recover	Vaccine	Glad
Glee	Elated	Ecstatic	Laugh	Smile	Merry	Thrilled
Love	Respect	Hero	Jolly	Favourable	Good	Better
Best	Amaze	Excellent	Thank	Bright	Encourage	Bravo

Table 5 Score matrix

Tweet type	Score
Fearlessness	1
Neutral	0
Fear	−1

Table 6 Dataset range table

Dataset	Training tweets	Testing tweets
USA	80,000	20,000
India	80,000	20,000
Brazil	25,814	6,454

3.2.5 Assigning Scores to the Tweets Using Word List Algorithm

Three different scores such as 1, 0, and −1 were assigned corresponding the word/words containing three different dictionaries fearlessness, neutral and fear respectively. I was decided to give 1 for all types of positive fearless tweets, 0 for neutral sentiment tweets and −1 for negative/fear related tweets on the topic of 2020 pandemic of COVID. Table 5 shows the scores with respect to three dictionaries. Figure 5 shows the tweet summary from the three countries after pre-processing: (a) Brazil (b) India (c) USA.

3.2.6 Primitive Fear Analysis

At first a rudimentary fear analysis was done using simple word count of words from the fear dictionary present in each individual tweets. The graphs of the fear level in the tweets of users in three different countries has been illustrated in Fig. 6.

Figure 6(a) describes the level of fear in each user's tweets from Brazil. The score is higher with respect with the higher count of words from the generated fear dictionary. Figure 6(b) shows the level of fear in each user's tweets from India. The score is higher with respect with the higher count of words from the generated fear dictionary. Figure 6(c) demonstrates the level of fear in each user's tweets from USA. The score is higher with respect with the higher count of words from the generated fear dictionary.

(a)

	Tweets	Scores
0	This is a project of great importance for the ...	1
1	Illegal miners had invaded indigenous lands at...	-1
2	em Piripiri, Piaui, Brazil	1
3	Brazil coronavirus death toll reaches	-1
4	As the death toll of climbs in Brazil, many fe...	-1
...	...	...
27185	Asi con EL ASESINO UNIVERSAL	1
27186	Please, I'm Brazilian, President Bolsonaro is ...	-1
27187	Pega a viso! em Recreio Dos Bandeirantes, Rio ...	1
27188	If Brazil takes a serious economic blow becaus...	1
27189	Quando acabar a quarentena todo mundo vai tira...	1

27190 rows × 2 columns

(b)

	Tweets	Scores
0	To my Asians partners are you listening	1
1	is Indias first vaccine against Covid-19. The ...	1
2	Hyderabad based pharmaceutical firm has develo...	1
3	So tired of "America failed...covid" narrative...	-1
4	COVID-19 LIVE	India's st vaccine candidate '...	1
...	...	...
74046	Unable to reschedule my Air India flight that ...	1
74047	If you want a change, you have to take a decis...	1
74048	K'taka to continue corona tests on dead patien...	-1
74049	: : : : : I	1
74050	India me...dunya me sabse zyada Covid Cases aa...	1

74051 rows × 2 columns

(c)

	Tweets	Scores
0	Trump ignores Covid-19 risk in renewed attack ...	-1
1	. The Coronavirus numbers are worse now - Japa...	-1
2	After months of recovery I can report that I a...	1
3	Ya veremos cmo avanza lo de esta nueva cepa o ...	-1
4	? is off today? THOUSANDS rally in Brooklyn fo...	-1
...	...	...
85882	In a few days, we'll see a big competition bet...	1
85883	Is that an upward turn in the ??? What happene...	1
85884	They are falling behind on DNA testing (Covid-...	-1
85885	Usa - Cina: aumenta la tensione. Le news -)	1
85886	Nach Austrittsdrohung der - einigen sich die M...	-1

85887 rows × 2 columns

Fig. 5 The tweet summary from the three countries after pre-processing: **a** Brazil, **b** India, **c** USA

3.2.7 Sentiment Classification Using Word List

A complex word list was used to classify tweets into three categories: Fearlessness, Neutral and Fear. Figure 7(a) exhibits the what percentage of overall tweets are fear related, neutral or fearless related from USA. Figure 7(b) describes the what percentage of overall tweets are fear related, neutral or fearless related from India, and same has been shown in Fig. 7(c) for Brazil.

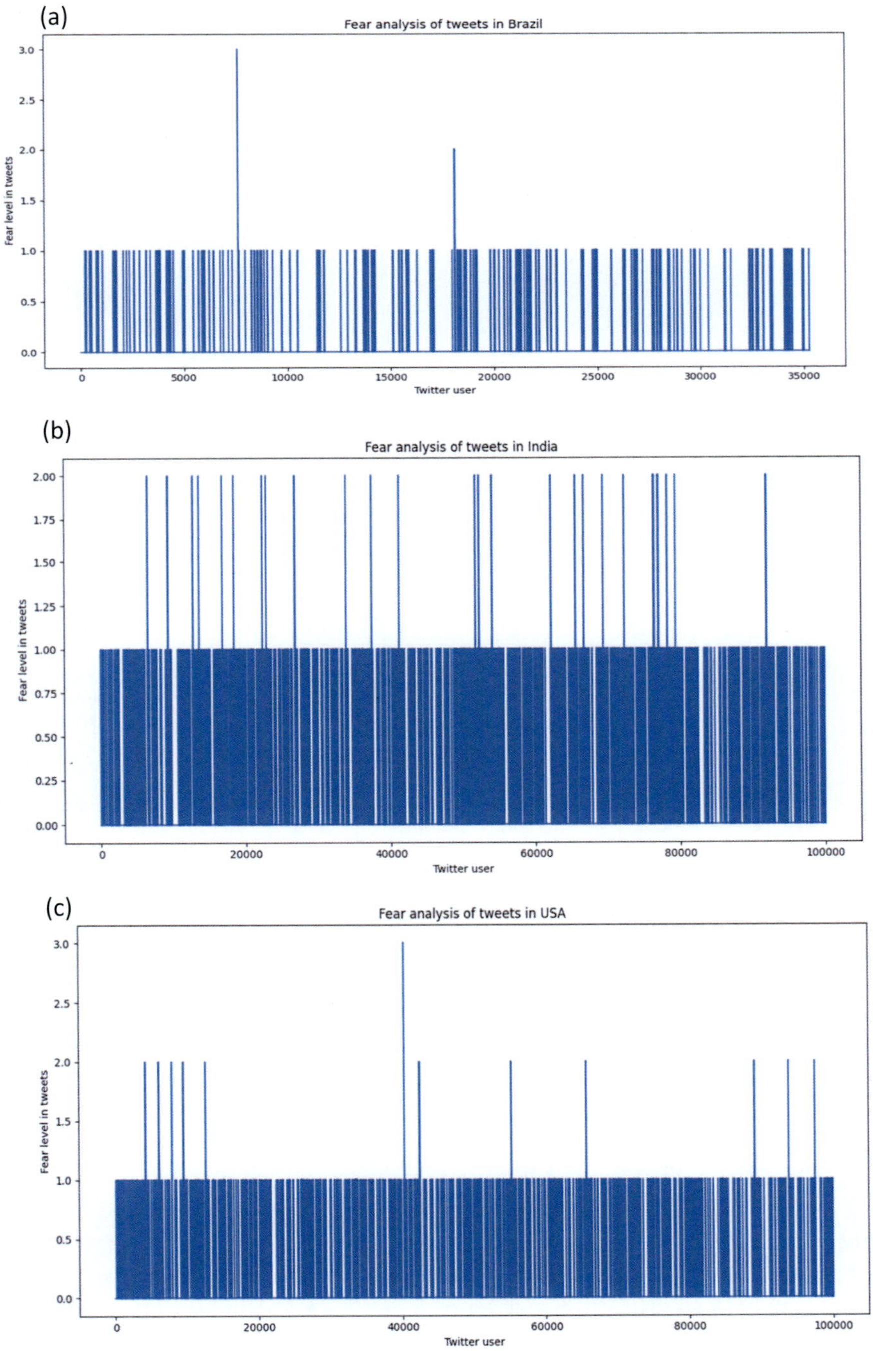

Fig. 6 The fear level analysis in the tweets of users in three different countries: **a** Brazil, **b** India, and **c** USA

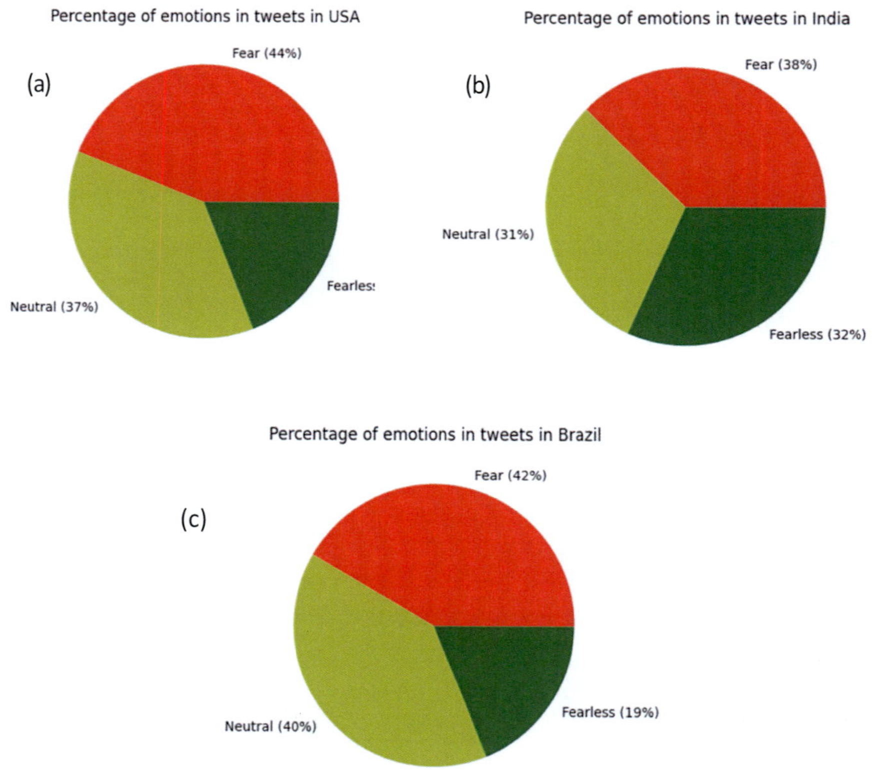

Fig. 7 The percentage of all three types of tweet emotions present in the three different countries

3.2.8 Sentiment Analysis Algorithm

In this section the three different types of machine learning statistical methods used for analysis and predicting the sentiments in the tweets from the three different countries are discussed.

3.2.8.1 Multinomial Naïve Bayes Based Classification

Naïve Bayes algorithm plays an important role in supervised learning for classification. This method assumes a probabilistic model and it agrees to uncertainty about the model by determining probabilities of the outcomes [24, 25].

Naïve Bayes computes explicit probabilities for hypothesis and it is also very robust for noise data. Naïve Bayes classifiers most significant in text classification and multiclass problems have higher success rate as compared to other algorithms on sentiment analysis. The Bayes rule for the class C_i ε C (set of classes) on the tweets T is defined as in Eq. (1):

$$P(C_i|T) = \frac{P(T|C_i)P(C_i)}{P(T)} \tag{1}$$

The max posteriori for classification by Naïve Bayes is defined as the most likely class C_p is calculated as for all C_i in C as mentioned in Eq. (2):

$$C_p = argmax(P(T|C_i)P(C)) = argmax(P(x_1, x_2, \ldots, x_n)|C_i)P(C)) \quad (2)$$

by dropping denominator of Eq. (1) and documents are divided into features x_1, $x_2,\ldots,x_n$ in feature set X.

It is assumed that the features are independent and therefore, multinomial distribution is best suited for the final classification for the class C_i as mentioned in Eq. (3):

$$C_{NB} = argmax(P(C_i)\prod\nolimits_{x \in X} P(x|C)) \quad (3)$$

3.2.8.2 Logistic Regression based classification

LR is the function used at the base of the method, the logistic function. The LR function was developed by statisticians to define properties of population growth in ecology, rising quickly and maxing out at the carrying capacity of the environment. It's an curve that can take real valued numbers and map it into a value between 0 and 1, excluding the limits [27, 28]. The multiple binary regression model is defined as mentioned in Eq. (4):

$$\text{logit}(X) = \frac{1}{1 - e^{-X\beta}} \quad (4)$$

where, π is probability of an observation in the category of Y variable, generally called the "success probability." β represents the constants and x represents the independent variable. For a tweet dataset of size n, the likelihood for a binary logistic regression is given by:

$$L(\beta; y, X) = \prod_{i=1}^{n} \pi^{y_i}{}_i \, (1 - \pi_i)^{1-y_i} \quad (5)$$

From Eq. (5), this yields the log likelihood as mentioned in Eq. (6):

$$\ell(\beta) = \sum_{i=1}^{n} \left[y_i X_i \beta - \log(1 + exp(X_i \beta)) \right] \quad (6)$$

3.2.8.3 Support Vector Machine (SVM) Based Classification

SVMs are a set of supervised methods which are used for regression, classification and outlier's detection. There are several advantages of support vector machines to design this kind of problem, firstly, it's very effecting in higher dimensional spaces, and secondly, still very where number of dimensions is greater than the number of samples. Uses a subset of training points in the decision function (called support vectors), so it is also memory efficient [26].

Let us assume the tweets be x and the scores be y in pair relationship $(x_1, y_1), (x_2, y_2), ...(x_n, y_n)$. Any hyperplane can be written as the set of points x as mentioned in Eq. (7):

$$\vec{w} \cdot \vec{x} - b = 0 \tag{7}$$

b represents the bias parameter in Eq. (7)

Objective function for the linear SVM model is determined as:

$$\frac{1}{2} w^T w + C \sum\nolimits_{i=1}^{l} \xi_i \tag{8}$$

w indicates the decision hyperplane, C is the parameter of the problem that defines how soft the margin should be and ξ_i is the slack variable in Eq. (8).

Linear SVM has been used to solve multiclass classification with ultra large datasets. One to many classifications is used for the sentiment classification.

4 Discussion on Results

In this part, the positive, neutral, negative sentiments which have been discussed also some graphical representations have been illustrated. Python 3 under Anaconda framework was used in this. Mostly due to its fluid and simple syntax and the vast amount of machine learning libraries and modules available such as scikitlearn, Pandas, Matplotlib, NumPy for analysis of the sentiments in tweets [18, 25].

For the results achieved, 80% of the datasets for training and 20% of the dataset for test purposes were used.

4.1 Sentiment Polarity

In this part, the chart shows the sentiment polarity over the twitter data for the data collected for the time period specified before [15]. The data has been shown by namely a few keywords like COVID-19, Coronavirus, Fear, Bravery, etc. It can be seen that the number of tweets for each case is shown separately for the three different countries. However, the positive and negative sentiments of users regarding the situation was found as the tweets which were analysed provided such results [20]. Units are measured in percentage [4].

Figure 8(a) illustrates the sentiments levels of all the tweets from USA. The tweets with score less than zero and plotted in red are classified as fear related tweets. Tweets in yellow with score zero and plotted in yellow are classified as neutral tweets and

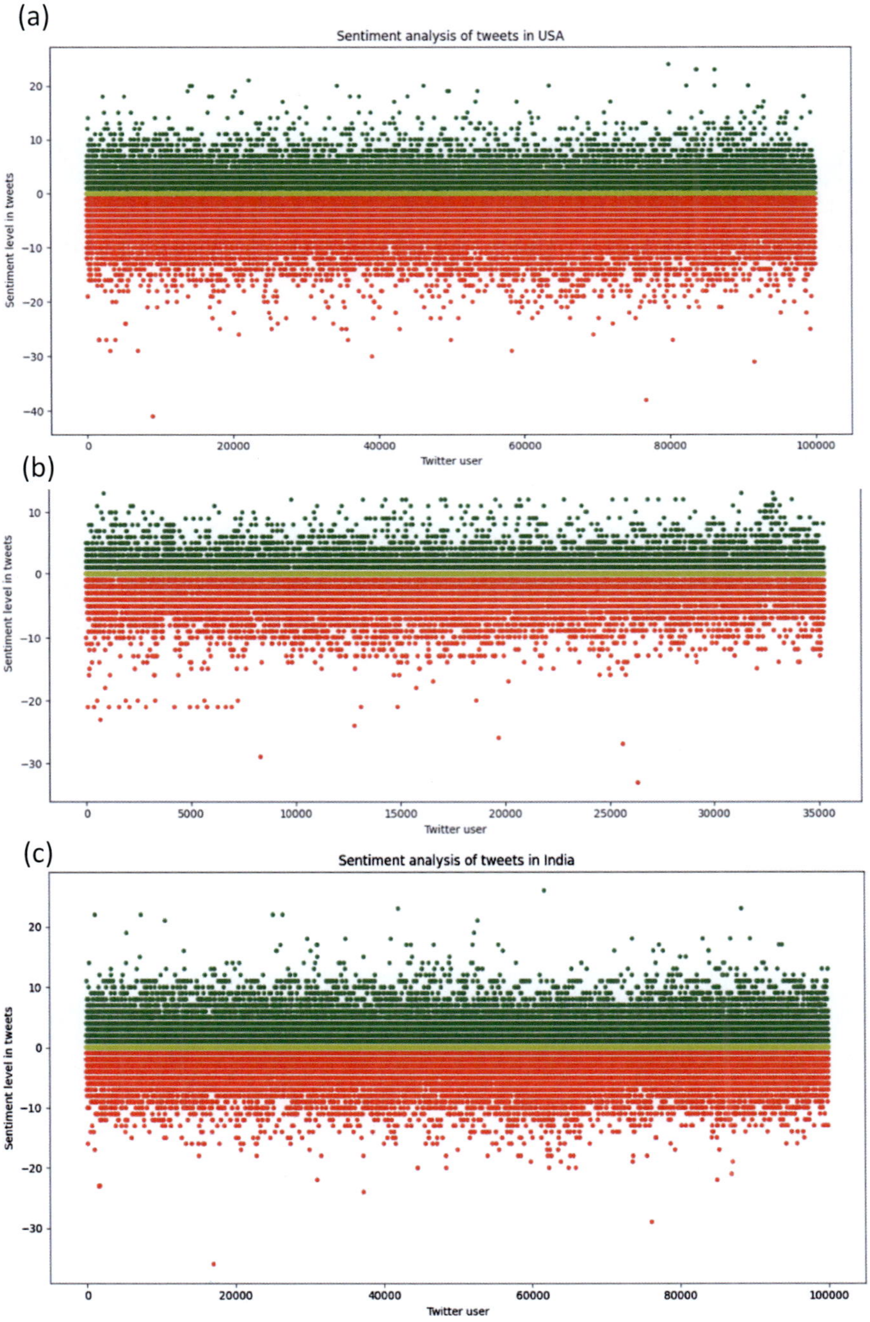

Fig. 8 Shows the sentiment polarities (fearlessness, neutral and fear) among the three countries

finally the tweets with score greater than zero and plotted in green colour are classified as fearlessness tweets. Figure 8(b) describes the sentiments levels of all the tweets from Brazil. The tweets with score less than zero and plotted in red are classified as fear related tweets. Tweets In yellow with score zero and plotted in yellow are classified as neutral tweets and finally the tweets with score greater than zero and plotted in green colour are classified as fearlessness tweets. Figure 8(c) exhibits the sentiments levels of all the tweets from India. The tweets with score less than zero and plotted in red are classified as fear related tweets. Tweets In yellow with score zero and plotted in yellow are classified as neutral tweets and finally the tweets with score greater than zero and plotted in green colour are classified as fearlessness tweets [16].

From the graphs illustrated in Fig. 8, it becomes clear that the twitter users in Brazil took this pandemic situation negatively. It can be seen clearly that the people of Brazil have fear in their tweets. The sentiments of the people of brazil showed that in some cases even more than 30% of the tweets were analysed as negative. The graph shows that while the users in Brazil were affected badly but the users in India were even worst hit as seen in the graph that the distribution is denser. One of the reasons of the graph being dense is that the number of tweets analysed for India were 1,00,000 whereas the Brazilian tweets were 35,000. Still the Negative side of the graph prevails in the case of India with USA being the least negative on the graph and as per numbers are concerned. It's also seen that many tweets were as such which made them look positive and the positiveness of tweets is the most in USA, followed by India and then Brazil. After judging the positive and negative tweets the number of neutral tweets were also found to be dominant as mostly people don't express their positiveness or negativizes on Twitter but here the neutral data also showed that many tweets were related to government, politics and other matters of general concern rather than being positive or negative [23, 24]. The types of tweets found in the three countries were relatively of the same type but the variation came when it was noticed that their opinions which can't be used in this analysis because the work is on real data and not on data which were opinions and could have been done but not actually done. In this regard the subjective nature of the analysis will be discussed in the next part.

4.2 *Subjectivity*

In the perspective of the twitter users, there are many subjective as well as objective point of views. There are three Pie-charts given below which show the numbers in percentage that how many tweets were neutral, positive or negative in this regard. There are many tweets which describe the user's enthusiasm and love in this pandemic due to various reasons like work from home, environmental betterment and clean weather, recovery form virus, news for vaccine etc. while there are others who are scared because of the fast spread, social distancing and lack of jobs.

4.3 Comparison of Models:

Three machine learning models namely 'Logistic Regression (LR)', 'Multinomial Naïve Bayes' and 'Support Vector Machine (SVM)'was used. Each model was applied separately for three different countries [19]. Figure 9 shows three bar graphs representing the accuracy of three models in predicting the classes of tweets.

Figure 9(a) describes the difference in accuracy between the three models namely 'Logistic Regression (LR)', 'Multinomial Naïve Bayes' and 'Support Vector

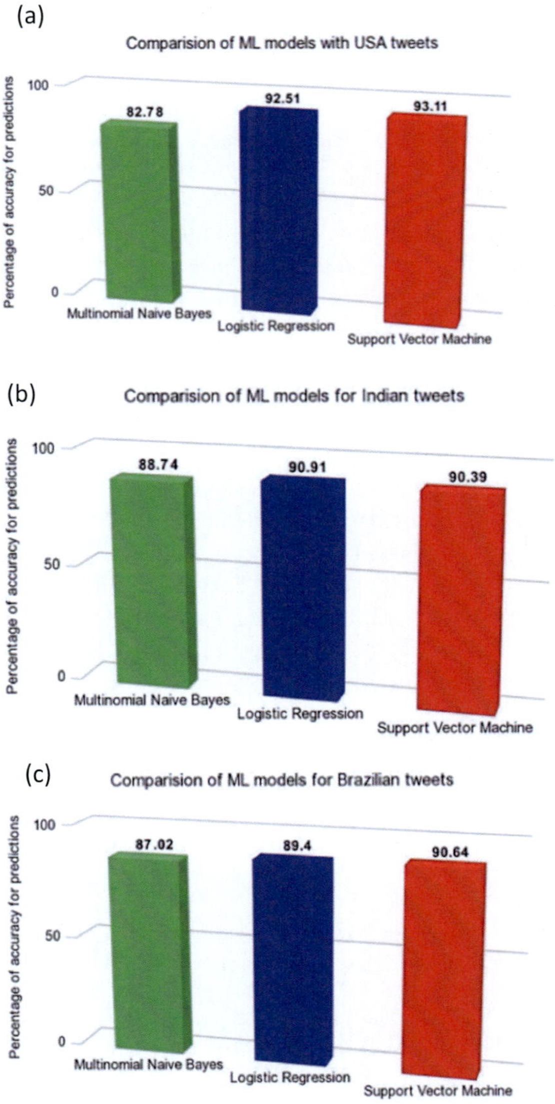

Fig. 9 The three bar graphs display the comparison of three models Naïve Bayes, Logistic Regression and SVM

Machine (SVM)'to for tweets from citizens of USA. Figure 9(b) demonstrates the difference in accuracy between the three models namely 'Logistic Regression (LR)', 'Multinomial Naïve Bayes' and 'Support Vector Machine (SVM)'to for tweets from citizens of India. Figure 9(c) shows the difference in accuracy between the three models namely 'Logistic Regression (LR)', 'Multinomial Naïve Bayes' and 'Support Vector Machine (SVM)'to for tweets from citizens of Brazil.

5 Conclusion

Here sentiment analysis was applied on the twitter data collected by us using the GetOldTweets3 Module worldwide in regard to the COVID-19. The data was collected during the time when the pandemic was the most widespread and that to over a wide time span. The time span was 01-04-2020 to 30-06-2020 and it was extracted using GetOldTweets3. The keywords which were used are "COVID-USA", "COVID-INDIA", "COVID-BRAZIL", "COVID-19 USA", "COVID-19 India", "COVID-19 Brazil". These keywords were used for sentiment polarity and sentiment analysis which fetched good results as discussed above. The results showed us about the type of tweets being neutral, positive and negative. These are mentioned in their accurate numbers above. The approximate number of tweets analysed were amounting to 2.3 lakh from three different countries. From this analysis a conclusion can be drawn about people's reaction from day-to-day posts and tweets on the social media platforms, in this case Twitter in particular. This data regarding COVID-19 outbreak helps understand how people, different organisations political or physical and the government broadcast these situations. In this analysis only unigrams were used for pre-processing of the collected data.

In future, other social media platforms like Facebook, LinkedIn which are also good sources of data may be considered. Hence the analysis can also be done on data from those platforms. There are many other algorithms and models for the analysis except the ones used hence they could also be tested to find the accuracy and analysis. Bigrams and trigrams will also be included for the pre-processing of the data. This type of pandemic may also occur in future and hence applying this as a model can also be helpful which can be done in future.

References

1. J. Samuel, G.G.M. Nawaz Ali, M.M. Rahaman, E. Esawi, Y. Samuel: COVID 19 Public sentiment insights and machine learning for tweets classification. Information **11**(6), 1–3 (2020)
2. K.H. Manguri, R.N. Ramadhan, P. Rasul, M. Amin, Twitter sentiment analysis on worldwide COVID-19 outbreaks. Kurdistan J. Appl. Res. **5**, 54–55 (2020)
3. N. Aguilar-Gallegos, L.E. Romero-García, E.G. Martínez-González, E.I. García-Sánchez, J. Aguilar-Ávila, Datasets on dynamics of coronavirus on Twitter. Data Brief **30**, 105684 (2020)

4. C. Ordun, S. Purosatham, E. Raff, Exploratory analysis of COVID-19 tweets using topic modelling, UMAP and Di-graphs (arXiv, 2020), pp. 1–2
5. R.J. Medford, S.N. Saleh, A. Sumarsono, T.M. Perl, C.U.J. Lehmann, An "infodemic": leveraging high-volume twitter data to understand public sentiment for the COVID-19 outbreak. Kurdistan J. Appl. Res. **7**(7), 2–3 (2020)
6. M. Alhajji, A.A. Khalifah, M. Aljubran, M. Alkhalifah, Sentiment analysis of Tweets in Saudi Arabia regarding governmental preventive measures to contain COVID-19 (Preprints, 2020), pp. 1–5
7. C.K.L.S. Pastor, Sentiment analysis of Filipinos and effects of extreme community quarantine due to Coronavirus (COVID-19) pandemic. J. Crit. Rev. **7**, 2–4 (2020)
8. N.K. Rajput, B.A. Grover, V.K.J. Rathi, Word frequency and sentiment analysis of Twitter messages during Coronavirus pandemic (arXiv, 2020), pp. 1–3
9. M. Ra, B. Ab, S. Kc, COVID-19 outbreak: tweet based analysis and visualization towards the influence of coronavirus in the world. Gedrag Organisatie Rev. **33**(02), 8–9 (2020)
10. A. D. Dubey, Twitter sentiment analysis during COVID19 outbreak. Kurdistan J. Appl. Res., 5–6 (2020)
11. J. Zhang, Y. Yang, Robustness of regularized linear classification methods in text categorization, in Proceedings of the 26th Annual International ACM SIGIR Conference on and Development in Information Retrieval, Special Interest Group on Information Retrieval (2003), pp. 190–197
12. E. Boiy, P. Hens, K. Deschacht, M.F. Moens, Automatic sentiment analysis in on-line text, in *ELPUB 2007* (2007), pp 349–360
13. K. Shu, A. Sliva, S. Wang, J. Tang, H. Liu, Fake news detection on social media: a data mining perspective. ACM SIGKDD Exp. Newsl. **19**, 22–36 (2017)
14. C. Makris, G. Pispirigos, I.O. Rizos, A distributed bagging ensemble methodology for community prediction in social networks. Inf. Multidisp. Digit. Publishing Inst. **11**, 199 (2020)
15. N. Heist, S. Hertling, H. Paulheim, Language-agnostic relation extraction from abstracts in Wikis. Information **9**, 75 (2018)
16. W. He, H. Wu, G. Yan, V. Akula, J. Shen, A novel social media competitive analytics framework with sentiment benchmarks. Inf. Manage. **52**, 801–812 (2015)
17. M.J. Widener, W. Li, Using geolocated Twitter data to monitor the prevalence of healthy and unhealthy food references across the US. Appl. Geogr. **54**, 189–197 (2014)
18. A. Kretinin, J. Samuel, R. Kashyap, When the going gets tough, the tweets get going! an exploratory analysis of tweets sentiments in the stock market. Am. J. Manage. **18** (2018). https://doi.org/10.33423/ajm.v18i5.251
19. M. De Choudhury, S. Counts, E. Horvitz, Predicting postpartum changes in emotion and behaviour via social media, in *Proceedings of the SIGCHI Conference on Human Factors in Computing Systems* (2013), pp. 3267–3276
20. Z. Wang, X. Ye, M.H. Tsou, Spatial, temporal, and content analysis of Twitter for wildfire hazards. Nat. Hazards **83**, 523–540 (2016)
21. J. Samuel, R. Kashyap, A. Kretinin, Going where the tweets get moving! an explorative analysis of tweets sentiments in the stock market, in *Proceedings of the Northeast Business & Economics Association* (2018)
22. M.M. Skoric, J. Liu, K. Jaidka, Electoral and public opinion forecasts with social media data: a meta-analysis. Information **11**, 187 (2020)
23. R.P. Kaila, A.V.K. Prasad, Informational flow on Twitter–corona virus outbreak–topic modelling approach. Int. J. Adv. Res. Eng. Technol. **11**(3), 128–134 (2020)
24. S. Geman, D. Geman, Stochastic relaxation, Gibbs distributions, and the Bayesian relation of images. IEEE Trans. Pattern Anal. Mach. Intell. **6**, 721–741 (1984)
25. S.K. Ghosh, S. Dey, A. Ghosh, Knowledge generation using sentiment classification involving machine learning on e-commerce: Int. J. Bus. Anal. **6**(2), 74–90 (2019)
26. A. Ortigosa, M.J. Martin, R.M. Carro, Sentiment analysis in Facebook and its application to e-learning. Comput. Hum. Behav. **31**, 527–541 (2014)
27. T. Pranckevicius, V. Marcinkevicius, Comparison of naive Bayes, random forest, decision tree, support vector machines, and logistic regression classifiers for text reviews classification. Balt. J. Mod. Comput. **5**, 221 (2017)

28. W. Ramadhan, S.A. Novianty, S.C. Setianingsih, Sentiment analysis using multinomial logistic regression, in *Proceedings of the 2017 International Conference on Control, Electronics, Renewable Energy and Communications (ICCREC)*, Yogyakarta, Indonesia, 26–28 September 2017 (2017), pp. 46–49
29. T. Chen, Xu. Ruifeng, Y. He, X. Wang, Improving sentiment analysis via sentence type classification using BiLSTM-CRF and CNN. Exp. Syst. Appl. **1**, 228–239 (2017)
30. M.M. Mironczuk, J. Protasiewicz, W. Pedrycz, Empirical evaluation of feature projection algorithms for multi-view text classification. Exp. Syst. Appl. **1**, 110–121 (2019)
31. A. Groß-Klußmann, S. König, M. Ebner, Buzzwords build momentum: global financial twitter sentiment and the aggregate stock market. Exp. Syst. Appl. **1**, 184–197 (2019)
32. C.W.S. Chen, S. Lee, M.C. Dong, M. Taniguchi, What Factors Drive the Satisfaction of Citizens on Governments' Responses to COVID-19? (JMIR Publications, 2020)
33. R. Chandrasekaran, V. Mehta, T. Valkunde, E. Moustakas, in *Topics, Trends, and Sentiments of Tweets About the COVID-19 Pandemic: Temporal Infoveillance Study* (JMIR Publications, 2020)
34. Mansoor, M., Gurumurthy, K., Anantharam, R.U., Badri Prasad, V.R.: Global sentiment analysis of COVID-19 tweets over time. Cornell Univ. J. **2** (2020)

A Role of Emerging Technologies in the Design of Novel Framework for COVID-19 Data Analysis and Decision Support System

S. Rama Krishna and Mohammad Sirajuddin

Abstract The rise and spread of the COVID-19 new coronavirus infection surprised the world with a significant impact on the lives of billions of people. Human life is endangered due to the severe spread of pandemics such as COVID. Many frontline warriors are working round the clock around the globe in defense intending to minimize the spread of this pandemic hence minimize mortality. Scientists are working hard to find remedial medicine and vaccine to launch their products to save mankind. Before we reach this stage of protection policymakers are trying alternative solutions. During this time, the changing trends in science and technology can help in fighting against the pandemic spread. This chapter addresses the challenging areas in the effective management of the COVID crisis. This chapter illustrates various novel methods developed by using technological inventions all around the globe, along with their merits, and demerits. This chapter explains the proposed framework which can help policymakers in the detection and prevention of COVID by using machine learning algorithms and other scalable technologies. The proposed framework includes both manual best practices with an adaption of technology change, such a hybrid solution significantly reduces time and efforts. This framework can also help the policymakers in taking the decisions faster and timelier managed on the go. It will trace, track, and predicts the impact with the help of novel technologies. Technologies included in this framework will help in identifying the COVID personal, preparation of contact tracing, identifying the containment zones, generation of schedules of the quarantine periods, the work schedule of frontline warriors, and assessing the required materials in controlling the outspread of virus infection.

Keywords COVID-19 · Prediction model · Machine learning · Bigdata · IoT · Cloud computing · Framework for contact tracing · Decision support system · Quarantine schedule

---

S. Rama Krishna (✉)
Department of C.S.E, Bapatla Engineering College, Bapatla, Andhra Pradesh, India

M. Sirajuddin
Department of C.S.E, Koneru Lakshmaiah Education Foundation (KLEF), Guntur, Andhra Pradesh, India

J. Nayak et al. (eds.), *Understanding COVID-19: The Role of Computational Intelligence*,
Studies in Computational Intelligence 963, https://doi.org/10.1007/978-3-030-74761-9_14

1 Introduction

WHO was informed of an unknown virus by China on December 31, 2019 [1, 7]. It was identified as nCov-19 a novel virus in the corona family after the reporting of a cluster of cases in Wuhan on January 12, 2020. This virus quickly started to spread in and around China and then to multiple countries [2, 3]. COVID-19 becomes an epidemic, and it is observed highly contagious [4, 5]. WHO identified the need for a public emergency with the COVID-19 and declared it on January 30, 2020. Figure 1 shows that among a total 1, 276, 000 confirmed corona cases, there was a recovery 264, 000 people, and 69,500 deaths were reported [6].

The research observations mentioned in [5] says that 41% of patients were suspected to have SARS-CoV-2 due to the transmissions related to the hospitals. It was observed that the patients with fever who visit the clinics during the period are highly dangerous. It may result in cross-infection in the patients and will raise a high-risk situation. A densely populated region of outpatient will allow massive spreading of this coronavirus. It was suggested to screen the patients and isolate them for the treatment to prevent cross infections and spread of COVID 19. Effective outpatient maintenance by separating the clinical area into COVID screening, common fever screening, and pediatric non-COVID screening areas with a strict isolation system was proposed in the literature [1]. In the article, it was also suggested that all the hospital support staff have a personal protection system. The next Sect. 2, explains the diagnosis procedures and safety measures opted by various countries to minimize the outspread of viruses.

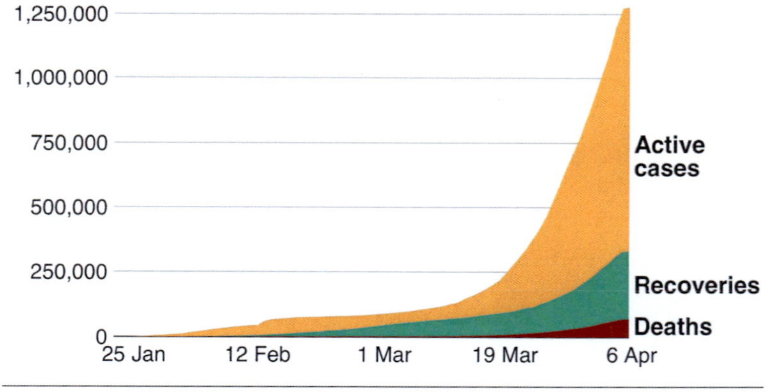

Fig. 1 Mortality and recovery rate in coronavirus case [6]

2 Diagnosis Procedures and Safety Measures

This section throws light on the various diagnosis procedures and safety measures to be considered for minimizing the outspread of corona virus.

2.1 Diagnosis Procedures and Symptoms of COVID-19

Diagnosis is the key component in fighting against pandemic like COVID-19. This section explains various tests performed for diagnosis of COVID-19. The following diagnosis methods are extensively used by the medical clinics.

2.1.1 Viral Test

This test is performed to investigate the existence of the virus in the body. RT-PCR test belongs to this category. This test is used for testing for current infection.

2.1.2 Antibody Test or Antibody Blood Test Isolation

This test is used for checking the presence of antibodies in the given blood sample. The presence of antibodies indicates the presence of a virus. This test takes more time to give result either positive or negative.

2.1.3 CT Scan Isolation

Another diagnosis approach is to check the condition of the lungs through a CT scan. This test can also be used to check the oxygen levels of a patient.

People suffering from COVID-19 have various symptoms like: *Cough, Difficulty in breathing, Fever, Headache, Sore Throat, Loss of taste or smell and Chills, etc.* In response to the accretive COVID-19 cases and the dearth of clinical laboratories, many manufacturers are coming up with many rapid and easy-to-use test kits or devices to facilitate testing outside the laboratories.

2.2 Self-preventive Measures

This section emphasizes on the various preventive measures to be taken by individuals to reduce the possibilities of propagation of this virus in the community.

2.2.1 Usage of Masks

The elementary reason for the propagation of COVID-19 is spread of tiny respiratory droplets from corona affected patients to normal people during speaking, coughing, and sneezing. As per the report by doctors, the droplet size ranges from 5 to 10 μm. So getting in direct contact with corona affected patients will be the major source of outspread of this disease. To minimize this outspread, we must consider two fundamental safety measures: i) confine the direct contact with infected individuals through physical distance, ii) minimize the likelihood of transmission per contact by wearing appropriate masks in public [21]. As per CDC [3] and W.H.O, wearing of masks by individuals minimizes the spread of contaminated droplets, hence outspread of virus can be confined. In some countries like China, South Korea, Japan, etc. the virus has been controlled through the usage of face masks. Besides wearing masks, the recommended safety measures must also be followed by every individual to come out of this pandemic situation.

In this section, we list out the types of masks that can be used to prevent ourselves from getting into the trap of the coronavirus.

i) **N-95 Masks:** This is the most recommended and adequate mask because it can block up to 0.3 microns sized tiny particles that are harder to capture. Due to this reason, it is highly recommended to wear. They are made up of polyester and synthetic fibers, including tangled fibers that act as a filter to make it harder for particles to pass through. These masks are recommended for single use only. However, CDC suggests that masks can be reused up to five times through mask rotations and reprocessing strategies.
ii) **Surgical Masks:** These surgical masks come in several varieties but less effective than the N-95 masks. They are made of breathable, paper-like synthetic fabric that provides flexibility in wearing. These masks are designed for one-time use only.
iii) **Homemade Masks:** Due to the shortage of masks in the market, many people have shown interest in preparing masks at home through dense fabric that is breathable and does not allow particles to pass through. Various styles of making homemade masks have spread around social media which are helping common people in making their masks at home.
iv) **Comparison of Masks:** N-95 masks gives better results than the surgical and homemade masks because N-95 masks have better filtering capabilities than the other masks [21]. Both homemade masks and surgical masks filter the number of microorganisms. But as per the medical experiment, the median fit factor of the homemade masks is one-half of the surgical masks. The surgical masks are 3 times more effective than the homemade masks. Wearing masks is effective only when they are used in combination with frequent hand washing with alcohol-based sanitizers. Proper disposing of masks after usage is also crucial. As per W.H.O, usage of N-95 masks with respirators may lead to serious health issues. Unlike other masks, N-95 masks with respirators cut-off 20% of oxygen and causes excess inhalation of carbon dioxide due to which

many users get feeling of tiredness, nausea, and other breathing-related issues. More research is being done on solving the issues related to N-95 masks.

2.2.2 Quarantine and Isolation

During this COVID Pandemic, these are the two buzz words used in everyday conversation which reflects the strategies to reduce the outspread this virus. These two words are used interchangeably which refers to separating people in different ways to reduce the outspread of coronavirus. Quarantine means isolation of people who may not have symptoms but may be susceptible to the infection because of close contact with COVID patients. Such people are recommended to be in quarantine for 14 days because the incubation time for COVID ranges from 1–14 days. Unlike, quarantine, Isolation means separating virus-infected patients or someone with a confirmed case of coronavirus infection from healthy people to minimize the outspread of infection. COVID Patient is put into isolation until the patient becomes free of symptoms and tests negative for the virus.

These two strategies are widely adopted by all the COVID affected countries to minimize the spread of infection. However, these two strategies are not enough to protect ourselves during an epidemic but also self-responsibility is equally important. Every citizen of a country with positive COVID cases must follow social distancing, avoiding public gatherings, staying at home, and catching up with friends and family virtually instead of in person. The for safety reasons, the recommended social distance is from 2 meters to 6 feet from others.

2.2.3 Usage of Drugs

There is no approved vaccination or drug for coronavirus treatment till now. However few drugs that have been in usage for treating malaria, SARS, Ebola are used for coronavirus treatment. In this section, we explain a few medicines that are widely adopted by doctors in the treatment of COVID. The most commonly used drugs are hydroxychloroquine and chloroquine. These medications have been used for the treatment of malaria, rheumatoid arthritis, and lupus for many decades. These same drugs are giving good results in COVID treatment. All the doctors are looking at Convalescent plasma (CP) therapy to treat COVID patients. In this medication, a blood particle called plasma containing the virus-fighting antibodies, is collected from people who have recovered from the infection and imparted into patients with an active coronavirus infection. Some of the countries like the U.S are using Remdesivir, an antiviral that is delivered via intravenous (IV) infusion. Doctors in the U.S are trying this medicine on severely COVID affected patients [22]. Coronavirus vaccines are in development and are in clinical trials. There is no enough medical data at this point of time to justify that the hydroxychloroquine and chloroquine gives better results on COVID-19 infected patients. More research has to be done on this part to determine the correct drug and its dosage for effective treatment of the coronavirus.

As we know prolong usage of every drug has severe side effects. Similarly, the higher dosage and long-time usage of hydroxychloroquine and chloroquine have the following side effects [22] Irreversible visual changes, abnormal heart rhythm, nerve or muscle pains, low blood glucose, may lead to psoriasis and hormonal imbalance.

2.3 Preventive Measures Opted by Countries and Their Impacts

In this section, we outlined various strategies adopted by different countries to control the outspread of COVID-19. This section also explains the consequences that originated due to the adoption of various strategies to control the outspread of this virus.

2.3.1 Lockdown or Public Curfew

In response to the spread of the virus, many countries imposed restrictions on public gatherings Most of the countries adopted regular or weekend curfew to make their citizens stay at home to minimize the spread of the virus. Some of the countries like the U.K., Ireland, and China preferred lighter restrictions [31]. Whereas the U.S., Canada, Iran, France, Germany, Belgium, the Netherlands, Switzerland, Portugal, Brazil, Austria, and Israel did not impose any curfew, but they called on peoples to stay at home. All virus affected countries not only shuts all their schools and universities but also banned public meetings and mass gatherings to minimize the spread of this virus. Preventive measures like closing country borders, restrictions on transportations, lockdowns, etc. brought much of global economic activities to halt. As we know manufacturing companies and industries are the biggest sources of every country's economy. Such manufacturing companies and industries were locked to support lockdowns imposed by the countries due to this today's world is facing an economic crisis. This situation also accelerated job losses in various countries like China, the U.S, Australia, etc. However, the governments of various countries (like India) adopted supportive measures through various schemes for employees and employers, who were adversely affected by this outbreak.

2.3.2 Social and Self Isolation

Many countries implemented strict self-isolation rules, with fines or imprisonment for violating the rules.

2.3.3 Rapid Testing Kits and Medical Tools

Several countries except Netherlands and Brazil offering test and treatment to trace COVID-19 for free of cost. United states, Iran partly contribute to the expenditure incurred for treatment. Free face mask distribution is also helps to prevent the virus to reach the sensitive human areas like eyes, nose and mouth. Countries like Spain, Turkey, and Italy support for free distribution of these masks. Turkey also supported nearly thirty countries by its export of medical masks. United Kingdom, Italy, Balkan and Spain are also imported masks from turkey. During this pandemic countries like India, China and Russia also spread their humanity by helping other countries with medical equipments. Tools like HeadFIT suggested by United Kingdom defence ministry effectively helped to enhance mood and confidence through which it effectively managed to defend COVID [22]. RT PCR test kits proven to be most reliable and quick diagnosis method for COVID 19 in many countries.

2.3.4 Hospital Phone Booths

Hospital Phone booths were innovative testing procedure firstly launched in South Korea. It is a comprehensive protective sample collection technique first seen in the hospital in Seoul. They designed non-contact swab collection procedure as person will enter in glass walled chamber on a side and in another side hospital staff will collect the sample by using flexible rubber gloves. This procedure helps to collect the sample time significantly increased up to 10 times. This is also a scalable solution with easy to disinfect after swab collection as shown in Fig. 2. This procedure is adopted with minor or no changes in many counties.

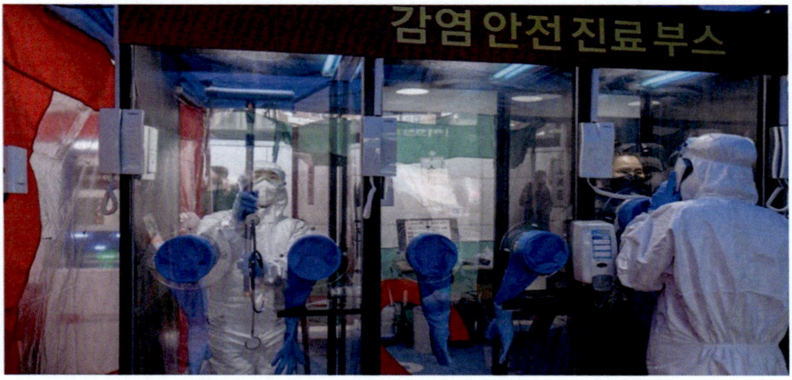

Fig. 2 Special booths for specimen collection [23]

2.3.5 Various Disinfectants in Use

Disinfectants are used to prevent and minimize the spread this coronavirus by cleaning the infected areas. Using the appropriate disinfectant is an important part of preventing and reducing the spread of this coronavirus. The Environmental Protection Agency (EPA) published a list of 83 approved disinfectants authorized to be used [25]. The most commonly used disinfectant is diluted solution of sodium hypochlorite or household bleach. But it is not clear whether bleach kills corona viruses on surfaces outside, or in the air. According to the survey report, bleach itself breaks down under ultraviolet light [26]. Usage of sanitizers and alcohol-based spirits are in wide usage. This massive usage of these sanitizers leads to the shortage of disinfectants in various countries and also given a platform for black markets to emerge. Many Asian and European countries have seen a rise in reports of illicit disinfectants in their markets. In particular, the lack of disinfectants and hand sanitizer created an opportunity for unscrupulous sellers. To solve this issue, many countries like India have permitted their local medical and research laboratories to manufacture quality disinfectants at marginal prices.

2.3.6 Plasma Therapy

Plasma Therapy is a low cost operating procedure. This takes shorter treatment time and even it is eco-friendly. The procedure functions like extracting plasma from the cured COVID patient's blood. After extraction of this plasma, it will be injected into the virus infected patient to grow sufficient antibodies in the defense of COVID. A compatible donor with a high amount of neutralizing body will help in this treatment. The study made during the period 23rd January 2020 to 19th February 2020 over ten patients out of which six male and four female enrolled to receive Plasma transfusion [27]. Their ages interquartile range from 45.0 years to 59.5 years where the median is 52.5 years. As per the doctor's advice up to 400 ml. plasma can be collected from each donor. This plasma can be utilized to treat two patients 200 ml. each. The study says that "CP therapy was well tolerated and could potentially improve the clinical outcomes through neutralizing viremia in severe COVID-19". This therapy does not approve as clinical practice though it is being experimented. It does not record or observed any adversary effects in its execution. In CP therapy two crucial aspects are one the dose size, and the second when this treatment has to start. The random clinical studies observed definite clinical benefits in this treatment [27]. India operated Convalescent Plasma Therapy on 49 years old patient in Delhi who was tested positive on April 4 successfully. After his successful recovery, it also experimented in other states and regions across the country.

2.3.7 Color-Based Zone Identifications

To take necessary action and preventive measures many countries like India divided the cities, districts, and states in terms of zones. Color-coded classifications of zones have been implemented to separate corona affected regions from the safe regions. Based on this classification, safe regions will be freed from lockdown and effective preventive measures can be implemented in regions with more number of corona positive cases. Based on the number of confirmed COVID cases cities/districts are classified into three categories as the red zone, orange zone, and green zone.

i) **Red Zone:** A city or district or state with more number of COVID confirm cases will be placed in this category. This category indicates an emergency or high alert area in which effective and immediate actions are needed to control the spread of the virus.
ii) **Orange Zone:** A city or district or state with less number of COVID cases will be placed in this category. A region will be put into this category only if few cases were found in the past with no increase in the number of positive cases further.
iii) **Green Zone:** Regions with no COVID-19 cases are considered as green zone regions. These category regions are safe however common preventive measures like social distancing, restricted public movement, limited transportation, etc. are allowed.
iv) **Containment Zone:** It refers to a specific area where positive cases of COVID-19 were found. The containment operations include sealing off the area with police barricades and prohibited entry and exit from this zone.

Based on this classification of zones, the government will implement strict rules in red and orange zones to minimize the spread of coronavirus. However, some relaxations are provided to the people or citizens in orange zones to improve from an economic crisis.

3 Related Work

This section explains how emerging technologies are used for handling this unpredicted emergency. We also outlined some recent advancements made in this direction. The role of AI, Robotics, Drones, Natural Language Processing (NLP), Machine Learning, Cloud Computing, and IoT are explained in this context.

3.1 Role of AI and Robotics

As we know that AI is the prominent technology that is ruling the world and also playing a central role in fighting against this virus. In this section, we demonstrated different aspects of using AI-based technology for minimizing the outspread of this harmful virus. The emerging trends in artificial intelligence and embedded technology should fill the technology gap to fight with the COVID.

i) **AI Bots**

In the severe concern to care for the COVID patient needs, there is a critical need to replace artificial intelligence bots to serve the roles of doctors, police, hospital supporting staff, diagnosis staff, and sanitization staff. AI-based chatbots give the response to the patient's queries related to the COVID. These chatbots can also be extended to provide medical guidance to the patients. These systems are implemented by using AI and NLP. Some popular chatbots are designed by using the dataset from WHO.

ii) **AI in Detecting COVID Patients Based on X-Ray and CT Scan**

Many scientists and researchers are proposing AI-assisted Screening for COVOD which is used for detecting abnormalities in the patient's lungs. These systems use a deep learning-based model for determining virus affected patients through lung scans. These systems not only reduce the burden on radiologists and physicians but also prevents radiologist from getting in direct contact with the patients during the test.

Shou Jin et al. [16] proposed AI-assisted CT imaging analysis for COVID-19 screening based on a deep learning model. This proposed model used 1,136 training cases (723 positives for COVID-19) from five hospitals, and achieved a sensitivity of 0.974 and specificity of 0.922 on the test dataset, which included a variety of pulmonary diseases." The AI system also highlighted all areas with legions, which allowed for a faster examination. This system was deployed in 16 hospitals and has performed over 1,300 screenings per day.

iii) **Robots for Relaxing Patients in Quarantine**

Lui et al. [8] proposed a mood booster Robo to help the people to fight against stress and mental instability during the quarantine and lockdown time in the hospitals and at homes. It helps people working as a companion during difficult times. The proposed method senses the mood of the user through audiovisual sensors. It generates customized fragrance through digital scent technology and relaxing music through the audio system and helps the users in elevating and boosting the mood.

iv) **Robots as a Medical Assistant**

Robots are being used as medical assistants [9, 10]. There is a severe impact on the front line fighters, such as doctors, nurses, and paramedical workers, during the defense of COVID patient treatments. This propounded work will help in monitoring the COVID patients and avoids the danger due to direct contact with them.

v) **Robots as Delivery Agents**

Li et al. [11] explained the use of the 3D delivery system using a UAV robot. These delivery systems will help to distribute the required food and medicine to the person anywhere and at any time. Research explains that The UAV will receive a request and delivers the needs. These systems will help in public distribution in coronavirus infected areas [12]. Asimov Robotics proposed robots for serving food and medicines to the patients in isolation wards. Also, these robots are equipped with video conferencing facilities to prevent healthcare staff from direct contact with patients. These robots not only distribute masks and sanitizers but also give awareness to people on how to prevent this pandemic. A self-driving Starship robot drops off deliveries in Emerson Valley, Britain. The robot goes right to people's door, eliminating the need for contact between people.

vi) **Robotic Ambulance**

Samani et al. [13] proposed an ambulance robot to assist the patients in remote areas by using AED to help patients with cardiac arrest. This propounded system works in both manual and semi-autonomous modes. Arif et al. [14] proposed a novel AMBUBOT based on AED for saving the lives of patients in the event of cardiac arrest.

vii) **Robotic Disinfector**

Robotic technology is used by china for disinfection purposes [15]. A hospital in Shenyang in china used a Patrol robot for checking the temperatures of people and also disinfecting large areas. Usage of these robots minimizes the demand for medical staff. Another popular city named Hangzhou in China is employing robots for disinfecting large areas. These robots are equipped with miniature tanks for disinfection purposes. They are controlled by remote controls. In addition to this, some disinfectant robots can also climb stairs for spreading disinfectants. One of such robots seen in Luoyang. Frequent usage of sanitizers is one of the preventive measures to protect ourselves from this virus. By considering this fact sanitizer dispensing robots are being used in China that enables people to use sanitizers. These robots are mobile and available for sanitization purposes especially on roads and in offices. UV light robots are used for disinfection purposes. These robots will do cleaning tasks by minimizing the cleaning staff and cleaning time.

viii) **Patrol Robots**

Robots are also equipped with necessary cameras, sensors, and alarm systems to patrol the streets to monitor the lockdown. Recently in Singapore, a robot dog has been deployed to patrol various places and to ensure that people are maintaining a physical distance. This robotic dog frequently broadcasts a message that reminds people to maintain physical distance. This robotic dog is shown in Fig. 3.

Fig. 3 Robot dog [24]

ix) **Robot Chefs**

One of the possibilities for the extreme spread of this virus is through food if it is prepared by a virus-infected person. To overcome this flaw robot chefs are being used in the kitchen for preparing foods without any human intervention and supervision. These robots will minimize the spread of the virus through food preparation and delivery.

x) **Robotic Automation for Production Boosting**

One of the issues experienced by the COVID affected countries is how to resume and stabilize productivity, especially in manufacturing companies. A survey by the American Chamber of Commerce Shanghai found nearly half of the 109 companies polled said that their biggest challenge in the coming weeks was to have enough staff to run the full production lines in factories. This reality will make ramping up robotic automation even more appealing to reduce costs, ensure continuity and productivity even if the country experiences another global pandemic or other shutdowns.

3.2 Drones

Drones are playing a crucial role in fighting against the COVID-19 pandemic. This section outlines the usage of drones in all possible ways to stop the outspread of this virus. Drones can be used as patrolling and law enforcement agents, delivery agents for supplying medical commodities, disinfectant sprayers in public spaces, and monitoring officials to identify COVID patients with elevated body temperature scanners to take necessary and immediate action. In this COVID-19 outbreak, robots and drones are proving to be an essential support for people in fighting against this pandemic situation.

3.3 Machine Learning

Many researchers are using ML and deep learning-based models for classifying corona affected and non-corona affected people. Deep learning algorithms are used in detecting corona affected patients based on the X-Ray and CT scan images. Also, this technology has opened up a new direction of research to control the outspread of COVID-19 as follows:

i) Identification and prediction of the impact of COVID patients.
ii) Clustering and identification of the red zones.
iii) Prediction and forecast of the essential needs, availability of resources, and medicine to help in the administrative decisions.
iv) Models to find financial and social impact prediction and assessment.

Feng Shi et al. [18] demonstrated how emerging AI technologies will strengthen imaging tools and automate the scanning procedure by protecting the radiologists from direct contact with patients. F. Rustam et al. [29] explained the importance of machine learning models in forecasting the COVID-19 confirmed cases, death rate as well as recovery rate. F. Petropoulos et al. [30] elucidated the forecasting of COVID-19.

3.4 Natural Language Processing

Lifang Li et al. [17] explained the importance of categorizing the situational information which commonly propagated in social media. They collected COVID related data from Sina Weibo and classified collected information into seven types of situational information. To accomplish this natural language processing-based classifier is used. This categorization of situational information helps authorities to understand the situation during a crisis or emergency. So that authorities can implement new strategies to bring the situation under control. Certain flaws were examined in this approach mainly, small data set is considered and limited classifiers were trained.

More research has to be done on the aforementioned points that would help take precautionary measures in the future.

3.5 Digital Learning and Internet Technologies

To support the lockdown announced by the countries, many entrepreneurs are coming up with digital learning and remote work tools. Most of the education institutes are conducting classes online for the students by creating virtual learning infrastructure. Many universities are showing interest in conducting online examinations and this demands newer and robust online examination software. During this lockdown, remote work tools are in high demand to allow employees of organizations to work from home safely. These tools should be implemented with the emerging technologies to provide seamless and reliable services which are also one of the challenging tasks.

3.6 Role of Cloud Computing in COVID-19 Pandemic

Cloud Computing technology has been at a decent pace over the past few years. To control the outspread of this virus, we need to collect the people's health-related information along with their location information continuously. For storing and analyzing this timely collected data cloud services are required. To ensure the security of people's health data stored in a cloud some specialized security schemes should be implemented.

Jianqing Liu et al. proposed a paradigm for statistical analysis of infectious diseases. To provide security to the collected data KOIPE security scheme was proposed. This propounded paradigm performs spatial clustering analysis of infectious diseases in a secure and privacy-preserving manner [19].

3.7 Internet of Things

IoT technology is in widespread use especially to implement smart city ideas. This adoption of IoT technology has opened up new research challenges more specifically in health care due to the unexpected COVID Pandemic. This context explains the role of IoT in health care. To automatically detect and control the spreading of this coronavirus the innovative IoT based applications are needed. Integration of emerging technologies like AI, Bigdata Analytics, Cloud Computing with IoT can leads to powerful automated tools for coming out of this unfortunate situation. In this direction, recently Mohammad M.N et al. [20] proposed an IoT-based smart helmet for detection and diagnosis of COVID-19. This propounded system is equipped with a thermal imaging system and facial recognition technology that automatically detects a person with corona symptoms and propagate this information to the authorities to take necessary actions [20]. One of the identified limitations of this technique is, it is difficult to wear a smart helmet for a long time. The devices used for health care should be small, flexible, and portable to use for people.

3.8 Identified Pitfalls

The following limitations are identified in the existing systems during our literature survey.

i) Lack of availability of large COVID dataset for high accuracy of classification and prediction.
ii) Usage of Robots and drones like technologies are expensive to implement especially in the developing countries where the population is more.
iii) Existing ML and Deep learning algorithms are trained and tested on very limited data.
iv) Use of AI-based schemes for detecting and predicting COVID-19 and its impact from the data collected using IoT devices.
v) Current research focuses on only one specialized domain to diminish the outspread of the virus. Integration of various domains like image processing, ML, NLP, IoT, etc. together will give a powerful solution to handle the current emergency.

3.9 Objectives

Based on the literature survey we have considered the following objectives for implementing the proposed framework.

i) Health Data Collection using IoT based portable devices.
ii) Massive health care data storage and security.
iii) Data Distribution to the authorities and doctors to take necessary actions.
iv) Use of AI-based schemes for detecting and predicting COVID-19 and its impact from the data collected using IoT devices.
v) Design of integrated framework to handle this pandemic situation.

4 Proposed Framework

In this paper, we proposed a novel framework to handle data during the COVID pandemic. To cope up with the changing needs during this pandemic situation, governments are trying to make us of innovative applications. This framework encompasses various emerging technologies in handling the COVID situation.

The proposed model helps in capturing the medical data of people. This framework provides the following functionalities.

i) Data Capture
ii) Data Preprocessing and Analysis
iii) Data Classification and Data Analytics
iv) Decision Making
v) Service Provisioning

This proposed framework captures the citizen's activity, contact, and medical data through IoT based gadgets, mobile G.P.S., health workers, robotic and clinical

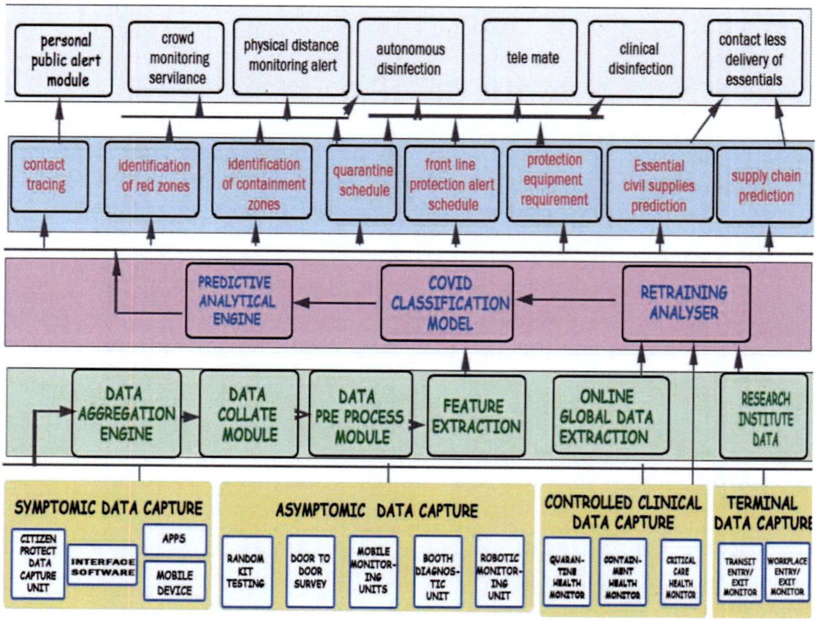

Fig. 4 Proposed framework

monitoring. This collected data is processed and analyzed to determine the possibility of getting infected by the virus.

The proposed framework is divided into five layers in abstract to perform some specific functionality and it is depicted in Fig. 4.

4.1 Layer 1: Data Capture Layer

This is a data capture layer which is responsible for capturing the required data for decision making using the technologies like IoT bands. Mobile technology, G.P.S, and robotic devices. Also, this layer has a provision of collecting data from clinical staff. It is used to collect symptomatic, asymptomatic citizen health travel and contact history, at the same time data obtained during the monitoring of clinical practices suspects, patients with COVID in quarantine, and clinical areas places. COVID patient detection, tracking of his activities, and predict the possibilities of COVID are captured in various ways. In the suggested model we propose four modules to capture the data as follows:

4.1.1 Symptomatic Data Capture Module

This module collects the data related to symptoms, travel, and contact information through IoT enabled wrist band (Protec. band). These can be operated in two modes, fully autonomous or mobile paired bands. Fully autonomous modules will capture the data and directly propagates to the data aggregation module. The pair able bands can be used in economy mode. These bands will sense the data and stores it when it is paired to the internet-enabled devices like a mobile phone with application control or free government internet gateways propagates the data to the data aggregator. Collected data contains body temperature, blood pressure levels, oxygen levels, heartbeat speed, and Cough.

4.1.2 Asymptomatic Data Capture

Some of the citizens do not exhibit the symptoms. These can become super spreaders or spread points. Such people need to be identified and traced quickly. So in our asymptomatic data capture module, we defined various techniques to collect the data from asymptomatic citizen which are explained as follows;

i) **Random Kit Based Testing:** Random Kits based on the serological procedure is used to collect data from citizens not having any COVID symptoms. The results of these random kit testing are recorded and propagated to the data aggregator [28].
ii) **Door to Door Survey:** This is a conventional method of data collection. Paramedical staff, health workers, and volunteers will perform door to door survey to capture people's medical data manually. They may also collect random samples from the citizens and send them to nearby COVID test centers.
iii) **Mobile Monitoring Units:** We also proposed mobile monitoring and diagnosis units in containment zones, making the required infrastructure available to the people to capture medical data.
iv) **Booth Diagnostic Unit:** This method will help to capture the data on the fly at junction points, petrol pumps, or waiting for centers so that citizens can participate voluntarily in giving samples with zero contact. In this data capturing approach, sample collection booths are organized for collecting blood samples from the people and they are tested for the presence of infection. This strategy is widely used by various countries for testing people.
v) **Robotic Monitoring Units:** Robots can be used for the collection of medical data or COVID test samples from the people in quarantine and hospitals. They will significantly help to reduce risk and improve the rest time for front line health warriors.

4.1.3 Controlled Clinical Data Capture

This module is used to capture the data related to the COVID positive or suspect patients. This module will help to understand how the clinical or quarantine procedures are impacting the citizen marked as COVID through the symptomatic or asymptomatic test procedures. This data will help in taking the decisions by the government to schedule the duties of frontline warriors, and monitor the essential requirements of the quality of drugs and their impact on the COVID patients during the quarantine, containment, and critical care times in the clinical or supervisory environments. This data capture majorly focuses on three areas discussed below.

i) **Quarantine Health Monitoring**

Quarantine is the strategic decision taken by the governments to ensure the minimization of spread with COVID suspected people through primary or secondary contact tracing. It will help in the decision making by studying the symptomatic behaviors of the COVID suspected patients. This data can be used to predict the dormant and propagation phases of the virus, strength, and spread of the virus. Hence necessary measures can be taken to prevent the community spread. In this module, during the quarantine period, periodic health monitoring information, symptom, and spread information can be gathered in quarantine centers.

ii) **Containment Health Monitoring**

The containment procedure will help to reduce the contact possibilities of the primary and secondary contact tracing. The study of the data collected in this module will help to understand the impact of public disinfection, identifying containment zones, and how much duration the containment to be kept. Data from people living in the containment zone is also collected and monitored.

iii) **Critical Care Health Monitoring**

Critical Care is said as the time duration for the COVID patient's clinical treatment as an inward patient in the COVID hospitals. This module will help to collect the information from the paramedical staff in the hospitals and used to analyze the data to understand the behavior of the COVID patient during critical times, the impact of usage of drugs, severity of the virulent nature. This data will be used to design the best clinical practices, identification of the drugs that impact during critical times, enumerate the essential clinical and pharmaceutical requirements.

4.1.4 Terminal Data Capture

Capturing the data will help to identify the spread of the virus in the community. This monitoring can be done to understand the procedures to be used to identify the symptomatic COVID patients in different activities in daily life. This will minimize the risk of the spread of the coronavirus by stringently monitoring the citizen's symptomatic data during various activities at the entry and exit terminals. This will help in taking the decisions to prepare exit policies of the lockdown decision by the government. In this approach, data is proposed to be collected at various entry and exit points of shopping malls, airports, railway stations, etc.

4.2 Layer 2: Data Processing and Analysis

Data processing is an essential technique needed for the extraction of quality data for policy and decision making. This layer takes the responsibility of extracting the essential data fuel which can be used for the analytical engine available in layer 3. Three of the modules; data aggregation engine, data collate module, and data pre-process modules will do the preprocessing. The roles and responsibilities of data processing modules are discussed below.

4.2.1 Data Aggregation Engine

This module helps to collect and aggregate the data obtained from the sources mentioned in Layer one. It receives data collected from symptomatic data capture, asymptomatic data capture, and terminal data capture to perform data aggregation. Because data collected from different strategies are stored in different formats and they are unstructured. The unstructured data scrubbed and formatted to make things ready for the data collate unit. Here we prefer the use of technologies like big data, cloud storage to handle the huge data capturing.

4.2.2 Data Collate Module

Data aggregated at Aggregation engine given as input to Collate module. The responsibility of the data collate module is to do dimensionality reduction and clean the data impurities. This module will help to make the data ready to input into the data preprocessing module. Here we propose the use of data cleansing techniques used in data mining for efficient data collation.

4.2.3 Data Pre-process Module

Data Preprocess is the key module to help in identifying and extracting the essential feature data to use as input to the analytical module in layer three. In these modules, major tasks such as dimensionality reduction, scaling, normalization techniques are used. We propose the use of mathematical methods along with data mining for preprocessing of the extracting the data fuel. This module will obtain data from two sources one from the data collate module second from external sources data. The

External data source includes two modules; the Online Global Data extraction module and research institute data will obtain the data from external resources.

4.2.4 Feature Extraction Module

After pre-processing task, the data will undergo a feature extraction module for identifying the potential significant features required in deciding the COVID patient.

4.2.5 Online Global Data Extraction Module

This module continuously monitors data around the globe we suggest technologies like web scraping and real-time data mining for extracting the essential data from the sources worldwide. Hence this module will help to understand the global virus behavior patterns.

4.2.6 Research Institutional Data

This module helps in the extraction of research updates from the different research units in real-time and used as input for improving the accuracy, identifying the features needed to be analyzed, the study on the usage of disinfects, and drugs and clinical guidelines framing. The data preprocess layer will enable the quality data extraction, transformation, and loading to the predictive analytical module mentioned in the next layer.

4.3 Layer 3: Predictive Analytics Layers

In this layer, citizen information is classified and confirmed as COVID or non-COVID patients in automated data analysis or tested procedures. Once this data is available, it will be forwarded to the predictive analytics module for the different critical decision making.

4.3.1 COVID Classification Module

We suggest using a classification algorithm with the classified dataset captured from the predictive analytic module for training. Achieving an accurate prediction from the captured data needs to be properly engineered. Feature extraction, intelligent real-time retraining module will make the model up to date and improve its performance in the classification. The machine learning model is built and trained by considering clinical data, research institute data, and online global data. The data obtained from the below layer two after pre-processing will undergo a feature extraction module for extracting the most relevant features required for classification model. After feature extraction, the resultant data is given as input to the disease classification module for training and prediction.

4.3.2 Retraining Analyzer

It identifies the need of retraining the machine learning model automatically. Whenever retraining of model is required this module gives an alert, so that retraining of model can be performed for better accuracy in results. This module trains the disease classification module periodically.

Preprocessed data is given as input to classification module. This module classifies whether the person is having COVID or not. If the person is having COVID then that person's data is sent to the predictive analytical engine and also to the personal public alert module to take necessary actions.

Once a citizen is classified as a COVID patient automatically this data will be forwarded to the predictive analytical engine.

4.3.3 Predictive Analytical Engine

Once this module receives data from the disease classification model regarding the COVID infected person, it immediately invokes several services that help in decision making and policy designing in real-time. Here we suggest the use of technology like Hadoop and big data analytics for doing predictive analytics.

4.4 Layer 4: Prediction Layer

A predictive analytical engine performs the following functionalities with the help of the components and modules present in this layer. These functionalities are outlined as follows.

i) Contact Tracing: It identifies the people to whom the infected person was contacted previously after the diagnostic of COVID disease. This module needs bigdata and cloud computing to trace and enumerate direct or indirect contact tracing.
ii) Quarantine Schedule: This schedule is automatically generated by the predictive analytical engine based on the received and processed patient's data from contact tracing.
iii) Identification of Containment zones: once the predictive analytical engine determines a person as a COVID patient then it automatically clusters regions

which are in 1 km radial distance of the patient's home into a containment zone.

iv) Scheduling the front line workers: This module will help to schedule front line protection staff required and where they are required report automatically by sending alerts so that the preventive measures can be taken to minimize the outspread of infection.

v) Protection equipment requirement: once contact tracing is completed and data is ready it is required to take preparatory steps to ensure proper protection between COVID and Non-COVID people. Hence this module extrapolates the requirement and needs for the protection equipment.

vi) Civil supplies prediction module: This module will help to enumerate the required civil supplies.

vii) The supply chain prediction: This module will take responsibility to predict the requirements in delivering the essential commodities and ensure the proper supply chain mechanism.

4.5 Layer 5: Service Layer

This layer includes several service modules that will directly help the people. In this layer, we also suggest how different technologies can help mankind from the COVID like pandemic situations.

i) Personal public alert module: this service helps to alerts the citizen about their diagnostic information, alerts to avoid direct contact with COVID affected persons by alarming their vicinity, and periodically sends the government measures for precautions. IoT based technology, the internet, cloud computing for massive computation, and mobile technology helps to deliver these services.

ii) Crowd monitoring and surveillance: Essential part of COVID prevention is to reduce contact with infected persons. To ensure physical distance among the citizen during the relaxation This service module helps to monitor crowds using drones, IP cameras, surveillance cameras, image or video processing, and machine learning models to identify crowd automatically and alert the citizens timely.

iii) Physical distance monitoring – Monitoring the physical distance in the workplaces, offices, malls, and public gathering is a tedious task. This job can be fulfilled with a physical distance monitoring service assisted with Robotic and drone technologies. This will help to ensure physical distancing in various places.

iv) Automatic disinfection is performed in infected regions by the robots. Robotic sanitizers can be employed to achieve this task.

v) The tele-mate module provides teleconferencing and video conferencing facilities to the people who are in quarantine so that they can communicate with other people or to medical staff.

vi) Clinical disinfection- In the hospitals and COVID nursing homes, there is a serious concern of widespread of the COVID virus. To prevent this virus spread the hospital frontline workers have to work round the clock. This tiresome job has to be simplified using technologies like artificial intelligence and machine learning, to disinfect clinical premises effectively.

vii) Contactless delivery of essentials can be achieved with the help of drones, especially in the containment zones. In these zones, a request is collected from people through mobile phones. Commodities are delivered by drones and robots.

5 Summary

In this chapter, we demonstrated the functionality and the use of the proposed framework in handling the pandemic situation. This framework simplifies and automates the policies which are being put into action by the various governments manually today to minimize the outspread of the virus. We explained how emerging technologies can play a vital role in implementing smart and automated applications that will analyze people's health data to predict the severity and level of virus outspread not only today but also in the future. Design of a single framework that will address all the issues of the pandemic situation is a challenging task. So, for simplicity, we modularize the proposed framework into various layers that enables us to use specialized techniques and technologies to implement the functionality of each layer efficiently. This layered framework will outperform in handling the various issues in the pandemic situation compared to the existing applications. This proposed framework can be evaluated and compared based on the performance parameters such as classification accuracy, quarantine schedule generation time, manpower schedule generation times, scalability, and security.

6 Conclusion

This chapter enlightens the roles of various recent trends in technologies that can help to handle COVID pandemic situations. The proposed framework will be useful in the detection and prevention of COVID by using machine learning algorithms. In the future, we address security and distributed storage related concerns with the help of efficient algorithms.

References

1. Y. Xiao, C. Tan, J. Duan, A. Wu, C. Li, An effective model for the outpatient management of COVID-19. Infect. Control Hosp. Epidemiol. **41**(8), 1–4 (2020)

2. C. Huang, Y. Wang, X. Li et al., Clinical features of patients infected with 2019 novel coronavirus in Wuhan, China. Lancet **395**, 497–506 (2020)
3. Situation update worldwide, European Center for Disease Control and Prevention website (2020), https://www.ecdc.europa.eu/en/geographical-distribution-2019-ncov-cases. Accessed 15 Mar 2020
4. Q. Li, X. Guan, P. Wu, et al., Early transmission dynamics in Wuhan, China, of novel coronavirus-infected pneumonia. N. Engl. J. Med. (2020)
5. D. Wang, B. Hu, C. Hu et al., Clinical characteristics of 138 hospitalized patients with 2019 novel coronavirus-infected pneumonia in Wuhan, China. JAMA **323**(11), 1061–1069 (2020)
6. Johns Hopkins university, https://coronavirus.jhu.edu/map.html. Accessed 6 Apr 2020
7. WHO, https://www.who.int/emergencies/diseases/novel-coronavirus-2019/events-as-they-happen. Accessed 6 April 2020
8. J.H. Lui, H. Samani, K.Y. Tien, An affective mood booster robot based on emotional processing unit, in *2017 International Automatic Control Conference*, (Pingtung, Taiwan, 2017), pp. 12–15
9. C.Y. Su, H. Samani, C.Y. Yang, O.N.N. Fernando, Doctor robot with physical examination for skin disease diagnosis and telemedicine application, in *2018 International Conference on System Science and Engineering (ICSSE)*, (IEEE, New Taipei City, Taiwan, 2018), pp. 28–30
10. S.C. Tsai, H. Samani, Y.W. Kao, K. Zhu, B. Jalaian, Design and development of interactive intelligent medical agent, in *IEEE Artificial Intelligence Virtual Reality (AIVR) Conference*, (Taichung, Taiwan, 10th-12th December 2018)
11. B.Y. Li, H. Lin, H. Samani, L. Sadler, T. Gregory, B. Jalaian On 3D autonomous delivery systems: design and development, in *2017 International Conference on Advanced Robotics and Intelligent Systems, ARIS 2017*, (IEEE, Taipei, Taiwan, 6–8 Sep 2017)
12. Y.W. Kao, H. Samani, S.C. Tasi, B. Jalaian, N. Suri, M. Lee, Intelligent search, rescue, and disaster recovery via the internet of things, in *2019 Global IoT Summit (GIoTS)*, (IEEE, Aarhus, Denmark, 17–21 June 2019)
13. H. Samani, R. Zhu, Robotic automated external defibrillator ambulance for emergency medical service in smart cities. IEEE Access J. **4**, 268–283 (2016)
14. M. Arif, H. Samani AMBUBOT: ambulance robot, automated external defibrillator robotic ambulance, in *16th IEEE International Conference on Advanced Communication Technology (ICACT)* (2014)
15. Buisinesinsider, (2 April 2020), https://www.businessinsider.com/robots-fighting-coronavirus-in-china-us-and-europe-2020-3?IR=T. Accessed 20 Dec 2020
16. https://www.medrxiv.org/content/medrxiv/early/. Accessed 20 Dec 2020
17. L. Li, Q. Zhang, X. Wang, J. Zhang, T. Wang, T.-L. Gao, W. Duan, K.K. Tsoi, F.-Y. Wang, Characterizing the propagation of situational information in social media during COVID-19 epidemic: a case study on weibo. IEEE Trans. Comput. Soc. Syst. **7**(2), 556–562 (2020)
18. F. Shi, J. Wang, J. Shi, Z. Wu, Q. Wang, Z. Tang, K. He, Y. Shi, D. Shen, Review of artificial intelligence techniques in imaging data acquisition, segmentation and diagnosis for COVID-19. IEEE Reviews in Biomedical Engineering, 16 April 2020
19. J. Liu, Y. Hu, H. Yue, Y. Gong, Y. Fang, A cloud-based secure and privacy-preserving clustering analysis of infectious disease. in *2018 IEEE Symposium on Privacy-Aware Computing* (2018)
20. M.N. Mohammad, H. Syamsudin, S. Al-Zubaidi, A.K. Sairah, R. Rusyaizila, E. Yusuf, Novel COVID-19 detection and diagnosis system using IoT based smart helmet. Int. J. Psychosoc. Rehabil. **24**(7), 2296–2303 (2020)
21. J. Howard, et al., Face Masks Against COVID-19: An Evidence Review (April 2020), https://doi.org/10.20944/preprints202004.0203.v1
22. https://www.goodrx.com/blog/coronavirus-treatments-on-the-way. Accessed 20 Dec 2020
23. https://guardian.ng/news/world/south-korea-dials-up-virus-testing-with-hospital-phone-booths/. Accessed 20 Dec 2020
24. https://cities-today.com/wp-content/uploads/2020/05/Spot-Bishan-park-1-002.jpg. Accessed 20 Dec 2020
25. https://www.epa.gov/pesticide-registration/list-n-disinfectants-use-against-sars-cov-2. Accessed 20 Dec 2020

26. https://www.chemistryworld.com/news/explainer-how-do-disinfectants-tackle-covid-19/4011546.article. Accessed 19 Dec 2020
27. K. Duan, B. Liu, C. Li, H. Zhang, T. Yu, J. Qu, M. Zhou, L. Chen, S. Meng, Y. Hu, C. Peng, M. Yuan, J. Huang, Z. Wang, J. Yu, X. Gao, D. Wang, X. Yu, L. Li, J. Zhang, X. Wu, B. Li, Y. Xu, W. Chen, Y. Peng, Y. Hu, L. Lin, X. Liu, S. Huang, Z. Zhou, L. Zhang, Y. Wang, Z. Zhang, K. Deng, Z. Xia, Q. Gong, W. Zhang, X. Zheng, Y. Liu, H. Yang, D. Zhou, D. Yu, J. Hou, Z. Shi, S. Chen, Z. Chen, X. Zhang, X. Yang, Effectiveness of convalescent plasma therapy in severe COVID-19 patients. Proc. Natl. Acad. Sci. **117**(17), 9490–9496 (2020). https://doi.org/10.1073/pnas.2004168117
28. https://en.wikipedia.org/wiki/COVID-19_testing
29. F. Rustam et al., COVID-19 future forecasting using supervised machine learning models. IEEE Access **8**, 101489–101499 (2020). https://doi.org/10.1109/ACCESS.2020.2997311
30. F. Petropoulos, S. Makridakis, Forecasting the novel coronavirus COVID-19. Plos One **15**(3), (2020). https://doi.org/10.1371/journal.pone.0231236
31. G.B. Ali Murat Alhas, aa.com.tr, AANEWS BROADCAST COMPANY, 20 04 2020, https://www.aa.com.tr/en/health/steps-taken-by-countries-in-fighting-COVID-19-pandemic/1812009. Accessed 3 Dec 2020

Gaussian Regression Models for Day-Level Forecasting of COVID-19 in European Countries

Umut Özkaya and Şaban Öztürk

Abstract Coronavirus (COVID-19) outbreak has reached a global disease and has begun to threaten all over the world. No vaccine or drug has been developed to treat the outbreak, still. According to the data of the World Health Organization (WHO), the COVID-19 virus can be easily transmitted through respiratory droplets and contact routes. States need to take necessary measures to control the outbreak. In this study, pandemic data for eight countries in the European region are analyzed between 22 January 2020 and 30 April 2020. These countries are Belgium, Netherlands, Germany, UK, Spain, France, Italy, and Turkey. The number of cases is taken into consideration in the selection of countries. Gaussian regression analysis is preferred for the COVID-19 outbreak analysis. A comparative analysis is performed using Ard-Exponential kernel function, Rational-Quadratic kernel function, and Squared Exponential kernel functions. The quasi-Newton algorithm is used to optimize regression models. Confirmed death and recovered case data are processed in the scope of the analysis. The first 90 days of the data are used for the training of Gaussian regression models. The last nine days are evaluated for prediction. Mean Absolute Error (MAE), Median Error (ME), Mean Squared Error (MSE), and Root Mean Squared Error (RMSE) error metrics are used for metric of performance. In general, it is seen that the Ard-Exponential kernel with Gaussian Regression (AEGR) method obtained high metric performances for all cases and countries.

Keywords Coronavirus · COVID-19 · Prediction · Gaussian regression · Quasi-Newton · Optimization

U. Özkaya (✉)
Electrical and Electronics Engineering, Konya Technical University, Konya, Turkey
e-mail: uozkaya@ktun.edu.tr

Ş. Öztürk
Electrical and Electronics Engineering, Amasya University, Amasya, Turkey
e-mail: saban.ozturk@amasya.edu.tr

J. Nayak et al. (eds.), *Understanding COVID-19: The Role of Computational Intelligence*,
Studies in Computational Intelligence 963, https://doi.org/10.1007/978-3-030-74761-9_15

1 Introduction

The new type of Coronavirus, called COVID-19 by the World Health Organization (WHO), threatens public health significantly [1]. COVID-19, which has various symptoms such as dry cough, dyspnea, fever, and bilateral lung infiltrates, spreads quite rapidly. Early diagnosis of this disease is vital for human life. In order to prevent the disease from spreading, quite high budgets are allocated to detection devices in the early period. Besides, effective treatment methods are still under investigation. The subject that many scientists agree is the need for vaccination. It is devoted to vaccination studies on an important budget [1]. Although COVID-19 has been associated with specific symptoms, these symptoms vary from person to person. For this reason, medical methods such as computer tomography (CT), X-ray imaging, and magnetic resonance imaging (MRI) are preferred [2]. It takes some time to acquire these images, interprets them by experts, and diagnose them. Considering the number of people in the COVID-19 outbreak, many sources are insufficient to carry out these procedures. Computer-aided diagnosis (CAD) methods are beneficial to support early diagnosis. In particular, they are used as an auxiliary diagnostic method or as a consultant system to ease the workload of experts [3].

Although CAD systems have been used effectively in the medical field for quite some time, their importance is better understood in this global epidemic. These systems, designed according to the needs of experts, can be used both for reducing workload and training. Thanks to the development of methods in computer science and increasing hardware strength, CAD systems that can produce better results day by day are used more effectively than in previous years. In the first years of CAD systems, tasks such as gray level difference, edge determination, and color detection were performed with quite basic methods [4]. The purpose of these early CAD systems was to make the specialist's job easier by pre-processing. Then, with the development of feature extraction and classification algorithms, the contribution of CAD systems has increased a little more. In this period, image features were extracted using methods such as local binary pattern (LBP), scale-invariant feature transform (SIFT), and speeded up robust features (SURF) [5]. These methods, called hand-crafted feature extraction methods, were highly dependent on the researcher's experience and new methods had to be calculated for each dataset. In addition, these hand-crafted CAD systems were able to make decisions similar to those made by experts. Today, CAD systems that can produce almost the same decisions as experts' decisions are used with fully automatic image processing methods [6]. In particular, it has become very easy for methods such as convolutional neural networks (CNN), auto-encoders (AE), dictionary learning to be trained in labeled data and make decisions like experts. These methods can be used in many cases, such as the training of inexperienced students, the decision support system, the second decision system, and the advisory system, by making successful decisions. Inspired by this success of CAD systems, many artificial intelligence (AI) researchers are developing various CAD systems to support the fight against COVID-19. The number of studies for the detection of COVID-19 has reached tens of thousands in just three months.

These studies generally focus on detection and classification. Segmentation, semantic segmentation, and prediction studies are rarer. Since it is not possible to explain all these studies in this study according to method and approach differences, they will be examined as classification, segmentation, and prediction tasks.

In the literature, two classes, generally COVID-19 and healthy, are used as classification tasks. Also, multi-class classification tasks are performed by adding classes such as Sars, pneumocystis-pneumonia to make classification tasks more challenging. Gozes et al. [7] utilized a deep learning approach to classify COVID-19 and normal image classes. They reached to 98.2% classification sensitivity. Hemdan et al. [8] proposed COVIDX-Net to classify 50 chest X-ray images (25 COVID-19 images and 25 normal images). According to their study, VGG19 and Dense Net models have better results. U car and Korkmaz [9] proposed a deep learning framework with Bayesian optimization-based Squeeze Net to classify COVID-19. They reached 98.3% classification accuracy. Narin et al. [10] used pre-trained CNN networks for classification of various pandemic diseases. Barstugan et al. [11] used gray level feature extraction methods. Then, they used support vector machines (SVM) to detect early phase detection of Coronavirus. Ozturk et al. [12] proposed a two-step data augmentation technique to eliminate the imbalance between classes. Then, the classification process was applied by reducing the number of features obtained from four feature extraction algorithms. In addition to these classification studies, many classification studies have been done and continue to be done [13–16]. In addition to the Classification studies, CT image segmentation studies also increased significantly. Rajinikanth et al. [17] proposed a manual framework to classify COVID-19 images using threshold filter, Harmony-Search-Optimization, and Otsu thresholding. Chen et al. [18] suggested the Residual Attention U-Net algorithm for the automated segmentation of COVID-19 regions. Rajinikanth et al. [19] used the Firefly Algorithm and Shannon Entropy-based multi-threshold and Markov-Random-Field (MRF) segmentation for better accuracy results. Qui et al. [20] presented the Attentive Hierarchical Spatial Pyramid (AHSP) model, a deep learning model called MiniSeg. Wu et al. [21] developed a novel segmentation method to perform real-time COVID-19 diagnosis. Their framework performs the segmentation only if the classification model reports positive. Fan et al. [22] proposed a deep learning framework called Inf-Net. Inf-Net aggregates high-level features with a parallel partial decoder.

The performance of some of the classification and segmentation studies in the literature is quite high. In addition to this high performance, studies focusing on processing time are also encountered. Instead of doing more detailed studies on these topics, it may be more useful to work on forecasting because the number of studies on forecasting in the literature is relatively small. Besides, the estimated figures are essential for the provision of necessary medical supplies and preliminary preparations. Elmousalami and Hassanien et al. [23] used time series models and mathematical formulation for day level forecasting models on COVID-19 cases. Rizk-Allah and Hassanien et al. [24] integrated integrates an improved interior search algorithm (ISA) with a multi-layer feed-forward neural network (MFNN) to predict the cases of COVID-19. Distant et al. [25] predicted new daily infections and total cases using the epidemiological dynamics of China by training an artificial neural network (ANN). Ahsan et al. [26] used Multilayer Perception and Convolutional Neural Network (MLP-CNN) for mixed COVID-19 dataset. Accuracy remained around f $94.6 \pm 3.42\%$, as they used very little data for training their models. Karthik et al. [27] proposed a study that includes CNN architecture. Their method has a F1-score of 97.20% and an accuracy of 99.80% on the X-ray dataset. Stübinger and Schneider et al. [28] forecasted the future spread of Coronavirus using lead-lag effects between different countries using a dynamic time warping approach. Ghazaly et al. [29] used Non-Linear Regressive Network (NAR) for the time-series analysis of COVID-19. Moftakhar et al. [30] suggested ANN and auto-regressive integrated moving average (ARIMA) models for forecasting of COVID-19 cases. Sannigrahi et al. proposed a spatial regression models to analyze COVID-19 cases of Europian countries [34]. Rath et al. [35] designed a multiple linear regression model to predict COVID-19 cases. Bent [36] tried to predict age of survivors from COVID-19 pandemics. Iyanda et al. [37] implemented a novel regression method based weighted and multi scale analysis to examine COVID-19 outbreak in 175 countries.

In this study, the analysis of confirmed, death, and recovered COVID-19 cases for eight different European countries is carried out with Gaussian regression techniques. Eight different countries are selected as Belgium, Netherlands, Germany, UK, Spain, France, Italy, and Turkey. These countries vary in COVID-19 case numbers, both due to different age ranges in their populations and differences in socio-cultural levels. This situation makes it challenging to predict cases of the country. At the same time, there are differences in the number of cases depending on the level of measures taken by countries against COVID-19. Gaussian regression is used to predict the course of COVID-19 in the next days. Three different kernel functions are used in the Gaussian regression technique within the scope of this study. These kernel functions are ard-exponential, rational-quadratic, and squared-exponential. The quasi-Newton algorithm is used as an optimization method in the training of the Gaussian regression technique. In the Gaussian regression analysis phase, 99 days of confirmed, death, and recovered data are used between 1/22/20 and 4/29/20. 90-day data for training and last 9-day data for prediction are evaluated. Recently,

This paper is organized to consist of 5 sections in total. Section 1 is defined as an introduction that includes general information about the COVID-19 and a literature review. Section 2 consists of dataset description, methodological background, and proposed method. Section 3 presents the obtained experimental results. There is a discussion in Sect. 4 and the conclusion in Sect. 5.

2 Materials and Methods

The following section describes the materials and work flow of the proposed model.

2.1 Dataset Description

This study uses dashboard information operated by the Johns Hopkins University Center for Systems Science and Engineering (JHU CSSE) for forecasting [31]. This dataset contains daily level information about affected cases, deaths, and recovery numbers. This time-series data is continuously updated, and the figures presented are cumulative. There is up-to-date information on the status of almost every country affected by COVID-19. The data presented in a table include a serial number, province or state, country or region, last update, abstract information, confirmed cases, deaths, and recovered rows and columns. With this information, table-based forecasting can be performed, or graphical forecasting can be done after the conversion to a graphic.

2.2 Methodological Background

Regression is used to estimate an output variable for a *D*-dimensional input variable. Gaussian processes regression (GPR) model, which is a nonparametric kernel-based probabilistic model, is usually simple to implement, easily adaptable to other problems, fully-probabilistic, and performing successfully for many problems [32]. Given a training set $TS = \{(x_i, y_i),\ i = 1, 2, \ldots, n\}$ with *n* pairs vector inputs. It is the aim of determining y_i parameters, which are output values by using input values x_i. A linear regression model with Gaussian noise using *TS* parameters is calculated as in Eq. 1.

$$y_i = f(x_i) + \varepsilon_i \text{ , where } \varepsilon_i \sim \mathrm{N}\left(0, \sigma^2\right) \tag{1}$$

where σ^2 represents error variance, and $f(x_i)$ is a latent function. When a Gaussian process (GP) is described as $\{f(x), x \in \mathbb{R}^d\}$ with *n* observations of $x_1, x_2, \ldots, x_n$, the joint distribution of the random variables is $f(x_1), f(x_2), \ldots, f(x_n)$. This means that every finite *x* value has a corresponding $f(x)$ vector. In this case, mean function $m(x)$ and covariance function $k(x,x')$ are used to define a GP. Gaussian distribution can be expressed as Eq. 2.

$$F_x \sim N\left\{m(x), k\left(x, x'\right)\right\} \tag{2}$$

Thus, $E(f(x)) = m(x)$ and $Cov[f(x), f(x')] = E[\{f(x)-m(x)\}\{f(x')-m(x')\}] = k(x,x')$ are a GP. Let's consider $f(x) \sim GP(0, k(x,x'))$, response y is modeled as Eq. 3.

$$P(y_i | f(x_i), x_i) \sim N\left(y_i | h(x_i)^T \beta + f(x_i), \sigma^2\right) \tag{3}$$

2.3 Proposed Method

Curve fitting or supervised learning techniques are generally preferred in prediction problems. Let's say x_i is the future situation of a time-series problem, it is desired that x_i value estimated as quite close to the real value. Regression process is performed by using kernel functions [32]. A kernel function is obtained between $f(x_i)$ and $f(x_j)$. In these functions, x_i and x_j expressions are vector with d × 1 dimensions. At the same time, the prediction value obtained for x_i affects prediction value obtained for the x_j value. Various types of functions can be used as kernel function $k(x_i, x_j)$. These functions are optimized with kernel parameters in the θ vector. It is necessary to update the kernel function expression to $k(x_i, x_j|\theta)$.

The parameters in the kernel functions take into account the standard deviation of the data (σ_f) and the characteristic length scale (σ_l). The distance of the input data according to the response values is defined by using σ_l. In regression, σ_f and σ_l expressions must be greater than zero. Within the scope of the proposed method, Gaussian regression technique with three different kernel functions is used in this study, respectively. These are Ard-Exponential kernel with Gaussian Regression (AEGR), Rational-Quadratic kernel with Gaussian Regression (RQGR), and Squared Exponential kernel with Gaussian Regression (SEGR). The ard-Exponential kernel function is specified used Eqs. 4 and 5.

$$k(x_i, x_j|\theta) = \sigma_f^2 \exp(-r) \tag{4}$$

$$r = \sqrt{\sum_{m=1}^{d} \frac{(x_{im} - x_{jm})^2}{\sigma_m^2}} \tag{5}$$

This kernel function uses the exponential length scale for each parameter. r is defined as Euclidean distance. Squared Exponential kernelfunction is defined as in Eq. 6.

$$k(x_i, x_j|\theta) = \sigma_f^2 \exp[-\frac{1}{2}\left[\frac{(x_i - x_j)^T (x_i - x_j)}{\sigma_l^2}\right] \tag{6}$$

Rational-Quadratic kernel function is represented as in Eq. 7.

$$k(x_i, x_j|\theta) = \sigma_f^2 \left(1 + \frac{r^2}{2\alpha\sigma_l^2}\right)^{-\alpha} \tag{7}$$

where α is scale mixture value for kernel function, r is specified as Euclidean distance.

In this study, the Quasi-Newton method is used for the optimization of kernel functions. In the Quasi-Newton problem, the zeros, local maxima, and minima are searched fora solution [33]. Newton method is used to obtain the stationary point of kernel functions. The gradient value at this point is equal to zero. In the Newton

Initialize J, x and B

While ($t<$Max number of Iteration)

Update $\mathbf{r}=\mathbf{Gx}-\mathbf{J}$

Calculate $\boldsymbol{\delta} \Leftarrow x - x^{old}$ and $\boldsymbol{\gamma} \Leftarrow r - r^{old}$

Update **B** (8)

Calculate $\Delta\mathbf{x}$ (9), $\boldsymbol{\alpha}$ (10) and $\mathbf{x}^{\mathbf{new}}$ (11)

t=t+1

end **while**

return $\mathbf{x}^{\mathbf{new}}$ and $\mathbf{B}^{\mathbf{upd}}$

Fig. 1 Pseudo code of quasi-Newton optimization

optimization method, Jacobian and Hessian matrices must be calculated to minimize objective function for high dimension optimization problems. The hessian matrix cannot be required for the optimization process by using the Quasi-Newton method. Instead, the solution is tried to be reached by analyzing consecutive gradient vectors. Hessian matrices should be used to solve multi-dimensional optimization problems. The Quasi-Newton method performs a low-grade update for prediction values. The pseudo-code of the Quasi-Newton algorithm is found in Fig. 1.

In the Quasi-Newton optimization method, the Hessian matrix (G) is updated with an iterative method, and an attempt is made to converge to a symmetric matrix, where J denotes the Jacobian matrix. B value is updated relative to the inverse of G in Eq. 8.

$$B = B^{old} + \left(1 + \frac{\gamma^T B^{old} \gamma}{\delta^T \gamma}\right) \frac{\delta \delta^T}{\delta^T \delta} - \frac{\delta \gamma^T B^{old} + B^{old} \gamma \delta^T}{\delta^T \gamma} \tag{8}$$

Step direction (Δx) is updated as in Eq. 9.

$$\Delta x = -Br \tag{9}$$

Line search parameter (α) and new fitting coefficients (x^{new}) are updated as in Eqs. 10 and 11.

$$\alpha = -\frac{\sum_{k} r_k \Delta x_k}{\sum_{k,l} G_{kl} \Delta x_k \Delta x_l} \tag{10}$$

$$x_k^{new} = x_k + \alpha \Delta x_k \text{ with } \min_{\alpha > 0} F(x + \alpha \Delta x) \tag{11}$$

The proposed framework is shown in Fig. 2.

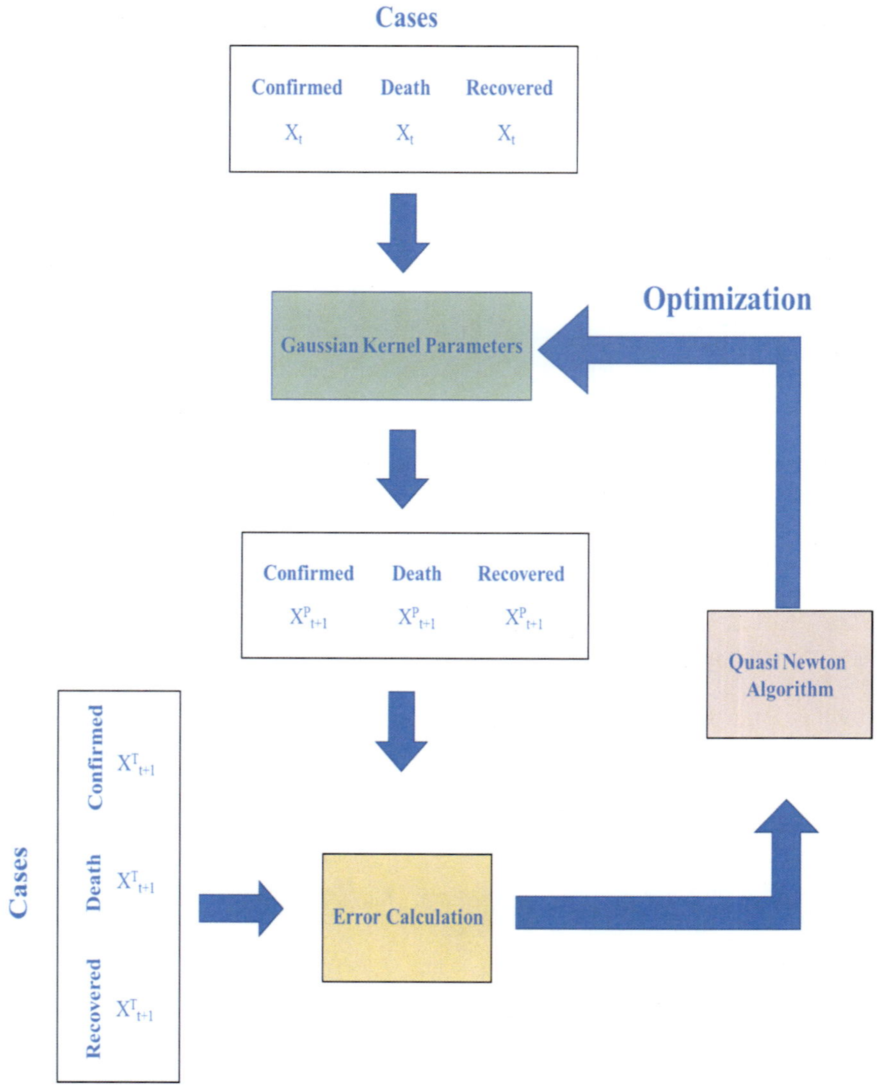

Fig. 2 The proposed framework

In the proposed framework, data of the previous day (X_t) is given to Gaussian regression function as the inputs. Data prediction is made for the next day (X^P_{t+1}) with the current kernel parameters. Error calculation is made between target data (X^T_{t+1}) and prediction data (X^P_{t+1}). Gaussian kernel parameters are updated with the Quasi Newton algorithm with taking into account the obtained error value. This process continues until a certain number of iterations as 75 iteration.

3 Experimental Results

In this study, the data from eight different European countries are analysed to form the COVID-19 outbreak spread model. Belgium, the Netherlands, Germany, UK, Spain, France, Italy, and Turkey are selected due to the rapid increase in the number of cases. Besides, four different error metrics are used for performance evaluation. These metrics are MAE, ME, MSE, and RMSE. Calculation of these performance evaluation metrics is performed between Eqs. 12-15.

$$MAE = mean_{i=1,n}|e_i| \tag{12}$$

$$ME = median_{i=1,n}|e_i| \tag{13}$$

$$MSE = median_{i=1,n}|e_i^2| \tag{14}$$

$$RMSE = \sqrt{median_{i=1,n}|e_i^2|} \tag{15}$$

Within the scope of Gaussian regression analysis, metric performance of confirmed, death and recovered cases for eight different countries are given in Tables 1, 2 and 3.

Comparative graphics about the forecasting of confirmed, death, and recovered cases are given in Figs. 3, 4, 5, 6, 7, 8, 9 and 10.

Table 1 Metric performance of gaussian regression methods for confirmed COVID-19 cases

Country	Proposed methods	MAE	ME	MSE	RMSE
Belgium	AEGR	185.33	136	66049	257
	RQGR	399.4444	335	216225	465
	SEGR	429.7778	432	425104	652
Netherland	AEGR	146.6667	180	46656	216
	RQGR	287.3333	250	236196	486
	SEGR	337.1111	337	423801	651
Germany	AEGR	1850.1	1656	7070281	2659
	RQGR	898.8889	873	201601	449
	SEGR	926.5556	901	239121	489
UK	AEGR	2427.2	2169	32092225	5665
	RQGR	16871	15306	1.6×10^9	40011
	SEGR	28535	25976	4.48×10^9	66995
Spain	AEGR	2274.4	1065	640000	800
	RQGR	3141.4	3177	8761600	2960
	SEGR	3150	3157	8567329	2927
France	AEGR	2081.7	1457	2468041	1571
	RQGR	9349.3	5433	303526084	17422
	SEGR	1100	1336	106276	326
Italy	AEGR	1364.6	1360	3356224	1832
	RQGR	1491.3	1485	373321	611
	SEGR	1428.2	1425	240100	490
Turkey	AEGR	428.77	370	769129	877
	RQGR	4992.3	5258	11316496	3364
	SEGR	3052.4	2708	51265600	7160

Table 2 Metric performance of gaussian regression methods for death COVID-19 cases

Country	Proposed methods	MAE	ME	MSE	RMSE
Belgium	AEGR	179	189	54289	233
	RQGR	276	255	416025	645
	SEGR	253.2222	133	524176	724
Netherland	AEGR	37.4444	34	3600	60
	RQGR	661.5556	700	1726596	1314
	SEGR	70.5556	53	32041	179
Germany	AEGR	110.2222	99	30276	174
	RQGR	1147.8	1087	4661281	2159
	SEGR	194.6667	126	246016	496
UK	AEGR	333	386	412164	642
	RQGR	4627.6	4653	86359849	9293
	SEGR	976.6667	978	1140624	1068
Spain	AEGR	323	296	209764	458
	RQGR	136.444	136	61009	247
	SEGR	147.333	128	92416	304
France	AEGR	282.6667	205	198916	446
	RQGR	3797	9696	43322724	6582
	SEGR	541.4444	411	1651225	1285
Italy	AEGR	136.2222	139	42025	205
	RQGR	475.6667	478	207936	456
	SEGR	477.1111	479	208849	457
Turkey	AEGR	16.8889	13	1681	41
	RQGR	368	334	703921	839
	SEGR	64.4444	66	20164	142

Table 3 Metric performance of gaussian regression methods for recovered COVID-19 cases

Country	Proposed methods	MAE	ME	MSE	RMSE
Belgium	AEGR	298.2222	267	216225	465
	RQGR	2061.8	2342	12931216	3596
	SEGR	3149.1	3576	33930625	5825
Germany	AEGR	3138.7	2997	28451556	5334
	RQGR	14346	12305	1.009×10^9	31767
	SEGR	37372	36894	4.9×10^9	70021
Spain	AEGR	3699.1	3078	35796289	5983
	RQGR	33740	34788	2.94×10^9	54249
	SEGR	1656.2	989	22071204	4698
France	AEGR	879.6667	755	2474329	1573
	RQGR	1907.7	2066	16475841	4059
	SEGR	1087	1086	1083681	1041
Italy	AEGR	683.3333	556	6260004	2502
	RQGR	5803.8	4456	315488644	17762
	SEGR	3977.3	3957	28143025	5305
Turkey	AEGR	7739	5408	417425761	20431
	RQGR	20947	19851	1.64×10^9	40614
	SEGR	22855	24721	2.052×10^9	45309

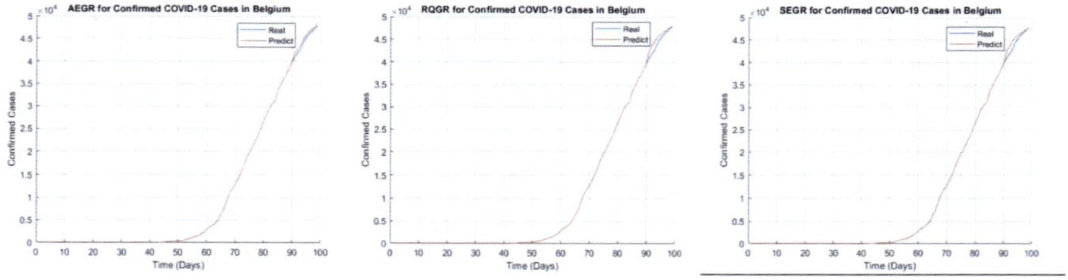

a) b) c)

Fig. 3 **a** AEGR prediction **b** RQGR prediction **c** SEGR prediction for confirmed COVID-19 cases in belgium

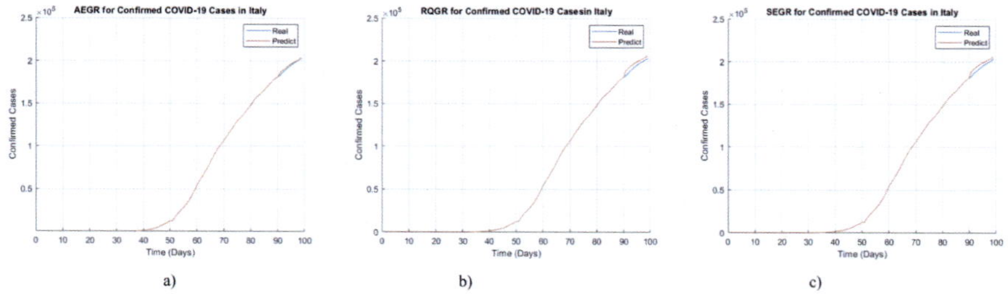

Fig. 4 **a** AEGR prediction **b** RQGR prediction **c** SEGR prediction for confirmed COVID-19 cases in Italy

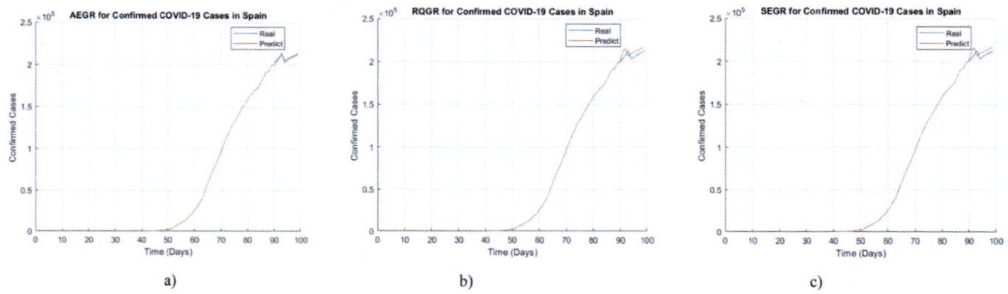

Fig. 5 **a** AEGR prediction **b** RQGR prediction **c** SEGR prediction for confirmed COVID-19 cases in Spain

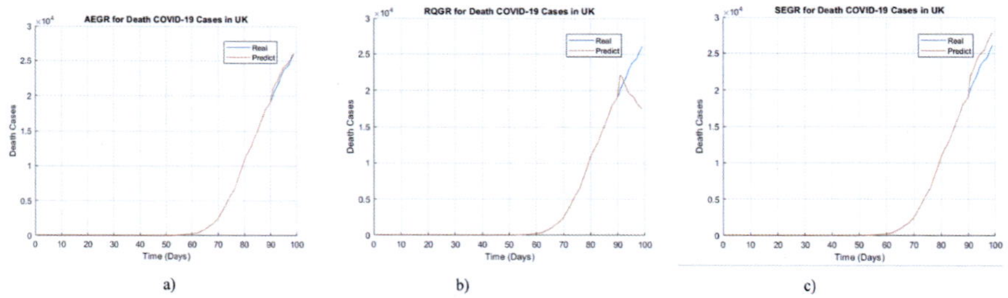

Fig. 6 **a** AEGR prediction **b** RQGR prediction **c** SEGR prediction for death COVID-19 cases in UK

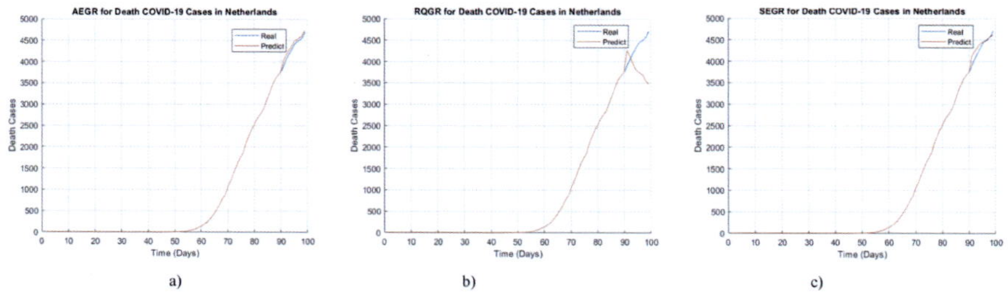

Fig. 7 **a** AEGR prediction **b** RQGR prediction **c** SEGR prediction for death COVID-19 cases in Netherlands

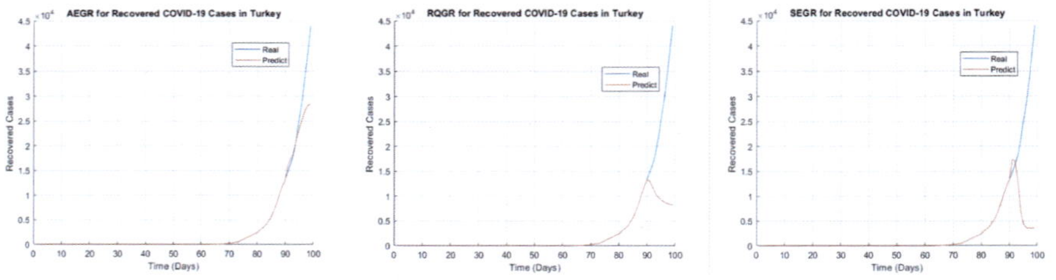

a) b) c)

Fig. 8 **a** AEGR prediction **b** RQGR prediction **c** SEGR prediction for recovered COVID-19 cases in Turkey

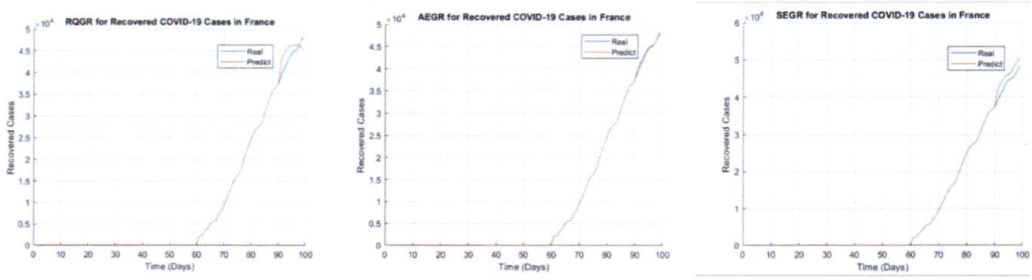

a) b) c)

Fig. 9 **a** AEGR prediction **b** RQGR prediction **c** SEGR prediction for recovered COVID-19 cases in France

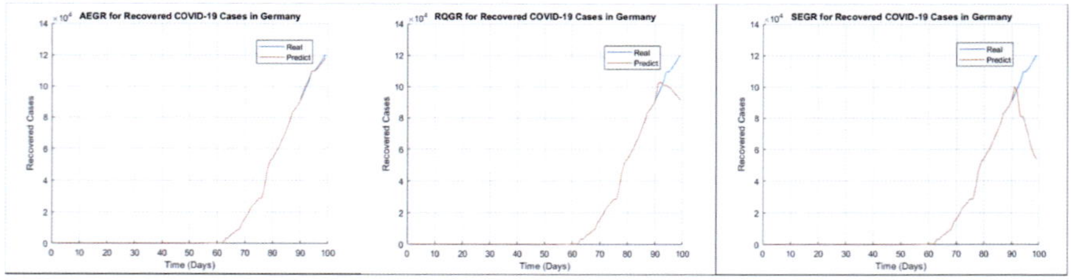

a) b) c)

Fig. 10 **a** AEGR prediction **b** RQGR prediction **c** SEGR prediction for recovered COVID-19 cases in Germany

4 Discussion

In COVID-19 data analysis, 90-day data is used for training in the Gaussian regression technique using different kernel functions. Although the 9-day data is used for the test phase, it is seen that the obtained results are quite promising. Numbers of confirmed, death, and recovered cases are predicted closely for selected countries.

Performance evaluation metrics vary depending on the number of confirmed cases, as in Table 1. The lowest metric value is obtained in the Netherlands with AEGR. These metric values are 146.6667 MAE, 180 ME, 46656 MSE, and 216 RMSE. The lowest performance for the Netherlands has belonged to the SEGR method with 377.1111 MAE, 337 ME, 423801 MSE, and 651 RMSE values. Another country with a lower number of confirmed cases than in other countries is Belgium. AEGR method shows the highest performance for Belgium with 185.33 MAE, 136 ME, 66049 MSE, and 257 RMSE values. As in the Netherlands, the SEGR method in Belgium has the lowest performance with 429.7778 MAE, 432 ME, 425104 MSE, and 652 RMSE metric values. Germany is one of the countries with a high number of confirmed cases. The RQGR method has the lowest error metrics. The RQGR method has obtained 898.8889 MAE, 873 ME, 201601 MSE, and 449 RMSE metric values. AEGR method obtains the best performance in the metric performance for the UK and Spain. AEGR method obtained metric values of 2427.2 MAE, 2169 ME, 32092225 MSE, and 5665 RMSE for the UK. The confirmed cases performances of the AEGR method in Spain are 2274.4 MAE, 1065 ME, 640000 MSE, and 800 RMSE. In the confirmed cases prediction for France, the SEGR method stands out with its metric performance of 1100 MAE, 1336 ME, 106276 MSE, and 326 RMSE. Italy and Turkey have more confirmed cases. AEGR obtained the best performance for both countries. AEGR method has 1364.6 MAE, 1360 ME, 3,356,224 MSE, and 1832 RMSE metric values for Italy, 428.77 MAE, 370 ME, 769129 MSE and 877 RMSE metric values for Turkey respectively.

Error metric values in Table 2 are quite low since the number of death cases is lower than the number of confirmed cases and recovered cases. It stands out with AEGR performance in prediction in the number of death cases. AEGR method for Belgium has 179 MAE, 189 ME, 54289 MSE, and 233 RMSE metric performance. AEGR method shows 37.4444 MAE, 34 ME, 3600 MSE, and 60 RMSE metric performance for the Netherlands. The prediction metric performance of the AEGR method for Germany is 110.2222 MAE, 99 ME, 30276 MSE, and 174 RMSE. The prediction metric performance of the AEGR method for the last nine days has 333 MAE, 386 ME, 412164 MSE, and 642 RMSE for the UK. In contrast to the AEGR method, which shows high estimation metric performance in other countries, the RQGR method stands out for Spain with 136.444 MAE, 136 ME, 61009 MSE, and 247 RMSE metric values. AEGR prediction for the number of death cases in France, Italy, and Turkey are over-performing considerably higher than other methods. Metric performance of AEGR method 282.6667 MAE, 205 ME, 198916 MSE, and 6582 RMSE for France; 136.2222 MAE, 139 ME, 42025 MSE, and 205 RMSE for Italy; 16.8889 MAE, 13 ME, 1681 MSE, and 41 RMSE for Turkey.

For recovered cases prediction, no update is made in the dataset for the UK and the Netherlands. Therefore, metric performance has not been evaluated. AEGR method shows the best performance in Belgium, Germany, France, Italy, and Turkey for recovered cases prediction. AEGR obtains metric performance with 298.2222 MAE, 267 ME, 216225 MSE, and 465 RMSE for Belgium; 3138.7 MAE, 2997 ME, 28451556 MSE, and 5334 RMSE for Germany; 879.6667 MAE, 755 ME, 2474329 MSE, and 1573 RMSE for France; 683.3333 MAE, 556 ME, 6260004 MSE, and

2502 RMSE for Italy; 7739 MAE, 5408 ME, 417425761 MSE, and 20431 RMSE for Turkey. SEGR method shows more successful performance for recovered cases prediction in Spain than the other techniques. It has a metric error, which is 1087 MAE, 1086 ME, 1083681 MSE, and 1041 RMSE.

Confirmed cases prediction of AEGR, RQGR, and SEGR methods for Belgium, Italy, and Spain are given in Figs. 3, 4 and 5. AEGR method has made a very successful prediction in the last nine days of test data. Primarily, the AEGR method has achieved a better convergence to the real data despite the fluctuations in the number of confirmed cases in the last nine days for Spain. Figures 6 and 7 include death cases prediction the UK and Netherlands for all methods. It achieves the best prediction performance by the AEGR method. Prediction for the number of recovered cases in Turkey, France, and Germany are given in Figs. 8, 9 and 10. AEGR method has made a more successful prediction for Turkey than other methods. However, the AEGR method could not fully converge the sudden increase in the number of recovered cases in the last nine days. SEGR method is more successful in predicting the number of recovered cases in France. RQGR and SEGR methods have shown a tendency to decrease in recovered cases prediction of the last nine days for Germany. The AEGR method has only converged very well to real data.

5 Conclusion

The COVID-19 outbreak, which is seen as a global threat, has begun to weaken all countries both economically and socially. Accurate prediction about the spread of the outbreak will help to control the outbreak more comfortably. A pandemic model for outbreak cannot be produced due to cultural differences and lack of data. The outbreak prediction of Gaussian Regression models is presented in this study. Also, it accomplished a successful prediction on the out breakin the long term. The effect of different kernel functions on outbreak prediction is also analyzed. Analyzes for selected eight different European countries to show that Gaussian Regression can be used as an effective forecasting tool, although COVID-19 displays different characteristics. In this study, Gaussian regression can be used as a robust time-series estimator. Comparative analysis of regression techniques for different countries is required to obtain the spread characteristic of the COVID-19 outbreak. The generalization capability of proposed methods for just one country may not be sufficient in global cases. Prediction on the number of confirmed, death, and recovered cases on a daily basis is a challenging task. Countries could take measures early thanks to the appropriate prediction model. Thus, the spread of the COVID-19 outbreak can be prevented more quickly. Consequently, it is suggested that Gaussian regression techniques should be developed for performance reasons in future studies.

References

1. C. Sohrabi, Z. Alsafi, N. O'Neill, M. Khan, A. Kerwan, A. Al-Jabir, R. Agha, World Health Organization declares global emergency: a review of the 2019 novel coronavirus (COVID-19). Int. J. Surg. **76**, 71–76 (2020)
2. Y. Li, L. Xia, Coronavirus disease 2019 (COVID-19): role of chest CT in diagnosis and management. Am. J. Roentgenol. **214**, 1–7 (2020)
3. L. Wang, Z.Q. Lin, A. Wong, Covid-net: a tailored deep convolutional neural network design for detection of covid-19 cases from chest x-ray images. Sci. Rep. **10**(1), 1–12 (2020)
4. R.P. Würtz, T. Lourens, Corner detection in color images by multiscale combination of end-stopped cortical cells, in *International Conference on Artificial Neural Networks* (Springer, Berlin, Heidelberg, 1997), pp. 901–906
5. D. Ni, Y.P. Chui, Y. Qu, X. Yang, J. Qin, T.T. Wong, P.A. Heng, Reconstruction of volumetric ultrasound panorama based on improved 3D SIFT. Comput. Med. Imaging Graph. **33**(7), 559–566 (2009)
6. U. Ozkaya, S. Ozturk, M. Barstugan, Coronavirus (COVID-19) classification using deep features fusion and ranking technique, in *Big Data Analytics and Artificial Intelligence Against COVID-19: Innovation Vision and Approach. Studies in Big Data*, vol. 8, ed. by A.E. Hassanien, N. Dey, S. Elghamrawy (2020), https://doi.org/10.1007/978-3-030-55258-9_17
7. O. Gozes, M. Frid-Adar, H. Greenspan, P.D. Browning, H. Zhang, W. Ji, E. Siegel, Rapid ai development cycle for the coronavirus (covid-19) pandemic: Initial results for automated detection and patient monitoring using deep learning ct image analysis (2020), arXiv preprint arXiv:2003.05037
8. E.E.D. Hemdan, M.A. Shouman, M.E. Karar, Covidx-net: a framework of deep learning classifiers to diagnose covid-19 in x-ray images (2020), arXiv preprint arXiv:2003.11055
9. F. Ucar, D. Korkmaz, COVIDiagnosis-net: deep bayes-squeezenet based diagnostic of the coronavirus disease 2019 (COVID-19) from X-Ray images. Med. Hypotheses **140**, (2020)
10. A. Narin, C. Kaya, Z. Pamuk, Automatic detection of coronavirus disease (covid-19) using x-ray images and deep convolutional neural networks (2020), arXiv preprint arXiv:2003.10849
11. M. Barstugan, U. Ozkaya, S. Ozturk, Classification of Coronavirus (COVID-19) from X-ray and CT images using shrunken features (2020), arXiv preprint arXiv:2003.09424
12. Ş. Öztürk, U. Özkaya, M. Barstuğan, Classification of coronavirus (COVID-19) from X-ray and CT images using shrunken features. Int. J. Imaging Syst. Technol. **31**(1), 5–15 (2020). https://doi.org/10.1002/ima.22469
13. L. Sun, Z. Mo, F. Yan, L. Xia, F. Shan, Z. Ding, H. Yuan, Adaptive feature selection guided deep forest for covid-19 classification with chest ct. IEEE J. Biomed. Health Inform. **24**(10), 2798–2805 (2020)
14. A.M. Hasan, M.M. Al-Jawad, H.A. Jalab, H. Shaiba, R.W. Ibrahim, A.A.R. Al-Shamasneh, Classification of Covid-19 coronavirus, pneumonia and healthy lungs in CT scans using Q-deformed entropy and deep learning features. Entropy **22**(5), 517 (2020)
15. C. Butt, J. Gill, D. Chun, B.A. Babu, Deep learning system to screen coronavirus disease 2019 pneumonia. Applied Intelligence, vol. 1 (2020)
16. P.K. Sethy, S.K. Behera, Detection of coronavirus disease (covid-19) based on deep features, vol. 2020030300. Preprints (2020)
17. V. Rajinikanth, N. Dey, A.N.J. Raj, A.E. Hassanien, K.C. Santosh, N. Raja, Harmony-search and otsu based system for coronavirus disease (COVID-19) detection using lung CT scan images (2020), arXiv preprint arXiv:2004.03431
18. X. Chen, L. Yao, Y. Zhang, Residual Attention U-Net for Automated Multi-Class Segmentation of COVID-19 Chest CT Images (2020), arXiv preprint arXiv:2004.05645
19. V. Rajinikanth, S. Kadry, K.P. Thanaraj, K. Kamalanand, S. Seo, Firefly-Algorithm Supported Scheme to Detect COVID-19 Lesion in Lung CT Scan Images using Shannon Entropy and Markov-Random-Field (2020), arXiv preprint arXiv:2004.09239
20. Y. Qiu, Y. Liu, J. Xu, MiniSeg: An Extremely Minimum Network for Efficient COVID-19 Segmentation (2020), arXiv preprint arXiv:2004.09750

21. Y.H. Wu, S.H. Gao, J. Mei, J. Xu, D.P. Fan, C.W. Zhao, M.M. Cheng, Jcs: An explainable covid-19 diagnosis system by joint classification and segmentation (2020), arXiv preprint arXiv:2004.07054
22. D.P. Fan, T. Zhou, G.P. Ji, Y. Zhou, G. Chen, H. Fu, L. Shao, Inf-net: automatic COVID-19 lung infection segmentation from CT scans. IEEE Trans. Med. Imaging **39**(8), 2626–2637 (2020). https://doi.org/10.1109/TMI.2020.2996645
23. H.H. Elmousalami, A.E. Hassanien, Day level forecasting for Coronavirus Disease (COVID-19) spread: analysis, modeling and recommendations (2020), arXiv preprint arXiv:2003.07778
24. R.M. Rizk-Allah, A.E. Hassanien, COVID-19 forecasting based on an improved interior search algorithm and multi-layer feed forward neural network (2020), arXiv preprint arXiv:2004.05960
25. C. Distante, I.G. Pereira, L.M.G. Goncalves, P. Piscitelli, A. Miani, Forecasting Covid-19 Outbreak Progression in Italian Regions: A model based on neural network training from Chinese data (2020). medRxiv
26. M.M. Ahsan, E. Alam, T. Trafalis, P. Huebner, Deep MLP-CNN model using mixed-data to distinguish between COVID-19 and Non-COVID-19 patients. Symmetry **12**(9), 1526 (2020)
27. R. Karthik, R. Menaka, M. Hariharan, Learning distinctive filters for COVID-19 detection from chest X-ray using shuffled residual CNN. Appl. Soft Comput. **99**, 106744 (2020)
28. J. Stübinger, L. Schneider, Epidemiology of coronavirus COVID-19: forecasting the future incidence in different countries, in *Healthcare*,vol. 8, no. 2 (Multidisciplinary Digital Publishing Institute 2020), p. 99
29. M.N. Ghazaly, M.A. AbdelFattah, A.A. Abd El-Aziz, Novel coronavirus forecasting model using nonlinear autoregressive artificial neural network. Int. J. Adv. Sci. Technol. **29**(5), 1831–1849
30. L. Moftakhar, S.E.I.F. Mozhgan, M.S. SAFE, Exponentially increasing trend of infected patients with COVID-19 in iran: a comparison of neural network and ARIMA forecasting models. Iran. J. Pub. Health **49**, 92–100 (2020)
31. E. Dong, H. Du, L. Gardner, An interactive web-based dashboard to track COVID-19 in real time. Lancet. Infect. Dis **20**(5), 533–534 (2020)
32. C.K. Williams, C.E. Rasmussen, *Gaussian Processes for Machine Learning*, vol. 2, no. 3 (MIT Press, Cambridge, 2006), p. 4
33. R. Haelterman, Analytical study of the Least Squares Quasi-Newton method for interaction problems. PhD Thesis, Ghent University. Retrieved 2014-08-14, 2009
34. S. Sannigrahi, F. Pilla, B. Basu, A.S. Basu, A. Molter, Examining the association between socio-demographic composition and COVID-19 fatalities in the European region using spatial regression approach. Sustain. Cities Soc. **62**, 102418 (2020). https://doi.org/10.1016/j.scs.2020.102418
35. S. Rath, A. Tripathy, A. Ranjan Tripathy, Prediction of new active cases of coronavirus disease (COVID-19) pandemic using multiple linear regression model. Diab. Metab. Syndr. Clin. Res. Rev. **14**(5), 1467–1474 (2020). https://doi.org/10.1016/j.dsx.2020.07.045
36. B. Flyvbjerg, The law of regression to the tail: how to survive Covid-19, the climate crisis, and other disasters. Environ. Sci. Policy **114**, 614–618 (2020). https://doi.org/10.1016/j.envsci.2020.08.013
37. A.E. Iyanda, R. Adeleke, Y. Lu, T. Osayomi, A. Adaralegbe, M. Lasode, N.J. Chima-Adaralegbe, A.M. Osundina, A retrospective cross-national examination of COVID-19 outbreak in 175 countries: a multiscale geographically weighted regression analysis (January 11-June 28, 2020). J. Infect. Publ. Health **13**(10), 1438–1445 (2020). https://doi.org/10.1016/j.jiph.2020.07.006

Computational Intelligence Based Hybrid Hyperparameter Tuned Prediction Techniques for COVID-19 Epidemiological Data

S. Dhamodharavadhani and R. Rathipriya

Abstract The aim of this work is to examine the appropriate predictive model for COVID-19 epidemiological data in the Indian population using Computational Intelligence based Hyper parameter Tuned Regression Techniques (CI-HTR). The development of the COVID-19 Epidemiological Data Prediction (EDP) model for India is proposed by using CI-HTR such as Gaussian process regression (GPR) model and shallow neural network models. For that purpose, this analysis uses a time series data set that includes daily cumulative COVID-19 epidemiological data. In order to increase forecast accuracy, hybrid models are constructed by merging Hyper parameter tuned regression models with Non-Linear Auto Regressive (NAR) neural network. Experiential research in this work reveals significant differences between the different types of CI-HTR models for the COVID-19 fatality prediction. The performance metrics like Root Mean Squared Error (RMSE), and the correlation factor called 'R-Value', are used to estimate the proposed methodology for COVID-19 dataset. It has been found that the hybrid GPR prediction model is the superior to the other models for the COVID-19 EDP.

Keywords Computational intelligence · Hyper parameter tuned prediction techniques · Support vector regression · Gaussian process regression · COVID-19 epidemiological data

1 Introduction

Severe Acute Respiratory Syndrome Coronavirus 2 (SARS-CoV-2) causes COVID-19. The first case was officially reported on 31st December 2019 in China. At present, there is no valid treatment or preventative measures to counter the COVID-19 epidemic. However, the best prevention strategies to control its spread are social distancing and self-health practices by individuals [1–5]. In [6], authors are suggested SNN based COVID-19 model was offers better accuracy. Jiang et al. [7] estimate the COVID-19 mortality rate based on the age of the population. They are: 4.5%

S. Dhamodharavadhani · R. Rathipriya (✉)
Department of Computer Science, Periyar University, Salem, India

J. Nayak et al. (eds.), *Understanding COVID-19: The Role of Computational Intelligence*, Studies in Computational Intelligence 963, https://doi.org/10.1007/978-3-030-74761-9_16

for those under 70, 8.0% for those aged 70–79, 14.8% were over 80 years of age. Furthermore, people over the age of 50 with health problems such as diabetes, heart disease, and Parkinson's disease were classified as high risk. It was clear from the works [8–11] that prediction methods based on computational intelligence are the best performers for health-related predictions. Nsoesie et al. [12] have reviewed a number of computational intelligence based prediction techniques such as regression models, Bayesian network, ARIMA forecasting model, etc. for the various domains. It provided a clear insight of computational intelligence techniques and forecasting methods. Gaussian process (GP) models require typically be situated used extensively happening engineering-built optimization applications [11]. A combination of GP regression (GPR) and adaptive neuro-fuzzy inference system is used for predicting the level of groundwater [12]. In [13], a comprehensive study was conducted to assess the viability of various surrogate models including GPR models via simulation-based optimization methods. Finally, the authors reported that GPR models and their ensemble perform better for forecasting tasks. In [14–19], Generalized Regression Neural Network (GRNN) model, Radial Basis Function neural network (RBFNN) were built as a new computational method for the field of infectious disease prediction. It has been observed from these works that standard neural network models can be considered as a decision-making tool or as a predictive tool with high predictive accuracy.

Generally, data mining and machine learning models are used to predicting the COVID-19 [3, 20–22]. Deep learning techniques are also used to detect the confirmed COVID-19 disease from X-ray dataset [23]. Pandemic prediction could be made on the basis of different variables including the influence of natural aspects, infection rate, confinement effect, age, gender and more. In these researches, the different methods and variables being used for prediction and their impacts have been widely studied [24–26]. Multivariate evidence revealed that its epidemics condition with COVID-19 has an optimistic and important impact on the desire of consumers to shop online [27, 28] analyzed clinical study time–series data besides SARS-CoV-2 viral illness, the common cause for COVID-19, to attempt discovers the high specificity of its clinical diagnosis and use a Bayesian approach [29–31].

GPR model was evaluated for the prediction of reservoir porosity and monthly groundwater level fluctuations. The predicted results revealed that the GPR model provided more accurate predictions than other models like ANFIS, GRNN [32, 33]. Authors of [34] introduced an advanced Random Forest model helped by the AdaBoost method. They assessed COVID-19 and given a report on a worldwide pandemic [5]. Kotwal et al. [13] conducted an extensive review to assess the mathematical models for COVID-19 prediction in India based on their predictive accuracy. They recorded that the accuracy of the short-term forecast was good, but the long-term forecasts were uncertain. The main purpose of all these works is to integrate epidemiological data and computational intelligence techniques for better understanding of the characteristics and impact of virus which will enable the health officials to make accurate decisions and a definitive plan of action.

The objective of this chapter is to develop hybrid predictive models based on computational intelligence techniques such as Gaussian process regression and shallow neural network models to predict COVID-19 fatality in India. The main contributions in this chapter are listed as follows:

- Computational intelligence techniques like GPR, and Shallow Neural Network models namely GRNN and RBFNN were compared for the prediction of COVID-19 Epidemiological Data.
- Non-Linear Auto Regressive (NAR) neural network is hybridized with neural network and GPR models to enhance the prediction quality standard prediction models.

The remaining chapter is arranged according to they follow. Section 2 details the computational intelligence techniques for Covid-19 Fatality rate fore casting for India. The outcomes and analysis of this study are given in Sect. 3. This work is summarized in Sect. 4 with potential future research.

2 Computational Intelligence Based Prediction Techniques

This section describes the techniques used in the Computational Intelligence appraoches

2.1 Hyperparameter Optimization (HO)

Hyperparameter optimization is used to find the optimal parameters for a given regression technique, which returns the best output on the given dataset based on the training model [6, 15]. It is mathematically represented in Eq. (1)

$$H^* = argmin_{o \in O} f(o) \tag{1}$$

Where f(o) is an objective function to identify the minimal RMSE value, H* denotes the hyperparameter set that provides the minimal RMSE value, and o denotes any real value based on the solution space. The important factor is that HO is costly inter ms of computational time, but it provides greater predictive accuracy.

2.2 GPR

Gaussian process is a machine learning technique [33] used to make uncertain predictions. It's also described as a finite discrete random variable collectively transferred in the Gaussians [15, 34]. These random variables represent the value for a function f(x) at input x in regression problems [35]. It is denoted as {f(x): x ∈ X}, mean function μ(x) and the covariance function k (x,xj) therefore it can be shown as in Eq. (2)

Table 1 GPR model-based hyperparameter optimization

GPR model-based hyperparameter optimization
1. Initialization of hyperparameters 2. Repeat until specified iteration reached 2.1 State the objective function according to the problem domain. 2.2 Specify the selection criteria (i.e. Minimal RMSE) for evaluating hyperparameters to be selected next from the surrogate model. 2.3Updatethe surrogatemodelby(score,hyperparameter)pair.

$$f(\cdot) \sim GP(\mu(\cdot), k(\cdot, \cdot)) \quad (2)$$

A covariance function as defined in Eq. (3) represents the covariance between random variable pairs in Gaussian process.

$$K_{ij} = k(x_i, x_j) = \alpha \exp\left\{\frac{||x_i - x_j||^2}{2\sigma_1^2}\right\} \quad (3)$$

Where σ1 = Characteristic length scale, α = Signal variance. Table 1 shows the hyperparameter optimization steps.

2.3 GRNN

Generalized Regression Neural Network (GRNN) belongs to the RBFNN category [6, 36]. The configuration of GRNN with two layers remains comparatively simple besides fixed. The first is the sequence, and the second is a summation [37]. In the pattern layer, when input is passed over each component, it creates a connection between the input and the response and is stored within the component. As a result, the number of components is equivalent to the number of individual values in the training set. In each pattern unit, a Gaussian PDF is applied to the network input, so that it is defined as Eq. (4).

$$\Theta = EXP\left[-0.5 * (A-t)`(A-t)/\left(\sigma\hat{}2\right)\right] \quad (4)$$

Here, Θ represents pattern layer's output, A is the origin, t is the training vector, and σ denotes "smooth parameter" or "smoothing factor" [6].

2.4 RBFNN

RBFNN [6] is a type of static neural network model whose activation or transfer function is given in Eq. (5) [38, 39]. It is a three-layer feed-forward (FF) neural network. Layer one is linear which transmits the input signal separately. Layer two is not linear, it uses Gaussian functions [40]. Layer three integrates a linear representation of the Gaussian productions. The weights between the hidden layer and the output layer are shifted during preparation.

$$f(x) = \frac{1}{\sqrt[\sigma]{2\pi}} e^{-\frac{(x-\mu)^2}{2\sigma^2}} \tag{5}$$

Where x is the input, μ is the mean, and σ is the standard deviation.

2.5 Nonlinear Autoregressive Neural Network (NAR)

The N online arau to regressive neural network is a type of neural network that is suitable designed for estimating future input variable values [6, 40, 41, 46]. The NAR Network helps to estimate future values based on the time [42]. Through its historical precedent it supported the use of are-feeding mechanism, where a certain predicted value would serve as feedback for new prediction satmo read vanced times [43–46].

$$y(time_{series}) = f(y(time_{series} - 1), \ldots, y(time_{series} - d)) \tag{6}$$

The Eq. (6) represents the predicted series y($time_{series}$) for given 'd' past values of y($time_{series}$).

2.6 RMSE

RMSE is calculated as follows: Find the differences of the first predicted and actual COVID-19 death cases and apply the square function to these values. Finally, find the square root of it [6, 47]. Equation (7) shows the mathematical formula of RMSE:

$$RMSE = \sqrt{\frac{1}{n}\sum_{i=1}^{n}(Predicted\ Death\ Case_i - Actual\ Death\ Case_i)^2} \tag{7}$$

where n is the number of death cases in the dataset.

2.7 Correlation Factor (R-Value)

It measures the strength of linear correlation that exists among predicted and actual values of COVID-19 death case for India [6, 45–47]. Equation (8) shows the mathematical representation of R-Value.

$$R = \frac{\sum_{i=1}^{n}(a_i - \overline{a_i})(b_i - \overline{b_i})}{\sqrt{\sum_{i=1}^{n}(a_i - \overline{a_i})^2}\sqrt{\sum_{i=1}^{n}(b_i - \overline{b_i})^2}} \quad (8)$$

where a denotes actual or real value, b denotes predicted value, $\overline{a}$ denotes the mean actual value, $\overline{b}$ denotes the mean predicted value, and n is number of death cases in the dataset.

3 Description of Data and Research Methodology

This section describes the dataset used and the functionality of the proposed approach.

3.1 Dataset Description

A COVID-19 dataset has been taken from the official website for prediction of the death cases in India. The daily COVID-19 data for the period up to August 2020 is taken as a database and used for model development [48]. Figure 1 showed the breakdown of COVID-19 confirmed cases in India up to August 2020. It is observed that COVID-19 fatality rate is very low compared to the recovered rate. But, still Indian government work on strategy to reduce COVID-19 fatality rate. To support this, CI-HTRM based prediction system is designed and developed in this chapter.

3.2 Proposed Methodology: CI-HTRM

In this work, a two-step approach to COVID-19 fatality rate prediction is provided. The first step is to develop a computational intelligence-based hyperparameter tuned prediction model using GRNN, GPR and RBFNN [49–51]. The second step is to use the NAR neural network to predict the error of these models. Finally, to improve mortality prediction accuracy, the results of CI-HTRM are linked to the predicted error. The work flow of the proposed methodology is given in Fig. 2. The explanation of the proposed CI-HTRM is as follows: First, the COVID-19 database must be pre-processed and divided into three sets, namely the training, testing, and verification

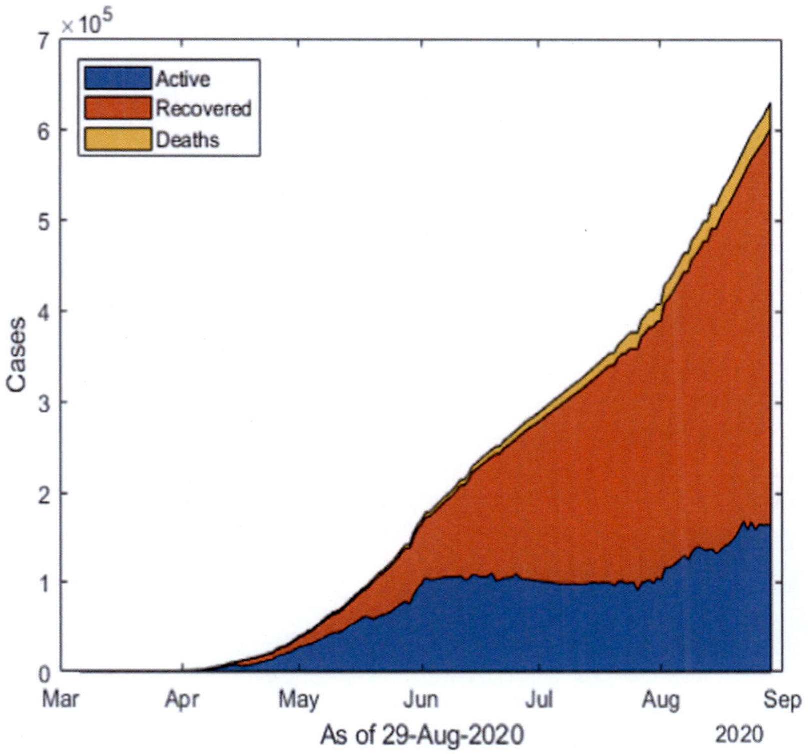

Fig. 1 Breakdown of confirmed cases in India

dataset. Second, build the CI-HTRM with the required parameter setup as specified in Table 2 and train it using the training dataset. Third, evaluate the proposed CI-HTRM using the test and verification database. Choose the best model as the learned model. Fourth, predict COVID-19 death cases for India and identify associated residuals. Fifth, build the hybrid CI-HTRM using the NAR-NN model for the time series COVID-19 dataset. Finally, combine the predicted COVID-19 death cases and the predicted residual to obtain the best predictive accuracy.

The Hybrid CI-HTRM is designed and developed using the following steps:

3.3 Algorithmic Steps for Hybrid CI-HTRM

1. COVID-19 time series data is pre-processed to remove the missing and noisy data.
2. Initialization of hyperparameters for CI-HTRM.
3. Define the model parameters for shallow neural network models.

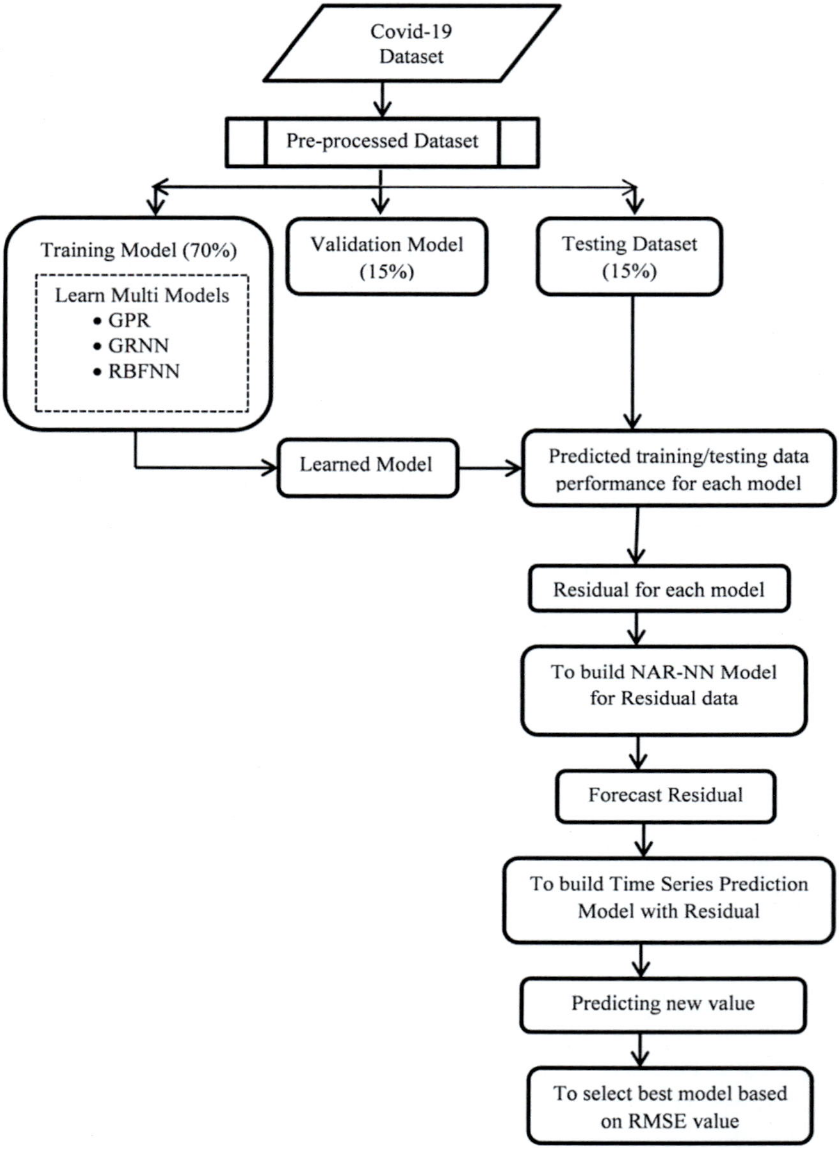

Fig. 2 Workflow of hybrid CI-HTRM

4. Predict the number of death cases ($Pred_{Covid}$) for 'n' period ahead using standard CI-HTRM
5. The residuals of CI-HTRM is inputed into NAR-NN model and predict the residual values (Pred_err) for the CI_HTRM individually.
6. Sum the Pred_errwith $Pred_{Covid}$ to get the new death cases $NewPred_{Covid}$.
7. $NewPred_{Covid}$ is returned as the output.

Table 2 Model parameters setup

Model parameters	GPR model	GRNN model	RBFNN model	NAR-NN model
Hidden layer (HL)	N/A	Fixed architecture	Fixed architecture	Fixed architecture
Number of neurons in HL	N/A	10–15	10–15	10–15
Training algorithm	N/A	Bayesian regularization	Bayesian regularization	Bayesian regularization
Kernel functions	Exponential	N/A	N/A	N/A
Hyperparameter	Covariance(σ) mean (μ)	Spread/ smoothing factor(σ)	Spread/ smoothing factor(σ)	N/A
Key performance indicator	RMSE	RMSE	RMSE	MSE

Table 3 The performance metrics value for GRNN, RBFNN and GPR models

Models	RMSE	R-value
GRNN	103.66	0.999983
RBFNN	97.51	0.999985
GPR	20.67	0.999999
Hybrid_GRNN	24.61	1.000000
Hybrid_RBFNN	20.14	1.000000
Hybrid_GPR	4.65	1.000000

Table 2 lists the model parameter setup for GPR, GRNN, RBFNN and NAR-NN models in the column 1 to 4 respectively. Figure 2 illustrates the working principles of the hyperparameter tuning used with the GPR model, the GRNN model, and the RBFNN model to achieve the minimum RMSE value for these models, respectively.

4 Experimental Results and Discussion

This section presented the results of an empirical study with a discussion of outcomes. Table 3 displays the values of RMSE and R-Value for GRNN, RBFNN and GPR models for COVID-19 death cases prediction. While comparing these models, it is seen that the GPR model has low RMSE value of 4.54 and high R-Value of 1. This indicates that the GPR model is a better model compared to in comparison to GRNN and RBFNN for the Indian COVID-19 dataset. The optimal spread value (σ) of GRNN and RBFNN is 4 and 1.38, respectively. Table 4 displays the values of

Table 4 Hyperparameter value for GPR

SigmaM	SigmaF	Sigma
1008.5	79.1	2.6

hyperparameters for Gaussian Process Regression (GPR) models.

Table 5 displays the predicted COVID-19 death cases for all three standards models using the timeseries COVID-19 dataset for India. The hybrid models developed in this work are a combination of the standard prediction models and NAR-NN based error forecasting model. The predicted COVID-19 death cases are presented in Table 6 for hybrid prediction models. It is known from the COVID-19 death cases predicted using standard and hybrid prediction models that there were no significant differences in the predicted values as there were no major changes in the RMSE values of these samples.

Figure 3 shows the predicted plot against the actual COVID-19 death cases of the GPR model. X-axis indicates the day of infection with COVID-19, and Y-axis shows the predicted COVID-19 death values in (10^4). The prediction using GPR model implies that the number of per day COVID-19 death cases remains constant even after reaching 65000 death cases.

Figure 4 shows the predicted plot against the actual COVID-19 death cases of the GRNN model. X-axis indicates the day of infection with COVID-19, and Y-axis shows the predicted COVID-19 death values in (10^4). The GRNN model shows the bell shaped curve for the predicted death cases, which has an increasing pattern.

Figure 5 shows the predicted plot against the actual COVID-19 death cases of the RBFNN model. X-axis indicates the day of infection with COVID-19, and Y-axis shows the number of predicted death cases. The RBFNN model shows the steep bell shaped curve for the death cases prediction. After 260th day of infection, number of death cases started to decrease. Figure 6 is the visual representation of RMSE based comparison of the CI-HTRMs. It is noted that GPR based HTRM is the performed better in comparison to other prediction models.

Table 5 Predicted COVID-19 death cases using standard models

Predicted date	GPR	GRNN	RBFNN
31-Aug-20	63390	64393	64632
1-Sep-20	63386	65355	65794
2-Sep-20	63382	66314	67024
3-Sep-20	63378	67271	68335
4-Sep-20	63373	68224	69738
5-Sep-20	63369	69174	71247
6-Sep-20	63365	70121	72875
7-Sep-20	63361	71065	74634
8-Sep-20	63356	72004	76538
9-Sep-20	63352	72940	78596
10-Sep-20	63348	73871	80820
11-Sep-20	63344	74799	83215
12-Sep-20	63339	75722	85784
13-Sep-20	63335	76640	88528
14-Sep-20	63331	77555	91440
15-Sep-20	63327	78464	94509
16-Sep-20	63323	79369	97718
17-Sep-20	63318	80270	101046
18-Sep-20	63314	81166	104464
19-Sep-20	63310	82056	107941
20-Sep-20	63306	82943	111441
21-Sep-20	63301	83824	114928
22-Sep-20	63297	84700	118367
23-Sep-20	63293	85572	121721
24-Sep-20	63289	86438	124959
25-Sep-20	63284	87300	128056
26-Sep-20	63280	88157	130989
27-Sep-20	63276	89008	133742
28-Sep-20	63272	89855	136305
29-Sep-20	63268	90696	138672
30-Sep-20	63263	91532	140844
1-Oct-20	63259	92364	142824
2-Oct-20	63255	93190	144620
3-Oct-20	63251	94010	146239
4-Oct-20	63247	94826	147695
5-Oct-20	63242	95636	148998
6-Oct-20	63238	96441	150161

(continued)

Table 5 (continued)

Predicted date	GPR	GRNN	RBFNN
7-Oct-20	63234	97240	151197
8-Oct-20	63230	98034	152117
9-Oct-20	63225	98823	152934
10-Oct-20	63221	99606	153659
11-Oct-20	63217	100383	154301
12-Oct-20	63213	101155	154870
13-Oct-20	63209	101921	155375
14-Oct-20	63204	102681	155822
15-Oct-20	63200	103436	156220
16-Oct-20	63196	104184	156573
17-Oct-20	63192	104927	156887
18-Oct-20	63187	105664	157167
19-Oct-20	63183	106395	157418

4.1 Discussion

This study has exploited three different computational intelligence based prediction model and its hybrid version to forecast the daily death cases of the COVID-19 in India up to 19th October 2020. First, CI-HTRM generally provides the best prediction accuracy when COVID-19 data is within the range of trained COVID-19 dataset. In particular, the prediction of the future number of death cases inevitably contains time values out of the range of the train data.

In such case, CI-HRTM may be suitable for short-term assessment several days in advance. The results also confirmed the same. Therefore, CI-HRTM could be used as a time series prediction model for the COVID-19 database. This is clear from the following:

- The average RMSE value of the standard version of GPR, GRNN and RPFNN is 73.94
- The average RMSE value of the hybrid version is 16.46.
- More specifically, GPR-based regression models work best in this regard. The hybrid GPR model is considered the best model because its RMSE value (4.65) is much lower than the others. This is because it captures the forecasting pattern even from small datasets. However, this result depends on the hyperparameter value and the characteristics of the data.

Table 6 Predicted COVID-19 death cases using hybrid models

Predicted date	Hybrid_GPR	Hybrid_GRNN	Hybrid_RBFNN
31-Aug-20	63398	64406	64669
1-Sep-20	63356	65394	65770
2-Sep-20	63378	66312	67045
3-Sep-20	63329	67291	68313
4-Sep-20	63371	68221	69746
5-Sep-20	63516	69176	71227
6-Sep-20	63362	70119	72873
7-Sep-20	63357	71061	74614
8-Sep-20	63353	72001	76530
9-Sep-20	63349	72937	78577
10-Sep-20	63345	73869	80807
11-Sep-20	63341	74796	83195
12-Sep-20	63336	75719	85769
13-Sep-20	63332	76638	88508
14-Sep-20	63328	77552	91423
15-Sep-20	63324	78462	94490
16-Sep-20	63319	79367	97700
17-Sep-20	63315	80267	101027
18-Sep-20	63311	81163	104446
19-Sep-20	63307	82054	107922
20-Sep-20	63302	82940	111422
21-Sep-20	63298	83821	114909
22-Sep-20	63294	84698	118348
23-Sep-20	63290	85569	121701
24-Sep-20	63286	86436	124940
25-Sep-20	63281	87297	128037
26-Sep-20	63277	88154	130970
27-Sep-20	63273	89006	133723
28-Sep-20	63269	89852	136286
29-Sep-20	63264	90694	138653
30-Sep-20	63260	91530	140825
1-Oct-20	63256	92361	142805
2-Oct-20	63252	93187	144600
3-Oct-20	63248	94008	146220
4-Oct-20	63243	94823	147676
5-Oct-20	63239	95633	148979

(continued)

Table 6 (continued)

Predicted date	Hybrid_GPR	Hybrid_GRNN	Hybrid_RBFNN
6-Oct-20	63235	96438	150142
7-Oct-20	63231	97237	151177
8-Oct-20	63226	98031	152098
9-Oct-20	63222	98820	152915
10-Oct-20	63218	99603	153640
11-Oct-20	63214	100380	154282
12-Oct-20	63210	101152	154851
13-Oct-20	63205	101918	155356
14-Oct-20	63201	102678	155803
15-Oct-20	63197	103433	156201
16-Oct-20	63193	104182	156554
17-Oct-20	63189	104925	156868
18-Oct-20	63184	105661	157148
19-Oct-20	63180	106392	157399

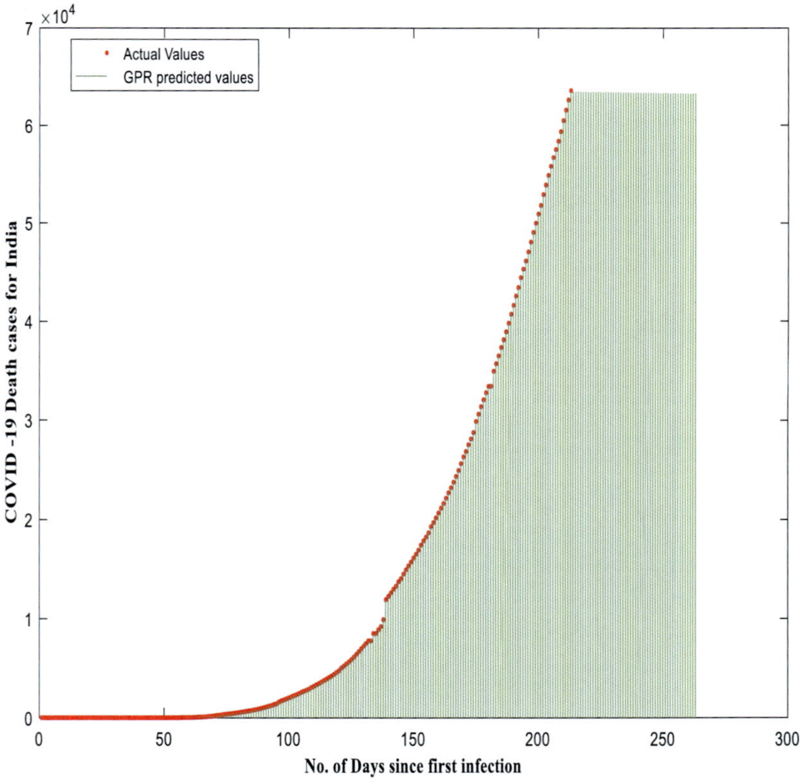

Fig. 3 Plot for GPR model

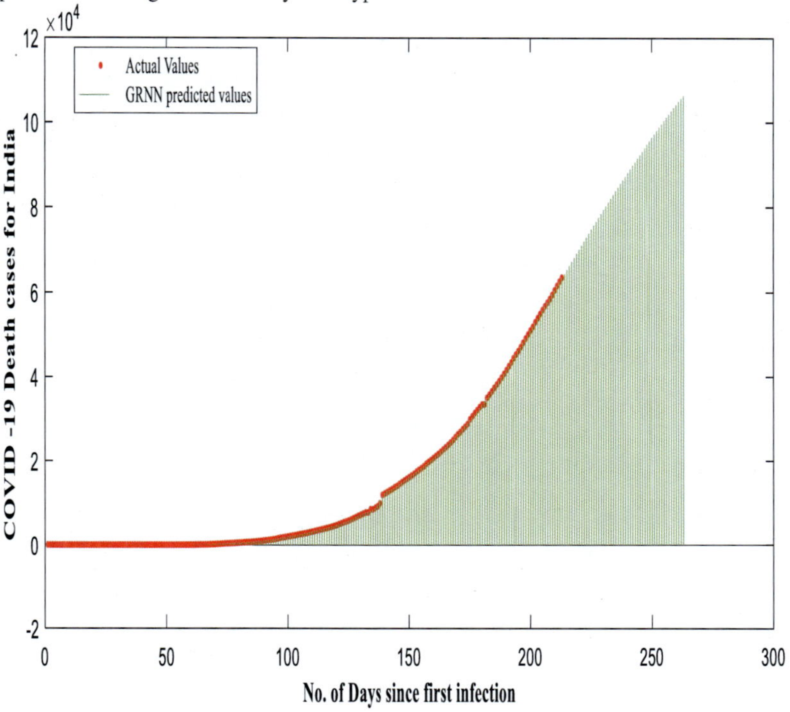

Fig. 4 Plot for GRNN model

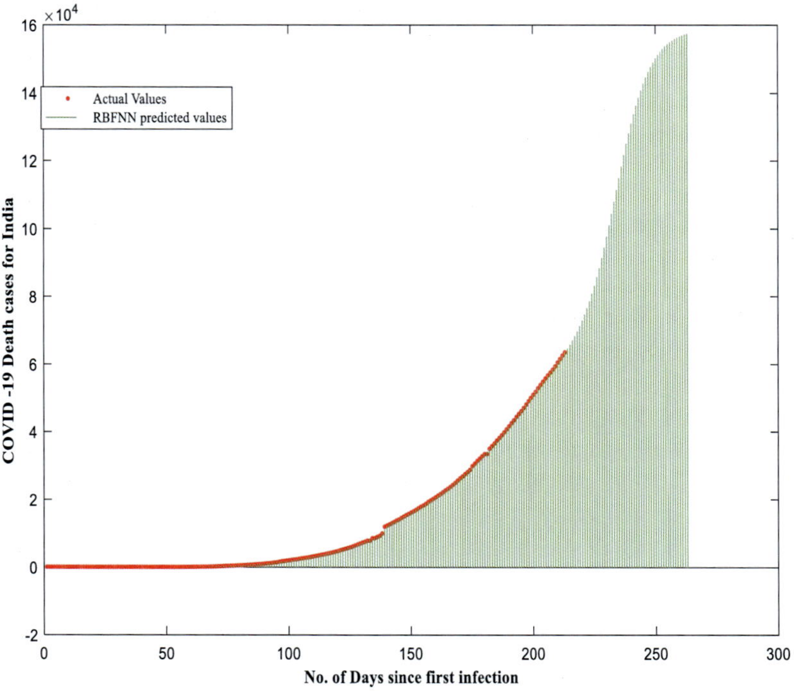

Fig. 5 Plot for RBFNN model

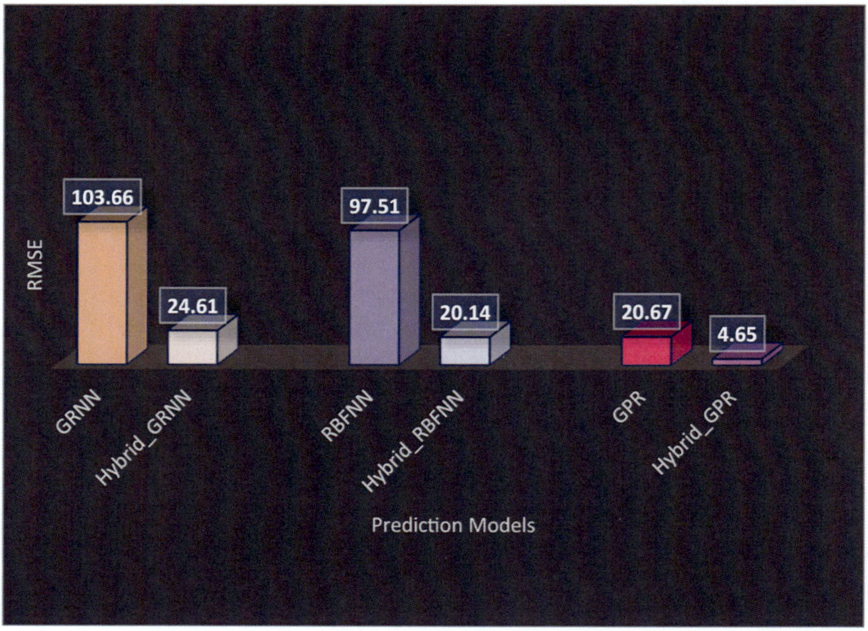

Fig. 6 RMSE value of CI-HTRM

5 Conclusion

This chapter proposed a methodology that hybridized the CI-HTRM with the NAR-NN model to achieve higher predictive accuracy in the prediction of COVID-19 death cases. The NAR-NN model was used to predict errors that should be added with the predicted death cases value. The Gaussian Process Regression (GPR) model has yielded a relatively good result in terms of optimized predicted values for death cases in the COVID-19 epidemiological data. According to the COVID-19 fatality prediction for India using CI-HTRM, the COVID-19 death cases are increasing gradually, but it depends on other environmental conditions, socio-economic factors, and socio-behavioral factors. This finding purely depends on the time series analysis of the Indian COVID-19 dataset. In future, the proposed CI-HTRM is hybridized with other computational intelligence techniques to enhance the prediction accuracy. Epidemiological data are highly sensitive, so predictive accuracy plays an important role in decision and policy formulation.

References

1. WHO| Novel Coronavirus – China. WHO. Archived from the original on 23 January 2020. [Retrieved 1 February 2020]. WHO| Novel Coronavirus – China. WHO. Archived from the original on 23 January 2020. [Retrieved 1 February 2020]
2. E. Simpson, I. Gurevych, Scalable bayesian preference learning for crowds. Mach. Learn. **109**(4), 689–718 (2020). https://doi.org/10.1007/s10994-019-05867-2
3. R. Sujath, J.M. Chatterjee, A.E. Hassanien, A machine learning forecasting model for COVID-19 pandemic in India. Stoch. Env. Res. Risk Assess. **34**, 959–972 (2020)
4. C. Iwendi, A.K. Bashir, A. Peshkar, R. Sujatha, J.M. Chatterjee, S. Pasupuleti, O. Jo, COVID-19 patient health prediction using boosted random forest algorithm. Front. Publ. Health **8**, 357 (2020)
5. S. Chauhan, Comprehensive review of coronavirus disease 2019 (COVID-19). Biomed. J. **43**(4), 334–340 (2020)
6. S. Dhamodharavadhani, R. Rathipriya, J.M. Chatterjee, COVID-19 mortality rate prediction for india using statistical neural network models. Front. Publ. Health **8**, 441–441 (2020)
7. A.C.C. Huang, C.G. Huang, C.T. Yang, H.C. Hu, Concomitant infection with COVID-19 and mycoplasma pneumoniae. Biomed. J. **43**(5), 458–461 (2020)
8. P. Deprez, P.V. Shevchenko, M.V. Wüthrich, Machine learning techniques for mortality modeling. Eur. Actuarial J. **7**(2), 337–352 (2017). https://doi.org/10.1007/s13385-017-0152-4
9. D. Hainaut, A neural-network analyzer for mortality forecast. ASTIN Bull. **48**(02), 481–508 (2018). https://doi.org/10.1017/asb.2017.45
10. S. Dhamodharavadhani, R. Rathipriya, Region-wise rainfall prediction using mapreduce-based exponential smoothing techniques, in *Advances in Intelligent Systems and Computing Advances in Big Data and Cloud Computing* (2018), pp. 229–239
11. S. Dhamodharavadhani, R. Rathipriya, Novel COVID-19 mortality rate prediction (MRP) model for india using regression model with optimized hyperparameter. J. Cases Inform. Technol. (JCIT) **23**(4), 1–12 (2021). https://doi.org/10.4018/jcit.20211001.oa1
12. D.K. Roy, B. Datta, Trained meta-models and evolutionary algorithm based multi-objective management of coastal aquifers under parameter uncertainty. J. Hydro Inform. **20**(6), 1247–1267 (2018)
13. B.H.A. Pilon, J.J. Murillo-Fuentes, J. Paulo, C.L.D. Costa, R.T.D.S. Júnior, A.M.R. Serrano, Gaussian process for regression in business intelligence: a fraud detection application, in *Proceedings of the 7th International Joint Conference on Knowledge Discovery*, vol. 3 (2015), pp. 39–49
14. J.M. Wang, D.J. Fleet, A. Hertzmann, Gaussian process dynamical models for human motion. IEEE Trans. Pattern Anal. Mach. Intell. **30**(2), 283–298 (2008). https://doi.org/10.1109/tpami.2007.1167
15. C.E. Rasmussen, Gaussian processes in machine learning, in *Advanced Lectures on Machine Learning Lecture Notes in Computer Science* (2004), pp. 63–71
16. K. Ažman, J. Kocijan, Dynamical systems identification using Gaussian process models with incorporated local models. Eng. Appl. Artif. Intell. **24**(2), 398–408 (2011). https://doi.org/10.1016/j.engappai.2010.10.010
17. M.E. Huber, Recursive gaussian process: on-line regression and learning. Pattern Recogn. Lett. **45**, 85–91 (2014). https://doi.org/10.1016/j.patrec.2014.03.004
18. C.C. Aggarwal, Radial basis function networks. Neural Netw. Deep Learn. **5**, 217–233 (2018)
19. F. Martínez, F. Charte, A.J. Rivera, M.P. Frías, Automatic time series forecasting with GRNN: a comparison with other models, in *Advances in Computational Intelligence Lecture Notes in Computer Science* (2019), pp. 198–209
20. D. Halbersberg, M. Wienreb, B. Lerner, Joint maximization of accuracy and information for learning the structure of a Bayesian network classifier. Mach. Learn. **109**(5), 1039–1099 (2020). https://doi.org/10.1007/s10994-020-05869-5

21. A.S. Albahri, R.A. Hamid, J. k Alwan, Z.T. Al-qays, A.A. Zaidan, B.B. Zaidan, et al., Role of biological data mining and machine learning techniques in detecting and diagnosing the novel coronavirus (COVID-19): a systematic review. J. Med. Syst. **44**(7), 44 (2020). 10.1007/s10916-020-01582-x
22. M. John, H. Shaiba, Main factors influencing recovery in MERS Co-V patients using machine learning. J. Infect. Publ. Health **12**(5), 700–704 (2019). https://doi.org/10.1016/j.jiph.2019.03.020
23. I.D. Apostolopoulos, T.A. Mpesiana, Covid-19: automatic detection from X-ray images utilizing transfer learning with convolutional neural networks. Phys. Eng. Sci. Med. **43**(2), 635–640 (2020). https://doi.org/10.1007/s13246-020-00865-4
24. G.R. Shinde, A.B. Kalamkar, P.N. Mahalle, N. Dey, J. Chaki, A.E. Hassanien, Forecasting models for coronavirus disease (COVID-19): a survey of the state-of-the-art. SN Comput. Sci. **1**(4), 1–1 (2020). https://doi.org/10.1007/s42979-020-00209-9
25. L.J. Muhammad, M.M. Islam, S.S. Usman, S.I. Ayon, Predictive data mining models for novel coronavirus (COVID-19) infected patients' recovery. SN Comput. Sci. **1**(4), 1–1 (2020). https://doi.org/10.1007/s42979-020-00216-w
26. K. Li, D. Chen, S. Chen, Y. Feng, C. Chang, Z. Wang, Predictors of fatality including radiographic findings in adults with COVID-19. Respir. Res. **21**(1), 1–10 (2020)
27. H.V. Nguyen, H.X. Tran, L. Van Huy, X.N. Nguyen, M.T. Do, N. Nguyen, Online book shopping in vietnam: the impact of the COVID-19 pandemic situation [published online ahead of print, 2020 May 10], (Publishing Research Quarterly, 2020), pp. 1–9, https://doi.org/10.1007/s12109-020-09732-2
28. I. Burstyn, N.D. Goldstein, P. Gustafson, Towards reduction in bias in epidemic curves due to outcome mis- classification through Bayesian analysis of time-series of laboratory test results: case study of COVID-19 in Alberta, Canada and Philadelphia, USA. BMC Med. Res. Methodol. **20**(1), 20 (2020). https://doi.org/10.1186/s12874-020-01037-4
29. Z. Vokó, J.G. Pitter, The effect of social distance measures on COVID-19 epidemics in Europe: an interrupted time series analysis. GeroScience (2020), 10.1007/s11357-020-00205-0
30. K. Tsioufis, C. Chrysohoou, M. Kariori, I. Leontsinis, I. Dalakouras, A. Papanikolaou et al., The mystery of "missing" visits in an emergency cardiology department, in the era of COVID-19, a time-series analysis in a tertiary Greek General Hospital. Clin. Res. Cardiol. **109**, 1483–1489 (2020). https://doi.org/10.1007/s00392-020-01682-1
31. N. Hatami, S. Ahi, A. Sadeghinikoo, M. Foroughian, F. Javdani, N. Kalani et al., Worldwide ACE (I/D) polymorphism may affect COVID-19 recovery rate: an ecological meta-regression. Endocrine **68**(3), 479–484 (2020). https://doi.org/10.1007/s12020-020-02381-7
32. S. Asante-Okyere, C. Shen, Y.Y. Ziggah, M.M. Rulegeya, X. Zhu, Investigating the predictive performance of gaussian process regression in evaluating reservoir porosity and permeability. Energies **11**(12), 3261–3261 (2018). https://doi.org/10.3390/en11123261
33. A. Dahl, E.V. Bonilla, Grouped gaussian processes for solar power prediction. Mach. Learn. **108**(8–9), 1287–1306 (2019). https://doi.org/10.1007/s10994-019-05808-z
34. D. Zou, Y. Cao, D. Zhou, Q. Gu, Gradient descent optimizes over-parameterized deep ReLU networks. Mach. Learn. **109**(3), 467–492 (2020). https://doi.org/10.1007/s10994-019-05839-6
35. S. Saha, et al., Distributed prognostic health management with Gaussian process regression, in *2010 IEEE Aerospace Conference* (IEEE, 2010)
36. S. Seyedzadeh, F.P. Rahimian, I. Glesk, M. Roper, Machine learning for estimation of building energy consumption and performance: a review. Vis. Eng. **6**(1), 1–20 (2018). https://doi.org/10.1186/s40327-018-0064-7
37. M. Petković, D. Kocev, S. Džeroski, Feature ranking for multi-target regression. Mach. Learn. **109**(6), 1179–1204 (2020). https://doi.org/10.1007/s10994-019-05829-8
38. A.J. Al-Mahasneh, S. Anavatti, M.G.A.M. Pratama, Applications of general regression neural networks in dynamic systems. Digital Systems (2018), 10.5772/intechopen.80258
39. N. Sharma, H. Om, Usage of probabilistic and general regression neural network for early detection and prevention of oral cancer. Sci. World J. **2015**, 1–11 (2015). https://doi.org/10.1155/2015/234191

40. R. Richman, M.V. Wüthrich, A neural network extension of the Lee–Carter model to multiple populations. Annals of Actuarial Science, pp. 1–21 (2019), 10.1017/s1748499519000071
41. S. Dhamodharavadhani, R. Rathipriya, Enhanced Logistic Regression (ELR) model for big data, in *Handbook of Research on Big Data Clustering and Machine Learning Advances in Data Mining and Database Management* (2020), pp. 152–176, 10.4018/978-1-7998-0106-1.ch008
42. N.F. Grinberg, O.I. Orhobor, R.D. King, An evaluation of machine-learning for predicting phenotype: studies in yeast, rice, and wheat. Mach. Learn. **109**(2), 251–277 (2020). https://doi.org/10.1007/s10994-019-05848-5
43. J. He, C. Du, F. Zhuang, X. Yin, Q. He, G. Long, Online Bayesian max-margin subspace learning for multi-view classification and regression. Mach. Learn. **109**(2), 219–249 (2020). https://doi.org/10.1007/s10994-019-05853-8
44. S. Kolb, S. Teso, A. Dries, L.D. Raedt, Predictive spreadsheet autocompletion with constraints. Mach. Learn. **109**(2), 307–325 (2020). https://doi.org/10.1007/s10994-019-05841-y
45. S. Dhamodharavadhani, R. Rathipriya, Variable selection method for regression models using computational intelligence techniques, in *Handbook of Research on Machine and Deep Learning Applications for Cyber Security Advances in Information Security, Privacy, and Ethics* (2020), pp. 416–436
46. S. Dhamodharavadhani, R. Rathipriya, Forecasting dengue incidence rate in tamil nadu using ARIMA time series model. Machine Learning for Healthcare, pp. 187–202 (2020)
47. S. Sivabalan, S. Dhamodharavadhani, R. Rathipriya, Arbitrary walk with minimum length based route identification scheme in graph structure for opportunistic wireless sensor network, in *Swarm Intelligence for Resource Management in Internet of Things*, pp. 47–63 (2020), 10.1016/b978-0-12-818287-1.00006-1
48. JHU, CSSEGISandData/COVID-19 (27 May 2020), https://github.com/CSSEGISandData/COVID-19/tree/master/csse_covid_19_data. Accessed 5 Sep 2020
49. S. Namasudra, S. Dhamodharavadhani, R. Rathipriya, Nonlinear neural network based forecasting model for predicting COVID-19 cases. Neural Process Lett (2021)
50. R. Devipriya, S. Dhamodharavadhani, S. Selvi, SEIR model for COVID-19 epidemic using delay differential equation. J. Phys: Conf. Ser **1767**(1), 012005 (2021)
51. S. Dhamodharavadhani, R. Rathipriya, COVID-19 mortality rate prediction for India using statistical neural networks and gaussian process regression model. Afri. Health Sci. **21**(1), 194–206 (2021)

Combating COVID-19 with Advanced Computational Intelligence Techniques

Computational Intelligence, Machine Learning and Deep Learning Techniques for Effective Future Predictions of COVID-19: A Review

K. Aditya Shastry and H. A. Sanjay

Abstract As per the World Health Organization (WHO), Coronaviruses represents a huge virus family that creates diseases in humans/animals. The newly discovered coronavirus is known as Covid-19 (Cov-19). In December 2019, this virus broke out in Wuhan, China causing massive havoc worldwide. The design, development, theory and application of standards related to computation form the Computational Intelligence (CI) methods. Conventionally, the 3 key components of CI are the Artificial Neural Networks (ANN), Fuzzy System (FS), & Computation related to Evolution (EC). Lately, techniques like chaotic systems, support vector machines (SVM), etc. have been included into the CI techniques. Machine Learning (ML) enables systems to automatically learn without being programmed explicitly. Deep Learning (DL) represents a family of ML techniques based on ANN. A great potential has been observed while applying CI, ML, and DL techniques for predicting Cov-19. In this regard, the key objective of this chapter is to present an extensive review to the readers on how CI, ML, and DL techniques can be utilized to effectively predict Cov-19. The chapter deals with the review of the different CI, ML, and DL techniques such as ANN, FS, and EC that have been applied for Cov-19 prediction. The application and suitability of CI, ML, and DL techniques for screening and treating the patients, tracing the contacts along with Cov-19 forecasting is discussed in detail. A discussion of why certain CI, ML, and DL techniques are useful for the Cov-19 prediction is also presented.

Keywords Covid-19 · Computational intelligence · Machine learning · Deep learning · Prediction · Analysis

K. A. Shastry (✉) · H. A. Sanjay
Department of Information Science and Engineering, Nitte Meenakshi Institute of Technology, Yelahanka, Bangalore 560064, Karnataka, India
e-mail: adityashastry.k@nmit.ac.in

H. A. Sanjay
e-mail: sanjay.ha@nmit.ac.in

J. Nayak et al. (eds.), *Understanding COVID-19: The Role of Computational Intelligence*, Studies in Computational Intelligence 963, https://doi.org/10.1007/978-3-030-74761-9_17

1 Introduction

A vast category of viruses known as Coronaviruses are present that cause illness in a moderate to severe fashion among humans and animals. Numerous such viruses are found to create infections related to respiratory varying from the common cold to more critical illnesses. Cov-19 belongs to the family of Coronaviruses that has been discovered recently. It was unknown till December 2019, during which it broke out in Wuhan, China. Currently, it is causing great havoc throughout the globe. Fatigue, Fever & dry cough are some of the common symptoms caused by Cov-19 that are currently discovered [1]. Additionally, some of the less frequent Cov-19 symptoms comprise of pains & aches, congestion in the nose, headache, throat pain, not able to smell or taste, skin rashes, and diarrhea along with color loss in toes & fingers. Some of these symptoms become hard to detect since they may be very mild and really start to become severe over a course of time. Another critical problem in detecting Cov-19 is that certain people may exhibit very mild symptoms or none. These people can serve as carriers of this disease and this has led to its rampant spread. Persons who don't exhibit any symptoms are called as asymptotic [2]. Furthermore, majority of the persons (around eighty percentage) are found to recover from Cov-19 without undergoing any treatments in hospitals. However, 1 of 5 persons who have acquired this illness can become critically sick and develop severe breathing difficulty. Such individuals must undergo hospitalization immediately. The persons who are found to be at risk of Cov-19 are elder persons & individuals possessing inherent medical issues such as high BP, problems related to lungs & heart, diabetes, or cancer. But a study has revealed that even a normal person may get critically sick because of Cov-19. However, the chances are considered to be rare. Research also indicates that any person irrespective of age/gender experiencing breathing difficulty with fever/cough need to undergo medical treatment almost immediately without wasting time. WHO has suggested recommending Cov-19 affected persons to call the health care providers (HCP) who can provide suitable directions to the patient [3, 4].

The spread of Cov-19 has been classified into 4 stages depending on the kind of spread and the related duration of time. Each country applied several techniques to prevent this spread comprising of lockdown (stay at home), covering mouth and nose with masks, restrictions on travel, avoiding large gatherings, cleaning hands using sanitizers, along with sanitization of locations. Unnecessary movement of citizens was restricted in certain countries. This has led to serious impact to economy as small & medium scale industries have been adversely impacted. As can be viewed, currently the GDP has dropped around the world quite suddenly. Nowadays, the only treatment available is isolation of the infected person who is provided certain common drugs that can assist him/her in recovery. Since the vaccine is not yet available for Cov-19, death rate has surged quite drastically around the world. The solitary isolation of persons has led to severe depression among sensitive individuals leading to even suicide [4, 5].

In developing countries like India, the normal life has been adversely impacted since prevention includes stringent measures such as lockdown. Initially, the confirmed cases were due to the persons returning from overseas. Subsequent local spread took place leading to community spread. The average age of the infected person in India is 39. Reasonably, individuals in the age span between 21 & 40 are considered to be infected more. Hence, effective prediction of Cov-19 is very much essential for reducing the spread of this twenty-first century pandemic [6, 7]. The paradigms of computation comprising of design, development, theory, and application form the CI. Classically, ANN, FS, and EC were the critical components of CI. Recently, other paradigms motivated by nature have also been incorporated in CI. Hence, CI denotes a field which is ever evolving and is comprised of techniques ranging from EC to ANN. Successful intelligent systems comprising of games, and cognitive systems are being built by employing CI techniques. Lately, ML and DL techniques have become highly popular. They have grown up to be a key component in AI. In turn, successful AI systems have been developed using CI, ML and DL methods [6, 7].

The challenges generated due to the Cov-19 pandemic can be effectively handled by CI, ML and DL. However, innovation alone cannot stop this pandemic. It requires the precision of humans also. Another critical aspect is that this pandemic may reveal several shortcomings of CI, ML and DL. The key working of CI, ML and DL is that it can identify numerous designs or patterns from training dataset. CI, ML and DL techniques have been popular among researchers in other application domains. They can be applied to novel situations by taking information from existing situations to improve the prediction accuracy of the CI, ML and DL techniques. In contrast, frameworks related to computer-based intelligence tend to change even when the setting changes marginally. Hence, the emergency associated with Cov-19 matches the fundamental requirement of CI, ML and DL that it represents a device, and its use is governed by the persons who use & structure it. Currently, this emergency requires that CI, ML and DL can be used to regulate the activity of humans & their associated development. Acquiring new training data under existing circumstances is one way of coping with this crisis. The subsequent choices will be guided by the capturing of any new segment of Cov-19 related data that are very much essential for leaders & CI, ML and DL frameworks. Research on Cov-19 prevention can progress rapidly only if the disease related data are shared freely and widely so that accurate predictions can be made. This will enable CI, ML and DL in aiding the Cov-19 prevention along with other relevant treatments. As mentioned earlier, the innovation can only help if the people who use it is responsible and innovative [6, 7].

Also, the inherent drawbacks of not only CI, but ML, and DL can be exposed by this Cov-19 emergency as these techniques require valid training dataset for proper data analysis. Favorable opinions are present regarding the implementation of CI, ML and DL methods in disease prediction. The trained CI, ML and DL models can be applied on test data that denote unknown data. This is the fundamental way in which the disease forecasting happens. Furthermore, the CI, ML and DL methods need to be adaptable to changing environments. This is very much true in the prevailing Cov-19 conditions where the situations are changing dynamically every day. This is very much true regarding the underlying assumptions of CI, ML and DL that represents a procedure for assessing unknown data from known data and requires the responsible use by the associated persons. During this current crisis, the appropriate utilization of the CI, ML and DL techniques are essential for the successful prediction of the disease. One method of handling this pandemic is the gathering of novel training data as and when they appear which can lead to better training of the CI, ML and DL models. Each portion of the new data about the present situation will lead to the proper selection of subsequent treatment choices. Hence, data sharing becomes highly essential as more data are collected and the CI, ML and DL models are trained, the more accurate the forecasting of Cov-19 can become [6, 7].

The key contribution of this chapter is that it provides an exhaustive survey of the CI, ML and DL techniques that were employed for Cov-19 prediction. This chapter focuses on the novel Cov-19 epidemic and how the modern CI, ML and DL technology were recently employed to solve the challenges during the outburst. A comprehensive review of works using CI, ML and DL techniques applied to handle the Cov-19 pandemic are demonstrated. The survey also presents discusses why certain CI, ML and DL techniques have been more effective for handling Cov-19 prediction. The chapter can assist researchers, and policymakers in tackling Cov-19 pandemic more effectively.

The rest of this chapter is organized as follows. The 2nd section deliberates on the various CI techniques that may be utilized for Cov-19 prediction. The 3rd section describes some significant case studies for Cov-19 prediction using CI, ML and DL. The 4th section discusses the justification of why certain CI, ML and DL techniques are better compared to others. The chapter ends with Conclusion and Future Scope.

2 Case Studies: CI Techniques for Cov-19 Detection

This section discusses popular CI techniques applied for Cov-19 prediction. Figure 1 demonstrates the CI and non-CI based approaches for Cov-19 prediction.

Figure 1 depicts the generic technique of CI and non-CI based applications that assists medical experts to detect the COV-19 symptoms [8]. As shown in Fig. 1, in the CI based approach the symptoms of patients are matched with that Cov-19 using CI techniques. The infection is confirmed based on the samples gathered and the next treatment is determined. Based on this, the patient is either quarantined or admitted. The treatment & monitoring based on CI is then begun. This is followed by recovery phase. Once again, the Cov-19 test is conducted. If the patient tests positive, then he/she is quarantined else he/she is declared as cured. The key difference between the traditional method and CI based method is that instead of manually matching the Cov-19 symptoms CI techniques may speed up this process. Subsequently, accuracy may also be increased in CI based methods since the traditional technique may be prone to human errors.

ANN, FS, & EC inspired by nature constitute the commonly utilized techniques of CI. The Sects. 2.1 to 2.3 explore these techniques along with their usage for predicting Cov-19.

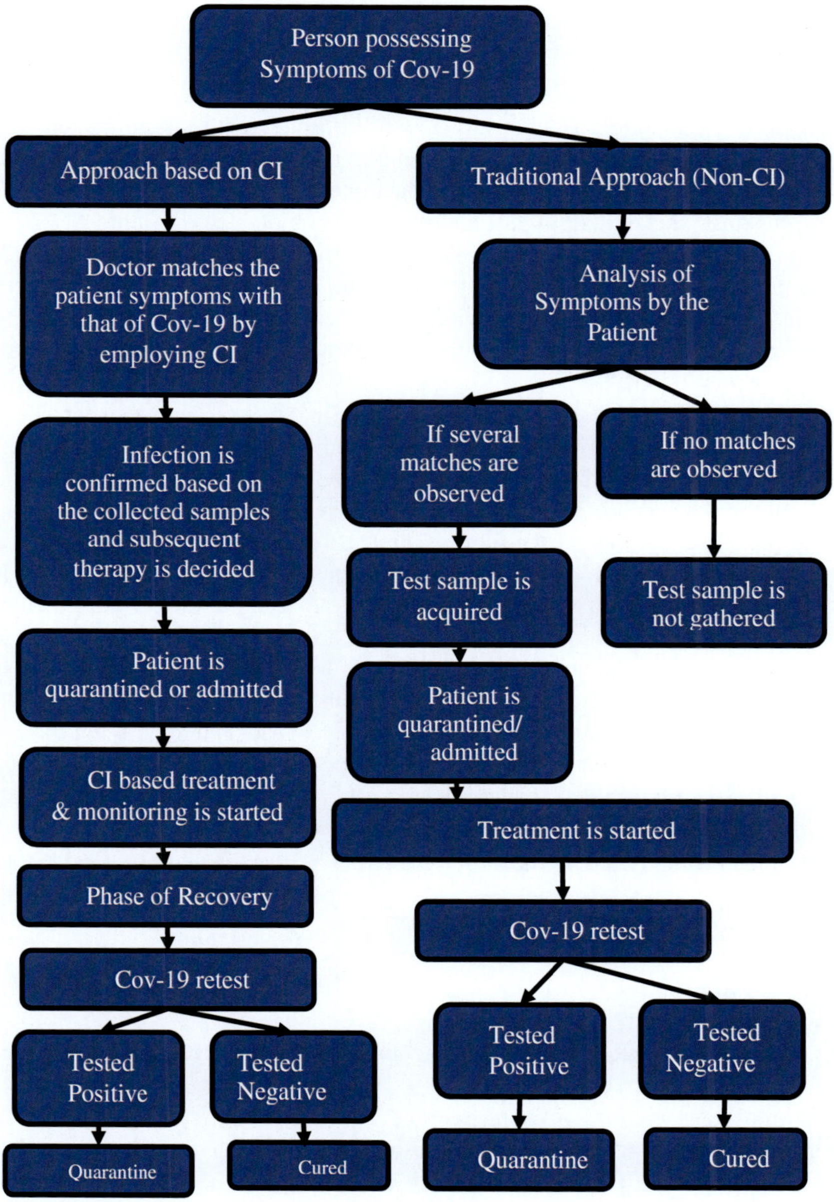

Fig. 1 Technique of CI & Non-CI based applications for Cov-19 prediction [8]

2.1 ANN for Cov-19 Prediction

Large datasets can be effectively analyzed by employing ANN. They aid in extracting meaningful patterns along with accurate forecasts & related trends. ANNs are being widely utilized in the engineering as well as the medical domain. ANN denotes a system for processing data comprising of several elements called neurons that are similar to the neurons in the biological system. ANNs have exhibited remarkable success in exploring the nonlinear & interaction effects from training/experimental/real-time data. For the purpose of forecasting the technique of curve fitting based on ANN is one among the widely utilized methods of artificial intelligence. ANN are usually comprised of 3 layers called the input, hidden & the output layers [9, 10].

The work [9] deals with the Cov-19 prediction in which the output layer consists of the number of day-to-day Cov-19 cases to be predicted and the input layer comprises of the number of days. The training data followed the pattern of the given predicted data in the ANN curve fitting method. This allowed the authors to make forecasts about the data behavior in the future. The least squares method was employed to generate the polynomial curve. Through the training process, the network adjusts the weights for minimizing the errors between the actual & the forecasted outputs. Random varying of weights is done to arrive at the optimal network that has the least error. Backpropagation algorithm was utilized for training the network. Forecasting was done on extrapolated datasets. Comparison between the real time data of WHO and the predicted values was carried out. Neurons in each interconnected layer with the previous & next layer neurons with synaptic weights. Moreover, each neuron of the hidden & output layers is added with a bias term. Each neuron output is computed by the sum of weights present in the neuron inputs in the preceding layer plus its bias. Each neuron output in the output/hidden layer is estimated using the Eq. 1.

$$O_j = f(I) = f\left(\sum\nolimits_{i=1}^{n} w_{ij}x_i + b_j\right) \tag{1}$$

Where,

- w_{ij} denotes the related weights of the jth neuron of the layer and the ith neurons of the previous layer
- b_j represents the bias of the jth neurons
- n signifies the total number of neurons of the previous layer
- f is the suitable transfer function that was employed

Figure 2 shows the architecture of the two-layered feed-forward ANN used in the work [9].

MATLAB's curve fitting tool was employed by the authors of [9] for building the network. The algorithm employed the technique of gradient descent to obtain the optimal solution by minimizing the mean square error (MSE) among the forecasted and the actual outputs demonstrated in Eq. 2.

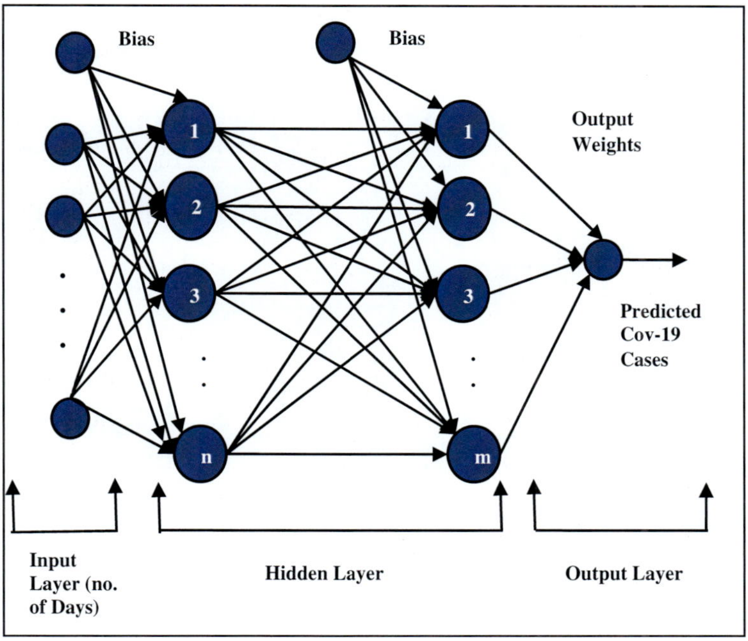

Fig. 2 ANN architecture for Cov-19 prediction [9]

$$E_p = \frac{1}{2}\sum (t_{pj} - O_{pj})^2 \tag{2}$$

Where,

- E_p denotes the error for the pth vector of presentation
- t_{pj} signifies the predicted output of the jth output neuron
- O_{pj} represents the actual output of the jth output neuron

Subsequent step includes the determination of error at the hidden neurons. These errors are then backpropagated to the values of weights connected among the hidden layer neurons & the input layer neurons. Likewise, the errors at the output neurons are backpropagated to the values of weights connected among the hidden and the output layer neurons employing Eqs. 3 and 4.

$$Error\delta_{pi} = \left(t_{pj} - O_{pj}\right)O_{pj}\left(1 - O_{pj}\right) \tag{3}$$

$$Error\delta_{pi} = \left(t_{pj} - O_{pj}\right)O_{pj}\sum \delta_{pi}W_{ki} \tag{4}$$

Equations 3 & 4 represent the output & hidden neurons respectively. The updating of Weights is performed by employing Eq. 5.

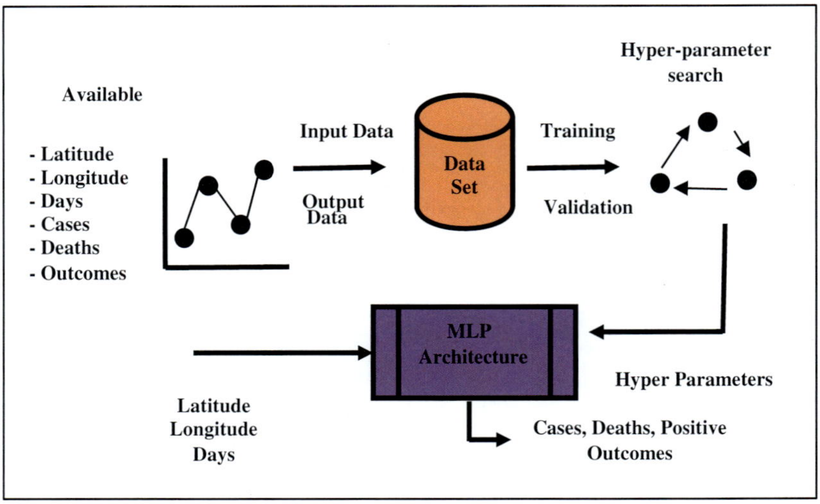

Fig. 3 ANN modelling for Cov-19 prediction [10]

$$\Delta W_{ji}(n = 1) = \eta(\delta_{pi} O_{pi}) = \alpha \Delta W_{ji}(n) \quad (5)$$

where, η denotes the learning rate parameter and α signifies the momentum factor.

A class of ANN is known as Multilayer Perceptron (MLP). Generally, the MLP term refers to feedforward ANN. However, sometimes MLP strictly refers to networks comprising of several perceptron layers with threshold activation. It contains 3 layers called input, hidden & output layers. The neurons in the hidden and output layers take advantage of nonlinear activation functions. Backpropagation is employed by the MLP. The fundamental difference between MLP and a linear perceptron is the presence of multiple layers & the activation function which is non-linear in nature. Non-linear data can be separated [10–12].

Figure 3 depicts the overall picture of the modeling process of ANN for the forecasting of Cov-19 [10]. The procedure of data transformation, data modeling, along with tuning of the MLP using multiple combinations of parameters is illustrated in Fig. 3.

In the Fig. 3, data is gathered first. A segment of the data is utilized for training. Remaining data segment is employed for testing the different parameter combinations of the MLP. The best architecture of MLP is then found out. This trained model is finally employed to forecast the future cases.

The publicly available repository governed by the Centre of John Hopkins University (JHU CSSE) [13] was utilized by the authors [10] for experimentation purpose. For training the MLP, data is reordered to create a set of inputs & outputs. The data collection date along with the location's longitude & latitude is added for each case. Conversion of the date into number of days is done in the dataset.

Every record in the dataset holds information regarding the number of patients (how many people have contracted the disease, recovered from the illness or death due to the sickness) at a specific place, on a given day since the 1st registered case. Input data is comprised of longitude, number of days & latitude. The output data constitute the number of patients present in every category. In this fashion, rearrangement of time series dataset is made for making it suitable for MLP training.

Authors [10] utilized 3-Layer MLP. The number of dataset inputs formed the input layer neurons [14]. The output layer is comprised of the number of patients predicted as having Cov-19. This number corresponded to the sole neuron in the output layer. MLP is built on computing the neurons values in an existing layer by summing the weighted outputs of the preceding layer, connected to the neuron [14, 15]. The activation function represents the weighted input sum that will be propagated as inputs to the next layer to activate the neurons in the subsequent layer. This activation takes place by mapping the input to output either in a direct fashion or by employing particular limits such as sigmoid, or tanh or by eliminating irrelevant values using ReLU activation function. The ReLU activation function eliminates the values which are negative and maps only the positive values in a straightforward manner [16]. Initially, random weights are allotted to every connection between the layers. These weights are then updated by employing the backpropagation process to minimize the total error of the network [17].

The MLP can predict only a single value. Hence if the problem requires several output values, multiple models have to be employed in a modular manner. Separate training of models implies all architectures can be tested, providing a better probability of choosing an efficient prediction model for each task. In the work [10] authors made use of 3 distinct trained MLPs. One MLP was used for predicting number of Cov-19 infected patients, 2nd MLP was employed for predicting number of patients who had recovered from Cov-19 and the 3rd MLP was utilized for predicting the number of patients who were deceased due to Cov-19.

Accurate outcomes were achieved by utilizing ANN for Cov-19 diagnosis because of the following reasons [9]:

- For handling large datasets, ANN has been observed to perform well because of its capacity to analyze, demonstrate patterns, prediction, etc. It has show tremendous promise in engineering as well as medical domain.
- ANN is able to analyze the non-linear association between the data attributes which is very much pertinent in Cov-19 datasets.
- The key benefit of ANN is that is does not comprise of complex computations & can be employed by several countries to assist in regulating the Cov-19 cases centered on the events/trends of other countries that are able to regulate/control the cases.
- The ANN method of curve-fitting is able to deliver improved prediction of the cases that are rising centered on the trends of China & South-Korea which are positively non-linear in nature and were able to successively control the Cov-19 cases.
- ANNs possesses the capacity to train the dataset of any country based on the trends of other countries.

2.2 Fuzzy Systems (FS) for Cov-19 Prediction

FL represents a logic where the values of truth can lie amid 0 and 1 including 0 and 1. It is a logic founded on multiple values. It is utilized to deal with the partial truth concept where the truth values can lie between the values of completely true representing 1 and completely false representing 0. FS are the structures centered on the techniques of FL that is inclined towards processing information in which using the classical set theory & binary logic become impossible to use [18].

Uncertainty forms a major factor while estimating the risks associated with healthcare facilities due to Cov-19. Fuzzy Inference System (FIS) has found to be an effective solution to handle uncertainties that are associated with assessment of healthcare facilities [19]. The FIS is prevalent in the domain of methods related to FL. The FIS includes 3 core concepts viz. The membership functions (MF), operations related to FS, and the procedures of inference [20]. FIS can be broadly partitioned into 4 segments such as: Fuzzification, weighting, estimation of procedures linked to inference & de-fuzzification.

Adaptive Neuro Fuzzy Inference System (ANFIS) has exhibited enhanced performances in prevalent applications related to prediction of time series data and other prediction problems [21]. It basically combines the features of ANN and FL that lead to considerable flexibility in assessing the nonlinearity present in datasets [22].

As mentioned before the FL and ANN are linked in the ANFIS model. It employs the IF–THEN rules for generating the mapping between input & output. The ANFIS model is illustrated in Fig. 4. The ANFIS model links the FL and ANN [22]. Figure 4 depicts the ANFIS model where, x & y are layer-1 inputs.

The O_{1i} is the layer-1 output of node I computed as shown in Eq. 6.

$$O_{1i} = \mu_{A_i}(x), i = 1, 2, O_{1i} = \mu B_{i-2}(y), i = 3, 4 \tag{6}$$

$$\mu(x) = e^{-\left(\frac{x-\rho_i}{\alpha_i}\right)2} \tag{7}$$

where μ (shown in Eq. 7) denotes the functions of Gaussian MF that will be generalized. Ai and Bi describe the membership values of μ. α_i and ρ_i denote the parameters set.

The layer-2 output is computed as shown in Eq. 8:

$$O_{2i} = \mu_{Ai}(x) X \mu_{Bi-2}(y) \tag{8}$$

The layer-3 output is computed using Eq. 9

$$O_{3i} = \overline{w_i} = \frac{w_i}{\sum_{i=1}^{2} w_i^{'}} \tag{9}$$

The layer-4 output is computed using Eq. 10:

$$O_{4,i} = \overline{w_i} f_i = \overline{w_i}(p_i x + q_i y + r_i) \quad (10)$$

where, $r_i,\ q_i,\ and\ p_i$ are the output parameters of node i.

The layer-5 output is estimated using Eq. 11.

Layer-5 comprises of 1 node and its outcome is estimated in Eq. 11:

$$O_5 = \sum_i \bar{w_i}\ f_i \quad (11)$$

Figure 5 depicts the forecasting of Cov-19 as per the authors [22]. The FPASSA-ANFIS model proposed by [22] employed the enhanced FPA for training the ANFIS model in order to choose its optimal parameters. Five layers constitute the FPASSA-ANFIS model. The confirmed Cov-19 cases (historic data) form the input features of Layer-1. The predicted values form the Layer-5. The optimal weights for Layers 4 and 5 are chosen in the training phase of the model. Formatting of input data in a time series format is done initially. The authors [22] have taken the autocorrelation function (ACF) for determining patterns hidden in the data that reveal the correlation between data points divided by the numerous time lags. They considered features with 5-lags i.e. ACF > 0.2. The data was divided into 75% as training dataset and the rest 25% as the test dataset.

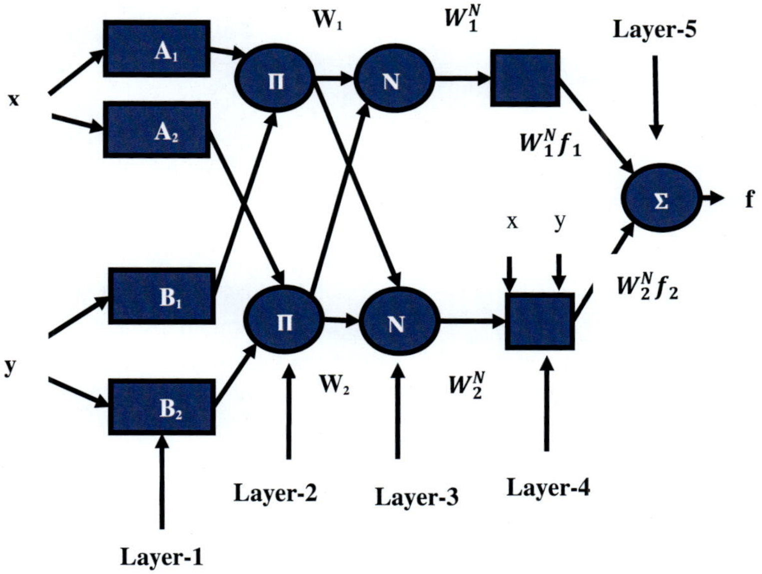

Fig. 4 ANFIS model structure [22]

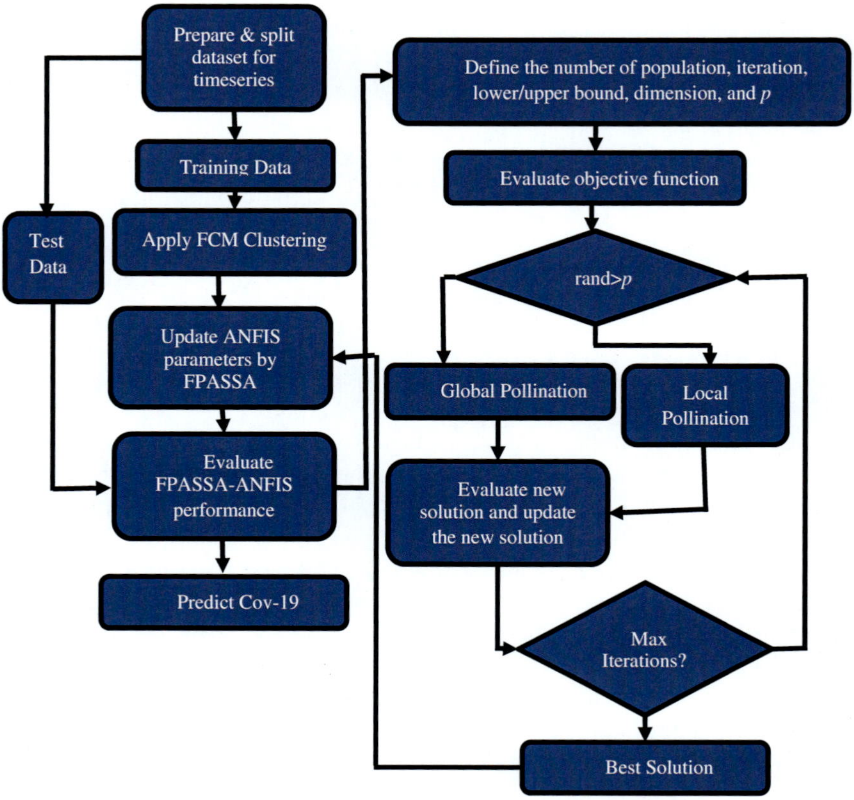

Fig. 5 ANFIS model to predict Cov-19 [22]

The quantity of clusters is determined by the fuzzy c mean (FCM) method to create the ANFIS model. For training the ANFIS model, the FPASSA-ANFIS model employed the improved FPA for parameter optimization. Five layers are reflected in the FPASSA-ANFIS model similar to the standard ANFIS model. Input features of layer 1 denote the Cov-19 cases. Forecasted values are generated by the layer 5. During the phase of learning, the best weights between layers 4&5 are chosen.

After the training phase ends, the phase of testing begins that produces the optimal answer to generate the outcome. The proposed method's performance is assessed by associating the real data and forecasted data by employing the performance metrics. Lastly, the FPASSA model was able to forecast the number of Cov-19 patients for the next day in China based on the previous data [22].

FS provided better accuracy for Cov-19 detection due to the following reasons: [19, 20, 22]

- FS perform better for predicting Cov-19 as they possess the ability to deal with the vagueness that persists in medical domain by creating a tangible model. This model will be able to perform decisions in the existence of uncertainty, inaccuracy & incomplete environments.

- The FL method used in FS can establish accuracy that is usually considered as unclear. Similarly, FS can create an unambiguous boundary which was not clear before.
- By utilizing FS, the boundary of Cov-19 symptoms such as tiredness, fevers which are vague can be made clear & quantifiable.
- Fuzzy methods are effective for handling conditions of illness which are vague in nature like in Cov-19.
- ANFIS method has been found to be effective for Cov-19 diagnosis since it merges the gains of both ANN and FIS. This technique ensures that no additional knowledge for modelling & training the system is required.
- The FS are typically useful in conditions where the scenarios are complex comprising of patterns which behave in a non-linear manner.

2.3 Evolutionary Systems for Cov-19 Prediction

Optimization problems can be resolved using tools such as evolutionary algorithms (EA) [23]. The EA was inspired by biological processes that search all the potential solutions in the solution space. It is formulated as an optimization problem [24–26]. This section deliberates the application of two evolutionary algorithms in the Cov-19 prediction. In [23], they were utilized to assess the parameters of the regression models that were developed for Cov-19 prediction.

2.3.1 Genetic Algorithm (GA)

Evolution concept motivated the birth of GAs which are subset of computational models [27]. Here, the problem is expressed similar to the chromosome data structure. GAs make use of “Potential Solutions/Candidate Solutions/probable hypothesis” for a problem formulated by the chromosome data structure (CDS).

GA preserves key data located in these CDS by employing “Recombination Operators” to the CDS [28–31]. In several instances, GAs is used as algorithms for optimizing objective functions. However, wide range of applications make use of GA [30, 32]. The GA implementation normally begins with the generation of chromosome population in a random fashion and is bounded above & below by the problem features. The evaluation of the generated chromosomes is done in the next step. The chromosomes that can generate optimal solutions are chosen to generate offspring’s in the subsequent generations. Measurement of the “goodness degree” of a result is done by the solutions of the current candidates in the population [33–37]. Figure 6 describes the outline of the GA algorithm.

The authors [22] utilized GA for identifying the parameters of the regression models [37]. The population-number was chosen as 300 & the maximum generation

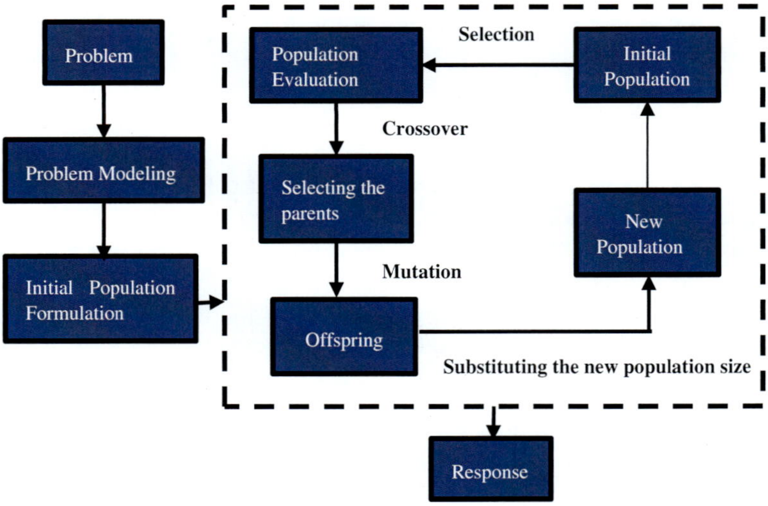

Fig. 6 GA algorithm [22]

was selected as 500 based on multiple experiments. These values were choses as they yielded the least error based on the cost function. The Mean Square Error formed the cost function between the actual and predicted values as shown in Eq. 12.

$$MSE = \sqrt{\frac{(E_s - T)^2}{N}} \quad (12)$$

Where,

E_s – Estimated values.

T – Target Values.

N – Number of Data.

2.3.2 Particle Swarm Optimization (PSO)

PSO was designed in 1995 by Kennedy & Eberhart [38] for resolving optimization problems. It is essentially a search technique. The group movement of birds searching for food inspired the development of PSO. It assumes that only 1 piece of food existed in the search space. Every PSO solution is termed as a particle that represents a bird in the real-world. Every particle's value is computed using a competency function that grows as the particle comes near to the target. Here, the target refers to the food that the birds are looking for in the real-world. The motion of a particle is governed by its velocity. Particle movements proceed in the direction where the final solution can be found by tracing the optimal particles in the present state [38–40].

The PSO algorithm basically follows the collective behavior of the birds. It is founded on the code of collective intelligence of the birds that are searching for the food. Apart from birds, this type of collective intelligence can also be observed in fishes where the fishes move cooperatively to search and hunt for food. Even when they are hunted, they try escaping from the predator by using collective intelligence [41–44].

Based on these points, the PSO comprises of the following basic steps:

i. Primary population is established & evaluated in the 1st step.
ii. In the 2nd step, the best personal & collective memories are determined.
iii. The updating of position and speed happens in the 3rd step.

The algorithm goes back to step-ii) if the stopping conditions are not satisfied. It is built on the population of the problem and hence is unlikely to be trapped in a local minimum. It basically works on probable rules and not on rules that are definite in nature. Hence, PSO denotes an algorithm that follows random optimization and can search for solutions in complex areas. So, PSO is more adaptable & robust in nature when compared to traditional optimization techniques. The target functions are nondifferential in nature since the PSO uses the result of information to lead the search.

The initial population has little to do with the chosen search route. The algorithm starts randomly and finally converges to the optimal solution. It is adaptable and regulates the balance between the global & local search spaces. This flexibility allows the PSO to converge to an optimal solution and enhance the capacity of the search. These features distinguish among GA & PSO and other evolutionary algorithms [39, 43, 45]. The authors [22] applied PSO for identifying the parameters of the regression models that were developed for Cov-19 prediction. The number of population & iterations were chosen as 1000 & 500 respectively using trial & error procedure for lessening the cost value.

ES provided accurate results for Cov-19 detection due to the following reasons [23]:

- EA forms effective technique for resolving problems related to optimization via intelligent techniques.
- The EA techniques like GA, PSO are based on population which makes it unlikely to be stuck in a local minimum. It operates as per the feasible rules and not rules which are obvious.
- EA algorithms are based on random initialization which can investigate areas related which are complex & undetermined. This has made EA algorithms more adaptable & robust when compared to traditional techniques for Cov-19 detection.
- EA handles target functions which are non-differential since EA utilize the outcome of data such as the index of performance/target function to monitor the search space in the problem area. The chosen search route response does not rely on the initial population. EA algorithms can start from anywhere and converge to an optimal answer which makes them effective for Cov-19 prediction.

- EA possesses higher adaptability for controlling the stability among the global & local search spaces. This distinctive feature of EAs resolves the issue of inaccurate convergences & enhances the search capability.
- All these features contribute to the effectiveness of EAs for Cov-19 prediction when compared to traditional methods.

3 ML and DL Techniques for Cov-19 Detection

This section demonstrates certain real-world case studies of where the ML and DL techniques have been employed for the effective prediction of Cov-19 [46]. CI technology is utilized to enhance the forecast accuracy for screening contagious & non-contagious illnesses [47]. The first expert system called MYCIN was devised in the year 1976 [48]. Around 450 rules gathered from health experts were used by MYCIN for treating patients with infections caused by bacteria. Antibiotics were also suggested. Health experts can employ this system for performing clinical related decisions [49]. Current literature has revealed that ML and DL technology has been effective in assisting medical experts during several pandemic outbreaks such as SARS, EBOLA, HIV, COVID_19 [50–58] and illnesses like cancer, diabetes, and heart related diseases that are not contagious [59–66].

3.1 Methods Based on ML & DL for Treating and Screening Cov-19 Patients

The detection of any disease at an early stage is crucial to saving lives [67, 68]. This needs to be more critical for diseases like Cov-19 that spread fast. Not only, does the early disease detection reduces cost but also allows the quicker diagnosis of the illness. Expert system development for healthcare aids the recognition, management & screening of Cov-19 in a cost-effective fashion when compared to the classic methods. ML and DL techniques are utilized to combine the screening & diagnosis process of the Cov-19 patient by using Computed Tomography (CT) with radio imaging, X-Ray, & data of blood samples collected in clinics.

Along with the classical screening & diagnosis, the medical experts can utilize the images from X-ray & CT scans to improve the process of screening & treatment. Nevertheless, due to the high surge of Cov-19, moderate performance is being shown by these devices. In view of this, the work in [69] has demonstrated the potential of ML and DL tools by proposing a novel technique using deep Convolutional Network that is able to diagnose Cov-19 in a quicker fashion.

The authors [69] have demonstrated that the ML and DL based expert system exhibited noteworthy performance on 1000 + CT images of 100 + patients infected with Cov-19 & 86 patients infected with pneumonia. This study suggested that the

DL model Convolutional NN (Resnet-101) was the superior model when compared to the ML models SVM & RF models. Recently, DL is being employed as an auxiliary tool for the automatic detection of Cov-19 with a higher accuracy [70]. This model utilizes the X-ray images of 100 + patients along with 400 + pneumonia cases. The DL technique attained more than 90% accuracy for binary classification and greater than 80% for the multiclass classification. The successful multiclass classification proved that the expert system may be effectively applied to aid radiology during the process of screening in an accurate and quicker way.

Also, SVM was employed as the model for classifying the features [71]. With the aid of SVM, the researchers identified 4 significant combinations of medical attributes based on the lab, clinical, and demo-graphic features utilizing the percentages of protein, and the age of patients. This model was proven to be robust and effective in forecasting patients suffering from severe conditions. The results demonstrated that the SVM model performed with AUROC of 0.9996 on the training dataset and 0.9757 on the test dataset. The survival analysis along with the cox-multivariant regression analysis exhibited that the model can effectively act as an auxiliary tool for medical experts.

In [72], the researchers assessed 253 samples of blood from Wuhan, and observed 11 discriminatory features that could act as a tool for differentiating Cov-19 from other diseases. This could be effectively used by medical personnel for quicker diagnosis. The RF algorithm was able to extract the 11 distinguishing features and performed with an accuracy of 95.95% and specificity of 96.97%. The authors have also deployed the related tools for aiding healthcare personnel.

3.2 Contact Tracing (CT) Applications for Tracking the Cov-19 Infected Persons

Contact Tracing (CT) forms the subsequent critical step after an individual is diagnosed with Cov-19. This is very much essential for stopping the disease from spreading rampantly. As per WHO, the spread of this infection occurs from one person to the other via droplets, nose discharges, saliva, and through close contacts [73]. Hence, CT becomes a vital step for breaking the chain of the transmission of the Cov-19 virus from person to person [74]. CT comprises of the recognition & management of people who were recently in contact with the infected Cov-19 person. This becomes very much essential for preventing further spread of the pandemic. Normally, the process of CT must identify all the primary & secondary contacts of the infected person within 14 days of the exposure. Systematic application of the CT procedure can effectively disrupt the chain of Cov-19 transmission. This can diminish the scale of the pandemic spread and hence bring down the number of infections that could occur if the spread is not controlled.

Keeping these points in mind, several countries have applied CT through mobile apps using digital technology like Global Positioning System (GPS), Bluetooth, social graph, etc. The CT procedure that uses digital technology operates faster than the manual process of CT. The gathering of individual personal data is done through these digital apps. This data can then be analyzed by CI tools for tracing a person who may be exposed to the novel Coronavirus because of the recent contacted chain.

The works have proven that the combining of CI, ML, and DL tools with the process of CT can effectively increase the detection rate [75]. Graph theory was applied on data associated with infectious animal epidemics disease that comprised mainly of shipping data amid every farm. The generated graph was employed for CT by utilizing its properties. Furthermore, resultant graphs had a probable forecast influence on the amount of infections that could occur. Still, certain limitations remain while addressing the issues related to privacy, data control, data security and scenario. Some of the countries are trying to overcome these challenges. For instance, "Israel" has made it compulsory for all its citizens to use mobile data to handle the Cov-19 pandemic [76]. However. Some apps related to CT were found to violate privacy laws prevailing in the country and hence were reported as unsafe for use [77]. But still these CT apps provide a valuable supporting tool that complements the manual tracing process.

3.3 ML and DL Technology in SARS-CoV-2 Prediction

In [78], the authors devised an innovative technique founded on Stacking ensemble with Support Vector Regression (SVR) for predicting Cov-19 before 6 days to the infection. They applied the model in 10 states of Brazil for forecasting the collective Cov-19 positive cases. They augmented the short-term prediction process to make the medical professionals & the administration aware of the impending pandemic [78]. The authors of [79] demonstrated a novel technique based on XGBoost for prediction of Cov-19 on the datasets related to clinical and mammographic factors. They observed that out of the 75 clinical attributes & samples of blood test,4 biological attributes resulted in a higher accuracy of 90% for Cov-19 prediction into the general, severe, and mortality rate categories [79].

Additionally, high value in single lactic dehydrogenase contributed to classify patients into the intensive medical care category. The degree of LDH was associated with several diseases linked to respiratory such as pneumonia, asthma, and bronchitis. The developed model incorporated decision rule for rapid forecasting of persons who were most likely to get infected with Cov-19. The authorized patients were then provided with intensive care which would probably reduce the infection rate. A time series forecasting technique based on DL algorithm for the LSTM network was devised in Canada [80]. The model predicted that the endpoint of this virus in Canada would be around the month of June 2020. This prediction was found to be accurate as the number of cases in Canada decreased considerably around June 2020.

In [81], ARIMA model and the wavelet-based prediction model were combined to develop a prediction model in real time. The model was able to perform short term prediction of the Cov-19 is numerous countries like UK, France, India, South

Korea, and Canada. This aided the medical experts and policymakers to alert these countries of an impending pandemic.

4 Discussion

This segment deliberates the comparison of various CI, ML and DL models employed in case of Screening & treatment, prediction, and vaccination of Cov-19.

For the Screening & treatment of Cov-19, the above studies provide enough proof of how expert systems can be effectively applied for the screening of Cov-19 patients. The key objective of the work was to diagnose the disease early and accurately predict the outcome. Primary identification of the pandemic helps in reducing the disease spread and, in the process, save more lives. This will also save additional medical expenses that may be caused due to the delayed diagnosis of the patients. Most of the above works made use of single classification algorithms on solitary data. However, hybrid classification techniques when applied on multiple or hybrid databases comprising of data associated with clinical, mammographic & demographic may deliver better predictions when compared to individual methods on single databases. This can be because each data type may deliver diverse views regarding the infected persons and be more helpful practically.

With regards to prediction of Cov-19, majority of the works employed deep learning algorithms and is found to have more potential, robust, and advance among the other learning algorithms. As per the studies presented for Cov-19 prediction, for the Brazilian scenario [78], the SVR & stacking-ensemble provided better results when compared to the ARIMA, Cubist regression, Random Forest, and Ridge regression models. This was because, they were able to learn the nonlinearities inherent to the evaluated epidemiological time series. In [79], the XGBoost technique obtained a higher accuracy when compared to RF, and Logistic Regression for predicting the mortality rate of Cov-19 patient. This was because of its higher interpretability, and less overfitting. In [80], DL using LSTM provided higher accuracy for Cov-19 prediction since it is more suitable for time series data. Also, it can effectively capture non-linearities existing in the Cov-19 datasets. The LSTM technique obtains better results on smaller sized datasets when compared to statistical techniques without any parameter selection assumptions. In [81], hybrid ARIMA-WBF strategy performed better on the European datasets when compared to the single ARIMA and WBF models. However, result comparison with DL techniques was not done. From the exhaustive survey conducted, it can be concluded that there is no single best technique for predicting Cov-19. The accuracy relies on the number and type of the datasets utilized for experimentation.

With regards to Vaccination of Cov-19, the works have employed several techniques related to CI along with ML and DL methods. These methods have addressed the problems faced in the classical/traditional techniques. They inferred that the combination of docking application with the CI, ML, and DL was more effective in discovering the reusability of the prevailing drugs for Cov-19 treatment. This procedure also made the procedure of developing drugs a less expensive process as existing drugs can be utilized to new pandemics also. The associated risks would also be minimized as the existing drugs would have previously been tested previously. The time for discovering new drugs can be drastically reduced during this emergency by using CI, ML, and DL techniques for exploring existing drugs.

5 Conclusions and Future Scope

In recent times, the Cov-19 pandemic has risen as the major challenge adversely impacting the health, economy of several countries globally. Development of forecasting techniques that can predict the disease accurately is crucial in delivering valuable insights of the spread & to study the impact of this pandemic. Because of the uncertainty that inherently exists in the Cov-19, classic models of epidemiology have demonstrated lower accuracy for long term forecasting. This chapter presented the introduction to the CI, ML and DL techniques that could be effectively applied for Cov-19 prediction. An extensive survey of the different techniques of CI, ML and DL such as ANN, FS, ANFIS, Evolutionary Systems (GA, PSO) that were applied for the pandemic prediction is discussed. The application of the CI, ML and DL techniques for screening, treatment, CT, prediction, and drug discovery for Cov-19 along with their comparison is also described.

However, certain limitations have also been observed. The prevailing emergency needs an enhanced technique possessing higher precision and accuracy for screening & forecasting the Cov-19 by examining the different features of the infected/suspected patients. Furthermore, most of the models used for Cov-19 prediction are single classifiers or single predictive models. Hybrid or ensemble models can be utilized and have better chances of accurate prediction since they incorporate the benefits of individual models.

In view of these limitations, in future, certain improvements are suggested. For doing long term forecasting HPC models have to be enhanced and the future research must do a comparative survey on several CI, ML, and DL models for individual countries. The global models cannot be used in the cases of pandemics since the nature of the outbreaks vary significantly from one country to the other. As can be noted from different studies, there is very less chance of an outbreak being replicated elsewhere. Even though assessing the maximum infected patients is difficult, the estimation of the proportion amid the number of deaths and the number of infected patients is critical. This ratio gives the mortality rate which is very much significant for accurately estimating the number of patients which in turn lead to better allotment of beds in ICUs. In future research, the prediction of mortality rate becomes very crucial for countries for planning new medical facilities.

References

1. V. Chamola, V. Hassija, V. Gupta, M. Guizani, A comprehensive review of the COVID-19 pandemic and the role of IoT, drones, AI, blockchain, and 5G in managing its impact. IEEE Access. **8**, 90225–90265 (2020)
2. S.K. Saxena, S. Kumar, V.K. Maurya, R. Sharma, H.R. Dandu, M.L.B. Bhatt, in *Current Insight into the Novel Coronavirus Disease 2019 (COVID-19),* ed. by S. Saxena Coronavirus Disease 2019 (COVID-19). Medical Virology: From Pathogenesis to Disease Control (2020), pp. 1–8
3. S.K. Saxena, S. Kumar, V.K. Maurya, R. Sharma, H.R. Dandu, M. Bhatt, in *Current Insight into the Novel Coronavirus Disease 2019 (COVID-19).* Coronavirus Disease 2019 (COVID-19): Epidemiology, Pathogenesis, Diagnosis, and Therapeutics (Springer, Singapore 2020), pp. 1–8
4. WHO: World Health Organization (2020). https://www.who.int/emergencies/diseases/novel-coronavirus-2019/question-and-answers-hub/q-a-detail/q-a-coronaviruses Accessed 5 July 2020
5. S. Kaushik, S. Kaushik, Y. Sharma, R. Kumar, J.P. Yadav, The Indian perspective of COVID-19 outbreak. Virusdisease **31**(2), 1–8 (2020)
6. R. Sujath, J.M. Chatterjee, A.E. Hassanien, A machine learning forecasting model for COVID-19 pandemic in India. Stoch. Env. Res. Risk Assess. **34**, 959–972 (2020)
7. J.M. Keller, D. Liu, D.B. Fogel, in *Introduction to Computational Intelligence.* Fundamentals of Computational Intelligence: Neural Networks, Fuzzy Systems, and Evolutionary Computation (Wiley-IEEE 2016), pp. 1–4
8. R. Vaishya, M. Javaid, I.H. Khan, A. Haleem, Artificial intelligence (AI) applications for COVID-19 pandemic, diabetes & metabolic syndrome. Clin. Res. Rev. **14**(4), 337–339 (2020)
9. S.K. Tamang, P.D. Singh, B. Datta, Forecasting of Covid-19 cases based on prediction using artificial neural network curve fitting technique. Global J. Environ. Sci. Manage. **6**, 53–64 (2020)
10. Z. Car, S.B. Šegota, N. Anđelić, I. Lorencin, V. Mrzljak, Modeling the spread of COVID-19 infection using a multilayer perceptron. Comput. Math. Meth. Med. **2020**, 1–10 (2020). https://doi.org/10.1155/2020/5714714
11. Center for Disease Control and Prevention, Coronavirus Disease 2019 (COVID-19) - How It Spreads. https://www.cdc.gov/coronavirus/2019-ncov/prepare/transmission.html Accessed 25 July 2020
12. M. Roser, H. Ritchie, Coronavirus Disease (COVID-19). Our World in Data, Oxford Martin. https://ourworldindata.org/coronavirus Accessed 26 July 2020
13. John Hopkins University & Medicine (2020). https://coronavirus.jhu.edu/map.html Accessed 27 July 2020
14. J. Heaton, I. Goodfellow, Y. Bengio, A. Courville, *Deep Learning.* ISBN: 0262035618. Genetic Programming and Evolvable Machines 19 (2017) (The MIT Press 2016), p. 800. https://doi.org/10.1007/s10710-017-9314-z
15. C. Bishop, Pattern Recognition and Machine Learning (Springer 2011), p. 738
16. A. Nader, D. Azar, in *Searching for Activation Functions Using A Self-Adaptive Evolutionary Algorithm.* Proceedings of the 2020 Genetic and Evolutionary Computation Conference Companion (ACM, Mexico 2020), pp. 145–146
17. C.M. Bishop, in *Pattern Recognition and Feed-Forward Networks.* The MIT Encyclopedia of the Cognitive Sciences, vol 13(2) (MIT 1999)
18. P. Prokopowicz, J. Czerniak, D. Mikołajewski, Ł. Apiecionek, D. Ślęzak, Theory and applications of ordered fuzzy numbers. Stud. Fuzziness Soft Comput. **356** (2017)
19. A. Adwibowo, Fuzzy logic assisted COVID 19 safety assessment of dental care (2020). https://doi.org/10.1101/2020.06.18.20134841, PPR: PPR178593
20. L.-Y. Wei, A hybrid ANFIS model based on empirical mode decomposition for stock time series forecasting. Appl. Soft Comput. **42**, 368–376 (2016)
21. J.S. Jang, ANFIS: Adaptive-network-based fuzzy inference system. IEEE Trans. Syst. Man Cybern. **23**(3), 665–685 (1993)

22. M.A.A. Alqaness, A.A. Ewees, H. Fan, M.A.E. Aziz, Optimization method for forecasting confirmed cases of COVID-19 in China. J. Clin. Med. **9**(3), 674 (2020)
23. S.F. Ardabili, M.O.S.A.V.I. Amir, P. Ghamisi, F. Ferdinand, A.R. Varkonyi-Koczy, U. Reuter, T. Rabczuk, P.M. Atkinson, COVID-19 outbreak prediction with machine learning. Mathematics **8**, 1–138 (2020). https://doi.org/10.1101/2020.04.17.20070094
24. T. Bäck, *Evolutionary Algorithms In Theory And Practice: Evolution Strategies, Evolutionary Programming, Genetic Algorithms* (Oxford University Press Inc., USA, 1996).
25. K. Deb, *Multi-Objective Optimization Using Evolutionary Algorithms,* vol. 16. (John Wiley & Sons, 2001)
26. E. Zitzler, K. Deb, L. Thiele, Comparison of multiobjective evolutionary algorithms: empirical results. Evol. Comput. **8**(2), 173–195 (2000)
27. P. Ghamisi, J.A. Benediktsson, Feature selection based on hybridization of genetic algorithm and particle swarm optimization. IEEE Geosci. Remote Sens. Lett. **12**, 309–313 (2014)
28. S. Ardabili, B. Beszedes, L. Nadai, K. Szell, A. Mosavi, F. Imre, in *Comparative Analysis of Single and Hybrid Neuro-Fuzzy-Based Models for an Industrial Heating Ventilation and Air Conditioning Control System.* Proceedings of the International Conference on Computing and Communication Technologies (RIVF) (2020), pp. 1–6. https://doi.org/10.1109/RIVF48685.2020.9140753
29. S. Ardabili, A. Mosavi, A.R. Várkonyi-Kóczy, in *Advances in Machine Learning Modeling Reviewing Hybrid And Ensemble Methods.* International Conference on Global Research and Education. (Springer, Cham, 2020), pp. 215–227
30. H. Mühlenbein, M. Schomisch, J. Born, The parallel genetic algorithm as function optimizer. Parallel Comput. **17**, 619–632 (1991)
31. D. Whitley, A genetic algorithm tutorial. Stat. Comput. **4**, 65–85 (1994)
32. J. Horn, N. Nafpliotis, D.E. Goldberg, in *A Niched Pareto Genetic Algorithm for Multiobjective Optimization.* Proceedings of the First IEEE Conference on Evolutionary Computation. IEEE World Congress on Computational Intelligence, Orlando, FL, pp. 82–87 (1994)
33. C.R. Reeves, A genetic algorithm for flowshop sequencing. Comput. Oper. Res. **22**, 5–13 (1995)
34. G. Jones, P. Willett, R.C. Glen, A.R. Leach, R. Taylor, Development and validation of a genetic algorithm for flexible docking. J. Mol. Biol. **267**, 727–748 (1997)
35. S. Ardabili, A. Mosavi, A.R. Varkonyi-Koczy, in *Advances in Machine Learning Modeling Reviewing Hybrid And Ensemble Methods.* Engineering for Sustainable Future, 1–10. https://doi.org/10.1007/978-3-030-36841-8_21
36. C.R. Houck, J. Joines, M.G. Kay, A genetic algorithm for function optimization: a Matlab implementation. NCSUIE-TR-95–09. 95, pp. 1–10 (1995)
37. D. Whitley, T. Starkweather, C. Bogart, Genetic algorithms and neural networks: optimizing connections and connectivity. Parallel Comput. **14**, 347–361 (1990)
38. J. Kennedy, R. Eberhart, in *Particle Swarm Optimization.* Proceedings of Proceedings of ICNN 1995-International Conference on Neural Networks, pp. 1942–1948 (1995)
39. R. Poli, J. Kennedy, T. Blackwell, Particle swarm optimization. Swarm Intell. **1**, 33–57 (2007)
40. M. Clerc, Particle Swarm Optimization, vol. 93, (John Wiley & Sons, Hoboken, 2010)
41. P.J. Angeline, in *Using Selection to Improve Particle Swarm Optimization.* Proceedings of IEEE International Conference on Evolutionary Computation Proceedings. IEEE World Congress on Computational Intelligence, pp. 84–89 (1998)
42. J. Sun, B. Feng, W. Xu, in *Particle Swarm Optimization With Particles Having Quantum Behavior*. Proceedings of the Congress on Evolutionary Computation, pp. 325–331 (2004)
43. K.E. Parsopoulos, M.N. Vrahatis, Recent approaches to global optimization problems through particle swarm optimization. Nat. Comput. **1**, 235–306 (2010)
44. Q. Bai, Analysis of particle swarm optimization algorithm. Comput. Inform. Sci. **3**(1), 180–184 (2010)
45. K.E. Parsopoulos, M.N. Vrahatis, in Particle *Swarm Optimization Method in Multiobjective Problems.* Proceedings of the ACM symposium on Applied computing, pp. 603–607 (2002)
46. S. Lalmuanawma, J. Hussain, L. Chhakchhuak, Applications of machine learning and artificial intelligence for Cov-19 (SARS-CoV-2) pandemic: a review. Chaos, Solitons Fractals **139**, 1–6 (2020)

47. S. Agrebi, A. Larbi, Use of artificial intelligence in infectious diseases. Artif. Intell. Precis. Health, 415–438 (2020)
48. E.H. Shortliffe, Computer-based medical consultations. Mycin, 264 (1976)
49. N. Peiffer-Smadja, T.M. Rawson, R. Ahmad, A. Buchard, G. Pantelis, F.-X. Lescure, G. Birgand, A.H. Holmes, Machine learning for clinical decision support in infectious diseases: a narrative review of current applications. Clin. Microbiol. Infect. **26**(5), 584–595 (2020)
50. M.M. Barbat, C. Wesche, A.V. Werhli, M.M. Mata, An adaptive machine learning approach to improve automatic iceberg detection from SAR images. ISPRS J. Photogrammetry Remote Sens. **156**, 247–259 (2019)
51. H.-C. Li, G. Yang, W. Yang, Q. Du, W.J. Emery, Deep nonsmooth nonnegative matrix factorization network factorization network with semi-supervised learning for SAR image change detection. ISPRS J. Photogrammetry Remote Sens. **160**, 167–179 (2019)
52. R. Shang, L. Qi, L. Jiao, R. Stolkin, Y. Li, Change detection in SAR images by artificial immune multi-objective clustering. Eng. Appl. Artif. Intell. **31**, 53–67 (2014)
53. F. Gao, J. You, J. Wang, J. Sun, E. Yang, H. Zhou, A novel target detection method for SAR images based on shadow proposal and saliency analysis. Neurocomputing **267**, 220–231 (2017)
54. A. Colubri, M.-A. Hartley, M. Siakor, V. Wolfman, A. Felix, T. Sesay, J.G. Shaffer, R.F. Garry, D.S. Grant, A.C. Levine P.C. Sabeti, Machine-learning prognostic models from the 2014–16 Ebola Outbreak: data-harmonization challenges, validation strategies, and mHealth applications. EClin. Med. **11**, 54–64 (2019)
55. S. Choi, J. Lee, M.-G. Kang, H. Min, Y.-S. Chang, S. Yoon, Large-scale machine learning of media outlets for understanding public reactions to nation-wide viral infection outbreaks. Methods **129**, 50–59 (2017)
56. G. Nápoles, I. Grau, R. Bello, R. Grau, Two-steps learning of Fuzzy cognitive maps for prediction and knowledge discovery on the HIV-1 drug resistance. Expert Syst. Appl. **41**(3), 821–830 (2014)
57. U. Chockanathan, A.M. DSouza, A.Z. Abidin, G. Schifitto, A. Wismüller Automated diagnosis of HIV-associated neurocognitive disorders using large-scale Granger causality analysis of resting-state functional MRI. Comput. Biol. Med. **106**, 24–30 (2019)
58. M. Togaçar, B. Ergen, Z. Comert, COVID-19 detection using deep learning models to exploit social mimic optimization and structured chest X-ray images using fuzzy color and stacking approaches. Comput. Biol. Med. **121**, 1–12 (2020)
59. A.R. Vaka, B. Soni, Breast cancer detection by leveraging Machine Learning. ICT Express (2020) (In press)
60. S. Saxena, M. Gyanchandani, Machine learning methods for computer-aided breast cancer diagnosis using histopathology: a narrative review. J. Med. Imaging Radiat. Sci. **51**(1), 182–193 (2019)
61. T. Nazir, A. Irtaza, Z. Shabbir, A. Javed, U. Akram, M.T. Mahmood, Diabetic retinopathy detection through novel tetragonal local octa patterns and extreme learning machines. Artif. Intell. Med. **99**, 101695 (2019)
62. I. Kavakiotis, O. Tsave, A. Salifoglou, N. Maglaveras, I. Vlahavas, I. Chouvarda, Machine learning and data mining methods in diabetes research. Computat. Struct. Biotechnol. J. **15**, 104–116 (2017)
63. P. Sharma, K. Choudhary, K. Gupta, R. Chawla, D. Gupta, A. Sharma, Artificial plant optimization algorithm to detect heart rate & presence of heart disease using machine learning. Artif. Intell. Med. **102**, 1–14 (2019)
64. A. Karen Garate-Escamilla, A.H.E. Hassani, E. Andres, Classification models for heart disease prediction using feature selection and PCA. Inform. Med. Unlocked **19**, 1–11 (2020)
65. T. Liu, W. Fan, C. Wu, A hybrid machine learning approach to cerebral stroke prediction based on imbalanced medical dataset. Artif. Intell. Med. **101**, 101723 (2019)
66. C.M.J.M. Dourado Jr, S.P.P. da Silva, R.V.M. da Nóbrega, A.C. da S. Barros, P.P. R. Filho, V.H.C. de Albuquerque, Deep learning IoT system for online stroke detection in skull computed tomography images. Comput. Netw. **152**, 25–39 (2019)

67. R. Vaishya, M. Javaid, I.H. Khan, A. Haleem, Artificial Intelligence (AI) applications for COVID-19 pandemic. Diabetes Metab. Syndr. Clin. Res. Rev. **14**(4), 337–339 (2020)
68. T. Ai, Z. Yang, H. Hou, C. Zhan, C. Chen, W. Lv, Q. Tao, Z. Sun, L. Xia, Correlation of chest CT and RT-PCR testing in coronavirus disease 2019 (COVID-19) in China: a report of 1014 cases. Radiology **296**(2), 32–40 (2020)
69. A.A. Ardakani, A.R. Kanafi, U.R. Acharya, N. Khadem, A. Mohammadi, Application of deep learning technique to manage COVID-19 in routine clinical practice using CT images: results of 10 convolutional neural networks. Comput. Biol. Med. **121**, 1–9 (2020)
70. T. Ozturk, M. Talo, E.A. Yildirim, U.B. Baloglu, O. Yildirim, U. Rajendra Acharya, Automated detection of COVID-19 cases using deep neural networks with Xray images. Comput. Biol. Med. **121**, 1–11 (2020)
71. L. Sun, G. Liu, F. Song, N. Shi, F. Liu, S. Li, P. Li, W. Zhang, X. Jiang, Y. Zhang, L. Sun, X. Chen, Y. Shi, Combination of four clinical indicators predicts the severe/critical symptom of patients infected COVID-19. J. Clin. Virol. **128**, 1–6 (2020)
72. J. Wu, P. Zhang, L. Zhang, W. Meng, J. Li, C. Tong, Y. Li, J. Cai, Z. Yang, J. Zhu, J., M. Zhao, H. Huang, X. Xie, S. Li, Rapid and accurate identification of COVID-19 infection through machine learning based on clinical available blood test results. medRxiv 1–12 (2020). https://doi.org/10.1101/2020.04.02.20051136.
73. WHO: World Health Organization, 2020. Health Topic, Coronavirus disease overview. https://www.who.int/health-topics/coronavirus#tab=tab_1 Accessed 10 July 2020
74. WHO: World Health Organization, 2020. CT in the context of COVID-19. https://www.who.int/publications-detail/contact-tracingin-the-context-of-Covid-19 Accessed 10 July 2020
75. MIT: Covid Tracing Tracker - a flood of coronavirus apps are tracking us. Now it's time to keep track of them. https://www.technologyreview.com/2020/05/07/1000961/launching-mittr-covid-tracing-tracker/ Accessed 5 July 2020
76. C. Rorres, M. Romano, J.A. Miller, J.M. Mossey, T.H. Grubesic, D.E. Zellner, G. Smith, CT for the control of infectious disease epidemics: chronic wasting disease in deer farms. Epidemics **23**, 71–75 (2018)
77. BBC: Coronavirus: Israel enables emergency spy powers. https://www.bbc.com/news/technology-51930681 Accessed 14 July 2020
78. M.H.D.M. Ribeiro, R.G. da Silva, V.C. Mariani, L.D.S. Coelho, Short-term forecasting COVID-19 cumulative confirmed cases: perspectives for Brazil. Chaos, Solitons Fractals **135**, 1–10 (2020)
79. L. Yan, H.-T. Zhang, J. Goncalves, Y. Xiao, M. Wang, Y. Guo, C. Sun, X. Tang, L. Jing, M. Zhang, X. Huang, Y. Xiao, H. Cao, Y. Chen, T. Ren, F. Wang, Y. Xiao, S. Huang, X. Tan, N. Huang, B. Jiao, C. Cheng, Y. Zhang, A. Luo, L. Mombaerts, J. Jin, Z. Cao, S. Li, H. Xu, Y. Yuan, An interpretable mortality prediction model for COVID-19 patients. Nat. Mach. Intell. **2**, 283–288 (2020)
80. V.K.R. Chimmula, L. Zhang, Time series forecasting of COVID-19 transmission in Canada using LSTM networks. Chaos Solitons Fractals **135**, 1–6 (2020)
81. T. Chakraborty, I. Ghosh, Real-time forecasts and risk assessment of novel coronavirus (COVID-19) cases: a data-driven analysis. Chaos Solitons Fractals **135**, 1–10 (2020)

Early Prediction of COVID-19 Outcome: Contrasting Clinical Scores and Computational Intelligence Methods

Salvatore Greco, Alessandro Salatiello, Nicolò Fabbri, and Angelina Passaro

Abstract Triaging incoming patients is critical for an optimal allocation of hospital resources, especially during a pandemic, when these tend to be quickly depleted. A typical approach for predicting patients' outcomes relies on clinical scores such as the Charlson Comorbidity Index (CCI). CCI-based triaging is a reliable approach for estimating the mortality risk in the general patients' population. However, this score is not optimized for predicting mortality in specific populations such as the one represented by COVID inpatients, often the most represented population in the emergency department cohorts during the current pandemic. Motivated by this, this chapter describes the development of a new COVID-19-specific clinical score: The General Assessment of SARS-CoV-2 patients Score (GASS). The score builds on the clinical experience gained during the first phase of the pandemic, and it is based on both clinical and laboratory data. It was aimed at predicting the 30-day mortality outcome of hospitalized COVID-19 patients and showed markedly better accuracy than the CCI. Furthermore, this chapter introduces an additional predictive model based on a classical Computational Intelligence method. Specifically, it describes the development

S. Greco · A. Passaro (✉)
Department of Translational Medicine, University of Ferrara, Via Luigi Borsari 46, 44121 Ferrara, Italy
e-mail: psn@unife.it

S. Greco
e-mail: salvatore.greco@unife.it

A. Salatiello
Section for Computational Sensomotorics, Department of Cognitive Neurology, Centre for Integrative Neuroscience and Hertie Institute for Clinical Brain Research, University Clinic Tübingen, Otfried-Müller-Straße 25, 72076 Tübingen, Germany
e-mail: alessandro.salatiello@uni-tuebingen.de

N. Fabbri
Department of General Surgery, Ospedale del Delta, Via Valle Oppio 2, 44023 Lagosanto, Ferrara, Italy
e-mail: n.fabbri@ausl.fe.it

A. Passaro
Medical Department, University Hospital of Ferrara Arcispedale Sant'Anna, Via Aldo Moro 8, 44124 Cona, Ferrara, Italy

J. Nayak et al. (eds.), *Understanding COVID-19: The Role of Computational Intelligence*, Studies in Computational Intelligence 963, https://doi.org/10.1007/978-3-030-74761-9_18

and validation of a feedforward artificial Neural Network (NN) that automatically maps patients' clinical and laboratory data to a 30-day mortality-risk score. Critically, the NN-based method was shown to be more accurate at predicting 30-day mortality of COVID-19 patients than both the CCI and GASS scores. However, the intrinsic black-box nature of the NN-based method makes it hard to reach an intuitive understanding of the internal computations underlying its decision process. This might affect its general acceptance among clinicians, and lead them to prefer using the GASS score.

Keywords COVID-19 · Mortality prediction · GASS score · Neural network · LASSO score · Computational Intelligence

1 Introduction

The recent pandemic by SARS-CoV-2 (Severe Acute Respiratory Syndrome CoronaVirus 2), and its clinical expression, COVID-19 (CoronaVirus Disease 2019) have been afflicting the entire world since the beginning of 2020. If, on the one hand, the exponential increase in the number of daily hospitalizations has put on the ropes the entire healthcare system, on the other hand, this new pandemic has brought to light the potential of Computational Intelligence (CI) to radically change healthcare management.

SARS-CoV-2 is the 3rd member in the family of coronaviruses, after SARS-CoV and MERS-CoV (Middle East Respiratory Syndrome CoronaVirus) and its burst radically changed clinicians' approach to patients and contextually modified our behavior worldwide. COVID-19 spread rapidly worldwide and, as of December 2020, it has caused more than 66 million cases in 218 countries with more than 1.5 million deaths[1]; neither SARS-CoV, nor MERS-CoV could be compared with SARS-CoV-2 in terms of pathogenicity. If, on the one hand, SARS-CoV caused 8000 cases ca. with a global mortality rate of 10% in 2002, MERS-CoV caused 800 cases with a mortality rate of 38%[2]. Differences between the viruses were also found in the proportion of patients requiring hospitalization. Patients with SARS-CoV infection underwent hospitalization in more than 70% of cases against the 20% ca. of patients with SARS-CoV-2 infection; in 2009, the percentage of patients with pandemic influenza who underwent hospitalization was even smaller[3]. In January 2020, the World Health Organization (WHO) declared COVID-19 a sanitary emergence of international interest; in March of the same year, the WHO officially recognized the state of pandemic.

COVID-19, the clinical manifestation of SARS-CoV-2 infection, acted both as a booster, and as an amplifier of potential opportunities and concerns. The sanitary and economic consequences of the pandemic are now well known, and many other

---

[1] The Real Time Statistics Project, Worldometer, https://www.worldometers.info/.

[2] International Committee on Taxonomy of Viruses (ICTV), https://talk.ictvonline.org/.

[3] World Health Organization (WHO), https://www.who.int/.

negative aspects will be clearer in the next months or years. Among the opportunities, the pandemic has spurred the deployment of Machine Learning (ML) methods in the fight against its dire social, economic and clinical consequences; so much so that ML is now playing a key role. Figure 1 shows the exponential growth of the number of publications on coronaviruses recorded in response to the SARS-CoV-2 pandemic, together with the boost in the adoption of ML techniques in these works. This manifests the growing interest of the scientific community in using ML methods for basic and applied research on SARS-CoV-2, COVID-19 and their ramifications. ML is now part of the standard scientific toolbox within the medical community and its applications, nowadays, cover any aspect of clinical activities, from territorial medicine to hospital life.

The ever-increasing number of COVID-19 cases requires a fast development of effective treatments. The ability to automatically discover hidden patterns in large amount of data, and its potential to personalize treatments make ML a valid tool that can go a long way towards achieving such goals. Additionally, ML can be of great help to intercept positive cases among the general population and to assist in curbing social interactions before the infection takes place. Moreover, it can be potentially useful in diagnosing the infection, in educational campaigns against the diffusion

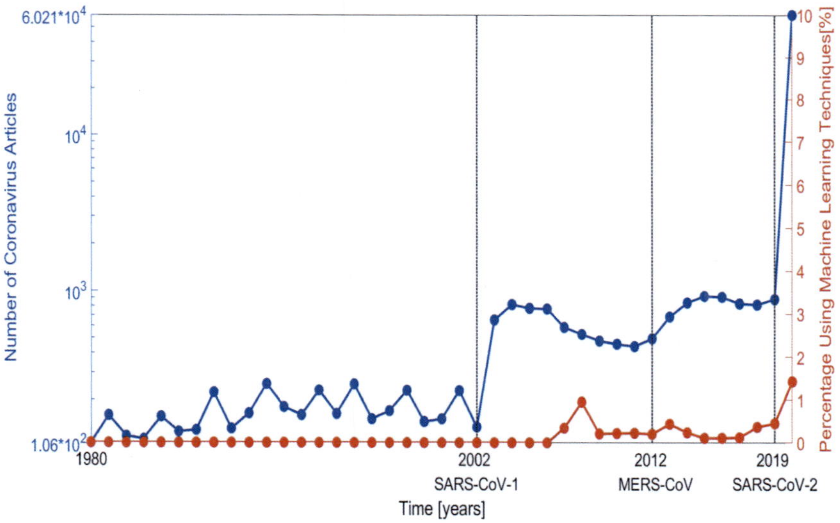

Fig. 1 Number of peer-reviewed publications on coronaviruses. The data were retrieved from PubMed on December 20, 2020. Coronavirus articles were retrieved using the query "coronavirus"; an estimate of the percentage of coronavirus articles adopting machine learning techniques was computed by selecting the publications that also mentioned one or more of the following keywords: "machine learning", "deep learning", "computational intelligence", and "artificial intelligence". The increase in coronavirus publications in response to SARS-CoV-2 (6711%) is more than an order of magnitude greater than that caused by SARS-CoV-1 (386.6%) and more than two orders of magnitude greater than that caused by MERS-CoV (37.9%). Almost 1.5% of coronavirus articles published in 2020 adopted ML-based methods; this represents a 233% increase with respect to the previous year (0.45%, 2019), and a 60% increase with respect to the previous peak (0.95%, 2008)

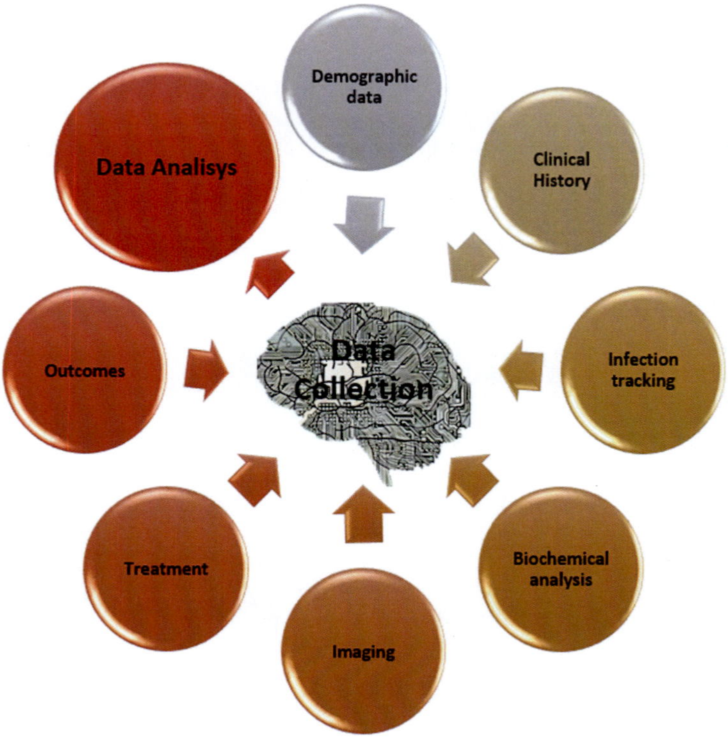

Fig. 2 Data entry flow and data analysis

of the virus, and in the development of new drugs. The areas of application of ML methods can be summarized as follows:

- Early detection and diagnosis
- Monitoring treatment
- Contacts tracing
- Projecting cases and mortality
- Developing treatments
- Reducing health care associated workload
- Identifying ways to prevent the disease

The laboratory infection diagnosis is usually performed with a pharyngeal swab and with virus-specific RNA detection and amplification with real time polymerase chain reaction assays (RT-PCR). Imaging and chest HRCTs can be useful for evaluating the interstitial lung involvement by COVID-19 even if, in many cases, the CT findings can be absolutely normal and the exam cannot help exclude the infection [1]. In this respect, the clinical symptoms of disease should be combined with clinical history of patients and with radiographic images to assess a quick treatment and to choose the inpatient best setting. Figure 2 summarizes the processes of data collection and data analysis in clinical settings.

Predictive models for the assessment of COVID patients were recently proposed worldwide, ranging from rule-based scoring systems to innovative machine learning models (e.g., deep learning). Having reliable predictive models during pandemics, in general, would help clinicians not only in their daily routine activities, but also in their first approach to patients. Triaging incoming patients is, in fact, of primary importance for an optimal allocation of hospital resources and this is especially critical during a pandemic, when these tend to be quickly depleted. A typical approach for predicting patients' outcomes relies on clinical scores such as the Charlson Comorbidity Index (CCI). Triaging patients with scores such as the CCI could be a reliable approach for estimating the mortality risk in the general patients' population. However, this score was not thought to be optimized for the prediction of the mortality risk of specific populations such as that represented by COVID inpatients.

This chapter will describe the development of two COVID-specific scores for the prediction of 30-day mortality. The first score is based on "on-the-ground" clinical experience, while the second score is based on a classical Computational Intelligence technique. It will be shown that even though the two scores were built using starkly different methods, they were both more reliable to predict 30-day mortality than the CCI index. Finally, this chapter will elaborate on the pros and cons of the proposed methods.

2 Related Work

A recent systematic review by Wynants et al. [2] analyzed 14,217 titles about prediction models: the review evaluated models for diagnosing SARS-CoV-2 infection in patients suspected to have it and for the prognosis of patients with the infection, but it also involved algorithms able to detect people in the general population at increased risk of SARS-CoV-2 infection or of admission to hospital.

The review recognized four different models able to identify people at higher risk of infection in the population, 91 models were found to be useful for detecting COVID-19, and 50 more models were found to be good at predicting disease outcomes (mortality, progression to the severe form of disease, ICU admission, ventilation, intubation, and length of stay). Many items were used for diagnosing the infection, such as age, body temperature and lymphocytes count; many others were used for predicting the prognosis of COVID-19 such as C reactive protein (CRP), creatinine blood levels, comorbidities or gender. It is important to highlight how critical the rapid detection of patients with the infection is, principally in the consideration of the false negativity of diagnostic tools such as the pharyngeal swabs [3] or the serological specific tests [4]. False-negative test could, in fact, delay treatment and increase the risk of viral transmission to others. In addition to this, not every patient can have rapid access to radiological examinations, making it essential to use predicting scores that could be reliable in diagnosing or in excluding disease, even without chest Rx or CT scans. Healthcare delivery requires the help of modern technologies belonging to the world of Artificial Intelligence to fight and plan against

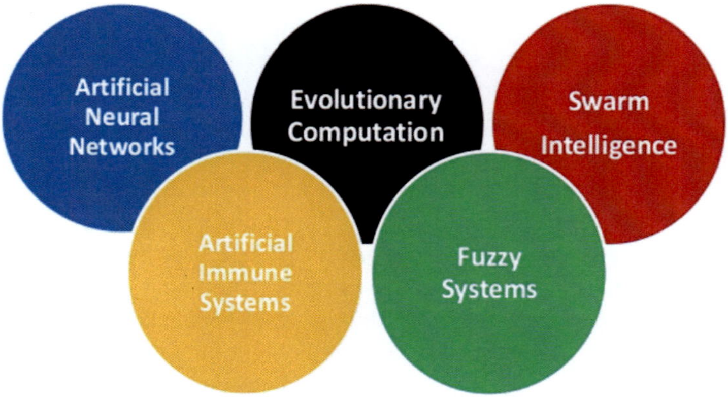

Fig. 3 Computational Intelligence: taxonomy proposed by Engelbrecht [25]

any kind of new illnesses. AI can quickly analyze patients' symptoms and signs, helping clinicians in their daily practice, from the time of viral detection to that of treatment (Fig. 3).

One of the most debated questions about SARS-CoV-2 concerns its origin, which has still to be established, even if some theories about it were already proposed in the literature. It is improbable, according to experts, that SARS-CoV-2 is related to a sort of laboratory handling of a related coronavirus: genetic observations unquestionably confirm, in fact, that SARS-CoV-2 did not result from any previously managed virus backbone [5]. It was hypothesized that SARS-CoV-2 could somehow derive from a natural selection in animal hosts (before a zoonotic transfer) or a natural selection in humans (following the zoonotic transfer).Computational intelligence has also found use in this area, helping researchers identify the most plausible origins of the virus, as well explained in the work by Nguyen et al. [6]. The results obtained from AI-enabled experiments that used clustering algorithms demonstrated that the observed SARS-CoV-2 viral genomes belonged to a cluster that included bats' and pangolins' coronavirus genomes.

This study, together with others, provided sufficient evidence that animals (such as bats and pangolins) were credible hosts for SARS-CoV-2. Many other aspects of disease are still unknown and AI and CI in particular, could help researchers in explaining them. After the diagnosis is made, it is needed the choice of the therapy that can ensure patients a better prognosis. At the beginning of the pandemic, different drugs were used for COVID-19 management, ranging from antibiotics to antivirals, to antimalarial drugs such as hydroxychloroquine or dexamethasone. To date, the only treatment demonstrating a lower mortality and, anyhow, only among patients with respiratory support, was dexamethasone, as summarized in the recent RECOVERY trial [7]. As a specific peculiarity, SARS-CoV-2 infection is known for inducing a pro-thrombotic state [8]: coagulopathy linked to COVID-19 can be clarified by the "two-way activation" theory, as already showed in the case of thrombocytopenia in critically ill patients (TICP) and the surrounding pro-inflammatory and thrombogenic response taking place when the endothelial damage occurs [9]. This reaction, then, is worsened by the boosted platelet activation and consumption thrombocytopenia. Differently from the characteristic coagulopathy and disseminated intravascular coagulation (DIC) profiles often seen during sepsis, patients with COVID-19 generally present unmodified coagulative and platelet profiles.

For all these reasons, treatments with low molecular weight heparins (LMWH) are often necessary to avoid thromboembolic events in patients with elevated serum levels of D-dimer and/or in a condition of reduced mobilization. Other treatments such as convalescent "hyperimmune" plasma and antivirals (such as remdesivir) have not demonstrated to be effective yet on reducing short-term mortality in hospitalized patients with COVID-19 [10, 11], even if other study protocols are currently being put in place to better evaluate the effects of these drugs on the virus.

Another point of discussion is represented by the lasting consequences of COVID-19: even if long-term sequelae of COVID-19 are still unknown, it is conceivable that some kind of chronic consequences can be encountered by patients who experienced the disease. This was thought on the base of past findings during SARS-CoV outbreak, as reported by Lam et al. in 2009 and by Ngai et al. in 2010 [12, 13] while many studies, based on radiology findings at different stages of illness and after patient's recovery, are being performed.

Last but not least, it is not to be underestimated the social impact of COVID-19 on patients: apart from the need for isolation by patients themselves, it is now clear how all their caregivers participate (not only emotionally) in facing this disease. For this, and for all other reasons described, much effort is dedicated not only to cares but also to prevention. COVID patients can meet different disease outcomes during the infection period and they strongly depend on clinical presentation, on patients' history and on how rapidly the first care is put in place.

Most of the studies focus on patients' risk of in-hospital mortality, on their mortality rate within an established period of time, on the need for intensifying cares (e.g., the choice of different settings of care such as the Intensive Care Units). In case of clinical recovery (meant before as time since the first positive swab detection until two consecutively negative and then as the time until one only negative swab and contextual absence of symptoms for at least 72 h), some studies calculated the mean time until negativized pharyngeal swabs to viral RNA detection that was about 28 days [14].

Another interesting aspect of disease regards the different responses by men and women: in particular, one Japanese study recently published by Takahashi et al. [15] reported higher plasmatic levels of cytokines such as IL-8 and IL-18 in male patients with higher induction of non-classical monocytes. Conversely, women had more robust T-cells activation. They found that a deprived T-cell response was inversely associated with patients' age and was contextually linked to worse disease outcomes in men, while this did not happen in women. Higher plasmatic levels of innate immune cytokines were associated, instead, with more severe illness in women, while it did not happen in men. It is still unclear why a gender difference in the immune response towards COVID-19 occurs, nor is it clear whether such difference can be overcome by a specific medical treatment. What is clear is that there is still a long way to go; using ML for creating prediction models on COVID-19 outcomes can surely help avoid negative disease outcomes and reduce patients' duration of in-hospital stay and hospitalization related costs.

3 Methods

This section details the methods adopted in this study. Specifically, Sect. 3.1 describes the data collection procedure; Sect. 3.2 describes a novel COVID-specific clinical score, named GASS (General Assessment of SARS-CoV-2 patients Score), introduced in this work for predicting 30-day mortality in COVID patients. Section 3.3 summarizes the phases taking to the development of a novel Computational Intelligence based model for the prediction of 30-day mortality. Finally, Sect. 3.4 compares the models' performance. The figure below (Fig. 4) provides a summary of the methods adopted in this study.

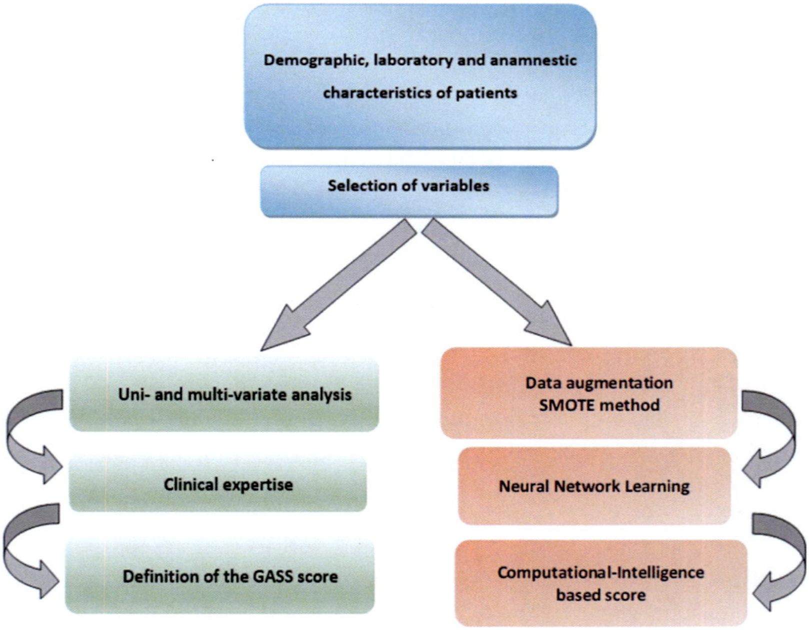

Fig. 4 Study design. Phases of definition of the clinical and CI-based score

3.1 Data Collection

The study discussed in this chapter took place in Italy, between March and June 2020. It was developed in two hospitals of Ferrara's territory; both were partially dedicated to COVID inpatients during the first phase of the pandemic. Ferrara is a province located in the eastern part of the Emilia-Romagna region of Italy, with a total population of approximately 350,000 inhabitants.

Italian experience with SARS-CoV-2 infections began in March: Italy was the first European Country to face a real sanitary emergency, after that the virus had already spread in Asia and in China, in particular.

During the period from the 13th of March to the 15th of June 2020, data from 499 patients, consecutively hospitalized into the two Ferrara's hospitals dedicated to COVID patients ("S. Anna" Hospital in Cona and "Ospedale Del Delta" in Lagosanto), were collected. The "S. Anna" Hospital approached the Coronavirus outbreak with a series of organizational measures. A specific, COVID-dedicated pathway was adopted, including an emergency department triage area, an infectious disease and a "suspect" observational unit, three internal medicine units, one pulmonology unit, and one intensive care unit.

The internal medicine departments in the "Ospedale del Delta" were converted to COVID departments and it was instituted a smaller medical department dedicated to the patients who required long-term cares.

All patients were diagnosed with SARS-CoV-2 infection: this was confirmed with a laboratory viral RNA detection at oro- and naso-pharyngeal swab. A period of two months ca. beyond the expected data was necessary to wait until full clinical recovery (during those months the chosen criterion was the double consecutive negativity of pharyngeal swabs to SARS- CoV-2 RNA detection), or to a different disease outcome (intensification of care, in-hospital mortality, 30-day mortality).

3.2 Clinical-Score-Based Prediction Model

For every patient, at hospital admission, it was calculated the Charlson Comorbidity Index (CCI), a well-known internationally validated score, commonly used in medical practice [16]; demographic and clinical data of the patients were reported electronically, while subject identifiers were removed before analyzing data, in order to maintain patients' anonymity. As necessary, informed consent was obtained after the nature and the possible consequences of the study had been fully explained.

Data analyses were performed by using SPSS 26.0 (IBM SPSS Statistics, IBM Corporation) software. A descriptive analysis of the population was carried out and results were expressed as absolute numbers, percentages, and mean $\pm$ SD. Data were evaluated retrospectively, paying attention to demographic, laboratory and anamnestic characteristics of patients. This was particularly useful for better understanding which variables were peculiar in COVID patients and which of them could somehow modify their prognosis.

The univariate analyses defined the differences between groups of survivors and deceased subjects and between patients who needed an intensification of care (meant as admission to Pulmonology departments or Intensive Care Units) and patients who did not; Chi-square test, Student t-test, and Mann–Whitney test were used as appropriate.

On the base of multivariate analyses, that showed strong relations with some of the variables studied (data not shown), it was developed a clinical score, the GASS Score (General Assessment of SARS-CoV-2 patients Score), hypothesizing its possible role in predicting COVID-19 outcomes (intensification of care, 30-day mortality, in-hospital mortality).

In Table 1 are summarized the descriptive analyses and the univariate analyses concerning the age of patients, their Charlson Comorbidity Index and the GASS score. All p values < 0.05 are considered statistically significant.

3.2.1 Choice of Variables

Age, gender (men, in the population described, needed more frequently an intensification of care), respiratory rate (RR), P/F ratio, lymphocytes count, XDP (D-dimer), a history of chronic kidney disease (CKI), evaluated glomerular filtration rate (eGFR), troponin I, brain natriuretic peptide (BNP) and C reactive protein (RCP) are the variables selected as determinant of the GASS score. The selection is based on expert opinion. Remarkably, all chosen variables are easily measured in a routine physical examination complemented with the collection and evaluation of venous and arterial blood samples.

3.2.2 Weighting the Variables

The variables were weighted according to their importance in modifying the chosen outcomes and it was assigned a different score for each of them.

Much attention was given to the age of patients, that usually modifies their prognosis and clinicians' approach (it was assigned a score ranging from 0 points for patients younger than 45 years to 4 points for very old patients, > 90 years); the same was made with PaO2/FiO2 ratio (P/F), one of the parameters collectible from arterial blood sampling that usually allows to make diagnosis of Acute Respiratory Distress Syndrome (ARDS): normal values of P/F did not give any adding point to the GASS score, while a value of P/F lower than 100 meant 3 points. Equally important was the contribute of eGFR (evaluated Glomerular Filtration Rate), whose scoring ranged from 0 (eGFR > 60 ml/min) to 3 points (eGFR < 30 ml/min). The minimum achievable GASS score was 0, while the maximum was 23 points. Table 2 summarizes the items of the GASS score.

Table 1 Patient demographics. The table reports average and standard deviation of the variables Age, Charlson Comorbidity Index (CCI) and GASS, of the whole dataset, and of subgroups defined by selected outcome measures (IoC: Intensification of Care, IHD: In-hospital Death, 30dD: 30-day death). The table also reports the *p*-values of two-sample t-tests for the difference of group means

	All (n = 499)	Range	IoC		IhD		30dD		IoC: Y vs. N (*p*-value)	IhD: Y vs. N (*p*-value)	30dD: Y vs. N (*p*-value)
			Yes (n = 129)	No (n = 370)	Yes (n = 126)	No (n = 373)	Yes (n = 125)	No (n = 374)			
Age (years)	72 ± 17	18–99	68 ± 12	73 ± 18	82 ± 10	68 ± 17	81.7 ± 10.3	68.2 ± 17.1	0.02	<0.001	<0.001
CCI (points)	2.0 ± 2.3	0–12	1.7 ± 2.4	2.1 ± 2.3	3.1 ± 2.3	1.7 ± 2.2	3.2 ± 2.4	1.7 ± 2.2	0.003	<0.001	<0.001
GASS (points)	7.0 ± 3.6	0–18	7.8 ± 3.1	6.8 ± 3.7	9.8 ± 3.3	6.1 ± 3.2	10.0 ± 3.3	6.1 ± 3.1	0.05	<0.001	<0.001

Table 2 GASS Score items. RR: respiratory rate; XDP: D-dimer; CKI: Chronic Kidney Injury; eGFR: Estimated glomerular filtration rate; BNP: Brain natriuretic peptide; CRP: C Reactive Protein

	0	1	2	3	4
Age (ys.)	< 45	45–60	61–75	75–90	> 90
Gender	Female	Male			
RR (apm)	< 20	20–30	> 30		
P/F	> 300	200–300	100–199	< 100	
Lymphocytes count (n/mmc)	≥ 1000	500–999	< 500		
XDP (mcg/ml)	< 0.5	0.5–2.0	> 2.0		
CKI	Moderate	Severe			
eGFR (ml/min)	> 60	45–60	30–44	< 30	
Troponin (ng/ml)	< 20	20–50	> 50		
BNP (pg/ml)	< 100	≥ 100			
CRP (mg/dl)	0–4.9	5.0–10.0	> 10.0		

The GASS score was considered for any patient and applied retrospectively to the whole population. After this first step, the population was further divided into three sub-groups according to the score calculated for each patient. Subjects with a GASS score ranging from 0 to 5 points belonged to the first group, while in the second one there were patients with GASS score of 6–10; the last group included patients with GASS scores > 10.

3.2.3 Statistical Analyses

Significant differences between groups were found for age, CCI and GASS Score: patients who underwent intensification of care were younger than patients who did not (68 ± 12 vs. 73 ± 18 years) and this can be explained by the eligibility for admission to a Pulmonology department or to the Intensive Care Unit (ICU): younger patients with fewer comorbidities are, in fact, usually more often candidate for admission to these medical departments because of their higher probability of a better prognosis. On the contrary, as it happened in our hospitals, older and comorbid patients, theoretically considered at higher risk of worse short-term prognosis, are usually treated in the internal medicine departments with the best therapy available at the time of admission.

As for in-hospital and 30-day mortality, strong differences were found for both CCI and GASS Score: in particular, as expected, patients who died within the 30th day since hospital admission or during the whole hospitalization had greater CCI and GASS score.

As for CCI, the mean scores of patients in the groups of in-hospital mortality vs. in-hospital survival were 3.1 ± 2.3 vs. 1.7 ± 2.2 points, respectively. Similar mean scores were calculated for the groups of patients who died within 30 days since hospital admission vs. patients who did not (3.2 ± 2.4 vs. 1.7 ± 2.2 points, respectively). All these differences were statistically significant (p value < 0.001).

This is to underline how the CCI score is somehow reliable in differentiating patients who will experience a worse prognosis from patients who will not.

Mean GASS scores were also different between groups with 9.8 ± 3.3 vs. 6.1 ± 3.2 points in the group of patients were in-hospital death was considered and 10.0 ± 3.3 vs. 6.1 ± 3.1 points in the group considering 30-day mortality. Statistical analyses showed that the three groups of patients with different GASS scores presented significant differences. In particular, the percentage of patients who needed an intensification of care was significantly lower than that of patients who did not need intensification of care in the group with GASS score 0–5 (24% vs. 38.4%, $p = 0.003$), and in the group with score 6–10 (56.6% vs. 46.2%, $p = 0.04$).

Significant differences were also found between the patients of the first group (GASS score 0–5) in terms of in-hospital mortality (9.5% vs. 43.4%, $p < 0.001$) and 30-day mortality (9.6% vs. 42.8%, $p < 0.001$). Opposite results, but similarly significant, were registered as for the patients of the third group (GASS score > 10) in terms of mortality (in-hospital and 30-day), as shown in Table 3.

A Chi-Square test confirmed that there the three groups of patients who presented different GASS scores were significantly different from each other: this happened for the need for intensification of care ($p = 0.01$), in-hospital mortality and 30-day mortality ($p < 0.001$ for both, data not shown).

The calculated sensitivity and specificity of the GASS score, measured using the "optimal" ROC threshold [19] were 42.11% and 85.92%, respectively.

These data, taken together, demonstrate how using a prediction score can effectively help clinicians make early decisions about patients. This has been happening in daily medical activities for decades: one of the best examples in this specific field is the Wells score, commonly used for predicting thrombosis in patients admitted to hospitals, or the SOFA score, for predicting probabilities of patients undergoing a septic state in case of infection.

3.3 Computational-Intelligence-Based Prediction Model

In addition to defining and evaluating a score based on medical expertise (i.e., the GASS score), this chapter introduces also a novel 30-day mortality risk prediction model based on a classical Computational Intelligence method, namely, a feedforward artificial neural network. As a baseline, this chapter also introduces a simpler linear prediction model based on LASSO logistic regression. All the analyses reported in this section were performed in MATLAB.

Table 3 Patient outcomes grouped by GASS score. The table also reports the *p*-values of Chi-Square tests for the associations between GASS score and outcome variables, for each risk class

GASS Score	Intensification of care (IoC)		In-hospital death (IhD)		30-day death (30dD)		IoC-GASS (*p*-value)	IhD-GASS (*p*-value)	30dD-GASS (*p*-value)
	Yes (n = 129)	No (n = 370)	Yes (n = 126)	No (n = 373)	Yes (n = 125)	No (n = 374)			
0–5	31 (24%)	142 (38.4%)	12 (9.5%)	162 (43.4%)	12 (9.6%)	160 (42.8%)	0.003	< 0.001	< 0.001
6–10	73 (56.6%)	171 (46.2%)	61 (48.4%)	182 (48.8%)	61 (48.8%)	185 (49.5%)	0.04	0.94	0.92
> 10	25 (19.4%)	57 (15.4%)	53 (42.1%)	29 (7.8%)	52 (41.6%)	29 (7.7%)	0.29	< 0.001	< 0.001

3.3.1 Data Consolidation

The success of machine learning methods might be negatively impacted by irregularities in the data. For this reason, before training any algorithm it is of primary importance to perform data consolidation. In this study, this procedure was applied to the complete database of health records composed of 499 patients and 151 variables. The first step was the removal of the variables that were not consistently measured in at least 95% of the patients and those that were not valid potential predictors (e.g., outcome variables, patient's metadata, etc.). This procedure took to a total of 30 potential predictors.

The data were further consolidated by selecting the 450 patients (89.64%) for which all the preselected variables were measured. Finally, the data records were sorted by hospitalization date and the first 360 patients (80%) were used to learn the prediction model and the remaining 90 patients (20%) as an independent test to estimate the generalization ability of the model. All continuous features were z-scored.

3.3.2 Data Augmentation

Because of the under-representation of the 30-day mortality class (24.3% vs. 75.7%) and the overall small size of the dataset, it is advisable to adopt a data augmentation technique. This ensures that the prediction model is not biased towards predicting the majority class. Specifically, in this study, this was accomplished by using the Synthetic Minority Over-sampling TEchnique (SMOTE) method [17]. The use of this technique led to a balanced dataset by generating artificial instances of patients belonging to the minority class.

3.3.3 Feedforward Neural Network Model

Artificial Neural Network models (NN) are among the most popular Computational Intelligence techniques. Their success stems from their favorable computational properties that endow them with the ability to approximate any measurable function to any desired degree of accuracy [18]. This motivated the application of this model to find a suitable solution to the problem introduced in this chapter, which boils down to finding a suitable function that maps patients features to 30-day mortality outcome.

To find a robust mapping that generalizes to independent test data, it is fundamental to optimize both the network architecture and the main training parameters. In this study, this was accomplished by performing an extensive hyperparameter search over all the combination of number of hidden units (h = 40, 80, …, 1280), number of layers (l = 1, 2, 10), and regularization strength ($\lambda = 0, 10^{-7}, 10^{-5}, 10^{-3}, 10^{-1}, 5*10^{-1}$). To avoid that the network models would overfit the data, a tenfold cross-validation scheme on the training data was adopted. All the networks were

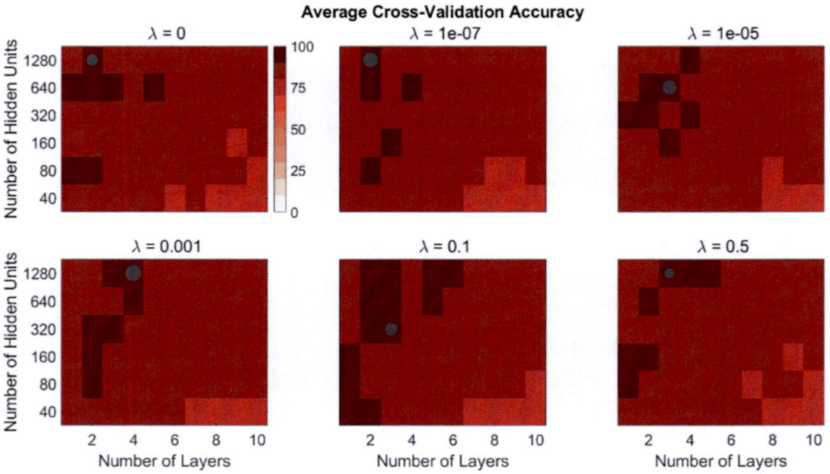

Fig. 5 Average tenfold cross-validation accuracy of the trained neural network models. Results are grouped in panels by the amount of regularization, determined by the parameter λ. Pixels are color-coded by accuracy: white indicates 0% accuracy, dark red indicates 100% accuracy. Gray dots denote the best classifiers for each λ. Dot size indicates the overall ranking of the classifier. The highest average cross-validation accuracy (92.5%) is reached by the neural network with 4 layers and 1280 hidden units, with $\lambda = 10^{-3}$ (bottom left panel)

trained using scaled conjugate gradient backpropagation [19]; training was automatically stopped whenever the fold validation error increased in 6 consecutive training iterations; cross entropy was the chosen cost function.

The results of this hyperparameter optimization procedure are reported in Fig. 5, which shows the average cross-validation accuracy for all the trained networks. This procedure selected a network architecture with 1280 hidden units distributed across 4 layers, with $\lambda = 10^{-3}$ (bottom left panel). This network achieved an average cross validation accuracy of 92.5%.

The selection of the final neural network model was based on a further optimization procedure, which aimed at finding the network with the highest validation accuracy for a random split of the training data and initialization of weights and bias parameters. This procedure selected a network with 93.06% validation accuracy. The corresponding training accuracy and test accuracy are 94.31% and 81.11%, respectively. The network's performance is further summarized in terms of area under the Receiver Operating Characteristic curve (ROC), sensitivity and specificity in Fig. 6.

3.3.4 LASSO Logistic Regression Model

As a baseline model, this work considered the linear logistic regression model optimized with LASSO regularization [20]. This method has the potential to avoid overfitting the training data and thus ensure a reasonable degree of generalization,

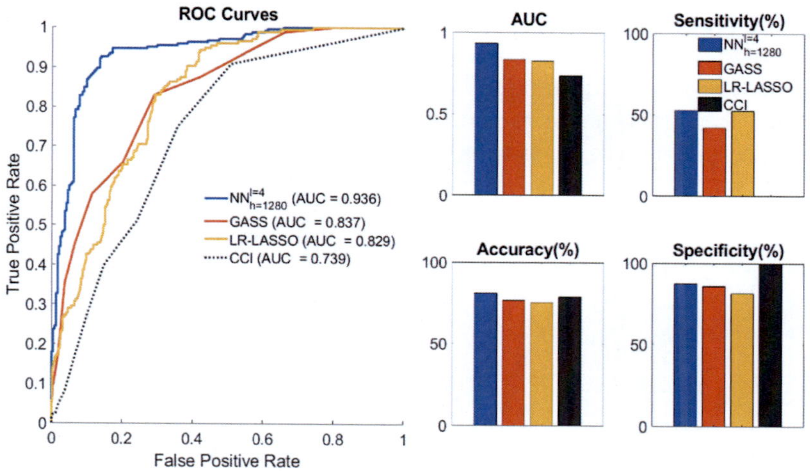

Fig. 6 ROC curves and summary statistics. NN: Feedforward Artificial Neural Network Model. LR-LASSO: LASSO Logistic Regression model; GASS: is the score based on physicians' expertise introduced in this work; CCI: Charlson Comorbidity Index, is a popular comorbidity index used to triage patients. Note that the sensitivity of the CCI score is 0%

by discarding redundant predictors. The parameter lambda, which determines the amount of regularization, was chosen by minimizing the tenfold cross-validation error using the MATLAB function lassoglm().

Out of the 30 preselected variables, this analysis selected only those that were better able to discriminate between 30-day mortality and 30-day survival when used within a linear multivariate model. The selected features are: ASL_{Inf}, *CKD*, *DEM*, *Cre* and *Age*. ASL_{Inf} is equal to 1 when the infection happened in an assisted living facility. *CKD* is equal to 1 if the patient had chronic kidney disease on hospital admission. *DEM* is equal to 1 if the patient has dementia. *Cre* and *Age* are continuous variables measuring the serum concentration of creatinine and the patient's age, respectively.

The variables selected with the previous method were used to fit a multivariate logistic regression model to the entire training data using the MATLAB function glmfit(). After fitting the model, it was computed the optimal operating point of the ROC curve [21] assuming identical false positive and false negative costs, and it was used as a discrimination threshold. The resulting logistic regression model is determined by the following Eq. (1), where p is the probability of dying within 30 days of hospitalization:

$$\begin{aligned} \text{In}\left(\frac{p}{1-p}\right) &= a_0 + a_1 * ASL_{Inf} + a_2 * CDK \\ &+ a_3 * DEM + a_4 * Cre + a_5 * Age \end{aligned} \quad (1)$$

This model was able to accurately predict the 30-day mortality both in the training data results (AUC = 0.8287, accuracy = 79.72%, sensitivity = 26.7%, specificity

= 96.88%), and, critically also in the independent test data (accuracy = 75.56%, sensitivity = 52.63%, specificity = 81.69%); these results are reported in Fig. 6.

In conclusion, in order to better evaluate the contribution of the variables present in the final model, the logistic regression model was fitted to the whole dataset (training data and test data). The final model, complete with the estimated coefficients is given by the Eq. (2):

$$\text{In}\left(\frac{p}{1-p}\right) = -6.11 + 0.488 * ASL_{Inf} + 0.315 * CDK$$
$$+0.77 * DEM + 0.215 * Cre + 0.055 * Age \quad (2)$$

This final analysis revealed that the model was able to explain a significantly large fraction of the data variance ($X^2 = 99.8$, p = $5.91*10^{-20}$) and that the variables *DEM* (t = 2.65, p = $8.12*10^{-3}$), *Age* (t = 4.88, p = $1.07*10^{-6}$), *and Cre* (t = 2.65, p = $2.23*10^{-3}$) can be considered as significant predictors of 30-day mortality.

Interestingly, the analysis of the odds ratios suggested that having dementia increases the odds of dying within 30 days of admission by over 200% (OR = 2.16); likewise, a unitary increase in serum creatinine concentration increased the odds of dying by 24% (OR = 1.24). Finally, a unitary increase in age increases the odds of dying by 0.06% (OR = 1.056).

3.4 Model Comparison

Figure 6 summarizes the prediction performance of all the predictive models introduced in this chapter. ROC curves and AUC measures are computed on the training data; accuracy, sensitivity and specificity are instead computed on the independent test set of 90 patients. The CCI index performed the worse: despite having a satisfying test accuracy (78.89%), its low sensitivity (0%) makes the score inadequate. Overall, the logistic regression model displayed prediction performance comparable to that of the GASS score, with the former being more sensitive (52.63% vs. 42.11%) and the latter more specific (81.69% vs. 85.92%).

Both models thus performed markedly better than the CCI. Finally, the neural network model was both more sensitive (52.94%) and specific (87.67%) and thus was generally more accurate (81.11%) than either the logistic regression model (75.56%) or the GASS score (76.67%).

4 Discussion

Triaging COVID-19 patients is fundamental to offer adequate clinical assistance under the constraint of the available hospital resources. In this work, it was shown that popular clinical scores such as the Charlson Comorbitidy Index (CCI) are not able to accurately predict the 30-day mortality of COVID patients. This suggests that the score cannot be used for triaging appropriately incoming COVID patients. Based on this finding, this chapter proposed a novel, COVID-specific, risk score based on on-the-ground clinical experience: the GASS score. The score is highly interpretable and was shown to be able to better predict 30-day mortality as compared to CCI. Additionally, this chapter has introduced a novel predictive model based on a classical Computational Intelligence (CI) method: a feedforward artificial Neural Network (NN). The NN showed better classification performance than either the CCI or the GASS scores.

One of the main strengths of a clinical score such as the GASS is its high interpretability and the possibility of being easily computed by clinicians with the data collected right after the patient's admission to the hospital. Conversely, such scores are often poorly specific (with a good sensitivity) or poorly sensitive (with a good specificity) because of the nature of clinical scores themselves.

The main strengths of the Computational-Intelligence-based predictive model introduced in this work are its high prediction accuracy and its proven reliability on independent test data. This method was, in fact, demonstrated to be accurate at predicting 30-day mortality outcome of even completely independent data, which were not used during its training. This suggests that the process can be quickly applied in triaging COVID patients to improve their chances of survival, contextually allocating in the best way hospital resources. Critically, a similar validation scheme would not have been possible for the GASS score, as this was built by using physicians' intuition, built leveraging on both on-the-ground first-hand experience with COVID patients and a critical analysis of all the available data. Prospective studies are already ongoing to deal with this current limitation.

On the other hand, the adoption of prediction scores based on Computational Intelligence techniques such as the one introduced here might be hampered by the difficulty in interpreting the rationale behind the classifications performed by the CI machine. Clinicians might in fact have little trust in a risk predicted by a CI machine that does not provide an intuitive explanation of its decision process; this might lead them to prefer adopting a highly-interpretable score such as the GASS, despite its lower accuracy.

However, the problem of interpreting the internal computations of CI methods such as ANNs is not unique to the clinical setting: it is a long-standing machine learning problem and is the subject of a great deal of promising research. For example, [22] proposes a method to generate, for each target class, the network input that leads to the maximum class score. The application of this method to the ANN presented in this work might reveal the vectors of patient attributes that lead to the highest and lowest 30-day mortality risk scores. This might lead to the identification of the most critical attributes for the network predictions and might therefore shed some light on the network's decision process. Alternative influential approaches are [23] and [24]. In [23], the authors go one step further and present a method to clarify the internal representation of the network's hidden units, based on a deconvolution network. Finally, in [24], the authors develop a powerful visualization tool based on intuitions gained by analyzing linear neural networks; the tool allows parsing the inputs into the sum of a signal and a distractor factor and is able to identify how much individual signal components contribute to the output class.

In conclusion, this chapter showed an example of the great potential of Computational Intelligence techniques in the fight against COVID-19. Events such as COVID-19 do not just represent a tremendous challenge to deal with: they also offer researchers the chance to develop cutting-edge tools that enable optimal patient care and allocation of hospital resources.

Acknowledgements The authors thank the International Max Planck Research School for Intelligent Systems (IMPRS-IS) for supporting Alessandro Salatiello.

References

1. M. Chung, A. Bernheim, X. Mei, N. Zhang, M. Huang, X. Zeng, J. Cui, W. Xu, Y. Yang, Z.A. Fayad, A. Jacobi, K. Li, S. Li, H. Shan, CT imaging features of 2019 novel coronavirus (2019-NCoV). Radiology (2020). https://doi.org/10.1148/radiol.2020200230
2. L. Wynants, B. Van Calster, G.S. Collins, R.D. Riley, G. Heinze, E. Schuit, M.M.J. Bonten, J.A.A. Damen, T.P.A. Debray, M. De Vos, P. Dhiman, M.C. Haller, M.O. Harhay, L. Henckaerts, N. Kreuzberger, A. Lohmann, K. Luijken, J. Ma, C.L. Andaur Navarro, M. Van Smeden, Prediction models for diagnosis and prognosis of covid-19: systematic review and critical appraisal. BMJ (2020). https://doi.org/10.1136/bmj.m1328
3. A.T. Xiao, Y.X. Tong, S. Zhang, False negative of RT-PCR and prolonged nucleic acid conversion in COVID-19: rather than recurrence. J. Med. Virol. (2020). https://doi.org/10.1002/jmv.25855
4. M. Lisboa Bastos, G. Tavaziva, S.K. Abidi, J.R. Campbell, L.P. Haraoui, J.C. Johnston, Z. Lan, S. Law, E. MacLean, A. Trajman, D. Menzies, A. Benedetti, F.A. Khan, Diagnostic accuracy of serological tests for covid-19: systematic review and meta-analysis. BMJ (2020). https://doi.org/10.1136/bmj.m2516
5. F. Almazán, I. Sola, S. Zuñiga, S. Marquez-Jurado, L. Morales, M. Becares, L. Enjuanes, Reprint of: Coronavirus reverse genetic systems: infectious clones and replicons. Virus Res. (2014). https://doi.org/10.1016/j.virusres.2014.09.006
6. T.T. Nguyen, M. Abdelrazek, D.T. Nguyen, S. Aryal, D.T. Nguyen, A. Khatami, Origin of novel Coronavirus (COVID-19): a computational biology study using artificial intelligence. bioRxiv preprint. https://doi.org/10.1101/2020.05.12.091397
7. P. Horby, et al.,Dexamethasone in hospitalized patients with Covid-19-preliminary report. New England J. Med. (2020)
8. M. Ackermann, S.E. Verleden, M. Kuehnel, A. Haverich, T. Welte, F. Laenger, A. Vanstapel, C. Werlein, H. Stark, A. Tzankov, W.W. Li, V.W. Li, S.J. Mentzer, D. Jonigk, Pulmonary vascular Endothelialitis, Thrombosis, and Angiogenesis in Covid-19. N. Engl. J. Med. (2020). https://doi.org/10.1056/nejmoa2015432
9. J.C. Chang, Hemostasis based on a novel "two-path unifying theory" and classification of hemostatic disorders. Blood Coag. Fibrinol. (2018). https://doi.org/10.1097/MBC.000000000000076
10. A. Agarwal, A. Mukherjee, G. Kumar, P. Chatterjee, T. Bhatnagar, Convalescent plasma in the management of moderate COVID-19 in India: an open-label parallel-arm phase II multicentre randomized controlled trial (PLACID Trial). MedRxiv (2020)

11. J.H. Beigel, K.M. Tomashek, L.E. Dodd, A.K. Mehta, B.S. Zingman, A.C. Kalil, E. Hohmann, H.Y. Chu, A. Luetkemeyer, S. Kline, D. Lopez de Castilla, R.W. Finberg, K. Dierberg, V. Tapson, L. Hsieh, T.F. Patterson, R. Paredes, D.A. Sweeney, W.R. Short, H.C. Lane, Remdesivir for the treatment of Covid-19—final report. N. Engl. J. Med. (2020). https://doi.org/10.1056/nejmoa2007764
12. M.H.B. Lam, Y.K. Wing, M.W.M. Yu, C.M. Leung, R.C.W. Ma, A.P.S. Kong, W.Y. So, S.Y.Y. Fong, S.P. Lam, Mental morbidities and chronic fatigue in severe acute respiratory syndrome survivors long-term follow-up. Arch. Intern. Med. (2009). https://doi.org/10.1001/archinternmed.2009.384
13. J.C. Ngai, F.W. Ko, S.S. Ng, K.W. To, M. Tong, D.S. Hui, The long-term impact of severe acute respiratory syndrome on pulmonary function, exercise capacity and health status. Respirology (2010). https://doi.org/10.1111/j.1440-1843.2010.01720
14. L. Cariani, B.S. Orena, F. Ambrogi, S. Gambazza, A. Maraschini, A. Dodaro, M. Oggioni, A. Orlandi, A. Pirrone, S.U. Renteria, M. Bernazzani, A.P. Cantù, F. Ceriotti, G. Lunghi, Time length of negativization and cycle threshold values in 182 healthcare workers with COVID-19 in Milan, Italy: an observational cohort study. Int. J. Environ. Res. Pub. Health (2020). https://doi.org/10.3390/ijerph17155313
15. T. Takahashi, P. Wong, M. Ellingson, C. Lucas, J. Klein, B. Israelow, J. Silva, J. Oh, T. Mao, M. Tokuyama, P. Lu, A. Venkataraman, A. Park, F. Liu, A. Meir, J. Sun, E. Wang, A.L. Wyllie, C. Vogels, A. Iwasaki, Sex differences in immune responses to SARS-CoV-2 that underlie disease outcomes. MedRxiv : The Preprint Server for Health Sciences (2020). https://doi.org/10.1101/2020.06.06.20123414
16. M.E. Charlson, P. Pompei, K.L. Ales, C.R. MacKenzie, A new method of classifying prognostic comorbidity in longitudinal studies: development and validation. J. Chronic Dis. (1987). https://doi.org/10.1016/0021-9681(87)90171-8
17. N.V. Chawla, K.W. Bowyer, L.O. Hall, W.P. Kegelmeyer, SMOTE: synthetic minority over-sampling technique. J. Artif. Intell. Res. 321–357 (2002). https://doi.org/10.1613/jair.953
18. K. Hornik, M. Stinchcombe, H. White, Multilayer feedforward networks are universal approximators. Neural Netw. **2**(5), 359–366 (1989)
19. M.F. Møller, A scaled conjugate gradient algorithm for fast supervised learning. Neural Netw. **6**(4), 525–533 (1993)
20. R. Tibshirani, Regression shrinkage and selection via the Lasso. J. Roy. Stat. Soc. Ser. B (Methodol.) (1996). https://doi.org/10.1111/j.2517-6161.1996.tb02080.x
21. R. Bettinger, Cost-sensitive classifier selection using the ROC convex hull method. SAS Institute, 1–12 (2003)
22. K. Simonyan, A. Vedaldi, A. Zisserman, Deep inside convolutional networks: visualising image classification models and saliency maps. arXiv preprint arXiv:1312.6034 (2013)
23. M.D. Zeiler, R. Fergus, Visualizing and understanding convolutional networks. In: European Conference on Computer Vision, pp. 818–833. Springer, Cham (2014)
24. P.J. Kindermans, et al., Learning how to explain neural networks: Patternnet and patternattribution. arXiv preprint arXiv:1705.05598 (2017)
25. A.P. Engelbrecht, *Computational Intelligence: An Introduction*, 2nd edn. (Chichester, UK: Wiley, 2007)

Enhanced Classification Using Restricted Boltzmann Machine Method in Deep Learning for COVID-19

R. Jayashree

Abstract Primary divulgence of the 2019 new coronavirus (COVID-19) is vital for ailment treatment and control. The recording of people who have a high possibility of getting COVID-19 is as important as monitoring COVID-19 patients during all phases of their disease. This chapter expanded to promote a multi-disciplinary approach to care for patients with COVID-19, including people who have high chances-of-getting infected, patients with mild, moderate, severe, and critical disease. A Restricted Boltzmann Machine with Dynamic Estimation of Distribution Algorithm (TwoRBM) constructs an interactive decision-making environment that helps in the classification of patients based on their symptoms and to identify people with a high chance of getting infected by a coronavirus. In this chapter, Convolutional Neural Networks (CNN) is used to categorize patients as COVID-19 disease-ridden or not using chest CT images. Wide-ranging researches are made in the view of the anticipated and the modest classification techniques on the Covid-19 dataset. Across-the-board study shows that the proposed archetypal produces more accurate classification.

Keywords CNN · Restricted Boltzmann Machine · TwoRBM

1 Introduction

The pandemic, novel coronavirus (COVID-19) got an outbreak at the end of the year 2019 and growing day-by-day as a contagious domestic epidemic where emergency critical decisions are in demand. Though various studies are under going on COVID-19, since the data are dynamic, prediction and classification are not easy tasks. Social distancing may reduce the growth of this epidemic drastically [1]. The previous research work on this describes the symptoms and environmental factors for getting infected [1]. Symptoms may emerge between 2 and 14 days after revelation

R. Jayashree (✉)
College of Science and Humanities, SRM Institute of Science and Technology, Kattankulathur, India
e-mail: jayashrr@srmist.edu.in

J. Nayak et al. (eds.), *Understanding COVID-19: The Role of Computational Intelligence*, Studies in Computational Intelligence 963, https://doi.org/10.1007/978-3-030-74761-9_19

Fig. 1 Period of revelation to the symptomless carrier

to the coronavirus. This epidemic spreads from one person to another and it may be asymptomatic initially. Figure 1 shows the period of revelation to the symptomless carrier of the deadly virus. Thus, it is important to identify the persons who are all in contact with an infected person even before the virus is detected through the test as isolating the infected person. The data from social networks like Facebook help to identify the activity of the infected person and the set of persons who were in personal contact with the person. Therefore, a cluster of persons who have great chances-of-getting infected can be effectively formed. Initially, this chapter proposes a model TwoRBM bestowing to the clustered past and social material of the persons with appropriate symptoms. After training the TwoRBM, dynamic estimation of the distribution provides the likelihood model to construct the critical features of the RBM.

The energy weight function representation of $COVID_{present}$ and $COVID_{absent}$ with social cluster information initially helps in decision making. This likelihood model is then pragmatic in Dynamic Estimation of Distribution Algorithm (DEDA) to effectually find the N people with the highest risk of getting COVID-19 from the current user. Forecasting models are uncertain in the results. Obtaining the most accurate results in the forecast helps in decision making which is vital in this critical situation. Premature decision making is dangerous in the early phase of a novel epidemic. The Deep Convolution Neural Networks (DCNN) can be implemented on three types of categorizing such as, 'normal', '$COVID_{absent}$', and '$COVID_{present}$'. The difference between 'normal' and '$COVID_{absent}$' is that 'normal' represents without any ailments and '$COVID_{absent}$' is the absence in coronaviruses but the presence of other types of pneumonia. The DCNN architecture VGG16 [2], is proficient and established on posteroanterior chest X-rays of patients [38] followed by a flat multi-layer perceptron. The classification performance evaluation is based on three factors: sensitivity, specificity, and accuracy. The remainder of the chapter is structured as follows. Section 2 describes the information related to preliminary work. Section 3 provides an overview of Bayesian Deep Learning and Restricted Boltzmann Machines. Construction of TwoRBM with DEDA and details of the CNN model, VGG16, are explained in Sects. 4 and 5 respectively. Section 6 presents the comparative experiments and corresponding analysis. The conclusions and the limitations of the proposed model are lastly followed in Sect. 7.

2 Literature Study

The earlier work on COVID-19 identification and classification is the fundamental motivation for this proposed model.

2.1 Factors/Symptoms of COVID-19

Lakshmi Priyadarsini and Suresh [1] prioritized human control epidemiological factors of COVID-19 which guides to take emergency steps. Hashmi and Asif [13] performed early detection of COVID-19 through symptoms. The symptoms get worse for some people around day 5 to 7 after exposure to the virus [11]. Often people start to get better or worse after day 7 [11]. The literatures are summarized in Table 1.

Table 1 Factors/symptoms of COVID-19

Source	Methodology	Findings
[1]	TISM approach	Epidemiological factors are under the control of human beings
[3]	Clustering and deep sequencing analysis	Study of different factors influencing the spread of coronavirus
[4]	Probability distributions	The amount of the spread of coronavirus is regulated through isolation
[5]	Statistical tests, regression model	Study on the risk factors of aged people during pandemic period
[6]	The Mann-Whitney test, linear regression and prism version 5	Predicts the geographical areas that are in a greater danger of considerable communal spread
[7]	Predictive method	Identify infected host through the social network activity
[8]	Statistical analysis using effective reproductive numbers	Study on the statistical relationship of virus transmissibility and the influence of environment
[10–12]	Symptom-based analysis	Recognized the common symptoms of viral infections

Table 2 Early detection of the disease

Source	Methodology	Findings
[13]	Linear regression analysis, Pearson and Spearman correlation coefficient	The chart, 'Hashmi-Asif', identify Covid-19 at early stages which decrease the spreading proportion
[14]	Sensitivity, specificity and statistical comparison	Detection of the outbreak of viral infection at early stage

Table 3 Classifications methods

Source	Methodology	Findings
[15]	Interactive estimation of distribution algorithm (IEDA)	Personalize computer-system search using Bayesian stimulated IEDA model
[16]	Forward neural network in the prediction model	Interactive decision method supports a decision-making system with optimization
[17]	A surrogate-assisted multi-objective evolutionary algorithm	The people who are interested or close in contact with a person (host) are identified using the Elo score
[19]	Interactive genetic algorithm	Study on a semi-supervised learning method in the interactive model
[20]	Parallel surrogate model-assisted	Enhanced mutation operators in the electromagnetic design
[21]	Surrogate-assisted meta-heuristic algorithms and Gaussian function	A high dimensional optimization problem is solved with relatively least computation expenses by articulating granularity within a surrogate model

2.2 Early Detection of the Disease

Early detection of viral infection through signs and symptoms decreases the progression rate. Common symptoms are grouped and rated between 1 and 4 based on the severity percentage [13]. Healthy persons with no symptoms are rated as 1 and persons with severe symptoms are rated as 4. Hashmi and Asif [13] presented a chart called as Hashmi-Asif Covid-19 (HAC) which detect epidemic disease 72 h earlier than the first symptoms arise. The literatures are summarized in Table 2.

2.3 Classifications Methods

People who are interested or close in contact with a person are calculated using the Elo score, which reduces or increases based on other people's activity related to this person in social media [17]. The literatures are summarized in Table 3.

Table 4 Prediction using social activities

Source	Methodology	Findings
[22]	ATRank and AUC on Amazon dataset	Study on a framework for integrating users' historical social network activities
[23]	recommender systems	Hybrid recommendation systems are more efficient than single recommendation technique-based systems
[24]	Multipartite hypergraph construction and ranking	Multi-partite hypergraph resolves E-commerce requests and are based on the group associations between different objects
[25]	Factorized weight interaction neural network model	Study on the model that provides the memorization and simplification benefits
[26]	Multi-layer perceptron, CNN	Accurate recommendation through deep learning-based model

2.4 *Prediction Using Social Activities*

Zhou et al. [22] presented a framework for integrating users' historical social network activities. Many interesting models provided users' social activity prediction methods [23, 24]. The literatures are summarized in Table 4.

2.5 *Restricted Boltzmann Machine*

Gao et al. [29] presented an analysis based on Boltzmann's function. Evolutionary algorithms (EAs) effectively solve optimization problems [31]. Searching efficiency is improved and computational cost is reduced with the surrogate management model based on RBM [30]. These literatures are summarized in Table 5.

2.6 *Deep Learning in COVID Dataset*

Cheng et al. [25] proposed a model based on wide and deep learning along with deep neural networks. Kim et al. [35] presented Convolutional Matrix Factorization (ConvMF) which is a novel context-aware personalization method of CNN with positive and negative categories to improve prediction accuracy. The study of posteroanterior chest X-ray and CT scan images helps in the early identification of viral infections using deep learning models [36, 42]. The literatures are summarized in Table 6.

Table 5 Restricted Boltzmann machine

Source	Methodology	Findings
[29]	Boltzmann's function-based analysis	The collective number of coronavirus deaths in China is predicted using Boltzmann's function
[31]	RBM and IEDA	Evolutionary algorithms provide the solution for optimization problems
[30]	Softmax RBM and EDA	Decision-making feature information enables dynamic modification in the probability model for the evolutionary advancement
[32]	Distribution algorithm based on RBM	Dynamic environment is tackled by Re-optimization approaches and they fail to consider the diversity of alternative solutions

Table 6 Deep learning in COVID dataset

Source	Methodology	Findings
[33]	dropout Restricted Boltzmann machines	Dropout considerably improved the performance of standard neural nets
[34]	RNN and long short-term memory (LSTM)	LSTM system is enriched through the incorporation of its extension network model
[35]	Deep learning-based recommendation	ConvMF Prediction accuracy for a novel context-aware personalization
[2]	Convolutional networks (ConvNets) and ImageNet Large-Scale Visual Recognition Challenge (ILSVRC)	The classification accuracy is maximum when depth of the network is high. Thus, ImageNet challenge dataset performance in enhanced using a ConvNet architecture with substantially increased depth
[38]	InceptionV3, VGG16, VGG19, Xception	VGG16 model perform better than all other model in CNN
[39]	ImageNet, ILSVRC	Regularized activation grading stabilizes the training of ResNet
[40]	DenseNet201, ImageNet and CNN	Study on DCNN with X-ray and CT images of COVID-19 infected patients
[37]	CNN—Xception, ResNet, and VGG-16	The VGG-16 model presented greater consistency and accuracy

2.7 *Dataset*

Data collection is very difficult in the COVID-19 pandemic since the data are dynamic and unpredictable [16–18]. The dataset sources with descriptions are given in Table 7 which are considered for initial classification using the TwoRBM model. The

Table 7 Dataset sources for TwoRBM

Source	Purpose	Description
[44, 45]	COVID-19 patient dataset with attributes	WHO provides guidance document for clinicians caring of COVID-19 infected patients
[41–46]	COVID and Non-COVID patients' chest X-RAY and CT images	Publicly available COVID-19-infected and non-infected patient's chest X-RAY and CT images
[48]	Information on COVID-19	Details of coronavirus symptoms, prevention and test measures
[49]	Up-to-date coronavirus confirmed, deaths and recovered data	Complete details of COVID-19 patients data based on geographical location
[50]	COVID-19 guidelines	Medicines and treatments with recommendations

COVID-19 infected and non-infected patient's chest X-RAY and CT images are available for research and study purpose that are utilized for the latter classification with DCNN in this chapter [41, 46]. In this chapter, a decision-making model is constructed initially with social network knowledge (Tables 1 and 4) using TwoRBM for each coronavirus present and absent possibility by fetching both the likelihood model of EA and the aptness function concurrently (Tables 3 and 5). VGG16 is used in the latter part of the model construction (Table 6). Finally, the proposed model is compared with the existing coronavirus early detection methods (Table 2).

3 An Overview of Bayesian Deep Learning and Restricted Boltzmann Machines

Bayesian inference aims to govern the posterior distribution, the distribution of archetypal parameters $\boldsymbol{\theta}$ that adapt to the data. Assuming d_i as the dependent variable, I_i as the independent variable, and ε_i as the measurement noise, $I_i = F(d_i,\theta) + \varepsilon_{i..}$ The posterior distribution, $P(\theta|\{d_i\},\{I_i\})$ in Bayes theorem is related to the likelihood. The estimation for a strange input is found by an average of the posterior. An ambiguity of archetypal forecasts is a voluntary derivative from Bayesian formalism. The Bayesian approximation is informal to state but pricey. The approximation of the posterior is also usually computationally expensive. The Bayesian approach not only predicts or classifies but also helps to calculate uncertainty. The uncertainty is important in safety–critical applications. Bayesian inference approximation is equivalent to neural network dropout, a way to regularize a neural network and limit over-fitting [28]. To minimize the loss, the last layer of the neural network (NN) is divided to give two outputs: the forecastedambiguity $\sigma_i{}^2$ and reliant on variable y_i. However, during the training period, each unit has a likelihood p of being set to 0. Finding parameters $\theta = (\Theta_1, \Theta_2, \cdots, \Theta_x)$ for a NN with x items is approximately equal to the posterior distribution $P(\theta|\{d_i\}, \{I_i\})$, where Θ_x is the limitation vector related with the xth unit. The dropout as a Bayesian Approximation proposes, a simple approach to quantify the neural network uncertainty [27]. It employs dropout during both training and testing. Random neuron dropping during training only reduces the network generalization error. Neural Network with dropout enabled during testing multiple feed-forwards, for the same input, generate multiple outputs [55]. Neurons randomly dropped only during training [33].

3.1 Restricted Boltzmann Machines

Markov random model is an undirected graph with connections between the variables, while Bayesian networks have directed connections. Boltzmann machines are a kind of Markov random model, a graphical model. The main advantage of undirected connections over a directed connection is that when there are many hidden variables it is easy to infer the hidden states associated with a set of visible states. Figure 2 displays a recurrent neural network with social network feedback [34]. Restricted Boltzmann Machines (RBMs), are two-layer generative neural networks that absorb a likelihood distribution over the inputs. They are a distinct class of Boltzmann Machine in that they have a constrained number of networks between visible and hidden units. Every node in the visible layer is related to every node in the hidden layer, but no nodes in the same cluster are associated. Train RBMs using the conflicting discrepancy learning procedure.

A Boltzmann machine defines a likelihood distribution over N random variables $r_1, r_2, \ldots, r_N$ such that

$$P(r_1, r_2, \ldots, r_N) \propto exp\left(\sum_i \theta_i r_i + \sum_{i,j} w_{ij} r_i r_j\right) \tag{1}$$

That is, the log-likelihood is linearly dependent on both individual variables r_i r$_\text{i}$ in addition to pairwise terms. Divide the random variables into two groups: visible variables $\{v_1, v_2, \ldots, v_{N_v}\text{v}_1, \text{v}_2, \ldots, \text{v}_{\text{Nv}}\}$ and hidden variables

Fig. 2 Recurrent NN with social network feedback

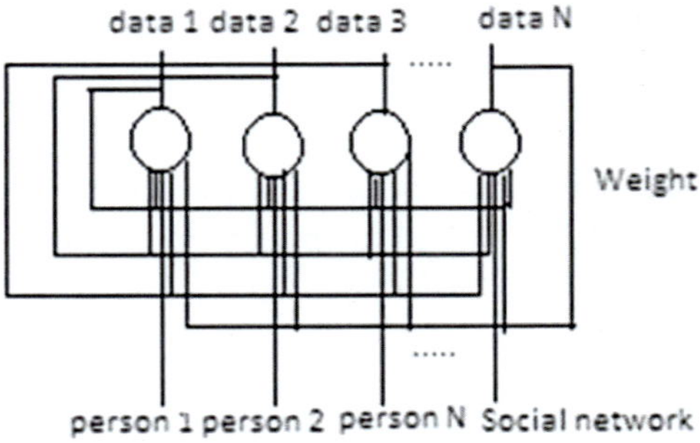

$\{h_1, h_2, \ldots, h_{Nh}\mathrm{h}_1, \mathrm{h}_2, \ldots, \mathrm{h}_{Nh}\}$ $\{h_1,h_2,\ldots,h_{Nh}\mathrm{h}_1,\mathrm{h2},\ldots,\mathrm{h}_{Nh}\}$. The restricted Boltzmann machine defines the following likelihood distribution from Eq. (1):

$$P(v_1, .., v_{Nv}, h_{1,..,}h_{Nh}) \propto exp\left(\sum_i \theta v_i v_i + \sum_i \theta h_i h_i + \sum_{i,j} w_{ij} h_i v_j\right) \quad (2)$$

The restricted Boltzmann machine (RBM) confines pairwise terms to those between a hidden variable and a visible variable. An undirected graphical archetypal with a visible and a hidden layer is known as RBM which is made for random variables that are either discrete or continuous. In this archetypal, there is no constrained dependency within a layer. In current years, RBMs have ascended to reputation due to their suitability with deep-learning. Construct a deep-learning model by considering one RBM's hidden layer(hl) as the second RBM's visible layer(vl). Thus, it is understood clearly that RBMs can encrypt multifaceted structures in data, producing them striking for supervised learning. A Boltzmann machine is a recurrent network with stochastic neurons and weights are symmetrical at the equilibrium [33]. The two groups of neurons are, visible and hidden. In restricted architecture, there are no connections within the visible group or hidden group. Network parameters are: Bias vector hidden units, $hu = [hu_1, hu_2, \ldots, hu_H]$; Bias vector visible units, $vu = [vu_1, vu_2, \ldots, vu_V]$; Connection weights, $W = \{w_{i,j}\}$. Network values are binary random vectors: $v = [v_1, v_2, \ldots, v_V]$ and $h = [h1, h2, \ldots, h_H]$. In RBM, the energy function is defined as $E_{Fun}(v, h) = --h^t Wv\text{-} vu^t v \text{ - } hu^t$ where probabilities to (v, h) based on Boltzmann distribution is

$$p(v, h) = \frac{exp(-Efun(v,h))}{z} \quad (3)$$

where $z = \sum_{v',h'} exp\,(-Efun(v', h'))$

4 Construction of TwoRBM

The TwoRBM model is to find COVID-19 positivity. Initially, the suspected persons for COVID-19 positivity, with or without symptoms, are represented as a cluster B. Let cluster A is the set of individuals who are all in physical contacts (directly or indirectly-identified easily using social network knowledge) with cluster B. Though cluster A is asymptomatic, they are the people with the threat of getting COVID-19 since this is a communal viral infection and, it spreads rapidly. These clusters A is further tested for COVID-19 positivity based on the leading symptoms and are classified either as positive ($\mathrm{COVID}_{present}$) or negative (COVID_{absent}). Social network knowledge is taken as a consideration only for clustering A. The architectural diagram of the proposed TwoRBM including the DCNN model is shown in Fig. 3. The 'Host' is a Coronavirus-infected person who is a part of cluster B. Initially the TwoRBM model is trained with training data to find COVID or non-COVID. Classification

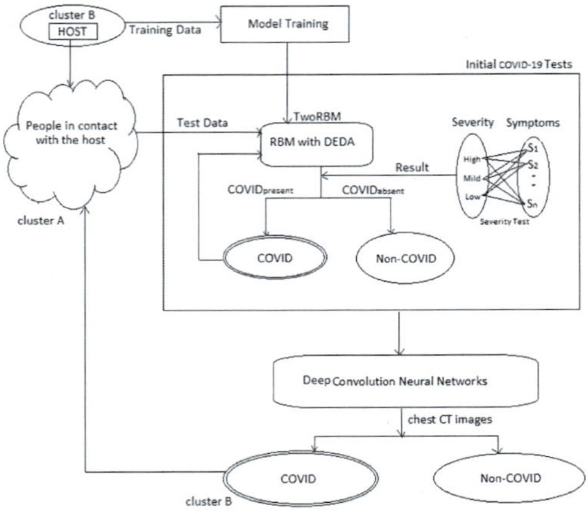

Fig. 3 Architecture of the TwoRBM and DCNN model

is performed based on the preliminary medical tests. Then, DCNN model classifies patients with greater accuracy based on chest CT scan. The severity of the leading symptoms of the given persons is considered as $COVID_{present}$ and other non-symptomatic or negative results as $COVID_{absent}$. Using the visible layer's outputs of the RBM, construct the likelihood archetypal of DEDA and the matching of the result will be estimated by energy functions weighting of both COVID and non-COVID models from Eqs. (2) and (3). The given user is estimated for the symptoms $\{s_1, s_2, \ldots, s_n\}$ where n is the number of indications with severity value. First, the TwoRBM build for user x in cluster A based on social group activities. Accordingly, the input layer of the $COVID_{present}$ and $COVID_{absent}$ have 'n' visible units. Let hidden layers of TwoRBM have 'm' units. Assuming the state of hidden components, the dynamic probabilities $P_{\theta l}$ and $P_{\theta dl}$ of the ith visible component in the $COVID_{present}$ and $COVID_{absent}$ models are considered with the following calculations as mentioned in Eqs. (4) and (5),

$$P_{\theta l} = p(a_{li} + \textstyle\sum^{m}{}_{j=1} W_{lij} h_{lj});\ v_{li} = 1|h_l \tag{4}$$

$$P_{\theta dl} = p(a_{dli} + \textstyle\sum^{m}{}_{j=1} W_{dlij} h_{dlj});\ v_{dli} = 1|h_{dl} \tag{5}$$

where (v_{li}, v_{dli}) and (h_{lj}, h_{dlj}) are the positions of the (i^{th} visible, j^{th} hidden) pairs in the $COVID_{present}$ and $COVID_{absent}$ respectively. $\theta_l = \{W_l, a_l, s_l\}$ and $\theta_{dl} = \{W_{dl}, a_{dl}, s_{dl}\}$ denotes the constraints of the $COVID_{present}$ and $COVID_{absent}$ respectively and $p(t) = \frac{1}{(1+exp(-t))}$ is the active function. After training the TwoRBM, the constraints (θ_l, θ_{dl}) have the leading symptom of the given user and denote a result based $COVID_{Present}$ and $COVID_{absent}$ likelihood on the given current user. Therefore, the TwoRBM user testing archetypal is more general to express the likelihood of COVID-19 in the given user. Pearson's correlation coefficient or the Pearson R statistical test is the best strength measuring test between the different variables of interest-based on the method of covariance. The requirement for the Pearson R statistical test is (1) parametric data, and (2) interval level variables, where interval is used with parametric tools in which distributions are predictable. The output value of the Pearson R statistical test isbetween –1 and 1 range. A maximum absolute value of the correlation coefficient specifies a robust relationship among variables. Find users with similar symptoms according to the Pearson correlation coefficient. Calculate Pearson similarity Psim (p_i, p_j) between user u_i and u_j as

$$Psim\left(p_i, p_j\right) = \frac{\sum_{k\in Ipi,pj}\left(Qpik - Q'pi\right)\left(Qpjk - Q'pj\right)}{\sqrt{\sum_{k\in Ipi,pj}(Qpik - Q'pi)2X\sum_{k\in Ipi,pj}(Qpjk - Q'pj)2}} \tag{6}$$

where $lp_{i,}$ p_j represents COVID/non-COVID test result for persons p_i and $p_{j,}Q_{pik}$ and Q_{pjk} are positive k_{th} symptoms by p_i and p_j, and Q'_{pi} and Q'_{pj} are the overall symptom results of persons p_i and $p_{j.}$ from Eq. (6), 'n' numbers of similar positive or negative users are carefully chosen. The testing function f(t) on symptom 't' is represented in Eq. (7) to evaluate the symptom severity through incorporating the trained TwoRBM along with the social information.

$$f(t) = \text{p}\ (\alpha \times social(t) + severity(t)) \tag{7}$$

where $social(t) = \sum_{j=0}^{n} Psim(\ p{,}p_j) \times Q_{ujt}$. Here social(t) specifies the weighted average matching score of 'n' comparable persons on symptoms 't' with positive or negative results. The 'σ' is the contributions of the social group which is a normalized function. The severity(t) represents the difference between $COVID_{present}$ and $COVID_{absent}$ symptoms' severity of TwoRBM on 't'. In detail severity(t) is governed by the energy function variance among COVID and non-COVID archetypal. The severity of the $COVID_{present}$ is more important and is contributed more to the calculation. The $COVID_{present}$ is a suitable model in a dynamic environment that can be applied readily in DEDA. The TwoRBM algorithm consists of the following contents,

4.1 Construct TwoRBM to Get More Consistent RBM and to Improve the Presentation of DEDA

Create cluster A using social network knowledge along with information collected from the infected patient. Analysis users from cluster A for symptoms. Evaluate and classify them into two groups depending on the severity of symptoms such as moderate or high as one group and low as another group. This classification helps to train TwoRBM to obtain $COVID_{present}$ and $COVID_{absent}$.

4.2 Emerging DEDA with the TwoRBM for Developing the Search List

Use the likelihood archetypal of DEDA to create a new list for the collaborative progressivity findings. Combine the trained TwoRBM model with communal data to estimate the severity of symptoms of the identified N user contacts.

4.3 Management of the TwoRBM Model

The COVID and non-COVID clusters are restructured grounded on the dynamically evaluated N user contacts. The average assessment error of the classification archetypal is evaluated and controlled to accomplish TwoRBM. If the error is higher than a given threshold (0.5), the model will be restructured along with the energy functions. The above process is repeated until discharge or death. Changeability in COVID-19 data and diversity in treatment/testing methods are the challenges in updating TwoRBM in a timely track.

The examined people are clustered depending on the severity of the vital symptoms of the COVID-19 positivity as shown in Fig. 3. Once the COVID-19 test is positive for N people, then they undergo further tests and are classified based on the positivity and severity of the result. The severity threshold represents the limit value in the symptom's severity.

5 Deep Convolutional Neural Network (DCNN)

Multi-layer perceptions are completely connected networks and DCNNs are its regularized versions formed by including extent amount of weights to the loss function. Each neuron in one layer is associated to all neurons in the next layer. This network factor makes them suitable for over-fitting data. In DCNN units are said to have spatially contiguous accessible fields when the hidden unit (HU) inputs layer κ-1 produces elements in its subset layerκ-1. Elements in layer $\kappa + 1$ have related connectivity with its underneath layers. To create a comfortable illustration of the data, each hidden-layer (hl) is a collection of various quality-maps. In the COVID-19 epidemic region, computed tomography (CT) of chest representation may be a considerably more reliable, purposeful, and fast procedure to categorize and estimate. In this chapter, DCNN is used to categorize the coronavirus infected persons as diseased or not based on CT images [51–54]. The VGG archetypal is observed as consecutive network architecture [2]. VGG-16 is a 16 layers DCNN. This network has rich quality representations for an extensive array of metaphors. The set-up has a metaphor with an input size of '224 $\times$ 224'. A backing package in toolbox model for the VGG-16 network is necessary for this deep learning. The unqualified archetypal does not involve the backing package. Grounded on the Visual identification Contest 2014 (ILSVRC2014) in 1000 classes DCNN reached 93% with topmost test accuracy, after training over 200 times on a dataset of more than 140 lakhs '224 × 224' images [38]. In this chapter, Adam optimizer [43] with pre-training weights similar to the ILSVRC2014 is executed over 200 times.

6 Experiments and Results

Using social network, fetch the information about the people who are all in contact with the infected person in past 15 days. Isolation or further tests/treatments is decided based on their estimated initial probabilistic severity score. Compatible matches are filtered and ordered, based on the severity of the symptoms of the users through partitioning-based clustering methods. The supreme common indications of coronavirus are shown in Fig. 4 [10–12]. Generally, the indications are mild in the beginning and may become vigorous later on. Symptoms are not necessarily present; some people may be infected with this deadly virus but may not have symptoms or may have very mild indications. In Fig. 5, 'Contact List' refers to the cluster of people who are highly vulnerable to get infected. That is, those who are in contact with the host in the recent 15 days. This group of people is identified through the social

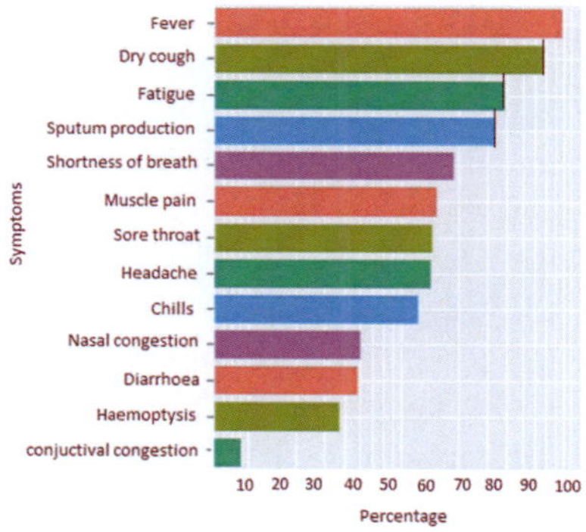

Fig. 4 Leading symptoms

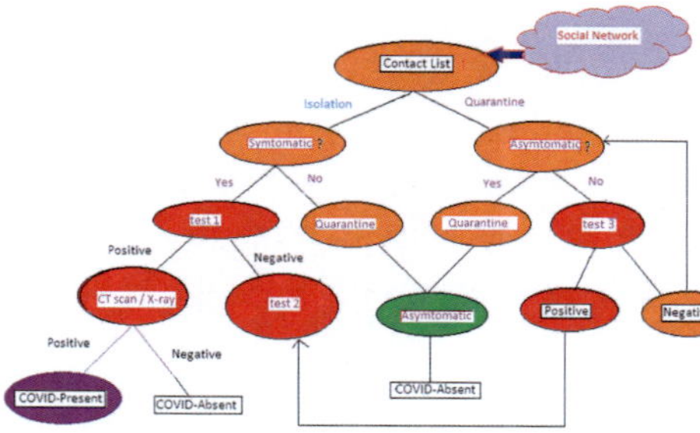

Fig. 5 Classification model

Table 8 Attributes as on dated July 23, 2020

Data	Attributes
Confirmed cases	CT_total_Positive_tested_count
People have been hospitalized	Hospitalized_cumulative
Cumulatively cases	$(COVID_{present}) + (COVID_{absent}) +$ withheld
People with positive results per day	Tested_ positive/day
Source of data	Data_source
Execution time of ETL	Prep_flow_runtime

network, which helps in fetching up information about a given person's activities and contacts (for example sharing photos and location on Facebook) [7, 47]. The people with no indication of the virus effect are set to quarantine for further observations. Mild or severe symptoms of infection need immediate testing and treatments accordingly. If the initial test shows the presence of infection, then a chest CT scan/X-ray is taken for the confirmation of $COVID_{present}$. The foremost considerable attributes are considered in the dataset [44, 45]. WHO provides guidance documents for clinicians caring from December 2019 for COVID-19 patients during all phases of their disease (that is, from screening to discharge) [44]. The dataset is accessed for experimental study in July 2020 (8 months dataset). Table 8 is the experimental attributes of the proposed work.

In the experimental datasets, the severity of each symptom is tagged as mild, moderate, or high depending on the severity of the symptoms of the given person. The persons are chosen in a random way from the datasets to carry out the experiments. In an orderly manner, evaluate training and testing datasets. That is, evaluate an initial 80% of all $COVID_{present}$ indications, and then test the remaining 20% training datasets. The COVID-19 data from the dataset [44, 45] is used for experiments. Feed the encryption of the symptoms into the TwoRBM. In the sample data, there are13 COVID-19 symptoms in category tags (Fig. 4). A user 'x' in the DEDA undergo a test for each symptoms' severity measures as $x = [x_1, x_2,.., x_{13}]$. Accordingly, the visible layer of the RBM-based COVID-19 severity archetypal has 13 units. Besides, the symptoms with the tag as 'high' in the training set are selected as the $COVID_{present}$ group while the symptoms with the tag as 'low' mark off as the $COVID_{absent}$ group. These groups are used to train the TwoRBM model. The trained TwoRBM symptoms severity model can represent the user's COVID-19 positivity/negativity results. The TwoRBM model demarcated in Eq. (7) is used in the evaluation of the severity of symptoms in the test-set. The terms calculated based on the classification for COVID-19 for the test dataset and archetypal are: tp_{c19} the percentage of coronavirus present that are properly categorized as coronavirus by the model, fp_{c19} the percentage of negative that are mislabeled as positive coronavirus by the model, tn_{c19} the percentage of negative that are correctly labeled as non-coronavirus, fn_{c19} the percentage of coronavirus present that are mislabeled as non-coronavirus by the model. In this work, 60 independent executions are achieved for each limitation setting in the TwoRBM algorithm. The precision measures the accuracy of the predictions and the recall measures efficiency in finding all the positives. A common metric in object accuracy measuring is Average Precision (AP). The AP calculates the average correctness value for recall value from 0 to 1. For COVID-19 data, precision and recall are calculated using the formulae, *Precision* = tp_{c19}*/(Total positive results)* and *Recall* = tp_{c19}*/(Total COVID-19 cases).* The very high correctness value for some given recall level is called as Interpolated Precision($\hat{I}$). Precisely, Interpolated Precision is the maximum correctness for any recall $\geq \hat{r}$. Overlapping of correctness boundary and the actual truth value is measured using intersect-over-boundary (ϕ). In COVID-19 datasets, the predefined ϕ threshold is 0.5. Average Precision finds the model accuracy for 13 symptoms,

$$AP = \int_0^1 \hat{I}(r)dr = \frac{1}{13}\sum_{i=1}^{13} \frac{i}{severity(i)} \tag{8}$$

Where severity(i) is the severity of the ith symptoms. Calculate Mean Average Precision (MAP) by taking the total of the $\hat{I}$ at N different recall levels. The AP is to confirm the correctness of the severity measure model. The MAP [18] is the average AP as shown in Eq. (8) of 'n' persons in the test dataset and stated as, $MAP = Avg\,(\sum_{k=0}^{n} AP_k)$, where AP_k is the AP of the kth person in the dataset. The algorithm contentment is reflected through MAP. Thus, a higher MAP value indicates higher algorithm contentment. An algorithm's effectiveness and reliability are measured through Hit Ratio (H_{Ratio}) [9]. The HRatio and MAP values are calculated to find the impacts of different numbers of HU in the TwoRBM model. The number of HU is related to that of category features which are determined as the input of the model. The ConvMF [55] is a classification algorithm based on CNN by integrating the $COVID_{present}$ and $COVID_{absent}$ results. Also, the TwoRBM algorithm with the communal data is compared in the experiments with average values of H_{Ratio} and MAP. Independently both the algorithms are executed on experimental datasets 15 times, and the average outputs are represented. The ConvMF and TwoRBM are studied on the metrics H_{Ratio} and MAP as shown in Table 9 (3 iterations for sample).

Table 9 Comparing the effectiveness of ConvMF and TwoRBM

Algorithm	ConvMF			TwoRBM		
Iteration no.	H_{Ratio}	MAP	Time in milliseconds	H_{Ratio}	MAP	Time in milliseconds
1	0.0256	0.589	123,670	0.055	0.639	903
2	0.0742	0.728	276,715	0.2149	0.934	215
3	0.0701	0.688	90,489	0.1171	0.705	137

Table 10 H_{Ratio} and AP values in DEDA and TwoRBM methods

DEDA		TwoRBM (RBM with DEDA)	
H_{Ratio}	AP	H_{Ratio}	AP
0.0128	0.815	0.0157	0.946
0.1571	0.855	0.181	0.971
0.0806	0.73	0.0677	1

6.1 Computational Cost Comparison

The representation of execution time enables to perform computational cost comparison. The average time is taken for execution in ConvMF and TwoRBM is 3434.793 and 7.297 in seconds respectively. The average (H_{Ratio}, MAP) of ConvMF and TwoRBM is (0.045707, 0.045707) and (1.65, 11.644) respectively. A fast algorithm with the highest H_{Ratio} and MAP is an efficient model. Thus, the proposed model is efficient than ConvMF which is shown clearly in Table 9.

6.2 DEDA

The TwoRBM model which is supported by DEDA is a rough distribution in the initial stage of development since social network user's activities are changeable and diverse. By enabling the connections among various social networks, the user's activities, and their close contacts in the past 15 days are gradually trackable. Updating the information about the patient on a timely basis is the effective evolution method to conceivably attain a precise and reliable TwoRBM model. Given the assessing time, the users connected to the $COVID_{present}$ or suspected patients are organized as per their timestamps in the dataset. The H_{Ratio} and AP are used as evaluation metrics for TwoRBM, which is based on RBM along with interactive DEDA. The average experimental outputs are calculated and show a better performance of TwoRBM in Table 10. The significance of the proposed TwoRBM model is measured using the Mann Whitney U test with the result p-value = 0.601508 for a 95% confidence interval. Though initially, TwoRBM classifies the COVID-19 dataset, Convolution Neural Networks (VGG16) is more suitable in identifying the presence of coronavirus through CT scan or X-ray images.

6.3 Experimental Setup

Python is a programming language that is used to train the anticipated deep learning archetypal in this experiment. Performed experiments on a Linux server with

Ubuntu 16.04 operating systemin Google Co-laboratory. Adaptive Moment Estimation (Adam) optimizer is utilized for pre-training DCNN models with random initialization weights. This helps in the field of deep-learning to attain quick optimal results. Split the dataset into two independent sets with 80% and 20% for training and testing respectively.

6.4 Dataset Features

The coronavirus infected and non-infected patients' chest images are applied in the experiment. Two sets of chest images are publicly available. They are, (1) *Chest X-Ray image* [41] Implement experiments using PyTorch on the available dataset which contains nearly 200 COVID-19 positive and 5000 COVID-19 negative images. The 'Augmentor' library in python is used for data augmentation. The dataset is divided into two categories such as $\mathrm{COVID}_{\mathrm{Present}}$ and $\mathrm{COVID}_{\mathrm{absent}}$. The number of ($\mathrm{COVID}_{\mathrm{Present}}$, $\mathrm{COVID}_{\mathrm{absent}}$) for training and test dataset is (84, 2000) and (100, 3000) respectively (2) *Chest CT image* [46]. The chest CT image shows a certain pattern for COVID-19 positive patients which is not detectable in X-ray images [46]. Therefore, in this chapter, a dataset of 379 COVID-19 positive and 463 COVID-19 negative patient's chest CT images are utilized for proposed model experiments. The Statistics of the test and training dataset for ($\mathrm{COVID}_{\mathrm{Present}}$, $\mathrm{COVID}_{\mathrm{absent}}$) are (173,168) and (88,64) respectively.

6.5 Deep-CNN for COVID-19

The imbalance problem produces skewed class sizes. That is, often classify images only as majority class which is an undesirable outcome. In the x-ray image dataset, this is a very common problem [42]. This problem is addressed and solved with the 'weight' parameter for classes $\mathrm{COVID}_{\mathrm{present}}$ and $\mathrm{COVID}_{\mathrm{absent}}$. The purpose of this parameter is to provide more weight (value) to the minority class, $\mathrm{COVID}_{\mathrm{present}}$ [36, 42]. In general, the 'weight' parameter for class $\mathrm{COVID}_{\mathrm{x}}$ is, $wt(COVID_{xt}) = constant_x(\sum_{x=0}^{2} S_x / 2S_x)$ where x = {present, absent}, S_x is the number of x-ray images and constant $_x$ is a constant variable. The best way to usethe 'weight' parameter is to represent the inverse class distribution present in the training dataset which is implemented in this experiment. The COVID-19 sensitivity remains high with short-range confidence intervals [38]. For the VGG16 model, evaluate the classification performance based on sensitivity, specificity, and accuracy on X-ray and CT chest images. The measure metric for each cluster is expressed with the subscripts *c19*forCOVID-19 as shown in Eqs. (9) to (11).

$$specificity_{c19} = tn_{c19}\frac{1}{tn_{c19} + fp_{c19}} \tag{9}$$

Table 11 Accuracy analysis of X-ray and CT chest images

Chest image	Model	Sensitivity	Specificity	Accuracy
X-ray	VGG16	52.71	92.56	79.34
CT-scans	VGG16	95.49	95.67	95.56

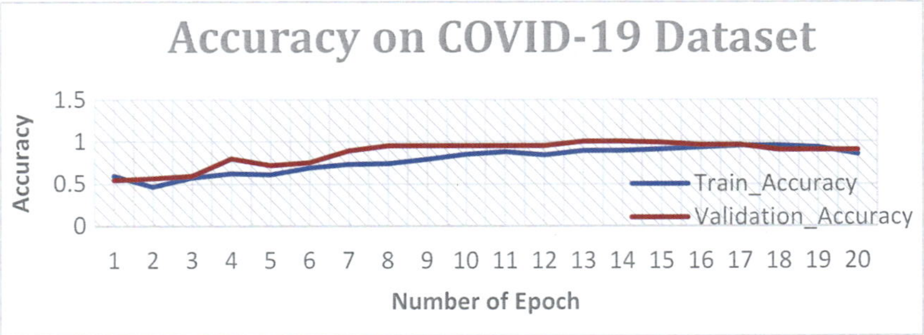

Fig. 6 Training and validation accuracy curves for COVID-19 dataset for Chest CT image

and

$$sensitivity_{c19} = tp_{c19}\frac{1}{tp_{c19} + fn_{c19}} \tag{10}$$

$$Accuracy_{c19} = \frac{tp_{c19} + tn_{c19}}{tp_{c19} + fn_{c19} + tn_{c19} + fp_{c19}} = \sum (specificity_{c19} + sensitivity_{c19}) \tag{11}$$

The COVID-19 detector in this section excludes other data such as geographical location, the density of population, temperature or humidity, etc., and obtain 95.56% $Accuracy_{c19}$ on the sample dataset based on CT image which is 16.22% greater than X-ray image dataset [51] as shown in Table 11 Thus, the proposed model is efficient in classifying patients under examination as COVID or Non-COVID using CT images. The training and validation accuracy curves are plotted using the chest CT image in Fig. 6 for the COVID-19 sample dataset. Pearson and Spearman's correlation provides a significant correlation of COVID-19 with its symptoms [13]. Detection of COVID-19 at an early stage enables the prevention of the spreading of disease in society. Sensitivity measures the efficiency of the prediction of $COVID_{present}$. The HAC chart performs linear regression to find the highest correlation [13]. This chart provides overall 95% sensitivity for symptomatic and 48.11% for asymptomatic cases for early identification, which is the cumulative sensitivity of all symptoms, and this enables 29.5% earlier detection of the viral diseases. The proposed model consists of 2 methods for classification. The first method for early detection along with social network knowledge and the second method is a DCNN with a sensitivity of 95.49%. Compare to the HAC chart which holds 71.55% as overll early detection, the cumulative precision of TwoRBM (from Table 6) with DCNN is 96.40%which is 24.85% higher, and results in 39.75% earlier detection of the disease. That is,

HAC diagnoses epidemic diseases 72 h earlier than existing testing methods and TwoRBM with DCNN can diagnose115 h earlier. Thus, the proposed model can detect the disease 10% better than the HAC chart.

7 Conclusion and Limitations

Two types of classification methods are implemented in this chapter for the COVID-19 dataset. First, the classifications of persons are carried out using the TwoRBM algorithm which combines the performance of RBM and DEDA. The proposed method utilizes the obtainable data from social networks and identifies the severity effect to enhance classification accuracy. The TwoRBM outperforms ConvMF in the HR and MAP. Thus, the TwoRBM model provides accurate symptom severity-oriented results for classifications compare to ConvMF. Further, the Mann–Whitney U test is accomplished on the sample dataset and obtained a 95% confidence interval which proves the efficiency of the proposed algorithm. Finally, the DCNN method is used on the sample chest CT images for classification after the preliminary test, since this the best model for image-based classification. The results are validated with an accuracy of 96.40%. The comparative study shows that HAC divulges coronavirus infection 3 days earlier whereas TwoRBM can divulge 5 days earlier than the current diagnosis system. The HAC chart only emphasis the preliminary stage of the endemic disease but the proposed model focuses on every stage of the disease, from asymptomatic to confirmation. TheVGG16 is larger in model size and thus vulnerable to slow training [2]. Corresponding sensitivity_{c19} and specificity_{c19} are extremely difficult in medical applications, especially infectious diseases such as COVID-19 which can be transmitted rapidly. The proposed method is only for research purposes and not suitable for the practical implementation in the medical field until further experiments are accomplished by the expertise.

Reference

1. S. Lakshmi Priyadarsini, M. Suresh, Factors influencing the epidemiological characteristics of pandemic COVID 19: a TISM approach. Int. J. Healthcare Manag., 1–10 (2020). https://doi.org/10.1080/20479700.2020.1755804
2. K. Simonyan, A. Zisserman, Very Deep Convolutional Networks for Large-Scale Image Recognition. arXiv.org, arxiv.org/abs/1409.1556 (2014)
3. C. Huang et al., Clinical features of patients infected with 2019 novel coronavirus in Wuhan, China. The Lancet, **395**(10223) (2020). www.thelancet.com/journals/lancet/article/PIIS0140-6736(20)30183-5/fulltext. https://doi.org/10.1016/s0140-6736(20)30183-5
4. J. Hellewell et al., Feasibility of controlling COVID-19 outbreaks by isolation of cases and contacts. The Lancet Global Health (2020). www.thelancet.com/journals/langlo/article/PII S2214-109X(20)30074-7/fulltext. https://doi.org/10.1016/S2214-109X(20)30074-7

5. C. Leung, Risk factors for predicting mortality in elderly patients with COVID-19: a review of clinical data in China. Mech. Age. Dev. **188**, 111255 (2020). https://doi.org/10.1016/j.mad.2020.111255. Accessed 12 May 2020
6. M.M. Sajadi et al., Temperature, Humidity and Latitude Analysis to Predict Potential Spread and Seasonality for COVID-19. Papers.Ssrn.com (Mar 2020). https://papers.ssrn.com/sol3/papers.cfm?abstract_id=3550308
7. P. Mahalle et al., Forecasting Models for Coronavirus (COVID-19): A Survey of the State-of-the-Art. (Apr 2020). https://doi.org/10.36227/techrxiv.12101547.v1. Accessed 20 Sept 2020
8. J. Wang et al., High Temperature and High Humidity Reduce the Transmission of COVID-19. arXiv:2003.05003 [q-Bio], arxiv.org/abs/2003.05003 (May 2020)
9. X. He et al. Neural collaborative filtering. In: Proceedings of the 26th International Conference on World Wide Web - WWW 2017 (2017), arxiv.org/pdf/1708.05031.pdf, https://doi.org/10.1145/3038912.3052569
10. "Study Confirms 'classic' Symptoms of COVID-19." ScienceDaily, www.sciencedaily.com/releases/2020/06/200624100047.htm. Accessed 11 Aug 2020
11. J. McBride, COVID-19 Symptoms: Day-by-Day Chart of Coronavirus Signs. Heavy.com (Apr 2020). https://heavy.com/news/2020/04/covid-19-symptoms-day-by-day-chart-of-coronavirus-signs/. Accessed 4 June 2020
12. Report of the WHO-China Joint Mission on Coronavirus Disease 2019 (COVID-19). Www.Who.Int, www.who.int/publications/i/item/report-of-the-who-china-joint-mission-on-coronavirus-disease-2019 (covid-19)
13. H.A.S. Hashmi, H.M. Asif, Early detection and assessment of Covid-19. Front. Med. **7** (2020). https://doi.org/10.3389/fmed.2020.00311. Accessed 29 June 2020
14. Y. Dai, J. Wang, Identifying the outbreak signal of COVID-19 before the response of the traditional disease monitoring system. PLOS Neglected Trop. Dis. **14**(10), e0008758 (2020). https://doi.org/10.1371/journal.pntd.0008758. Accessed 21 July 2020
15. Y. Chen et al., Personalized search inspired fast interactive estimation of distribution algorithm and its application. IEEE Trans. Evol. Comput. **21**(4), 588–600 (2017). Accessed 21 Aug 2020
16. L. Pan et al., A classification-based surrogate-assisted evolutionary algorithm for expensive many-objective optimization. IEEE Trans. Evol. Comput. **23**(1), 74–88 (2019)
17. Y. Tian et al., A surrogate-assisted multiobjective evolutionary algorithm for large-scale task-oriented pattern mining. IEEE Trans. Emerg. Top. Comput. Intell. **3**(2), 106–116 (2019)
18. D. Parra, S. Sahebi, Recommender systems: sources of knowledge and evaluation metrics. In: Advanced Techniques in Web Intelligence-2, pp. 149–175 (2013). https://doi.org/10.1007/978-3-642-33326-2_7. Accessed 6 Aug 2020
19. X. Sun et al., Interactive genetic algorithms with large population and semi-supervised learning. Appl. Soft Comput. **12**(9), 3004–3013 (2012). https://doi.org/10.1016/j.asoc.2012.04.021. Accessed 31 Oct 2020
20. M.O. Akinsolu et al., A parallel surrogate model assisted evolutionary algorithm for electromagnetic design optimization. IEEE Trans. Emerg. Top. Comput. Intell. **3**(2), 93–105 (2019)
21. J. Tian et al., Granularity-based surrogate-assisted particle swarm optimization for high-dimensional expensive optimization. Knowl. Based Syst. **187**, 104815 (2020). www.sciencedirect.com/science/article/abs/pii/S0950705119302904. https://doi.org/10.1016/j.knosys.2019.06.023. Accessed 11 Aug 2020
22. C. Zhou et al., ATRank: An Attention-Based User Behavior Modeling Framework for Recommendation (2017)
23. J. Lu et al., Recommender system application developments: a survey. Decis. Support Syst. **74**, 12–32 (2015). https://doi.org/10.1016/j.dss.2015.03.008. Accessed 25 May 2020
24. M. Mao et al., Multiobjective e-commerce recommendations based on hypergraph ranking. Inf. Sci. **471**, 269–287 (2019). https://doi.org/10.1016/j.ins.2018.07.029
25. H.T. Cheng et al., Wide & deep learning for recommender systems. In: Proceedings of the 1st Workshop on Deep Learning for Recommender Systems - DLRS 2016 (2016). https://doi.org/10.1145/2988450.2988454. Accessed 2 July 2020

26. M. Alfarhood, J. Cheng, DeepHCF: a deep learning based hybrid collaborative filtering approach for recommendation systems. IEEE Xplore (2018). https://ieeexplore.ieee.org/document/8614046. Accessed 21 Aug 2020
27. S. Ryu et al., A bayesian graph convolutional network for reliable prediction of molecular properties with uncertainty quantification. Chem. Sci. **10**(36), 8438–8446 (2019). https://doi.org/10.1039/c9sc01992h. Accessed 20 Jan 2020
28. Y. Zhang, A.A. Lee, Bayesian semi-supervised learning for uncertainty-calibrated prediction of molecular properties and active learning. Chem. Sci. **10**(35), 8154–8163 (2019). https://doi.org/10.1039/c9sc00616h. Accessed 13 July 2020
29. Y. Gao et al., Forecasting the cumulative number of COVID-19 deaths in China: a boltzmann function-based modeling study. Infect. Control Hosp. Epidemiol. **41**(7), 841–843 (2020). https://doi.org/10.1017/ice.2020.101. Accessed 1 Sept 2020
30. L. Bao et al., Restricted boltzmann machine-assisted estimation of distribution algorithm for complex problems. Complexity **2018**, 1–13 (2018). https://doi.org/10.1155/2018/2609014. Accessed 20 July 2020
31. L. Bao et al., Restricted boltzmann machine-driven interactive estimation of distribution algorithm for personalized search. Knowl. Based Syst. **200**, 106030 (2020). https://doi.org/10.1016/j.knosys.2020.106030. Accessed 20 July 2020
32. S. Peng et al., Estimation of distribution with restricted boltzmann machine for adaptive service composition. In: 2017 IEEE International Conference on Web Services (ICWS) (June 2017). https://doi.org/10.1109/icws.2017.23. Accessed 6 Aug 2020
33. N. Srivastava et al., Dropout: a simple way to prevent neural networks from overfitting. J. Mach. Learn. Res. **15**, 1929–1958 (2014). www.cs.toronto.edu/~hinton/absps/JMLRdropout.pdf
34. A. Sherstinsky, Fundamentals of recurrent neural network (RNN) and long short-term memory (LSTM) network. Physica D **404**, 132306 (2020). arxiv.org/pdf/1808.03314.pdf. https://doi.org/10.1016/j.physd.2019.132306. Accessed 18 Mar 2020
35. D. Kim et al., Convolutional matrix factorization for document context-aware recommendation. In: Proceedings of the 10th ACM Conference on Recommender Systems - RecSys 2016 (2016). https://doi.org/10.1145/2959100.2959165. Accessed 9 June 2020
36. S. Xu et al., CXNet-M1: anomaly detection on chest X-Rays with image-based deep learning. IEEE Access **7**, 4466–4477 (2019). https://doi.org/10.1109/access.2018.2885997. Accessed 16 May 2020
37. X. Xu et al., Deep Learning System to Screen Coronavirus Disease 2019 Pneumonia. arXiv: 2002.09334 [Physics] (Aug 2020). arxiv.org/abs/2002.09334
38. Deep Convolutional Neural Networks to Diagnose COVID-19 and Other Pneumonia Diseases from Posteroanterior Chest X-Rays. DeepAI (May 2020). https://deepai.org/publication/deep-convolutional-neural-networks-to-diagnose-covid-19-and-other-pneumonia-diseases-from-posteroanterior-chest-x-rays. Accessed 2 Aug 2020
39. C. Szegedy et al., Inception-v4, Inception-ResNet and the Impact of Residual Connections on Learning. arXiv:1602.07261 [Cs] (Aug 2016). arxiv.org/abs/1602.07261
40. A. Jaiswal et al., Classification of the COVID-19 infected patients using DenseNet201 based deep transfer learning. J. Biomol. Struct. Dyn., 1–8 (2020). https://doi.org/10.1080/07391102.2020.1788642. Accessed 26 Oct 2020
41. S. Minaee, Shervinmin/DeepCovid. GitHub (Oct 2020). https://github.com/shervinmin/DeepCovid. Accessed 21 Oct 2020
42. N.S. Punn, S. Agarwal, Automated diagnosis of COVID-19 with limited posteroanterior chest X-Ray images using fine-tuned deep neural networks. Applied Intelligence (Oct 2020). https://doi.org/10.1007/s10489-020-01900-3. Accessed 27 Oct 2020
43. D. Kingma, J.L. Ba, ADAM: A Method for Stochastic Optimization
44. COVID-19 Activity - Dataset by Covid-19-Data-Resource-Hub. Data.World, https://data.world/covid-19-data-resource-hub/covid-19-case-counts. Accessed 19 Aug 2020
45. Data.World. Data.World, https://data.world/login?next=%2Fcovid-19-data-resource-hub%2Fcovid-19-case-counts%2Fworkspace%2Ffile%3Ffilename%3DCOVID-19%2BActivity.csv. Accessed 23 July 2020

46. UCSD-AI4H. UCSD-AI4H/COVID-CT. GitHub (May 2020). https://github.com/UCSD-AI4H/COVID-CT
47. S.J. Fong et al., Composite monte carlo decision making under high uncertainty of novel coronavirus epidemic using hybridized deep learning and fuzzy rule induction. Appl. Soft Comput. **93**, 106282 (2020)
48. CDC. Coronavirus Disease 2019 (COVID-19). Centers for Disease Control and Prevention (Feb 2020). www.cdc.gov/coronavirus/2019-ncov/symptoms-testing/coronavirus-self-checker.html
49. Coronavirus. Www.Who.Int, www.who.int/health-topics/coronavirus/coronavirus#tab=tab_3
50. Infectious Diseases Society of America Guidelines on the Treatment and Management of Patients with COVID-19. Www.Idsociety.org, www.idsociety.org/practice-guideline/covid-19-guideline-treatment-and-management/
51. K. El Asnaoui, Y. Chawki, Using X-ray images and deep learning for automated detection of coronavirus disease. J. Biomol. Struct. Dyn. 1–12 (2020). https://doi.org/10.1080/07391102.2020.1767212
52. H. Ko et al., COVID-19 pneumonia diagnosis using a simple 2D deep learning framework with a single chest CT image: model development and validation. J. Med. Internet Res. **22**(6), e19569 (2020). https://doi.org/10.2196/19569. Accessed 20 Dec 2020
53. O. Gozes et al., Coronavirus Detection and Analysis on Chest CT with Deep Learning. arXiv: 2004.02640 [Cs, Eess] (Apr 2020). arxiv.org/abs/2004.02640. Accessed 21 Aug 2020
54. D. Singh et al., Classification of COVID-19 patients from chest CT images using multi-objective differential evolution–based convolutional neural networks. Euro. J. Clin. Microbiol. Infect. Dis. 1–11 (2020)
55. J. Zhao et al., Hybrid recommendation algorithms based on ConvMF deep learning model. In: Proceedings of the 2019 International Conference on Wireless Communication, Network and Multimedia Engineering (WCNME 2019) (2019). https://doi.org/10.2991/wcnme-19.2019.36. Accessed 11 Aug 2020

An Integrated GIS/GPS Based e-Governance Approach for Different Agencies to Fight Against COVID-19

Meghna Patel and Mitesh Patel

Abstract COVID-19 Coronavirus is now one of the most contagious diseases discovered recently and declared as global pandemic by World Health Organization (WHO). Emergency services are running at full capacity to keep citizens safe. Different agencies like hospitals, health centres, police departments, district administration etc. have to work hand to hand. The key to manage such an unprecedented and drastic situation with involvement of multiple agencies/authorities is effective use of technology. Understanding the spatial and time based dynamics of COVID-19 is useful for its mitigation, it helps to study the impact of the pandemic and can be used in decision making, planning and community action. Health geography highlights the interaction of public health offcials, affected actors and first responders to evaluate the disease propagation and predict new outbreaks. This chapter is an effort to evolve an integrated GIS/GPS based e-Governance platform which effectively used by all those agencies involved in control of the disease. It is useful to monitor and manage tracking of human movement, coordinate to health department and sanitary services, prepare proper scientific and socioeconomic analysis and progressive flow related to COVID-19 situation/spread in real time basis. The platform is having different modules for different agencies including patients for effective and timely management of the crisis with minimum overheads and ensuring minimum suffering to all agencies. It contains modules for Patients/Citizens, Police Department, Health Department, Hospitals, Corporations/NGOs, ground workers etc. All agencies will have real time GIS based dashboards, MIS reports and notification systems to deal

M. Patel (✉)
A.M. Patel Institute of Computer Studies, Ganpat University, Ganpat Vidyanagar, Kherva 384012, Gujarat, India
e-mail: meghna.patel@ganpatuniversity.ac.in

M. Patel
Space Applications Centre, Indian Space Research Organisation, Department of Space, Ahmedabad 380015, Gujarat, India
e-mail: mitesh@sac.isro.gov.in

J. Nayak et al. (eds.), *Understanding COVID-19: The Role of Computational Intelligence*, Studies in Computational Intelligence 963, https://doi.org/10.1007/978-3-030-74761-9_20

with the situation just in time. Hope its findings will inspire new reflections on the COVID-19 pandemic.

Keywords COVID-19 · e-Governance · GIS · GPS · Geo-fencing

1 Introduction of COVID-19

Many viral diseases influence our health. In recent time, a COVID-19 was detected in Wuhan, China in December 2019 and spread across the world. It has affected a huge part of global population [1, 2]. WHO announced COVID-19 as pandemic on 30th, Jan 2020. The disease is easily transmissible and infect humans having weak immunity [14, 15]. The present modern system of transportation transforms the world in a global village, which carries the rapid spread of COVID-19 virus [13]. WHO declared that COVID-19 fight could make or break the global war [3]. As a preventive measure Prime Minister of India, Shri Narendra Modi announced nation wide lockdown on 25th March, 2020. Following Table 1 shows the illustration of State wise status of COVID-19 in India.

Table 1 State wise status of COVID-19 in India as on 13th Oct, 2020 [12]

Name of State / UT	Active Cases		Cured/Discharged/Migrated		Deaths	
	Total	Change since yesterday	Cumulative	Change since yesterday	Cumulative	Change since yesterday
Andaman and Nicobar Islands	198	12	3770	6	55	
Andhra Pradesh	43983	2312	708712	5504	6256	32
Arunachal Pradesh	2940	49	9403	171	24	
Assam	28439	54	166039	1457	826	10
Bihar	10669	375	185911	1370	955	9
Chandigarh	1170	14	11898	111	192	1
Chhattisgarh	27421	73	116540	2769	1286	33
Dadra and Nagar Haveli and Daman and Diu	102		3064	5	2	
Delhi	20535	1166	284844	2975	5809	40
Goa	4465	191	33698	495	511	4
Gujarat	15414	281	133615	1442	3574	8
Haryana	10401	172	131228	1225	1592	13
Himachal Pradesh	2637	50	14690	219	251	1
Jammu and Kashmir	9992	474	72706	861	1333	11
Jharkhand	7776	391	84461	890	798	11
Karnataka	115795	4494	592084	12030	10036	70
Kerala	94473	1928	199634	7836	1025	22
Ladakh	961	19	4126	89	64	
Madhya Pradesh	14932	245	130721	1702	2645	21
Maharashtra	212905	8732	1281896	15656	40514	165
Manipur	2756	25	10707	203	93	2
Meghalaya	2434	44	5273	131	64	1
Mizoram	156	18	2046	36	0	
Nagaland	1409	150	5813	70	18	1
Odisha	23430	172	230192	2577	1040	18
Puducherry	4617	78	26555	264	565	2
Punjab	8576	699	112099	1234	3860	27
Rajasthan	21671	259	137848	1858	1665	15
Sikkim	384		2925	5	57	2
Tamil Nadu	43747	348	607203	5165	10314	62
Telengana	24208	306	189351	2009	1233	5
Tripura	3738	4	24623	220	317	1
Uttarakhand	6976	397	47609	678	762	15
Uttar Pradesh	38815	1204	393908	3342	6438	44
West Bengal	30604	368	262103	3155	5682	60
Total	838729	23124	6227295	77760	109856	706

Through international movements of people the COVID-19 crossed the boundary of India and spread into local population [2]. To prevent COVID spread in 1.3 billion people, Government of India has introduced different strategies and approaches for transmission control like nation wide lockdown, Do Gaj ki Doori, Vocal for Local, Aarogya Setu App etc. By imposing nation wide lockdown, Government has reduced the spread of the COVID-19 virus. Government has installed contact-less thermal scanners at airports, railway stations, bus stations and other public places. Insurance Regulatory and Development Authority (IRDA) has issued standard guidelines to all general, standalone and private health insurers to offer individual covid standard health policy called "Corona Kavach Policy". Government has encouraged and funded industries to expedite the COVID related research and development. Corporations and Urban Health Centres have setup tents to conduct rapid antigen test for free in the remotest location. As per the WHO estimate, there should be 3 beds per 1,000 people while India has only 0.7 beds per 1,000 people. Thus, to provide for more isolation wards, the Government converted more than 20,000 railway coaches into isolation wards. Reserve Bank of India has taken emergency measures by cutting different interest rates to record low to increase liquidity in the economy. United Nations (UN) and World Health Organization (WHO) has praised India's response to the pandemic as 'comprehensive and robust', terming the lockdown restrictions as 'aggressive but vital' for containing the spread and building necessary healthcare infrastructure. Oxford COVID-19 Government Response Tracker (OxCGRT) noted the government's swift and stringent actions, emergency policy-making, emergency investment in health care, fiscal stimulus, and investment in R&D of vaccine and drug and gave India a score of 100 for the strict response [21].

1.1 India's e-Governance in Pandemic

COVID-19 has put India's e-governance infrastructure and innovation to the test. Government of India has taken different initiatives in pandemic situation under the category of Government to Citizen services. This includes Aadhar based distribution of food supplies, transfer of funds under Direct Benefit Transfer (DBT) using Pubic Financial Management System (PFMS) and Aarogya Setu App for contact tracing, syndromic mapping and self-assessment. The Aadhaar based financial inclusion scheme, Pradhan Mantri Jan Dhan Yojana (PMJDY), brings low-income people into the banking system. This creates the backbone of a social security delivery system for social inclusion. The following Fig. 1 shows the illustration of Direct Benefit Transfer using Public Financial Management System during COVID-19.

Under the Jan Dhan Yojana, bank accounts linked with Aadhaar numbers are used to distribute direct benefit transfers, wages under the Mahatma Gandhi National Rural Employment Guarantee Act (NREGA) and other poverty alleviation and welfare delivery schemes, like the food and public distribution system. DBT played a major role in sustaining life of the under-privileged people of the society impacted by the COVID-19 crisis, helping millions in providing immediate relief in tiding

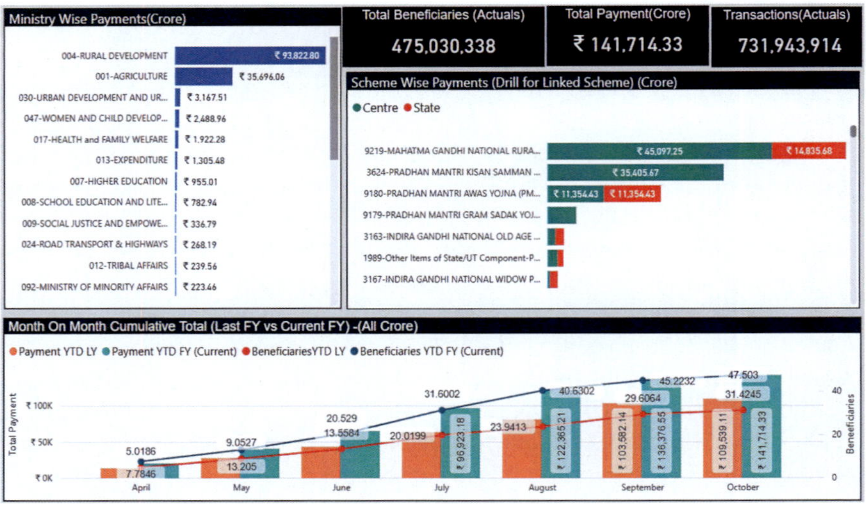

Fig. 1 Direct benefit transfer using public financial management system during COVID-19 [20]

over the turbulent period. Under 180 welfare schemes, the state governments using PFMS have disbursed Rs 9,217.22 crore to 4.59 crore beneficiaries between March 24 and April 17. Till 9th October, 2020, of the financial year 2020–21, total 47 Crore beneficiaries received DBT relief to the tune of Rs. 1,41,714 crores. Applications like UMANG, BHIM, and mPassport developed by the Government have transformed the way Government delivers to the citizens and citizens interact with the Government. Digital payment helps citizen to stay away from other people and observe social distancing. Unified Payments Interface, an instant real-time payment system developed by National Payments Corporation of India has gained popularity among people who aim to stay safe using digital payments. State Departments has also published a data on Open Government Data (OGD) Platform India—data.gov.in to bring the transparency into the system.

2 Overview

In the situations like pandemic or national emergency, agencies like central government, state government, district administration, municipal corporations, hospitals, police departments have to work in hand to hand. They all are having different softwares and procedures to manage the data related to affected geographical locations, citizens, medicine stocks, vacant bad etc. Because of heterogeneous development and scattered data it creates data inconsistency, redundancy, duplication of efforts and human errors. A central platform which caters to all the needs of different departments can provide effective MIS which helps all the stakeholders in decision making and resource management. An integrated e-governance platform with GIS based dashboard which compile and display the information in easily understandable format. This platform also communicates with other systems using APIs.

3 Objective of Proposed Work

Widespread of COVID-19 has socioeconomic impact on industries like Commercial, Education, Entertainment, Sports, Transport etc. An individuals are not loosing their lives but their job also [5]. To make the populations safe, the emergency facilities are running in their full capacities. Various agencies work hand to hand like police departments, hospitals, health departments, district administration etc. It is responsibility of each and every citizen to help the authorities to deal with this situation [4]. To manage the situation an integrated e-Governance platform which provides a common source of information for all agencies is the need of time.

All countries are influenced by COVID-19 situation and it has made an important effect on healthcare services and treatment systems. This pandemic has created the demand for various tools, technologies and applications which fulfills the requirement and solve various problems of different category of people within limited time frame. It has also creates a need for advanced technology which provides solution of essential medicines, healthcare management, checkup, detection, logistic, surveillance services with less involvement of human [6–8]. The proposed work is done to help in managing and controlling the spread of disease. It helps authorities to effectively detect and diagnose problems regarding COVID-19. In this work we discuss various modules which helps to deal with solving the problems of coronavirus. The system offers daily updates on infected people based different criteria like area, state, age, gender, medical history etc. The proper implementation of these systems can enhance communication and education regarding health of the public. This system gives various novel ideas and solutions to fight against medical emergencies at local and national level.

An effective e-governance system is needed for synchronous actions within citizens and various agencies. Gathering and Managing accurate information from various agencies is a difficult task. Incorrect information can put citizens in trouble and make them run from pillar to post for medical emergencies and logistics support. In this current scenario, centralized platforms can play a vital role for managing and exchanging the information which can be entered, used, verified and certified by different agencies like citizens, government departments, NGOs, Health Workers etc. The proposed work offers a GIS based information system and dashboard for tracking the coronavirus affected citizens. System provides the services to help agencies to manage the pandemic situation. The proposed system generates alerts regarding infected users and the close contact person who is infected with coronavirus using the GIS system. It also provides role based dashboard for different agencies which provides bird eye view of the situation. Dasboard also provides drill down MIS for critical information. Technologies like contact tracing, health QR, Geo-fencing is used as services in the proposed system.

4 Literature Review

Nowadays, GIS based technology is used in government, environment management, public safety, urban planning, management of natural resources, real estate industries, telecommunication, insurance companies, retail sector banking sector and many more. It is also used in mission like programs like Smart Cities, Bharatmala, Clean Ganga, Sagarmala, City Gas Distribution [9, 10].

Epidemiologists and different agencies like UNICEF, WHO, CDC used GIS based study and Map for taking decision and controlling the spread of Zika and Ebola viruses. To control and track disease like Dengue and Malaria, the institute NIMR (National Institute of Malaria Research) used a GIS based system. As well as, the Polio eradication, to reach in susceptible population areas to survey, analyse and map are used for vaccination using the GIS based system. It helped in finding out the children who unvaccinated, reasons behind it, to map healthcare centres in districts and so on [10].

Understanding of geography is needed to detect, understand and respond properly in virus outbreak. GIS helps researchers to map different parameters like demographic, geographic, previous history and origin of virus. It also helps into study spread pattern and intensity to take preventive steps, monitoring and control. Government agencies use GIS to analyse virus outbreak pattern in real-time bases by which can be identified the population at high risk. They can inform and take preventive measures for citizens like arranging healthcare facilities and medical support [16].

Geographic Information System (GIS) is used when location based information needs to be processed. It gives common ground for multiple agencies to collaborate using spatial analytics, mapping and location intelligence. Government agencies can map confirmed, active cases, recovered cases and can predict the future spread of virus [17].

Timeline maps are used to get the clear picture of how and where infection has spread over time. It also help to predict the future hot-spots. It can also be used with different demographic parameters like age, people with health issues etc. Interactive Maps, dashboards and story maps are effective ways to understand the needs of medical supplies, equipment, goods and services [10].

In current situation of pandemic, geo-spatial technology is gaining popularity globally. Technologies like contact tracing, syndromic surveillance, prediction of virus hot spots and geo-fencing are operationalized in very short duration. And more effective health care apps will be there shortly.

This kind of geo-spatial solutions are adopted world wide and people are accepting it very fast which was not known to people. It looks like this technology will gain popularity in the near future and people will soon use it effectively, but the question is, is the implementation of such an application which restricts and monitors a person's movement ethical?

4.1 Geospatial Technologies in COVID-19

Contact Tracing, Geo-fencing and virus surveillance dashboard are mainly used for controlling and monitoring COVID-19 outbreak [11].

4.1.1 Contact Tracing. To prevent further spread of virus and to track close contact of infected people, contact tracing app is commonly used. Before COVID-19, it was generally used to find disease carriers of tuberculosis, SARS, Ebola and MERS.

Till date vaccine is not developed and it takes time to develop herd immunity. In this situation contact tracing can play a key role. Implementation process needs to be automated and streamlined to make it effective. Countries like China, South Korea and India are affected first and they have implemented contact tracing apps in the starting phase. Australia, middle east countries and Europe have implemented afterwards [18, 19].

As far as technology is concerned, use of GPS based contact tracing is a good and cheaper solution in current situations. Application fetches location information with medium to high accuracy, and it also requires less efforts.

4.1.2 Quarantine Alert Systems. After it was confirmed that the virus is spreading by close contact with infected people, the government has started taking measures to restrict movement of COVID positive patients. Home quarantine or shift the patient to an isolation ward for two weeks is the common measure. China has implemented health code after lock down. There are other technologies to track the quarantine violators [18, 19].

Taiwan, Hong Kong and some states of India have implemented geo-fencing to detect quarantine violations in near real-time. In this, authorities set a virtual boundary around the place of quarantine. As soon as a mobile or device moves out of boundary, an alert is generated to authority. This solution is very effective and simple.

Kerala and Tamil Nadu have operationalised systems which determines device GPS signals to retrieve carrier's location while other states are in the process of implementation. Gujarat application is implemented which uses Geo-fencing and GIS mapping technologies to detect quarantine violation and to track movement of self-isolated persons. As such the government assured that information gathered is confidential and will be used for COVID-19 related issues only, but it still raises questions as it uses mobile services of private telecom companies.

Taiwan is using digital fencing based on proximity to the nearest mobile tower. This technology solves the privacy concern of citizens. Other countries like Indonesia, Italy and Australia are also showing interest in similar technology. Hong Kong is using Geo-fencing technology which preserves the privacy of citizens. When a person arrives into the city, he has to download the StayHomeSafe application and has to wear a smart bracelet. App continuously communicates with signals like Wi-Fi, cellular and Bluetooth and creates a unique code. If the application does not receive a signal then it generates an alert to the respective agencies about a person having moved out of a quarantined location. According to developers they have used these technologies instead of GPS because GPS is not effective in vertical buildings.

4.1.3 Coronavirus Maps and Dashboards. There are a number of initiatives on a global scale which shows near real-time insights of population, movement patterns which allows accurate prediction of virus spread, which helps authority to initiate action in a timely manner.

Companies like Apple, Google and Facebook developed dashboards with the use of data which they are collecting regularly without revealing personal data. Google and Apple use their map services for generating this kind of dashboard and prediction. Facebook Mobility Data Network uses location, census and satellite imagery. Facebook also introduced surveys on user symptoms to compile a country wide map of symptoms prevalence.

Biocomplexity Institute of the University of Virginia has developed surveillance dashboard which uses ArcGIS Living Atlas system to display new cases, recovered cases and death across the globe. Biocomplexity Institute has released the most sophisticated system. The researchers have decided to integrate disease modeling mechanism into GIS for epidemiological forecasting. Institute is also working on a dashboard to predict demand and shortage of medical equipment.

COVIDcast by Delphy Research Group is a good example of an interactive COVID-19 tracker map. It displays real-time information on cases and deaths as well as doctor visits, symptoms, and Google search trends. COVIDcast is an accurate solution for tracing and predicting the surges and falls in disease spread because the information is taken from authorized sources.

How the pandemic gave a platform for geospatial innovations Mostly, corona virus applications were integrated with geospatial technologies like GIS and GPS. Several geospatial solutions have been designed, developed and implemented to fight against the virus in this health emergency [18, 19].

4.2 The Argumentative

Governments and companies worldwide are investing to build solutions to stem the virus. In countries like South Korea, New Zealand, and China, corona virus apps have already been operationalised. But still people are raising concerns about software and systems fetching and using their personal data and location information. Several people are looking towards it as it violates the human rights.

Even disclosing GPS data without personal identity information can reveal the behavior or pattern of someone which can be harmful. It also raised concern about transparency of application. Some countries have open sourced the contact tracing application and geo-fencing solution so people can test and verify the application is working as claimed by the developer or not. Some applications also request users to fill personal information like name, address and mobile number for future correspondence.

In case of someone gains access to information, will create a huge controversy and will put the government into trouble. Countries knowing the implementation of it are hesitant to use contact tracing apps. They preferred to stay away until any privacy preserving protocols were introduced. But, it is not the fault of geo-spatial technology itself.

5 Proposed Research Methodology and Research Model

An integrated e-governance platform is a combination of workflows, services, user agencies and modules. In any application user agencies are the actual users who use the system. Workflows are predefined steps of the process which have to be followed for achieving a particular task. Workflow also includes the rule engine which can be applied to the process based on user inputs. To achieve a smooth operation of workflows different modules have been designed. Modules can be part of workflow or workflows can also be a part of a larger module. Services are the process which runs independently, there may be external services which the system communicates like services of other systems, hardware interfaces etc. The following Fig. 2 shows the illustration of integrated e-Governance platform.

5.1 User Agencies

Main reason to build any e-Governance system is to provide government services to all citizens. There are many users involved in the system like Citizens, Patient/Home quarantined persons, Municipal Corporations, Hospitals, Police Departments, NGOs, Health Departments, Monitoring agencies like State and Central departments. All these agencies will work together in the fight against COVID-19. Relevant information is shared across different user agencies as per need and requirement. Different user agencies have a number of roles to be played in the system.

5.1.1 Citizen: Citizens are a key user of the system. There are different roles which a citizen has to play or can take a benefit from the system.

A citizen can check if any infectious person is there nearby with the use of mobile/browser. System will fetch the GPS location of the mobile/device and send it to the server. Server scans the nearby locations of 100 m for any infected location or quarantined house. In case any infected person or location is there then it will show the location with a red marker on the map without revealing details of the infected person. Mobile application continuously scans the surrounding location in the background for any infected person and generates notification if it finds infected person nearby.

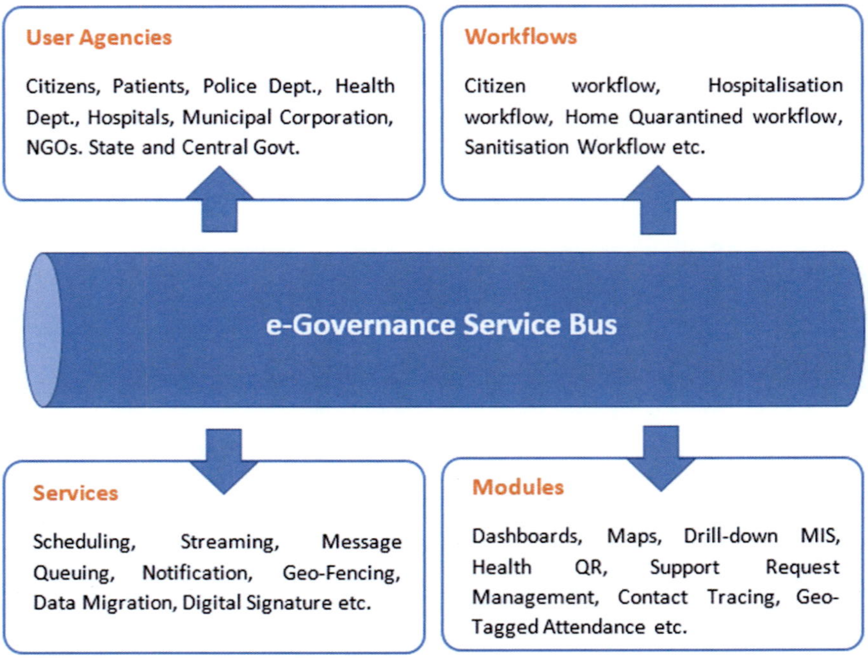

Fig. 2 Integrated e-Governance platform

If a citizen or his family member develops any symptoms of COVID-19 then he can submit the details to the authorities by filling up an online form with detailed information. Online form contains different information like name, age, contact number, any disease or comorbidity. After submitting the form all related agencies get notification regarding the citizen. All relevant agencies will get the information on their respective dashboards.

Anyone can submit the information about another person or neighbour who is having symptoms or having a travel history.

In a pandemic situation all person/vehicle movement is restricted. For movement a citizen has to submit a request to the corporation for movement due to emergencies or essential services. After approval, a letter is issued which can be shown to check posts. A Citizen can generate requests for e-pass for movement for particular date range and for particular location.

In the situation of national emergency or breakdown, there is a need for a huge workforce. People of different qualifications with different fields can play a major role in the system. A person can submit a request for volunteers online. Online form includes details like qualifications, age, willingness to work hours, work online or on-site etc.

5.1.2 Patient/Home Quarantine Person: As per the request from the citizen, worker of primary health centre visits the home of a citizen and if symptoms are found then he can suggest for COVID-19 test. Based on the test result, the health worker will flag the citizen as a Covid positive patient. He can suggest hospitalisation or home quarantine treatment.

Home quarantined people are not allowed to go outside for their daily needs. So he is always dependent on someone who helps him. For any requirement he has to wait for someone. Same way supporting agencies like health workers have to visit the house even if there is no need for support which also increases the chance of health workers getting infected. To resolve this issue, the patient module has different functionalities like, patients can send requests for medical help or logistic help. Requests for logistic support contain details like food supply, daily needs, sanitization requests etc. Medical support requests contain request types like regular medicine, medical emergency, need for doctor, need for nurse etc. Requests are then forwarded to different agencies based on request type and priority. Logistic support request is routed to corresponding agencies like NGO, Gram Panchayat, Police Station or Administrative authorities. Requests for medical help are routed to medical teams/hospital/health department/primary health center of a particular area on priority basis. Requests get automatically escalated to higher authorities/agencies if not resolved in a stipulated time frame.

Movement of patients is restricted and monitoring of his movement is a challenge for the police department. So, a module for home quarantined people is having location tracing functionality. It automatically monitors location of phone, in case a patient goes outside of the house then alerts are being generated and also forwarded to police officials for further action.

To help the police department, system automatically generates random attendance requests for home quarantined patients. Patient has to upload a geo-tagged selfie from the home. If geo-tagged location is far from home/quarantined location then application automatically notifies the authority. Application also uses technologies like face recognition/fingerprint recognition to increase the accuracy of the system.

5.1.3 Municipal Corporation/Panchayat: AMC officials or sarpanch can define quarantine zones by drawing a polygon in a particular area. Dashboard for corporations gives detailed picture of the quarantine zone.

Once a person is marked as Covid positive or area marked as quarantined zone the sanitization activities can be carried out. Reminders of next scheduled sanitization are generated automatically. System keeps a record of the area where sanitization has been done with a timestamp. Also, users can manage the inventory of sanitization materials like glows, masks, primary medicines etc.

A citizen can request e-pass for movement during lockdown. Municipal Commissioner or sarpanch can approve e-pass for movement of citizens for particular date and time with source and destination location to be travelled. Details about vehicles and passengers are also captured in this module.

5.1.4 Hospitals: Once a patient's report is positive, the health department makes a decision if he has to be shifted to hospital or not. Health workers feed information into the system and allot the hospital based on vacant beds. System automatically informs the hospital about the patient's condition and information like age, gender, health issues, covid test report etc. Based on the information, the hospital arranges the bed, medicines and informs the doctors.

Medical inventory management is very important in nationwide emergencies. Government agencies are putting their best efforts to monitor and manage the stock and needs of all hospitals. Central inventory management is very important at this time. Hospitals can enter the details about stock of medical equipment, medicines, PPE kits etc. Based on available stock, the system generates different inventory reports for all hospitals and also predicts the future needs.

Hospitals can enter details about occupied/vacant beds and details about allotment of bed to the patient. Details about patients and their health can be entered like Critical/Normal/Recovered etc. Allotment of bed and wards like General/ICU/Special etc. can also be entered. To bring transparency in the system, a realtime dashboard displays the appropriate and relevant details on the website publicly.

Once a patient is recovered and relieved from hospital, the same is informed to the local health centre so they can monitor the patient's health until he is fully recovered or completes his quarantined period.

5.1.5 Police Department: Police officials are doing endless efforts to maintain law and order in their respective jurisdiction. Technologies like medical QR codes, face recognition, e-pass and geo-fencing can help the officials. Police department defines the area on the map which they want to put under their surveillance.

The home quarantined person system automatically generates requests for geo-tagged selfie in which patient has to capture his photo with GPS location. System automatically run face recognition algorithm to check if uploaded photo is of same patient or not. It also checks the geographic location of the phone. If location is too far from the patient's quarantined location then system generated alerts for police officials. All data along with time stamp is being recorded.

Police officials can make video calls through the app. System records the details like the patient has answered the call or not. If not then police officials can set reminders for the next call.

With the use of a Geo-fencing module if any infected person is going out of his quarantined location then the alert will be generated for police.

For movement during lockdown, E-pass can also be verified using this system. QR code can help the officials to easily scan and verify the details. Police officials can cancel the pass if any citizen violates the rules. So, next time if he applies for e-pass then the corporation office can verify the information about the violation.

5.1.6 NGOs and Volunteers: NOGs and volunteers can help the government for logistic support. User can register into the system and can define the area in which he would like to serve. The request then forwarded to corporation or panchayat for approval. Once approved, they can get alerts about patients who need any logistic support.

Requests for volunteers can be sent to NGOs or corporations. They can give tasks to the volunteer according to their education, skills and preferences.

If a patient/quarantined person generates any request for logistic support then NGO or volunteer which is coming in the same area will get the notification about the logistic needs.

An NGO or volunteer can flag the request as serving so other NGOs will not be able to serve the same request. After the request is served, the user will flag the request as served.

Citizens who are served by NGO can also write reviews or give ratings based on experience. In case, the citizen is not satisfied with the service then he can escalate this to the higher authority. Which then will be forwarded to corporation or panchayat and can be resolved without much delay.

5.1.7 Health Department: Health department/primary health care center plays a vital role in the system. Health workers who visit door to door for testing people. Enters the details about any suspected COVID positive patient or citizens who is having symptoms can register the request for medical support. Then health workers conduct medical tests and also provide some basic precautionary medicine.

Requests for COVID test will be routed to the team of pathologists. After results come, the pathologist can flag the patient as +ve or -ve. Pathologists can also upload the test report for future reference.

Once a person is declared +ve by primary health centre, corporation will take decision, if patient will be home quarantined or shift him to hospital based on condition.

If he feels the need for COVID test then he will add detailed information like name, mobile number, address, gender, age, medical conditions etc. while registering the record system will also fetch geolocation and save. Details entered by health workers are the first authenticated source of information about the patient.

Health department gets the notification if any hospital needs medical support like medicines, PPE kits etc.

In the case of home quarantined person, the system then sets a quarantined period of 14 d. After 14 d the system schedules the appointment to conduct a test and confirm if the patient is fully recovered or not.

5.1.8 Monitoring Agencies (State/Central Government): Agencies like state and central government will have a central dashboard from where they can get the complete picture in a summarised manner.

Dashboard can display GIS based maps of affected areas. Timelines can be generated to give a picture about the spread of the disease. Drill through MIS can be used to give detailed view of each and every patient and hospitals.

Government agencies can also monitor if there is a demand for stock items like medicines, masks, medical equipments and PPE kits which are not fulfilled by the health department.

Any request by citizens, patients, home quarantined people, hospital etc. can be escalated if not fulfilled in a stipulated time frame.

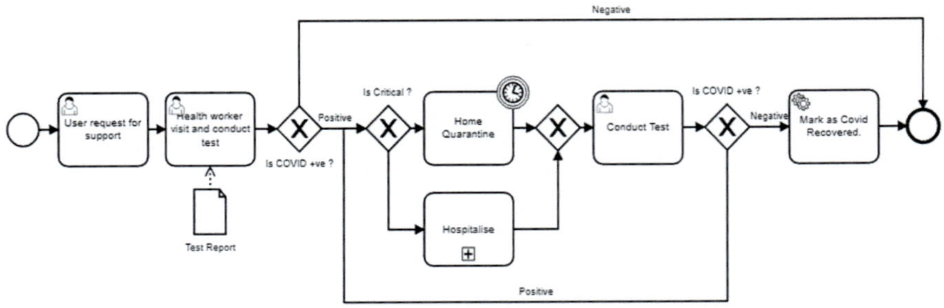

Fig. 3 Workflow

5.2 *Workflows*

Workflow is the sequence of steps to complete a particular task. It passes the data between different users and processes. It standardizes the process. Business Process Model and Notation (BPMN) is a flow chart method which models the business processes. It gives a clear picture of every business process which can be easily modify and deploy.

As an integrated system, there are number of workflows and rule engines which send information and documents from one department to another. It gives a standardized process across departments or user groups. The following Fig. 3. Shows illustration of Workflow.

Workflow for a citizen is described here for reference. As described one workflow divides into user tasks, processes, decisions, timer based events and documents.

5.2.1 Citizen Workflow is a workflow which consists of all processes from COVID test of a citizen who is suspected to be positive to relieve from hospital or when quarantined period is over. This involves different subflows like home quarantined, hospitalisation etc.

5.2.2 Home Quarantine Workflow starts with the patient found positive and is not critical so, health worker decides to suggest the home quarantine. It triggers a timer event of a quarantined period which is 14 d. It has subflows of logistics and medical request generation. If a request is not served in a stipulated time frame then it will escalate to higher authorities. A parallel process of sanitisation is also initiated which is fulfilled by corporations or panchayats.

5.2.3 Hospitalisation Workflow starts with health workers suggesting hospitalisation. He allocates the hospital based on vacant beds. Health worker also feeds the necessary information about the patient like name, age, whether the patient's condition is critical or not, details about disease like B.P, diabetes, asthma etc. He also uploads COVID test reports. System automatically generates the alert for the respective hospital. Hospital then prepares the medical team, medicines, PPE kits, arrange beds and prepare for further pathological reports. After admitting a patient, the system updates the dashboard of the website which shows statistics for the general public. During a period when a patient is hospitalised, staff updates inventory, patient's condition and experience details like doctor visit, ICU and pathology charges etc. Standard charges among all hospitals are calculated based on data entered. If it is a Government hospital the system automatically sends requests for medicines and PPE kits to the health department when it reaches re-order level.

5.2.4 Attendance Workflow defines the steps to mark attendance of home quarantined people. This workflow is divided into two processes. One is random attendance and one is from the police department. Random requests are generated by the system at any random time. This process sends notification alerts to quarantined people. Person has to capture his selfie from the mobile app. This process also captures location and time. System runs a face recognition algorithm and confirms the captured photo is of the patient. It also checks that the captured location is not away from a quarantined location. If the photo is not of a quarantined person or location is not within a permissible area. Then system sends an alert to the police officials with details. In case the user has not marked the attendance in a stipulated time frame then it also generates alert to the police officials.

5.3 Services

Services are programs which run in the background continuously to perform some specific tasks. It mostly runs as a OS service or background process. Services don't need any user intervention and it doesn't communicate with GUI directly. Any information is logged in log files or databases.

5.3.1 Data Migration service migrates data from different sources. Authenticated and accurate data is the key to success of any application. Base data of citizens can be migrated into the system from different sources like Aadhar, PAN, ration card, license records etc. This data is used by health workers to verify the details of particular citizen. It saves time of data entry and also gives accurate, authenticated and verified information.

5.3.2 Digital Signature services like eKYC, Aadhar enabled authentication, Digi locker can be used effectively. Digitals forms can be filled without going into time consuming processes like paper form for hospitalisation, discharge certificates, course of treatment done, insurance papers, test reports etc. it will also give fast, accurate and contactless form signing, storing and sharing platform.

5.3.3 Notification service generates notification for different agencies. Push notifications/SMSs can be sent to mobile devices or web sites. Notification can be sent on an event, on schedule or by user. Notification message pops up on mobile device. Users don't have to be in the app or using their device to receive them. System uses notification Services for all user groups like patients receiving notification for attendance, any infectious person nearby. Police officials receive notification if any patient is roaming out of 50 mtr radius of his home. NGO or volunteer notification if a user requests logistic support. Primary health centres receive notification if home quarantined people send requests for medical support.

5.3.4 Streaming service is used for streaming data like maps and GIS. Streaming data which is constantly changing like showing live GPS location of volunteers, health workers, patients etc. on map with interactive dashboards.
5.3.5 Scheduling service is used when a program has to execute at a particular time or at a particular interval like migration of data at particular intervals, sending messages at particular times, schedule backup to a remote location. Send reminders to user agencies like reminders for pathological tests, reminders to health workers to supply medicines to home quarantined people, reminders for corporation officials to sanitize particular areas. In some modules like citizen location data which is accumulated over the time can be deleted if not required after some days automatically.
5.3.6 Geo-Fencing service is a trust-but-verify approach. The service runs constantly in the background and checks if a quarantined person is roaming beyond 50 mtr area of quarantined location. As soon as he moves beyond a geo-fenced area, the system generates alerts for users as well as police officials.
5.3.7 Contact Tracing service continuously registers the owner's location data and transmits it to a database for analysis and contact identification. When someone they crossed paths with for an extended time is tested COVID-19 positive, the owner gets a notification, describing when and where the contact happened without disclosing personal details.

5.4 Modules

Modules are individual programs which contain business logic. It can be integrated with other modules, workflows or API to achieve a particular task. Loosely coupled modules are easy to develop, maintain and test.
5.4.1 Health QR Code is the colour coded QR code having details about the person and details. If a citizen is COVID positive then his QR codes colour will be RED, if he has come in contact with an infected person then colour will be ORANGE, if he has violated rules like moving out of quarantine zone while he is home quarantined then colour will be BROWN. If person is not in COVID positive or not in contact of infected person then colour will be black. This QR code is very effective, anyone can check other's QR code and can verify that a particular citizen is COVID positive or not. Authorized person can scan QR code and view the details like citizen's basic information, history of COVID infection or he details about rules violated during lock down etc.
5.4.2 Dashboard is an important module which combines the maps, markers, statistics, and notifications on a web page or mobile screen. Dashboard constantly communicates with different services like maps and notifications and updates the map at runtime. Dashboard updates at either on a particular time, on particular events or data changes.
5.4.3 Drilldown MIS allows users to view different layer of data by clicking on specific data element of web page in an application. Different MIS like statistics of quarantine zones, support requests can be displayed in the MIS, users can click on

elements and further details can be displayed. All reports can be exported in Excel or pdf format.

5.4.4 Support Request module allows different functions like logistic support request, medical support request etc. It also gives functionality like request escalation, in which requests are automatically forwarded to higher authorities if not fulfilled in a stipulated time frame. It is having APIs to give details about requests based on different parameters like area wise, category wise and criticality wise.

5.4.5 Maps module used to integrate display maps with GIS functionalities like map, zoom and panning in webpages and dashboards. Markers are used to point locations and containment zones.

5.4.6 Contact-Tracing module used to find the people who have came in contact with infected person. Geo-spatial queries can be executed to find the people who came in contact at particular time on same location.

5.4.7 Geo-attendance module allows user to capture the photograph which also store geographic location including latitude and longitude of quarantined location. Using this module, agencies can send a request to home quarantined people. Individual has to upload geo-tagged photo to the portal or mobile application. On server side face recognition algorithm will compare the uploaded photo to the photo captured at the time of health worker's visit. Geo-spatial query can determine if a photo is captured from home quarantined location or from different location. If home quarantined person has violated any of the rule then system generates an alert for agencies.

5.5 *Proposed Implementation Architecture*

To implement such a huge system, the architecture plays very important role. Architecture should be scalable and can be used in heterogeneous environment. The following Fig. 4 shows the illustration of Blueprint of Scalable Architecture.

5.5.1 Angular provides an effective way of communication and modularised platform to communicate between GUI and Restful APIs. Two way binding is a useful feature in GIS based applications where live data updates can reflect on a webpage synchronously. Angular with Material UI is a great combination for frontend development. Single Page Application (SPAs) can render pages very fast. Responsive web layout can easily adapt to different screen sizes like mobile, tablets, monitors and TVs.

5.5.2 Mobile Application is used when there are hardware dependencies like camera and GPS services and local storage for files and data processing.

5.5.3 Restful APIs are the endpoint which communicates with GUIs like mobile apps and websites, and Business logic like microservices.

5.5.4 Microservices are loosely coupled programs which work to achieve different tasks.

5.5.5 NodeJS is a cross-platform, open-source runtime used for building networking and server-side application. It used to build java script micro services. Architecture which includes applications like JSON API-based applications, I/O bound

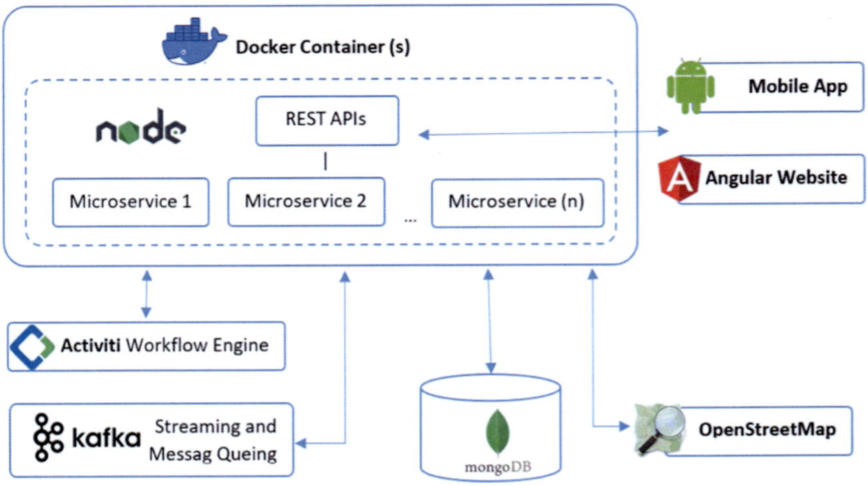

Fig. 4 Scalable architecture blueprint

applications, data streaming applications, single-page application and data-intensive real-time application can be developed with its reach libraries framework. NodeJS applications can be easily deployed into docker containers.

5.5.6 Docker Containers can be used to deploy APIs and microservices. These containers can be scaled up and scaled down as per the requirement across different physical machines or geographically spreaded locations.

5.5.7 Activi BPMN Workflow Engine can easily manage the complex business process and rule engines. Workflow engine-like activities can be integrated with microservices via APIs. Workflows can be easily modified flow diagrams like GUI.

5.5.8 Streaming and Message Queuing Services like apache kafka can be used to manage and monitor continuous streaming data like locations and notifications.

5.5.9 No SQL Database like MongoDB can be used considering the volume of data and OLTP kind of transactions.

5.5.10 Open Street Map is an open-source project which enables users to integrate maps into application. Interactive maps are a key for effective GIS. Location or person can be pinpointed with a map marker. Police, NGO, Health department can draw boundaries of areas which are coming under their jurisdiction with the use of polygon tools.

5.6 Proof of Concept

To demonstrate and study the feasibility of the proposed system. Some of the modules have designed, developed and tested as a proof of concept. Technologies used for

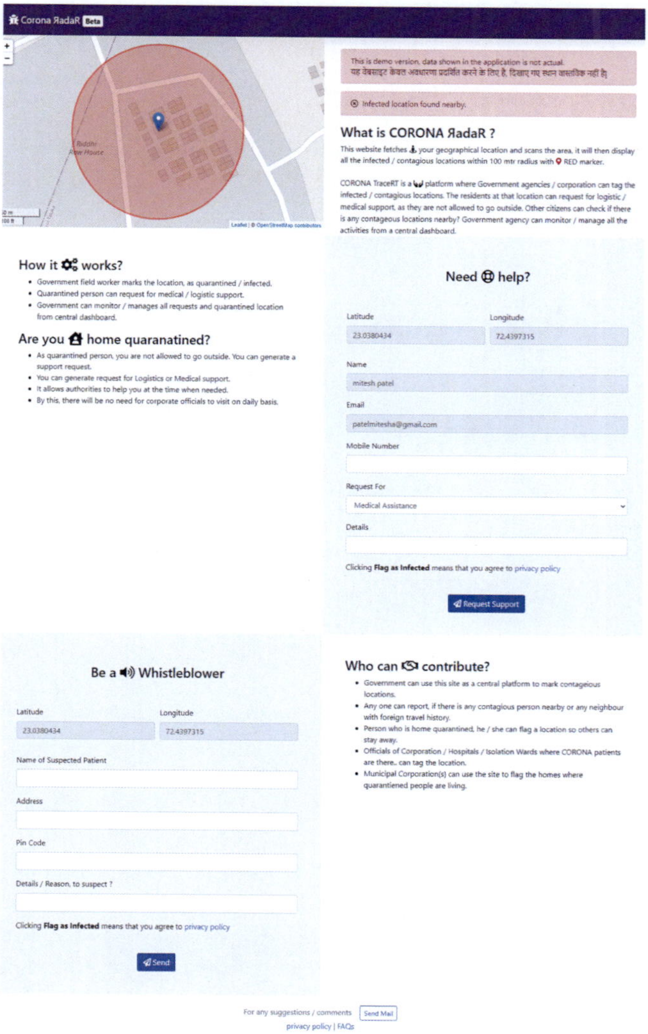

Fig. 5 Citizen module

development are the same as mentioned in implementation architecture. Following Fig. 5 shows the illustration of Citizen Module.

Citizen screen covers different functionalities like scan nearby area for infected person or quarantined location, support request generation etc. System automatically fetches the location of the device and scans the nearby locations if there is a infected person nearby then it will display red circle and generates an alert. Alert does not contains any sensitive information which reveals the identity of infected person. Citizen can generate the request for medical or logistic help. Citizen can also inform

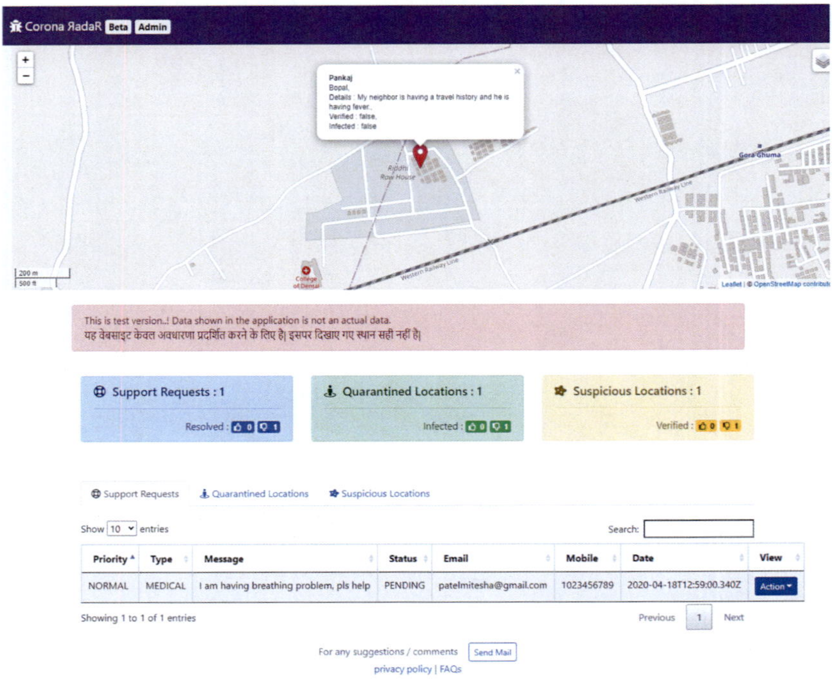

Fig. 6 Dashboard

the authority about other suspected person who is having travel history or who is not feeling well. Following Fig. 6 shows the illustration of Dashboard.

Dashboard shows all the information at a central place. Different agencies can have different view as per the requirement. It shows different information like Support Requests, Quarantined Person's location etc. All information is displayed in different layers in map. By clicking on map user can get the detailed information about the location. By clicking on data the location of request is updated in the map.

Significant benefits of System

- GIS based maps can be very useful when catering to different geographical location.
- Different alerts and notifications can be generated for different agencies.
- Daily reports can be generated with one click of a button.
- Data compilation can be very easy when all information is at a central location.
- Operations of different departments and citizens can be made smooth.
- Miscommunication and wrong information can be avoided because information is flowing through the verification process.
- Use of open source technologies to reduce cost and future enhancements.
- Administration can estimate the current situation of patients, medical supplies and logistic needs.

- Modules can be developed through an app in which, system can generate random alerts for a quarantined person to upload geo-tagged photos to verify he/she is not roaming around.

Utilization of information:

- Citizens - can get alerts, if any infected person is there nearby.
- Health department - can send medical support to a citizen. Department can also make an arrangement/quarantine facility in case of need.
- Corporation - can send a team for sanitization or other supports if anyone found CORONA positive.
- Police departments - can monitor the location so no one enters into the containment area.
- NGOs - logistics like food/medicines/transport can be given to quarantined/needy people with the support of NGOs or local management.
- Administration - information can be utilised based on hierarchy from localities to district, district to state administration and finally central authority can also monitor and manage all the different activities.

Challenges of the system

- Maintaining privacy is a big challenge when capturing and processing personal information, geographic location etc.
- Continuously capturing and processing geographic data like GPS location may consume more power of a device.
- Processing geo-spatial queries like scanning surrounding location for contact tracing is a challenging task.
- Design the database to handle large data.
- Handling lakhs of requests at a time and processing them like streaming services.
- Designing scalable and geographically distributed application architecture.

Limitation of the system

- Some of the modules are dependent on GPS sensor of mobile. User has to explicitly give the permission and it should be always switched on.
- Background services used for tracking the device consumes more battery.
- Citizen has to always carry mobile phone for different operations like geo-tagged attendance and QR code based operations.
- Phone should be connected to internet to communicate to the central server.

6 Conclusion

Geo-spatial application and integration with e-governance platforms is the demand of the current time. Integrated technology which can connect all government agencies like NGOs, corona warriors, Police, Hospitals and citizens can change the game and give an effective and collaborative solution to fight against the COVID-19. To achieve this goals, all agencies including citizens, government agencies have to work hand in hand. This kind of integrated platform can be developed module by module by and operationalised as per the need from village to district and state to central government level. Different modules catering to the different needs can exchange the information with each other using APIs. Roles and responsibilities can be assigned to different user or agencies. All activities can be monitored from central control room or in a geographically distributed departments. Platform can be open sourced or crowd sourced for faster development, where developers across the India can contribute to the project and government can use it without paying a huge cost to companies. This kind of development approach also brings transparency in the system. Data privacy and security can be achieved by deploying security at different levels like web security, network security, application security and data encryption. Applications can give the good performance in this kind of distributed architecture.

References

1. H. Lu, C.W. Stratton, Y.W. Tang, Outbreak of pneumonia of unknown etiology in Wuhan, China: the mystery and the miracle. J. Med. Virol. **92**(4), 401–402 (2020)
2. S. Kaushik, S. Kaushik, Y. Sharma, R. Kumar, J.P. Yadav, The Indian perspective of COVID-19 outbreak. VirusDisease **31**, 1–8 (2020)
3. V. Shenoy, S. Mahendra, N. Vijay, COVID 19 lockdown technology adaption, teaching, learning, students engagement and faculty experience. Mukt. Shabd. J. **9**(4), 698–702 (2020)
4. J.H. Yoo, The fight against the 2019-nCoV outbreak: an arduous march has just begun. J. Korean Med. Sci. **35**(4), e56 (2019)
5. M. Nicola, Z. Alsafi, C. Sohrabi, A. Kerwan, A. Al-Jabir, C. Iosifidis, R. Agha, The socio-economic implications of the coronavirus pandemic (COVID-19): a review. Int. J. Surg. (Lond. Engl.) **78**, 185 (2020)
6. M. Javaid, A. Haleem, Industry 4.0 applications in medical field: a brief review. Curr. Med. Res. Pract. **9**(3), 102–109 (2019)
7. M. Ienca, E. Vayena, On the responsible use of digital data to tackle the COVID-19 pandemic. Nat. Med. **26**(4), 463–464 (2020)
8. M. Javaid, A. Haleem, R. Vaishya, S. Bahl, R. Suman, A. Vaish, Industry 4.0 technologies and their applications in fighting COVID-19 pandemic. Diab. Metab. Syndr. Clin. Res. Rev. **14**, 419–422 (2020). https://doi.org/10.1016/j.dsx.2020.04.032
9. D. Maguire , ArcGIS: general purpose GIS software system, in *Encyclopedia of GIS*, ed. by S. Shekhar, H. Xiong (Springer, Boston, 2008), pp. 25–31. https://doi.org/10.1007/978-0-387-35973-1_68
10. How enterprises are using GIS to track Covid-19 impact - ET CIO (ETCIO.com, 26 March 2020), https://cio.economictimes.indiatimes.com/news/strategy-and-management/how-enterprises-are-using-gis-to-track-covid-19-impact/74821102. Accessed 21 Dec 2020
11. A. Koptelov, How geospatial tech helps fight COVID-19 right now (16 June 2020), https://www.itransition.com/blog/geospatial-tech. Accessed 21 Dec 2020
12. Ministry of Health and Family Welfare, Government of India, https://www.mohfw.gov.in. Accessed 21 Dec 2020
13. M.A. Shereen, S. Khan, A. Kazmi, N. Bashir, R. Siddique, COVID-19 infection: origin, transmission, and characteristics of human coronaviruses. J. Adv. Res. **24**, 91–98 (2020). https://doi.org/10.1016/j.jare.2020.03.005. ISSN 2090-1232

14. F. Zhou, T. Yu, R. Du et al., Clinical course and risk factors for mortality of adult inpatients with COVID-19 in Wuhan, China: a retrospective cohort study. Lancet **395**, 1054–1062 (2020). https://doi.org/10.1016/S0140-6736(20)30566-3
15. V. Raghupathi, An empirical investigation of chronic diseases: a visualization approach to medicare in the United States. Int. J. Healthc. Manag. **12**(4), 327–339 (2019)
16. A. Patkar, A.K. Berry, Awareness of COVID-19, preventive strategies for the same and factors affecting immunity in adult community dwellers with no medical background-a questionnaire based study. I. J. Health Sci. Res. **10**(7), 287–295 (2020)
17. I. Franch-Pardo, B.M. Napoletano, F. Rosete-Verges, L. Billa, Spatial analysis and GIS in the study of COVID-19. A review. Sci. Total Environ. **739**, 140033 (2020)
18. M.N.K. Boulos, E.M. Geraghty, Geographical tracking and mapping of coronavirus disease COVID-19/severe acute respiratory syndrome coronavirus 2 (SARS-CoV-2) epidemic and associated events around the world: how 21st century GIS technologies are supporting the global fight against outbreaks and epidemics. Int. J. Health Geogr. **19**(1), 8 (2020). https://doi.org/10.1186/s12942-020-00202-8. PMID: 32160889; PMCID: PMC7065369
19. M.E. Kretzschmar, G. Rozhnova, M.C. Bootsma, M. van Boven, J.H. van de Wijgert, M.J. Bonten, Impact of delays on effectiveness of contact tracing strategies for COVID-19: a modelling study. Lancet Pub. Health **5**(8), e452–e459 (2020)
20. National Informatics Centre, https://www.nic.in/blogs/direct-benefit-transfer-a-blessing-during-the-time-of-pandemic. Accessed 21 Dec 2020
21. Oxford COVID-19 Government Response Tracker (OxCGRT), https://www.bsg.ox.ac.uk/research/research-projects/coronavirus-government-response-tracker. Accessed 21 Dec 2020

Deep Learning for COVID-19: COVID-19 Detection Based on Chest X-Ray Images by the Fusion of Deep Learning and Machine Learning Techniques

Swati V. Shinde and Deepak T. Mane

Abstract The proposed book chapter focuses on development of an efficient AI based medical imaging solution for COVID-19 by leveraging the easily available COVID X-Ray Images (CXR). For this the experimentation with different deep learning and machine learning algorithms is performed. A convolution network (CNN) is one of the widely used deep learning algorithms used for medical imaging systems. In this chapter different variants of CNN are used. Data augmentation and dropout techniques are used to avoid overfitting. Among these different variants, CNN with ten convolutional layers and 15 epochs has given the best performance of 88.23% training 85.94% validation accuracy. This is followed by the use of Alexnet for feature extraction from CXR images and the extracted features are given as the input to different machine learning classifiers including Gaussian Naïve Bays, K-Nearest Neighbor (KNN), Support Vector Machine (SVM), Decision Tree (DT), Random Forest (RF), Gradient Boost, Ada Boost for classification of input images among the classes of Covid-19, No Finding, Pneumonia. Among these machine learning classifiers, SVM has given the best performance of 86% testing accuracy. Thus, the deep learning algorithms have proven to give satisfactory performance. This performance can still be improved with the help of more images and by fine tuning the pre trained models. In addition, this chapter highlights the importance of transfer learning, brief description about medical imaging.

Keywords CNN · KNN · SVM · DT · RF · Gradient Boost · Ada Boost

1 Introduction

Novel coronavirus (COVID-19) started spreading in China in 2019 and thereafter spread all over the world resulting into 26,332,484 positive cases and 870,033 deaths as on 3rd September 2020 and many cases are still in pipeline [1]. This has created

S. V. Shinde (✉)
Computer Engineering Department, Pimpri Chinchwad College of Engineering, Pune, India

D. T. Mane
Computer Engineering Department, Rajershi Shahu College of Engineering, Pune, India

J. Nayak et al. (eds.), *Understanding COVID-19: The Role of Computational Intelligence*,
Studies in Computational Intelligence 963, https://doi.org/10.1007/978-3-030-74761-9_21

a challenging situation in almost all countries by affecting mankind life and all business and educational sectors. Figure 1 shows the statistics of the total number of coronavirus cases and deaths in different countries [1].

In the challenging situation of COVID-19, all countries are facing a shortfall in healthcare resources including medical experts. In such situations, the use of computer assisted and Artificial Intelligence (AI) enabled automated systems are of great help. As the use of Artificial Intelligence and allied technologies including machine learning, deep learning, soft computing has given the breakthrough and better than human level performances for diagnosing many diseases, many researchers across the globe are experimenting these for diagnosing COVID-19.

Reverse Transcription Polymerase Chain Reaction (RT-PCR) test is the only screening method preferred currently for COVID-19 [2]; however it has the limitations in terms of labor-intensiveness, very specific, manual, need of specialized laboratories, specimen collection, time consuming and its sensitivity is highly variable [2].

Alternative screening methods are radiography examination including Chest X-Ray (CXR) and CT scan imaging. CXR images are easily available as compared to CT scan images and are inexpensive while CT scan images need advanced imaging equipment. Also CXR imaging provides the advantage of rapid triaging, availability and accessibility in many clinical sites and imaging centers and portability within an isolation room. Because of these advantages, CXR imaging has become a good alternative to RT-PCR. But to interpret these CXR images, radiology experts are

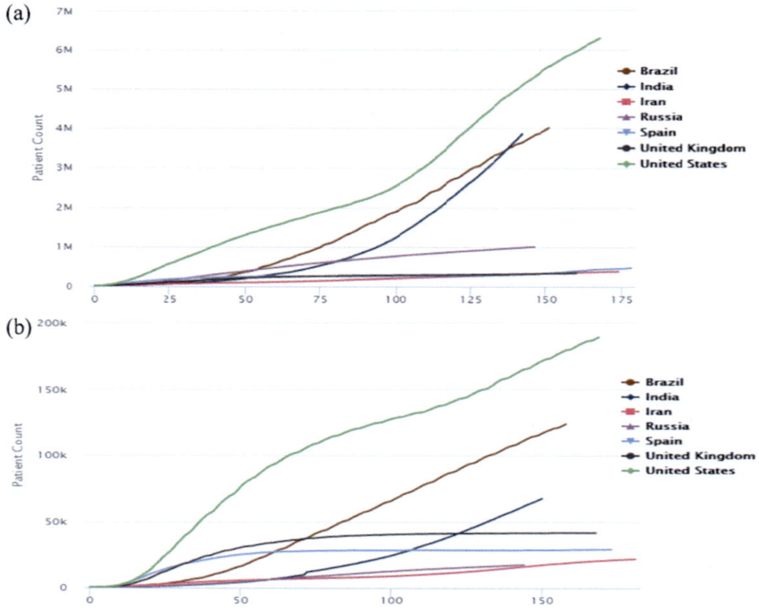

Fig. 1 Corona virus Worldwide Graphs (**a**) Cumulative number of cases vs number of days after 10, 000 cases (**b**) Cumulative number of deaths vs number of days after 100 deaths

needed and also these images cannot be completely interpreted by human sight. So AI based computer assisted automated systems play a very important role in analyzing these images and do accurate diagnostic predictions based on these images [4].

Importance of AI techniques in healthcare is witnessed through many competitions and challenges like Kaggle where winning models are tested against human-level expertise and proved to outweigh the human level performance. Although COVID-19 is a relatively younger and less explored disease, literature shows that many researchers have contributed to their best through many articles in all possible sub domains of COVID-19. But nearly all researchers have mentioned the most concluding and prevailing thing about this existing research is that due to the less data, less information, current situations, unavailability of drugs and other resources whatever the current conclusions they have presented cannot be generalized yet and it needs further introspection in coming times. This is because the study is performed on limited samples and it needs further investigation in the coming time with the increase in the data and other resources. So continuing research with the increasing resources is the need of the time. So our proposal presents a prospective automated solution for diagnosing COVID-19 and risk assessment of the patients based on Artificial Intelligence algorithms.

Since the 1990s, handcrafted features extracted from images are heavily used for developing many medical diagnosis systems. The success of these AI based automated diagnosis lies in the accurate extraction of handcrafted features from large amounts of training data. Afterwards, since 2012, deep learning started gaining a lot of popularity in medical applications because of its ability to learn optimal features from the data of given problems. In deep neural networks once an image is given as an input, it passes through layer by layer by transforming image into abstract level features to deeper level features and finally classifying these features into predefined classes.

The proposed book chapter focuses on development of an efficient AI based medical imaging solution for COVID-19 by leveraging the easily available CXR images. For this the experimentation with various machine learning algorithms and deep learning architectures will be performed in order to evaluate the suitability of AI for COVID-19 diagnosis.

The proposed system aims to develop an adaptive and hybrid ensemble system that makes accurate predictions about COVID-19. This system accepts radiological images and classifies them into predefined categories of COVID and non-COVID (Pneumonia, Normal, other viral infection). The proposed system has following features-

- The proposed chapter explains the experimentation details in incremental manner starting with basic CNN and upgrading it by varying its hyper parameters till it gives the best accuracy of 85.84%. In addition, alternative strategy of using CNN for feature extraction is demonstrated here wherein these extracted features are fed to different machine algorithms for further classification of COVID-19 images. Here the Alexnet is used for extracting the features from input images. So the proposed chapter demonstrates the novice researchers about use of pretrained

model for medical image analysis through feature extraction with the help of the use case of COVID Diagnosis through experimentation. Also this part of the experimental result explains the use of deep learning and machine learning algorithms together for medical image analysis.
- The AlexNet pretrained model has the complex architecture with 25 layers while proposed CNN models have 3, 5, 10 layers only which is the huge saving on computational complexity.
- The Alexnet is trained on large dataset of ImageNet. Similar performance is obtained using the CNN model by hyperparameter tuning through experimentation with 3, 5 and 10 layers. This will help novice researchers to reach to the optimal performance.
- AlexNet is trained on large ImageNet dataset so it is capable of extracting very raw features from the COVID images while proposed model is trained on COVID dataset so it is rightly tuned to classify the COVID images.
- Pretrained model along with different machine learning classifiers are giving good training accuracy but reduced test/val accuracy resulting in slight overfitting. While the proposed model is giving consistent training and test/val accuracy.

Thus the proposed system will be an efficient AI based medical imaging solution for COVID-19 by leveraging the easily available CXR images and by implementing the deep neural network design tailored for COVID-19 detection.

2 Related Work

In [3], authors have proposed the **COVID_MTNet architecture,** COVID-19 detection based on multi-task deep learning using transfer learning and NABLA-N approach of image segmentation for both CXR and CT scan images. Detection model-Inception Residual Recurrent Convolutional Neural Network has given the testing accuracy of 84.67% for CXR images and of 98.78% for CT scan images. Also the author has presented a novel approach of quantitative analysis of the infected region in X-Ray and CT scan images.

In [4], the deep convolution network namely 'COVID-Net' tailored specifically for COVID-19 diagnosis based on CXR images is suggested and it is made publically available online as one of the open source networks for COVID-19 detection. In addition authors have also contributed for the dataset of 13,975 CXR images with the maximum number of COVID-19 images. This model uses the unique explainable approach and deeply dives into critical factors of COVID cases.

The paper [5] proposes the method namely 'DeepCOVIDExplainer' for automatic diagnosis of COVID-19 symptoms based on the CXR images and classifying them into pneumonia, COVID-19 and normal cases. For this author have used 16,995 CXR images which are pre-processed, augmented and classified by the ensemble of DenseNet, ResNet, and VGGNet. This is followed by highlighting the affected

Table 1 Literature review of recent methods for COVID detection

Ref	Task	Methodology	Architecture	Performance	Dataset	Limitation
[3]	COVID_MTNet:ection, Segmentation, Quantitative analysis of infected region	Data Augmentation Transfer Learning Covid X-ray Detection Covid X-ray Segmentation Covid CT-Detection Covid CT-Segmentation	Inception Residual RCNN-Detection NABLA-N-Segmentation	Testing accuracy: X-Rays 84.67% CT-98.78%	Pneumonia-5,216 https://github.com/ieee8023/covid-chestX-ray-dataset COVID-19- https://github.com/ieee8023/covid-chestX-ray-dataset	Model needs to be trained and tested with more COVID-19 samples
[4]	Classification of CXR images into COVID-19 and Non COVID-19 (viral infections, bacterial infections, and healthy patients)	Principled network design prototyping Machine-driven Design Exploration COVID-Net Network Architecture Implementation COVID-Net Auditing via Explainability	VGG-19 ResNet-50 COVID-Net	Accuracy-93.3% PPV-98.9% Sensitivity-91.0%	COVIDx Dataset 13,975 images	small sample size does not enable generalizable statements about the reliability of the method,

(continued)

Table 1 (continued)

Ref	Task	Methodology	Architecture	Performance	Dataset	Limitation
[5]	'DeepCOVIDExplainer' –COVID-19 ClassificationInterpretation Segmentation	Preprocessing By histogram equalization enhancement, unsharp masking of the original CXR input images and filtering Augmentation Classification normal, pneumonia, and COVID19 viral Highlighting class discrimination regions Obtaining Interpretations	DenseNet's, ResNet's, and VGGNet's	PPV-89.61% Recall-3%,	Total CXR images: 16,995 Total Patients: 13,808 Classes: Normal, pneumonia, and COVID-19	–
[6]	Classification of COVID-19 chest X-ray images	Transfer learning on three pre-trained models	ResNet50, InceptionV3 and InceptionResNetV2	ResNet50-98% InceptionV3-97 InceptionResNetV2-87	Data shared by Cohen GitHub repository + Chest X-Ray Images (Pneumonia) https: //www. kaggle. com/paultimothymooney/chest-xray-pneumonia	only limited number of COVID-19 CXR images for training the model
[6]	Classification of COVID-19 chest X-ray images	Deep feature extraction Class decomposition Transfer learning Evaluation and composition	Decompose, Transfer, and Compose (DeTraC) ResNet+ https://github. com/asmaa4may/DeTraCCOVId19	Accuracy–95.12% Sensitivity-97.91%, Specificity-91.87% Precision-93.36%	180-Normal CXR images- from JSRT and COVID-19 image data collection. https: //githubcom/ieee8023/ covid-chestxray-dataset. 2020;	–

(continued)

Table 1 (continued)

Ref	Task	Methodology	Architecture	Performance	Dataset	Limitation
[7]	Classification of CXR & CT images and segmentation	Data pre-processing and data splitting Data augmentation Training and classification dataset	Resnet50, Inception_V3, Inception_ResNet_V2, Xception, VGG16, VGG19, DenseNet201, and MobileNet_V2	Resnet50, Inception_Resnet_V2 and MobileNet_V2 have given the highest accuracy of more than 96%	5856 images Consisting of 4273 pneumonia and 1583 normal images	Improvement in performance by using more datasets, feature extraction –YOLO, Segmentation-U-Net
[9]	CT Scan Image Segmentation Classification	Segmentation of CT Scan Images	UNet++	a per-image sensitivity of 94.34%, specificity of 59 99.16%, accuracy of 98.85%, PPV of 88.37%, and NPV of 99.61% in retrospective dataset	Images: 46,096 anonymous images Patients: 106 admitted Classes: 51 of COVID, 19 of pneumonia and 55 control patients Hospitals: Renmin Hospital of Wuhan University	

region by using Grad-CAM++. The performance of this model in terms of positive predictive value (PPV) is 89.61% and recall is 83%.

In [6], for automatic diagnosis of COVID-19 using CXR images, three different models of pre-trained CNN namely ResNet50, InceptionV3 and InceptionResNetV2 are used for diagnosis of COVID-19 based on CXR images. ROC curve, Confusion matrix and fivefold cross validations are used for performance analysis. It has been observed that the ResNet-50 model has the highest accuracy of 98% while InceptionV3 and Inception-ResNetV2 have given the accuracy of 97 and 87% respectively.

In paper [7], Deep Convolutional Neural Network (DCNN) used for binary classification of pneumonia images is compared with fined tuned pre-trained models of Resnet50, Inception_V3, Inception_ResNet_V2, Xception, VGG16, VGG19, DenseNet201, and MobileNet_V2. These fine tune models are tested on CXR images and CT scan images with total 5856 images consisting of 4273 pneumonia and 1583 normal. Among these Resnet50, Inception_Resnet_V2 and MobileNet_V2 have given the highest accuracy of more than 96%.

In paper [8], deep features are extracted from CXR images using ResNet50 and fed to support vector machines (SVM) for detection of COVID-19. SVM is used for classification instead of deep networks due to less data availability. The classes considered are COVID-19, pneumonia and normal. The performance of the model is-Accuracy-95.38%, FPR-95.52%, F1 score-91.41%, MCC and Kappa—90.76%. The results are compared with traditional image processing methods combined with SVM like are local binary patterns plus SVM, Gray Level Co-occurrence Matrix (GLCM) plus SVM and histogram of oriented gradients plus SVM.

The comparative literature review of these existing methods is given in Table 1.

3 Introduction to Deep Learning

Machine Learning is the sub domain of the Artificial intelligence (AI) that focuses on the study of different algorithms to bring intelligence in the machines. Artificial Neural Networks (ANN) is one of the machine learning technique inspired by the functioning of human brain. Deep learning focuses on the ANN with many hidden layers that are capable of both feature extraction followed by the classification. Deep learning is mainly suitable for image and video processing applications.

Medical imaging is one of the important application area of deep learning wherein images in different modalities like X-Ray, MIR, CT Scan, Pathology are processed to diagnose the different diseases. This is proved to be a major shift from manual diagnosis towards automated diagnosis. The success of these AI based automated diagnosis lies in the accurate extraction of handcrafted features from large amount of training data. Convolution neural network (CNN) is the most preferred deep learning model for image classification.

Deep learning exists since long but it has gained momentum recently because of the following reasons:

1. Availability of Big Data through internet
2. Increase in computational power through parallel computations by GPUs
3. The efficacy of CNN and other deep neural networks in image modeling, language modeling and sequential data processing

Deep learning enables computers to see and recognize objects in images. This ability of deep learning is extremely useful to solve computer vision problems. Deep learning is not only useful in computer vision problems but also useful in Natural Language processing and speech processing applications.

Medical applications generate a lot of healthcare data containing significant information every day. Machine learning, particularly deep learning, is capable of integrating, analyzing and building the model to make predictions based on this data.

3.1 Advantages of Deep Learning

- Applicable to a wide range of application domains of speech processing and speech Synthesis, Natural language Processing, Gaming, Image Classification, object detection, Object Localization, Image segmentation, etc.
- Ability to extract optimal features thus by reducing the need of explicit feature engineering
- Its adaptability to new problems through transfer learning without losing earlier knowledge.
- Number of architectures with specific characteristics suited to different research problems like convolution neural networks, recurrent neural networks (RNNs), long short-term memory (LSTM), etc.

3.2 Disadvantages of Deep Learning

- Data hungry algorithms which demands a large amount of training data
- It is computational cost for training the model is high
- Considered as black box because it gives decisions but no interpretation behind decisions is available.
- Optimization for its architecture in terms of number of layers, training algorithm, different hyper parameters is needed

3.3 Importance of Deep Learning in Medical Image Analysis in Comparison with Machine Learning

- Machine learning algorithms cannot capture the complexity of many healthcare research problems.

- Machine learning algorithms greatly depend on hand craft features suggested by the medical experts.
- As the data varies from patients to patients and its comprehension varies with the experience of medical experts, making it more labor intensive and non-reliable.
- Earlier machine learning algorithms like K-Nearest Neighbors, Decision Trees, Logistic Regression, Support Vector Machines, etc. takes only raw image data into account without learning of hidden representations.
- Nowadays, with the advancements in image acquisition devices like X-Ray, CT and MRI scanners, high resolution images are produced which are tough to interpret by using machine learning algorithms.

4 Transfer Learning

Transfer learning is an important feature of deep learning techniques wherein model trained for one task is reused for its expansion to another task. Here the expansion term signifies that earlier capability of the model is further updated and not replaced for its use to another task.

In transfer learning, usually the deep learning model is initially trained on some large benchmark datasets having a variety of labels like ImageNet [10]. This trained model is called a pre-trained model and it is made available for further task specific training, called fine-tuning. Pre-trained model is capable of extracting generic features accurately from input images as it is trained on a diverse dataset and fine-tuned models are customized for a given task.

The intuition behind transfer learning is that the fine-tuned model gets the advantage of learned generic features of pre-trained model which otherwise is not possible only with task specific smaller dataset.

4.1 Importance of Transfer Learning

- Many times the situation arises where the distribution of the data with which model is trained is different than that of testing data. This might be due to variation in test data may be due to variation into different data acquisition instruments or due to any other reason. In such situations most of the existing statistical models need to be retrained from scratch using newly arrived data which is costlier or infeasible in some cases to collect the needed training data again and retrain the models. During such circumstances, transfer learning does the knowledge transfer between the different task domains without the need to retrain and by extending the model for additional data [11].
- Transfer learning is proved to be more efficient when data gets out-dated frequently [11].

- Transfer learning is useful in the cases where data of one geographical location may not have the same distribution as of another location.
- Transfer learning solves the problem of limited data to some extent and it is recommended as an effective solution to insufficient data. Pre-trained models extract generic features and only rely on task specific data for rest of the deeper features. This reduces the size of required training data and training time.
- Data annotation is also a complex and time consuming task in the healthcare domain. Transfer learning helps the annotation task with the use of pre-trained models and semi supervised learning.

Because of these reasons, transfer learning methods are gaining notable popularity in renowned venues of data mining (ACM, ICDM, PKDD, KDD, and IEEE, etc.), machine learning (ICML, ECML, NIPS, IJCAI, and, AAAI, etc.) and applications of machine learning and data mining (ACL, SIGIR, WWW, and, ACM etc.) [10].

4.2 When to Use Transfer Learning?

There are no hard and fast rules about when transfer learning might be used but some guidelines are given below [11]. These are grouped in two categories- 1. The factors that decide whether to use for transfer learning. 2. The factors that decide whether to retain transfer learning.

1. Following are the factors that decide whether to use transfer learning-
    - If enough training data is not available to train the network from scratch
    - If there exists a pre-trained network trained on similar task
    - If there exist a pre-trained network trained on different task but similar data
2. The factors that decide whether to retain transfer learning- In this case transfer learning serves as an optimization task that saves time and computational complexity. Here after evaluating the performance of transfer learning, it is decided whether to retain it or not [12].

Following are three possibilities defined in [13, 14] as shown in Fig. 2.

- **Higher start.** Early performance of the model with transfer learning is higher than without transfer learning.
- **Higher slope.** The rate of improvement in performance during training with transfer learning is steeper than without transfer learning.
- **Higher asymptote.** The converged performance of the trained model with transfer learning is better than that of without transfer learning.

Following Table 2 gives summary of different transfer learning models-

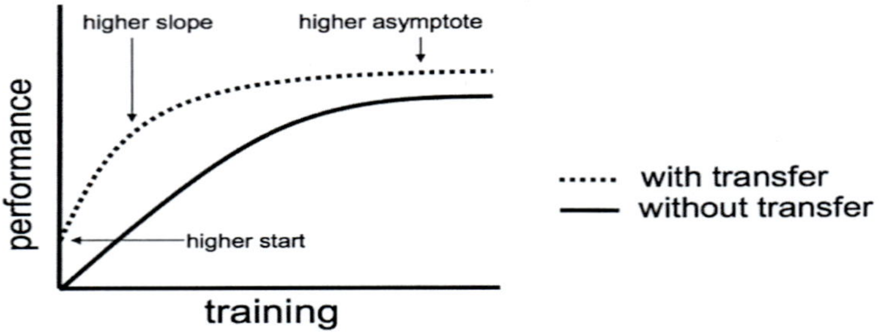

Fig. 2 Possibilities in transfer learning [13]

Table 2 Summary of deep learning models

Year	CNN architecture	Input size	No. of layers	Top-5 error rate	No. of parameters	Winning position	Developers
1998	LeNet5	32 * 32	8	–	60 thousand	–	YannLeCun et al.
2012	AlexNet	227 * 227	7	15.3%	60 million	1st	Alex Krizhevsky, Geoffrey Hinton, IlyaSutskever
2013	ZFNet	224 * 224	–	14.8%	–	1st	Matthew Zeiler and Rob Fergus
2014	GoogleNet	224 * 224	19	6.67%	4 million	1st	Google
2015	VGGNet	224 * 224	16	7.3%	138 million	2nd	Simonyan, Zisserman
2016	ResNet	224 * 224	152	3.6%	–	1st	Kaiming He

5 Medical Imaging

Medical imaging is one of the significant tools for disease diagnosis which shows anatomical structure of different body organs of human being. There are different modalities of medical images including Radiographs, Magnetic Resonance Imaging (MRI), Computer Tomography (CT), Positron emission tomography (PET), Skeletal Scintigraphy (Bone Scan), Histopathological, Endoscopy, Thermography, Nuclear Images, etc. These images provide clues for diagnosis and prognosis of different diseases and also decide the further treatment plan.

5.1 Medical Imaging Tools

- **ImageJ** [15], It is the java based, open source, user friendly biomedical image processing tool developed by National Institute of Health, USA. It is useful for quantification of images, automatic tumor foci detection, etc. ImageJ has thousands of plugins and scripts dedicated to large number of image processing task. It can be downloaded from https://imagej.nih.gov/ or https://imagej.net/.
- **CellProfiler is used** for quantitative analysis of biological images. This software is free and open-source. For using this software, no earlier expertise in programming is required. It converts image into grayscale for feature extraction followed by detection and segmentation of cell nuclei. Thus it extracts 772 features grouped into texture, geometry and holistic. It can be downloaded from https://cellprofiler.org/releases/ [16, 17].
- **Medseg**- Medseg is a free medical image segmentation tool for CT and MRI in AI applications. It is web based simple, fast with gaming and radiology-based controls. It imports data in the form CT and MRI images as.nii or.nii.gz into the tools and then performs automatic or interactive segmentation using AI models. It corrects the errors manually and downloads the finished segmented image. It can perform the interactive AI-based segmentation of CT images (Liver, spleen, Kidneys, thorax: lungs, pancreas, heart, spinal canal, liver tumors, abdominal lymph nodes) and MRI images (Liver, kidneys, spleen, prostate) [18].
- **RIL-Contour** (Radiology Informatics Laboratory Contour): This tool is for medical image annotation for and with deep learning. It facilitates the consistent annotation of large medical imaging datasets and supports multiuser workflows. This software provides two main windows namely dataset project viewer and the dataset annotation window. The dataset project viewer displays a list of the imaging files associated with a project. The dataset project viewer is user-friendly to work with large datasets through hierarchical file [19]
- **NiftyNet-** It is an open source CNN architecture based on TensorFlow, specially designed for researchers in medical image analysis and image-guided therapy. This allows using pre-trained models and building the solution to different medical imaging problems. Code: https://github.com/NifTK/NiftyNet [20]
- **TurtleSeg-** It is an interactive 3D image segmentation tool. Its main feature is Spotlight which is an automated system that guides users for giving inputs. It allows user to load a 3D image followed by using 2D Livewire to manually contour a sparse number of different slices. Based on this information, full 3D segmentation is built automatically. It can import and exports many 3D file formats like DICOM and STL files [21].

Following are some of the software's for analysis of IHC images-

- ImmunoRatio
- ImmunoMembrane
- Phenomics
- Visiopharm

Table 3 Different tools of medical image processing

Tool name	Functionality	Downloading source	Licensing
ImageJ	Quantification of images, automatic tumor foci detection	https://imagej.nih.gov/ https://imagej.net/	Free, Open Source
CellProfiler	Detection and segmentation of cell nuclei	https://cellprofiler.org/releases/	Free, Open Source
Medseg	Automatic or interactive segmentation of CT and MRI images	https://medicalsegmentation.com/	Free
RIL-Contour	Medical image annotation	https://gitlab.com/Philbrick/rilcontour	Free, Open Source
NiftyNet	Medical image analysis and image-guided therapy	https://github.com/NifTK/NiftyNet.org	Open source
TurtleSeg	Interactive 3D image segmentation	https://www.turtleseg.org/	Paid

- Aperio Digital Pathology
- TissueGnostics

Following is the list of some tools to medical imaging given in Table 3.

6 COVID-19 X-Ray Image Dataset

For the experimentation, the dataset of CXR images is used from the source [22]. This data is the collection from two different sources, one is of COVID-19 X-ray image dataset developed by Cohen JP [22] and other is of X-Ray image dataset of pneumonia developed by Wang et al. [23]. This dataset contains a total of 1125 images consisting of 500 images of pneumonia, 500 images of non-finding class while 125 images of COVID-19 class.

6.1 Dataset Details

The experimentation dataset is divided into a training set of 810 images, a test set of 225 images and a validation set of 90 images as given in Fig. 3. Some sample images in the dataset are given in Fig. 4.

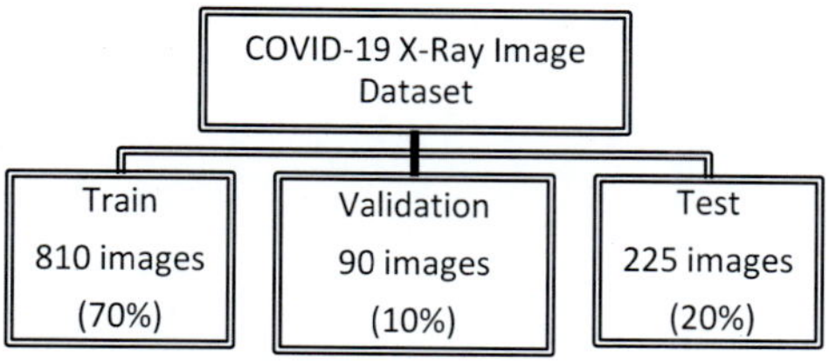

Fig. 3 Distribution of dataset among training, validation and testing set

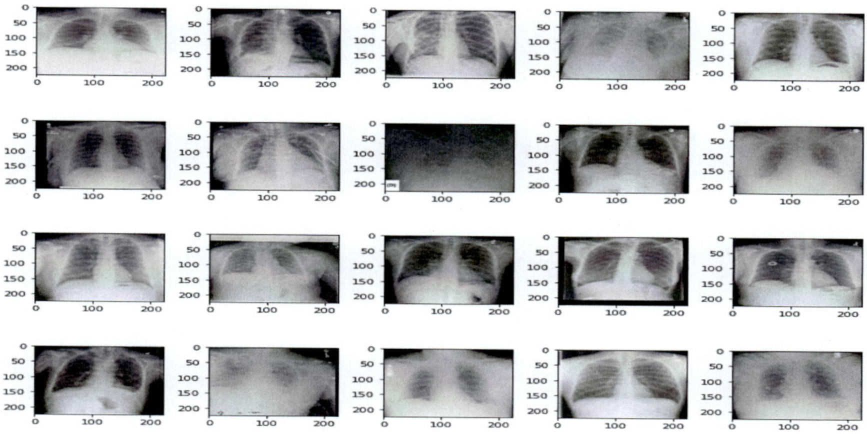

Fig. 4 Sample images of the COVID-19 dataset

7 Experimentation Details

This section represents the experimental details of CNN with different convolution layers.

7.1 CNN with Three Convolution Layers and Two Dense Layers

This section describes the structure of CNN with three convolution and two dense layers

a. **Data preparation**
The ImageDataGenerator class provided by tf.keras is used for reading images from disk, preprocessing them into proper tensors and converts these images into batches of tensors needed for training the network. The preprocessing in terms of rescaling the images and resizing them into required dimensions is done. Batch size of 32 and input image size of 224 * 224 is considered

Table 4 Hyper parameters used for the CNN

Convolution layers	Conv1: 16 filters of 3 * 3 size Conv2: 32 filters of 3 * 3 size Conv3: 64 filters of 3 * 3 size Activation Function: ReLu
Input size	224 * 224
Dense1 layer	512 nodes Activation function: ReLu
Dense2 layer	3 nodes Activation function: Sigmoid
Optimizer	Adam
Loss function	Binary cross entropy
Performance metric	Accuracy

Table 5 Layer wise output shape and number of parameters with three conv. layers and without augmentation

Layer	Output shape	#Parameters
Conv2D	224 * 224 * 16	448
MaxPooling	112 * 112 *16	0
Dropout	112 * 112 * 16	0
Conv2D	112 * 112 * 32	4640
MaxPooling	56 * 56 * 32	0
Conv2D	56 * 56 * 64	18,496
MaxPooling	28 * 28 * 64	0
Dropout1	28 * 28 * 64	0
Flatten	50,176	0
Dense	512	25,690,624
Dropout2	512	0
Dense	3	1539
Total parameters	25,715,747	

b. **Model Creation**

The convolution network model consisting of three blocks of convolution layers with max-pooling layers are used. The number of filters (with size 3 * 3) used in these convolution layers are 16, 32, 64 respectively. These blocks are followed by two dense layers with 512 and 3 nodes respectively. First dense layer uses 'relu' and second uses 'sigmoid' activation function. The hyper parameters of the model are given in Table 4 and layer wise output and number of parameters are given in Table 5.

c. **Model compilation and training**

The created model is compiled by using *ADAM* optimizer and *binary cross entropy* loss function with accuracy as a performance metric. This is because there are three

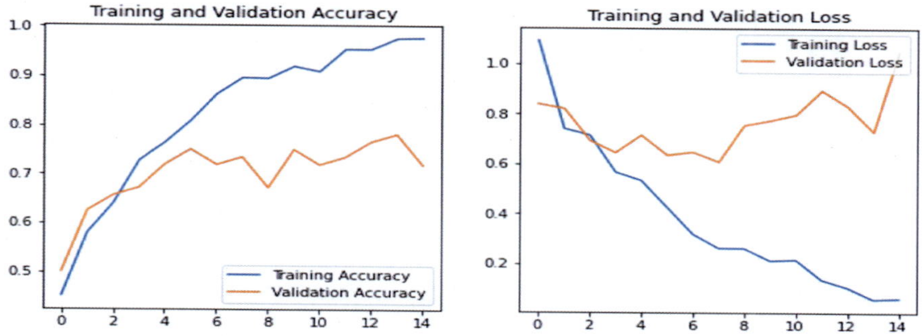

Fig. 5 Training and validation performance with three convolution layers

classes in the dataset. So three nodes are used in output layer with sigmoid activation function. Each node is giving the output as the probability of belonging to that class i.e. it is a binary decision in the context of each output node. So binary cross entropy loss function is used.

The compiled model is trained for 15 epochs on training data and its performance is evaluated on training and validation data. The Fig. 5 gives the visualization of model performance on training and validation data.

Figure 5 shows that the training accuracy is increasing linearly over time, whereas validation accuracy started falling after around 70% resulting into a wider margin with training accuracy. The model has given the training accuracy of 97.69% while validation accuracy of 71.88%. This is because the model is suffering from the drawback of overfitting. Due to overfitting the network has not generalized well and learned the noise present in the data. The overfitting can be removed by multiple ways like data augmentation and by adding dropouts to our model.

d. **Data augmentation**

Data augmentation is a remedy for overfitting problems which augments the dataset by using random transformations in order to increase the number of data samples. Data augmentation generates the more training data from the existing limited training data and this newly generated data helps in improving the generalization ability of the network by feeding it with variations in training examples. This can easily be done by passing different transformations as an argument to the ImageDataGenerator which takes care of applying these transformations during the training process. For this experimentation, the transformations applied are horizontal flip, random rotate and zoom augmentation.

- **Horizontal Flip-** For this, the argument 'horizontal_flip=True' is passed to the 'ImageDataGenerator' class. The single image is taken and it is passed five times in order to apply an augmentation to the same image five times as given in Fig. 6.
- **Random rotate-** For this, the argument 'rotation_range = 45′ is passed to the 'ImageData Generator' class which applies 45° of rotation randomly to the training examples as shown in Fig. 7.

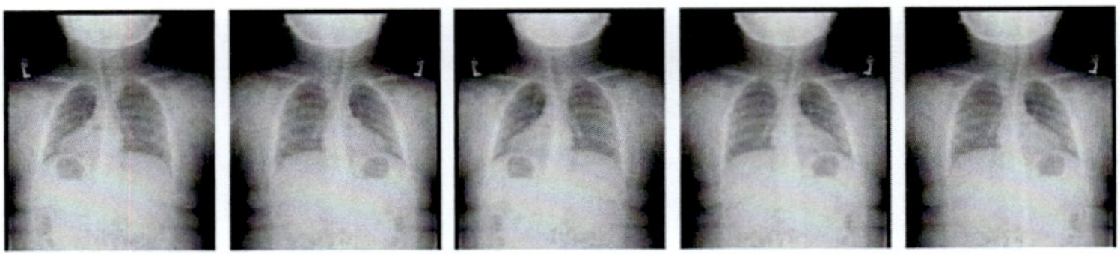

Fig. 6 Horizontally flipped training images

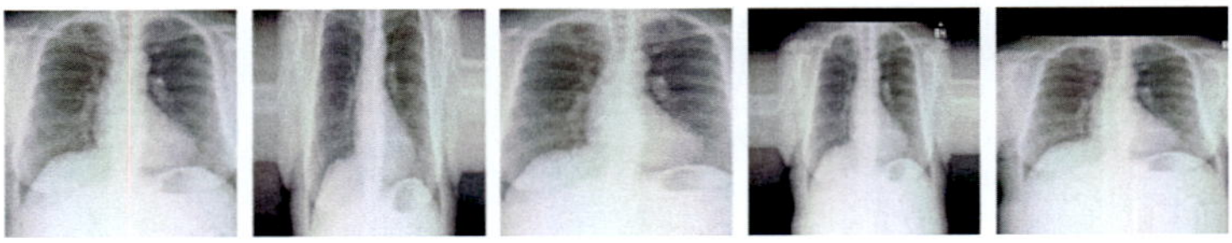

Fig. 7 Randomly rotated training images

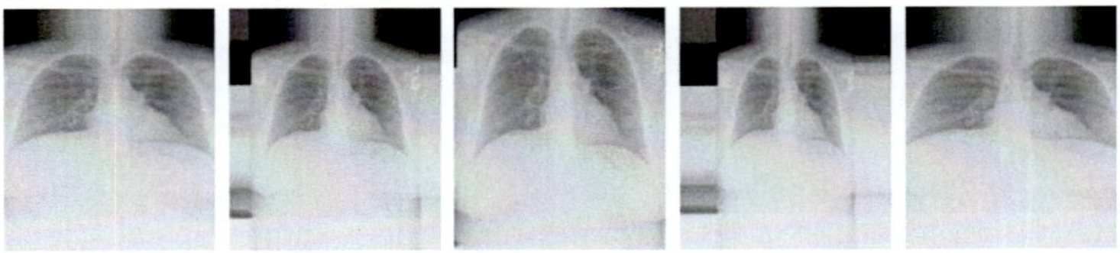

Fig. 8 Randomly zoomed training images

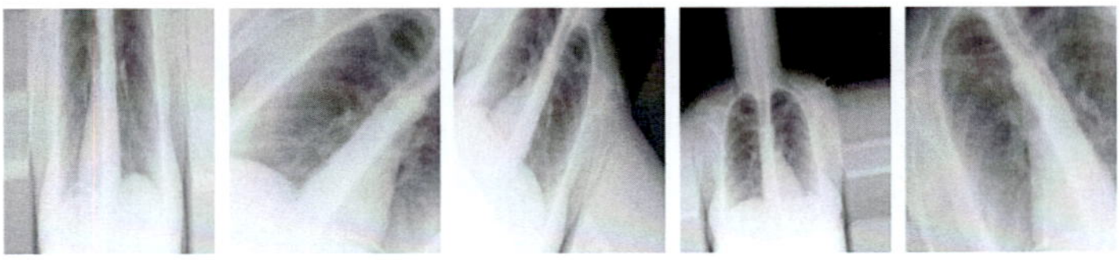

Fig. 9 All augmentation operations together

- **Zoom-** It applies a zoom augmentation to the dataset that randomly zooms training images up to 50% as in Fig. 8.
- **All augmentation operations together-** After applying all the previous augmentations including horizontal flip, 45° rotation, rescale, width shift, height shift, and zoom augmentation to the training images, images will look like in Fig. 9. These augmented images are given as input to the CNN described in Table 6.

e. Dropout-other technique to reduce overfitting

It is the simplest way to prevent network from overfitting. During training of the network, some randomly selected neurons are selected and they are dropped out from the training process i.e. their activations values are ignored in forward pass and in backward pass of error propagation their weights are not updated. This avoids the

Table 6 Layer wise output shape and number of parameters with three conv. Layers with augmentation

Layer	Output shape	#Parameters
Conv2D	224 * 224 * 16	448
MaxPooling	112 * 112 * 16	0
Dropout	112 * 112 * 16	0
Conv2D	112 * 112 * 32	4640
MaxPooling	56 * 56 * 32	0
Conv2D	56 * 56 * 64	18,496
MaxPooling	28 * 28 * 64	0
Dropout1	28 * 28 * 64	0
Flatten	50,176	0
Dense	512	25,690,624
Dropout2	512	0
Dense	3	1539
Total parameters	25,715,747	

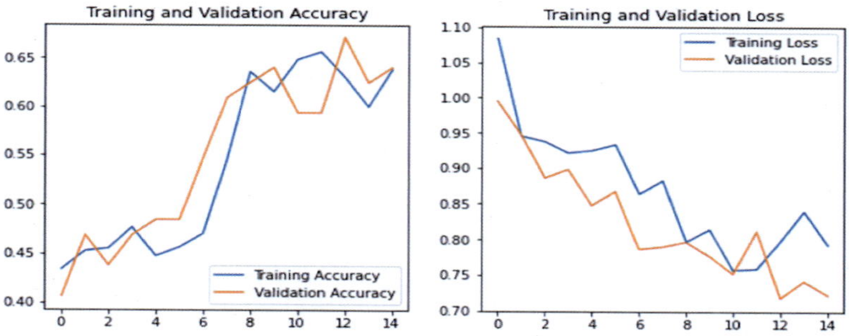

Fig. 10 Training and validation performance after augmentation

creation fragile models which are very much specialized to the training data. The dropout rate for a layer is specified in terms of fractional number such as 0.1, 0.2, 0.3, etc. and it means dropping out 10, 20 or 30% of the nodes randomly from the specified layer. Here, the dropout of 30% is applied to the first and last max pooling layers and of 10% to a fully connected layer of 512 nodes.

The augmentation and dropout strategies together have successfully removed the overfitting and given the performance of 63.88% training accuracy and 64.06% validation accuracy. The training loss of 0.7931 and validation loss of 0.7222 is given by the network. The performance is as shown in Fig. 10.

After increasing the number of epoch from 15 to 50, the same model has given the training accuracy of 68.38% and validation accuracy of 75%. The accuracy and loss are as shown in Fig. 11.

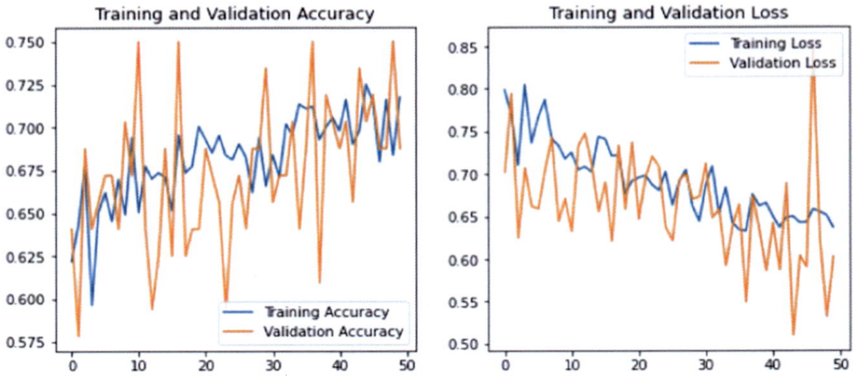

Fig. 11 Training and validation performance after augmentation

Here the complete visualization of layerwise activations of the model is given in the following plots as depicted in Fig. 12. The plot for every channel in each of our 8 activation maps is extracted and stacked the results in one big image tensor, with channels stacked side by side.

To put up it together

- The first layer is extracting the edge features by retaining all the information in the original image.
- The layers starting from the first layer going towards the end layers extract the more and more abstract features necessary to classify the image among classes.

7.2 *CNN with 5 Convolution Layers and Two Dense Layers with 50 Epochs*

After increasing the number of layers from three to five and after keeping number of epochs as 50, model has given the training accuracy as 68.25% validation accuracy as 71.88%. The plots of loss and accuracy for training and validation data are given in Fig. 13.

7.3 *CNN with 10 Convolution Layers and Two Dense Layers with 15 Epochs*

Convolution and max pooling layers are again increased from five to ten and epochs are reduced from fifty to fifteen. The structure of the CNN is shown in Table 7. This combination has given the best training and validation accuracy of 88.23 and

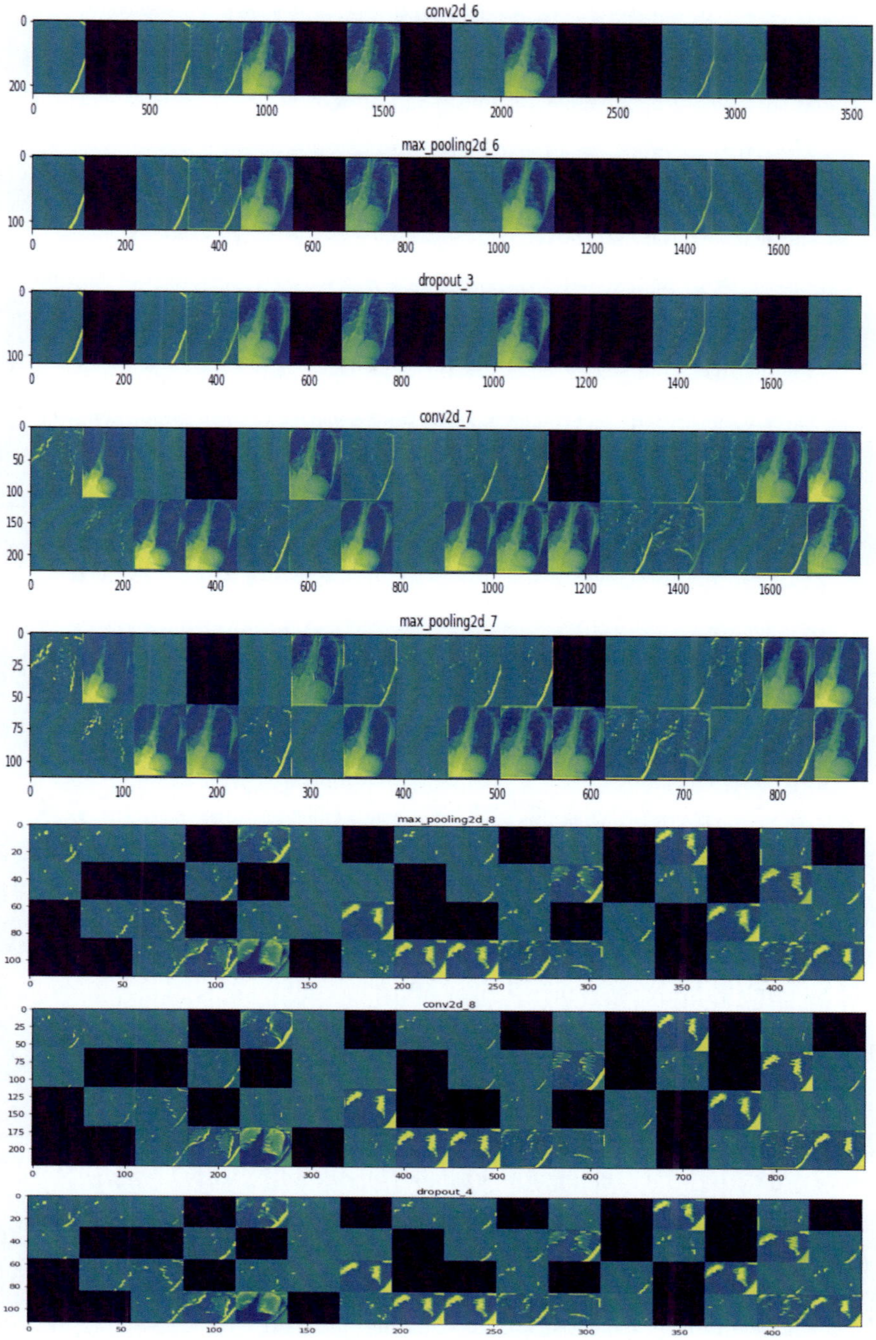

Fig. 12 Visualization of layerwise activations of the model

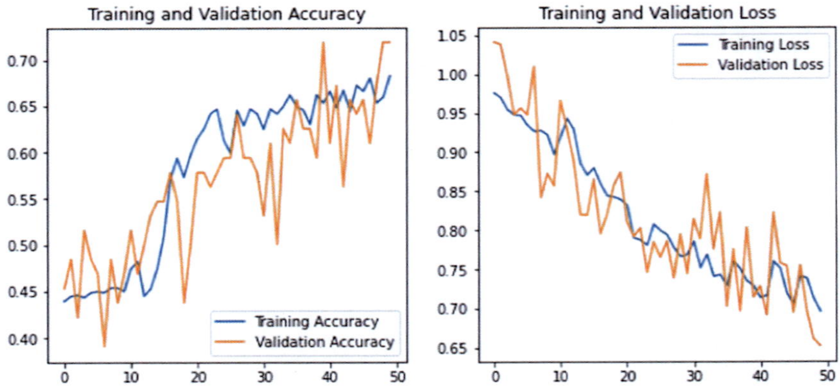

Fig. 13 Loss and accuracy plots for training and validation data with five convolution errors

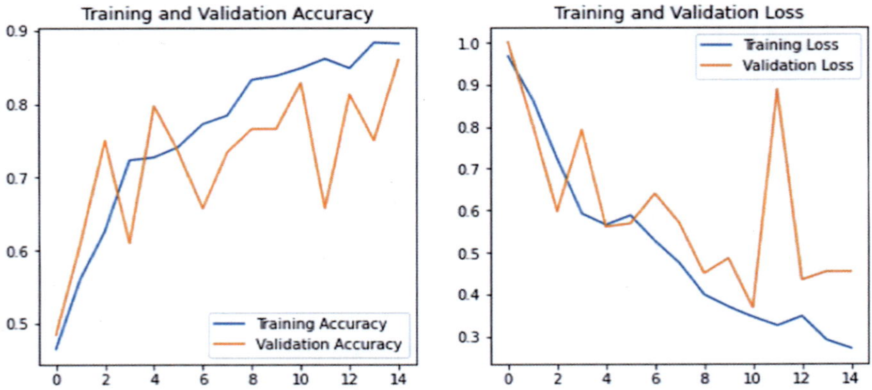

Fig. 14 Performance of CNN with ten convolution-pooling layers and fifteen epochs

85.94% respectively as given in Table 8. The plots of loss and accuracy for training and validation data are given in Fig. 14.

7.4 Fusion of Pretrained Convolution Neural Network for Feature Extraction and Different Machine Learning Algorithms for Classification

Feature Extraction Using Alexnet

Transfer learning is the most useful technique used nowadays to extract features from the dataset. It is also proven that it gives significant results as compared to the traditional feature extraction techniques [23]. Here the pretrainedAlexnet model is used to extract the features from the X-ray dataset and further they are provided to

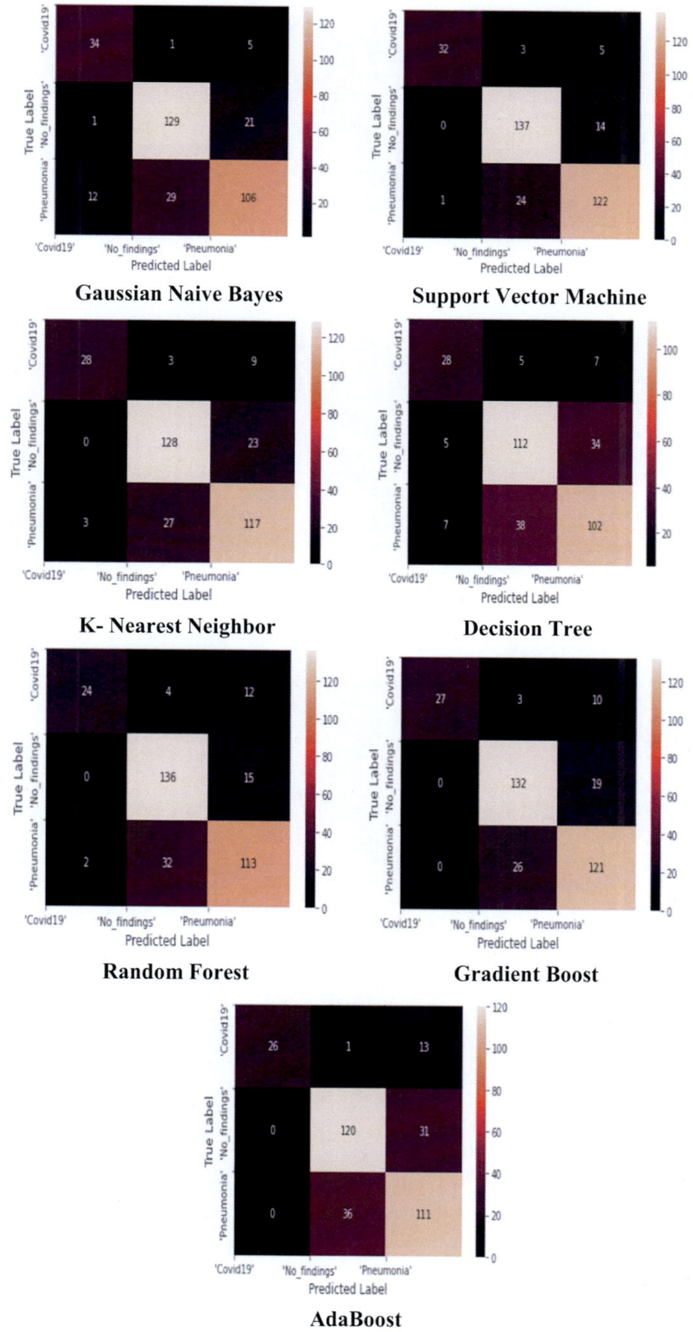

Fig. 15 Performance of machine learning classifiers in terms of confusion matrix

Table 7 Layer wise output shape and number of parameters with ten conv. layers

Layer	Output shape	#Parameters
Conv2D	224 * 224 * 16	448
Conv2D	224 * 224 * 32	4640
MaxPooling	112 * 112 * 32	0
Conv2D	112 * 112 * 32	9248
Conv2D	112 * 112 * 64	18,496
MaxPooling	56 * 56 * 64	0
Conv2D	56 * 56 * 64	36,928
Conv2D	56 * 56 * 32	18,464
MaxPooling	28 * 28 * 32	0
Conv2D	28 * 28 * 32	9248
Conv2D	28 * 28 * 64	18,496
MaxPooling	14 * 14 * 64	0
Conv2D	14 * 14 * 32	18,464
Conv2D	14 * 14 * 64	18,496
MaxPooling	7 * 7 * 64	0
Flatten	3136	0
Dense	512	1,606,144
Dense	3	1539
Total parameters	1,760,611	

Table 8 Training and validation performance with ten convolution-pooling layers and 15 epochs

Epoch no.	Loss	Accuracy	Validation loss	Validation accuracy
1	1.0033	41.72	1.0004	48.44
2	0.8912	55.75	0.8054	60.94
3	0.7693	58.26	0.5960	75.00
4	0.6088	71.89	0.7927	60.94
5	0.5562	73.63	0.5607	79.69
6	0.5637	77.56	0.5680	73.44
7	0.5269	76.27	0.6385	65.62
8	0.5170	75.29	0.5692	73.44
9	0.4126	82.78	0.4491	76.56
10	0.3677	84.03	0.4854	76.56
11	0.3229	86.06	0.3677	82.81
12	0.2832	89.48	0.8876	65.62
13	0.3438	84.18	0.4331	81.25
14	0.2817	88.44	0.4528	75.00
15	0.2775	88.23	0.4527	85.94

different Machine learning classifiers to classify the given Xray image to COVID 19, Pneumonia or No diseases class. The Alexnet [24] is the pretrained model consists of 25 layers. The detailed description about these layers is mentioned in Table 9.

As mentioned in the table, the input which is fed to the Alexnet must be of size 227 × 227 × 3. As our Xray database input size is different than this, hence in preprocessing the given input image is converted to 227 × 227 × 3 size. Once the input dataset is prepared as per the requirements, Alexnet pretrained model is applied and layer 'fc7' features are extracted and used for classification purpose. It can be easily understood based on the above table that there are total number of 4096 features which are extracted from each X-ray image using Alexnet model.

Below the algorithm is given which explains the detailed steps of the transfer learning mechanism used for the feature extraction here.

Algorithm: Feature extraction using deep learning pretrained model Alexnet

Start

1: for p := 1 →P (P is the total number of images in training dataset

1.1: Read p^{th} RGB covid Image.

1.2: Resize image to 227 X 227 X 3.

1.3: Load the pretrained 'Alexnet' model

1.4: Extract the features using the 'fc7' layer& store it row wise.

2: end for

3: Input feature vector to different Machine learning classifiers and calcu late its performance.

End

These features are applied to various machine learning classifiers and its performance analysis is shown in Fig. 15 and Table 10. Figure 15 depicts the confusion matrices of each of the classifiers. Table 10 depicts the performance in terms of Precision, Recall, F-Score, Support, Testing and Training Accuracy.

Based on the above Tables, it can be stated that the performance of the SVM, Random forest, Gradboost and Adaboost is better when applied on these deep learning features. Instead of extracting features of layer 'fc7' of Alexnet, further experimentations can also be carried out on the other layer features that can be investigated and experimented to identify the appropriate feature set that gives maximum results.

Table 9 Alexnet layers and its description

Sr No	Label	Layer	Description
1	'data'	Image Input	227 × 227 × 3 images with' zerocenter normalization
2	'conv1'	Convolution	96 11 × 11 × 3 convolutions with stride [4 4] and padding [00]
3	'relu1'	ReLU	ReLU
4	'norm1'	Cross Channel Normalization	Cross channel normalization with 5 channels per element
5	'pool1'	Max Pooling	3 × 3 max pooling with stride [2 2] and padding [0 0]
6	'conv2'	Convolution	256 5 × 5 × 48 convolutions with stride [1 1] and padding [2 2]
7	'relu2'	ReLU	ReLU
8	'norm2'	Cross Channel Normalization	Cross channel normalization with 5 channels per element
9	'pool2'	Max Pooling	3 × 3 max pooling with stride [2 2] and padding [0 0]
10	'conv3'	Convolution	384 3 × 3 × 256 convolutions with stride [1 1] and padding [1 1]
11	'relu3'	ReLU	ReLU
12	'conv4'	Convolution	384 3 × 3 × 192 convolutions with stride [1 1] and padding [1 1]
13	'relu4'	ReLU	ReLU
14	'conv5'	Convolution	256 3 × 3 × 192 convolutions with stride [1 1] and padding [1 1]
15	'relu5'	ReLU	ReLU
16	'pool5'	Max Pooling	3 × 3 max pooling with stride [2 2] and padding [0 0]
17	'fc6'	Fully Connected	4096 fully connected layer
18	'relu6'	ReLU	ReLU
19	'drop6'	Dropout	50% dropout
20	'fc7'	Fully Connected	4096 fully connected layer
21	'relu7'	ReLU	ReLU
22	'drop7'	Dropout	50% dropout
23	'fc8'	Fully Connected	1000 fully connected layer
24	'prob'	Softmax	Softmax
25	'output'	Classification Output	Crossentropyex with 'tench', 'goldfish', and 998 other classes

Table 10 Findings of different deep learning features on machine learning algorithms

Method	Findings	Precision	Recall	F1-score	Support	TrainAcc	TestAcc
Gaussian Naive Bayes	Covid-19	0.72	0.85	0.78	40	0.79	0.80
	No Finding	0.81	0.85	0.83	151		
	Pneumonia	0.80	0.72	0.76	147		
	Weighted Avg	0.80	0.80	0.79	338		
SVM	Covid-19	0.97	0.80	0.88	40	0.89	0.86
	No Finding	0.84	0.91	0.87	151		
	Pneumonia	0.87	0.83	0.85	147		
	Weighted Avg	0.86	0.86	0.86	338		
KNN	Covid-19	0.90	0.70	0.79	40	0.90	0.80
	No Finding	0.81	0.85	0.83	151		
	Pneumonia	0.79	0.80	0.79	147		
	Weighted Avg	0.81	0.81	0.81	338		
Decision Tree	Covid-19	0.70	0.70	0.70	40	1.0	0.71
	No Finding	0.72	0.74	0.73	151		
	Pneumonia	0.71	0.69	0.70	147		
	Weighted Avg	0.72	0.72	0.72	338		
Random Forest	Covid-19	0.92	0.60	0.73	40	1.0	0.80
	No Finding	0.79	0.90	0.84	151		
	Pneumonia	0.81	0.77	0.79	147		
	Weighted Avg	0.81	0.81	0.80	338		
Gradboost	Covid-19	1.00	0.68	0.81	40	1.0	0.82
	No Finding	0.82	0.87	0.85	151		
	Pneumonia	0.81	0.82	0.81	147		
	Weighted Avg	0.84	0.83	0.83	338		
Adaboost	Covid-19	1.00	0.65	0.79	40	0.79	0.76
	No Finding	0.76	0.79	0.78	151		
	Pneumonia	0.72	0.76	0.74	147		
	Weighted Avg	0.77	0.76	0.76	338		

8 Challenges and Future Research Perspective

The following are the challenges and further scope of work of the proposed model.

- As the deep learning methods are data hungry, the lack of sufficient training data sets is considered as an obstacle but this is changing nowadays because of adoption of clinical digitization and availability of datasets on publicly accessible repositories.

- Having large image datasets creates yet another challenge of accurate labelling or annotating of the medical images. As these medical images are large in size, they are processed in patches. This even needs the patch level annotations.
- Precise medical diagnosis can't rely only on medical image analysis, but it also needs useful information that is not present within the images like patient history, age, demographics, etc. So the system that integrates both imaging with clinical information is of utmost importance.
- Combining imaging with clinical features results in the challenges of balancing the large number of imaging features extracted by the deep learning network with the small number of clinical features to prevent the clinical features from being drowned out.
- Another data-related challenge in medical applications is class imbalance. Small proportion of the total patients is disease positive, which results in an imbalance dataset. So the development of deep learning systems that are capable at handling this class imbalance is another important area of research
- Deep learning algorithms seek the proper tuning of a large number of hyperparameters that the developer has to fix through repetitive experimentation. Reducing the dependency of these algorithms on various hyperparameters needs to be explored.
- Deep learning algorithms are proved to be accurate but considered as black boxes as the interpretability of their results needs to be obtained. Thus development of methods/techniques to obtain resembling interpretation of the results of deep learning algorithms is an active research area.
- Most of the research papers found in the literature focus on the supervised learning task by ignoring the unsupervised task. However unsupervised learning resembles more with human learning and follows the generative approach rather than discriminative approach of learning. Also unsupervised learning needs unlabelled data which is quite easier than obtaining the labelled data required for supervised tasks.
- Image quality affects the performance of deep learning algorithms, so the improvement in the performance of deep learning models in spite of low quality images needs to be devised.

In case of transfer learning across the domains, bridging the gap between source domain and target domain having different tasks, different feature spaces need to be filled by implementing heterogeneous transfer learning.

9 Conclusion

This chapter has demonstrated the importance of deep learning and machine learning for covid-19 detection. The deep neural network called the convolution network is trained on X-ray images of the patients. X-ray image based covid-19 test is helpful as it gives diagnosis results of the test in a very less time and that too in a contactless mode. X-ray machines are affordable and are available in almost every hospital.

So X-ray based covid-19 test is a useful solution to reduce the current testing load on RT-PCR testing. This chapter explains the experimentation details with CNN by varying its hyper parameters. The best accuracy obtained is 85.84%. In addition the experimentation is about extracting the feature values by using CNN and using different classifiers on for training these features. Here the Alexnet is used for extracting the features from input images. And these are passed to different machine learning classifiers and SVM has given the best performance.

References

1. https://www.worldometers.info/coronavirus/. Accessed 5 Sep 2020
2. A.I. Khan, J.L. Shah, M.M. Bhat, CoroNet: a deep neural network for detection and diagnosis of COVID-19 from chest x-ray images. Comput. Methods Programs Biomed. **196**, 105581 (2020). https://doi.org/10.1016/j.cmpb.2020.105581. ISSN 0169-2607
3. Md.Z. Alom, M.M.S. Rahman, M. Nasrin, T. Taha, V. Asari, COVID_MTNet: COVID-19 detection with multi-task deep learning approaches. arXiv preprint arXiv:2004.03747 (2020)
4. T. Gao, G. Wang, Chest X-ray image analysis and classification for COVID-19 pneumonia detection using Deep CNN (2020). https://doi.org/10.1101/2020.08.20.20178913
5. Md. Karim, T. Döhmen, D. Rebholz-Schuhman, D. Stefan, M. Cochez, O. Beyan, DeepCOVID-Explainer: explainable COVID-19 predictions based on chest X-ray images (2020)
6. R.M. Pereira, D. Bertolini, L.O. Teixeira, C.N. Silla, Y.M.G. Costa, COVID-19 identification in chest X-ray images on flat and hierarchical classification scenarios. Comput. Methods Programs Biomed. **194**, 105532 (2020). https://doi.org/10.1016/j.cmpb.2020.105532 ISSN 0169-2607
7. A. Idri, K.E. Asnaoui, Y. Chawki, Automated methods for detection and classification pneumonia based on x-ray images using deep learning. arXiv preprint arXiv:2003.14363 (2020)
8. P.K. Sethy, S.K. Behera, Detection of coronavirus disease (COVID-19) based on deep features (2020). https://doi.org/10.20944/preprints202003.0300.v1
9. O. Gozes, M. Frid-Adar, N. Sagie, H. Zhang, W. Ji, H. Greenspan, Coronavirus detection and analysis on chest CT with deep learning. arXiv preprint arXiv:200402640 (2020)
10. M. Raghu, C. Zhang, J. Kleinberg, S. Bengio, Transfusion: understanding transfer learning with applications to medical imaging, in *33rd Conference on Neural Information Processing Systems (NeurIPS 2019)*, Vancouver, Canada (2019)
11. https://ai.googleblog.com/2019/12/understanding-transfer-learning-for.html. Accessed 5 Sep 2020
12. https://towardsdatascience.com/a-comprehensive-hands-on-guide-to-transfer-learning-with-real-world-applications-in-deep-learning-212bf3b2f27a. Accessed 5 Sep 2020
13. QiangYang SinnoJialinPan, A survey on transfer learning. IEEE Trans. Knowl. Data Eng. **22**, 1345–1359 (2010)
14. https://machinelearningmastery.com/transfer-learning-for-deep-learning/. Accessed 2 Sep 2020
15. C.T. Rueden, J. Schindelin, M.C. Hiner, et al., ImageJ2: ImageJ for the next generation of scientific image data. BMC Bioinformatics 18–529 (2017). https://doi.org/10.1186/s12859-017-1934-z.PMID 29187165
16. C. McQuin, A. Goodman, V. Chernyshev, L. Kamentsky, B.A. Cimini et al., Cell Profiler 3.0: next-generation image processing for biology. PLOS Biol. **16**(7) (2018). https://doi.org/10.1371/journal.pbio.2005970s10
17. J. Xie, R. Liu, J. Luttrell, C. Zhang, Deep learning based analysis of histopathological images of breast cancer. Front. Genet. **10** (2019). https://www.frontiersin.org/article/10.3389/fgene.2019.00080, https://doi.org/10.3389/fgene.2019.00080.. ISSN 1664-8021
18. https://medicalsegmentation.com/. Accessed 2 Sep 2020

19. K.A. Philbrick, A.D. Weston, Z. Akkus et al., RIL-contour: a medical imaging dataset annotation tool for and with deep learning. J. Digit. Imaging **32**, 571–581 (2019). https://doi.org/10.1007/s10278-019-00232-0
20. E. Gibson, W. Li, C. Sudre, et al., A deep-learning platform for medical imaging. Comput. Methods Programs Biomed. **158**, 113–122 (2018). https://doi.org/10.1016/j.cmpb.2018.01.025. ISSN 0169-2607
21. P.F. Christ, et al., Automatic liver and tumor segmentation of CT and MRI volumes using cascaded fully convolutional neural networks. arXiv abs/1702.05970 (2017)
22. S. Rajaraman, S.K. Antani, M. Poostchi, K. Silamut, M.A. Hossain, R.J. Maude, S. Jaeger, G.R. Thoma, Pre-trained convolutional neural networks as feature extractors toward improved malaria parasite detection in thin blood smear images. PeerJ **6**, e4568 (2018). https://doi.org/10.7717/peerj.4568
23. A.T. Sapkal, U.V. Kulkarni, Comparative study of leaf disease diagnosis system using texture features and deep learning features. Int. J. Appl. Eng. Res. **13**(19), 14334–14340 (2018)
24. A. Krizhevsky, I. Sutskever, G.E. Hinton, 2012 AlexNet. Adv. Neural Inf. Process. Syst. 1–9 (2012)

Convolutional Neural Networks for COVID-19 Diagnosis

Vidyadevi G. Biradar, H. A. Sanjay, and H. C. Nagaraj

Abstract The COVID-19 infection has led to uncontrolled destruction to humankind. Health professionals and law and order bodies are the primary people facing challenges to control coronavirus crisis. However, there is a major responsibility on the researchers in investigating the possible Artificial Intelligence based aids that help in diagnosis of coronavirus infection at an early stage, mitigate the spread of coronavirus infection and bring peace and confidence in the people across the world. AI-driven tools show incredible potential to draw high precision interpretation from medical images and extend great support to doctors in decision making. Therefore, AI-powered support systems have been equipped as an integral part of health care center infrastructures. Chest-CT scans and X ray imaging for COVID 19 infection diagnosis aids in evaluating emergent conditions of patient is the necessary as RT-PCR test used for Coronavirus infection detection takes longer time. The examination of CT scan and X -Ray images from suspected patients with infection by an expert radiologist required careful observation and interpretation. With the outbreak in corona virus infection radiologists are subjected to extraordinary workload. Therefore, to reduce workload of radiologists it is necessary to develop an automated tool. In the present COVID 19 crisis there is emergency for development of an automatic tool which is helpful in quick analysis. The research outcomes of the state-of- the-art investigations in this direction indicate that COVID-19 detection models which are developed using the paradigm of deep convolutional neural networks have shown high hopes challenging COVID 19 crisis. In this chapter discusses, fundamentals of convolutional neural networks, transfer learning techniques, provides a list of COVID-19 dataset, COVID- 19 diagnosis systems based on DenseNet, ResNet

V. G. Biradar (✉) · H. A. Sanjay
Department of Information Science and Engineering, Nitte Meenakshi Institute of Technology, Bengaluru 560064, India
e-mail: vidyadevi.g.biradar@nmit.ac.in

H. A. Sanjay
e-mail: sanjay.ha@nmit.ac.in

H. C. Nagaraj
Department of Electronics and Communication Engineering,
Nitte Meenakshi Institute of Technology, Bengaluru 560064, India
e-mail: principal@nmit.ac.in

J. Nayak et al. (eds.), *Understanding COVID-19: The Role of Computational Intelligence*,
Studies in Computational Intelligence 963, https://doi.org/10.1007/978-3-030-74761-9_22

and SqueezeNet architectures, their comparison, ensemble techniques and research directions.

Keywords Deep learning · COVID-19 · Transfer learning · RT-PCR · Image augmentation

1 Introduction

The COVID-19 virus affects respiratory track and lungs leading to lungs inflammation and accumulation of fluids in sacks which is life threatening. There is no effective vaccine invented till now to treat this infection, the only the way to stop virus infection is staggered lockdowns, social distancing and use of personal safety kits etc. The COVID-19 pandemic pricked the consciousness of researchers to bear the responsibility to curb the infection. The nature of the infection has imposed responsibility not only on medical professionals, but tremendous pressure on researchers of other domains like data science and artificial intelligence [1].

It is observed by the radiologists that COVID-19 infections are observed in the lower sections of lungs. At present virus infection is diagnosed through RT-PCR-test which is costly, tedious and accuracy of the test is dependent on the quality of the materials used. The X-ray and CT-scan images of lungs are utilized to infer the details on infections. With the growing infection rate, radiologists are undergoing fatigue due to overwhelmed workload. It is difficult to confirm severity of infections through manual inspections. Therefore, it is the need of the hour to identify and evaluate an automated technology to diagnose COVID-19 infections through lung image analysis. The thrust of research for analysis of X-ray and CT-scan lung images is looking on artificial intelligence.

The deep learning paradigm which is under the umbrella of artificial intelligence is proven to be the promising technology in health care units. This technology not only helps in increasing turnover rate of corona infections detection but also helps in monitoring the infection progression. In the hospitals where AI assisted tools are used for COVID-19 diagnosis, the summary report on COVID-19 infections generated by artificial intelligence systems have matched the RT-PCR test reports including the description on ground glass opacities and pulmonary consolidation including lungs morphological changes. Therefore, there is demand for fine tuning of these systems by optimizing deep learning models to improve the performance further [2].

To challenge COVID-19 pandemic many research organizations have developed AI- tools for COVID -19 pneumonia detection from Chest-CT or X ray scans. Deep learning systems are utilized in health care units for screening, triage infection, and in monitoring the disease progression. Some organizations have released online tools, others are organizing crowd-sourced repositories of COVId-19 [4]. Infervision, releases AI based virus infection diagnosis tool known as InferRead CT Lung COVID-19. This tool is developed by exploiting existing pneumonia detection deep learning models [3].

Deep Learning methods comprises of plethora of algorithms that are motivated by the way brain works. A special type of deep learning networks is convolutional-neural- networks (CNNs), these networks are beating humans in various computer vision tasks and have become a natural choice for the analysis of chest X- rays for corona virus infection diagnosis.

The subsequent sections in this chapter discusses, fundamentals of convolutional neural networks, transfer learning techniques, provides a list of COVID-19 dataset, COVID- 19 diagnosis systems based on DenseNet, ResNet and SqueezeNet architectures, their comparison, ensemble techniques and research directions.

2 Convolutional Neural Networks for COVID-19 Diagnosis

CNNs are the sub-category of deep learning models which are useful in image classification tasks. This section gives a brief introduction on the typical architecture of CNNs, hyperparameters of CNN, steps to design a CNN and performance metrics for testing CNN models.

2.1 Convolutional Neural Networks

The characteristics of CNNs are like neural networks which are designed specifically for analyzing image data. The architecture of CNN has two parts; first part, extracts features from input image and the second part, carries out classification based on the features extracted by the first part. The first layer convolves input image with several filters and generates feature maps, which are passed through activation layer to yield normalized feature maps. These feature maps are reduced with pooling layer, the resulting feature maps are fed to subsequent convolutional layer in the model network and so on.

Types of layers in a typical CNN:

The basic layers in CNN are convolutional layer, pooling layer, Rectified Linear Unit (ReLu) and a fully-connected (FC) layer. The goal of the convolutional layer is to compute vital image features from the input image, the success in feature extraction depends upon the computation of filter weights that are used for convolution during CNN model training which are updated through back propagation algorithm. The number of features maps computed depends upon the number of filters/kernels with which image is being convolved. The pooling layer receives feature maps from preceding convolutional layer and this layer is placed in between two convolutional layers [5, 6]. The pooling layer brings reduction in feature set without losing significant information.

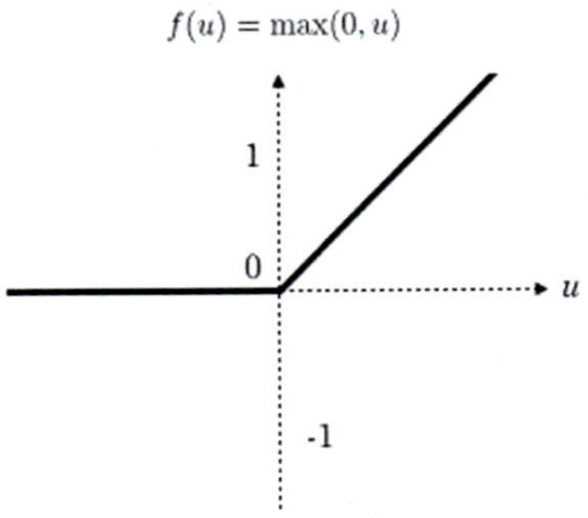

Fig. 1 Characteristics of ReLu activation function [7]

ReLU activation layer refers to a function [7] which is non - linear as shown in Fig. 1, which is represented as ReLu(x) = max (0, x), where x is the input pixel value, where, ReLu preserves positive values and replaces negative values with zeros.

The FC layer is the last layer in CNN, which is for classification, this layer receives a feature vector from the last convolutional layers. This layer classifies image into specified number of classes and returns a vector of probabilities which indicates the class of an image. The parameters of convolutional neural networks that can be varied during CNN design and experimentation are called as hyperparameters. The hyper parameters associated with CNN need to be fine-tuned to achieve better classification accuracy.

There are four hyper parameters for convolutional layer, i. No. filters/kernels for convolution, ii. The size of filters, iii. Stride value (1 or 2), iii. Padding, iv. Zero padding- padding border of the image to retain the original size of the image. There are two hyper parameters for pooling layer, they are i. size of the window, and ii. Stride value.

2.2 Hyperparameters

The pre-trained models can be reused for solving a novel problem, for this along with transfer learning, hyper parameters of convolutional neural networks need to be tuned to customize the CNN for the new task that leads to better classification accuracy [8, 9]. The hyperparameters are:

1. Learning rate

   - Fixed learning rate
   - Gradually decreasing/increasing learning rate,
   - Setting learning rates based on momentum
   - Adaptive learning rate

   Selection of a method for setting learning rate is based on the choice of optimizer such as Adam, SGD, Adagrad, AdaDelta and RMSprop.

2. Number of epochs: it indicates how many times the whole training image dataset are passed through CNN; this value must increase until a minimum difference between test error and validation error is observed.
3. Batch size: This is the numbers of samples fed in each training stage, small batch size is preferred. Generally, batch size in the range16–128 is a good choice.
4. Activation function: This induces non-linearity into the model, ReLu works well for CNN, other alternative functions include LeakyRelu, Elu, tanh and sigmoid etcetera.
5. Hidden layers and units: The optimal number of hidden layers needed depends upon the complexity of image dataset and it must be experimentally determined. Larger number of hidden layers adds to computational complexity and lower number leads to model underfitting. Therefore, it must be determined through experimentation.
6. Filter/kernel weights initialization: It is good to initialize weights of the filters with random numbers or uniform distribution.
7. Dropout: dropout drops neurons from the neural network randomly to avoid model overfit [10], it is desirable to set the value to 0.5. This indicates that 50% of the neurons are deactivated during training process.
8. Selection of optimal CNN hyperparameters: Grid search and random search techniques are used for searching optimal combination CNN hyper parameters as manual searching is tedious.
9. Visualization tools: It is necessary to understand the image features which are the focus of CNN for image classification task. This is very much essential to gain the trust in CNN functionality [11, 12].

2.3 *Steps for Design and Implementation of a CNN Model*

The Keras library provides functions to design and implement a CNN, however other platforms like fatsia and PyTorch can also be used.

- Define the model - The CNN may be sequential model or graph model and the popular one used in image classification is sequential model. In this step, add required number of layers to the model and specify number of filters/kernels, type of activation function, pool size and stride size etc.
- Compile the model-The hyper parameters like loss function, optimizer, learning rate and model performance metrics must be specified during compilation.
- Evaluation of performance - the CNN model is trained and tested.

Table 1 Confusion matrix

Predicted			
		Negative	Positive
Actual	Negative	True positive	False negative
	Positive	False negative	True positive

2.4 Performance Metrics for CNN Models

This section provides the performance metrics for testing of CNN models.

- **Confusion matrix** - it is used to interpret the accuracy of CNN model and its errors in classification task. Table 1 shows confusion matrix. Accuracy is indicated by the confusion matrix. Misguides the results interpretation when model is trained on imbalanced dataset. Therefore, evaluation on other metrics is required.
  The four parameters of confusion matrix are, True-positive (TP) - this indicates whether the model predicts the positive class accurately. The true-negative (TN) - this indicates whether the model predicts the negative classes accurately. The false-positive (FP) - represents whether model predicts positive classes incorrectly, false-negative (FN) - indicates whether model outcome has incorrectly predicted the negative classes.

- **Area under receiver operating curve (AUG)** [13]
  This metrics is used to evaluate the classification accuracy of CNN model in multi-class classification. This is measured for different threshold values which represents extent of separability between classes. Higher value of AUG is desirable, and it indicates better model prediction. The receiver operating curve (ROC) is the plot of TPR Vs. FPR. Is shown in Fig. 2. A guideline for classification of accuracy of CNN model from AUG values is: The range between (0.90–1) is satisfactory, (0.80–0.90) is good, 0(0.70–0.80) is fair, (0.60–0.70) is poor and (0.50–0.60) is not satisfactory.

- **Sensitivity /Recall/TPR (true positive rate)**
  It gives the result on the portion of positives which are correctly predicted. It is computed using below formula (1),

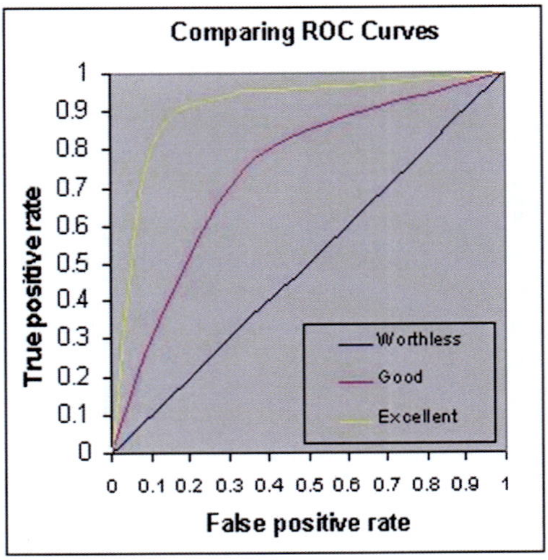

Fig. 2 AUG under receiver operating curve

$$Sensitivity = \frac{True\ positive(TP)}{TP + FN(false\,negative)} \tag{1}$$

- **Specificity**
  It gives the result on proportion of negatives which are correctly predicted. It is computed using below formula (2)

$$Specificity = \frac{TN}{TN + FP} \tag{2}$$

- **Precision**
  It is computed using below formula (3),

$$Precision = \frac{TP}{TP + FP(false\ positive)} \tag{3}$$

- **False Positive Rate (FPR)**

False positive rate is computed using formula (4),

$$False\ positiverate(FPR) = 1 - specificity \tag{4}$$

- **F1-Score**

This measure is important to balance precision and recall in case of uneven distribution of samples in the database. It is computed using the formula (5),

$$F1 - Score = 2 \times \frac{Precision - Recall}{Precision + Recall} \tag{5}$$

Relation Between Sensitivity, Specificity, FPR and Threshold
The sensitivity is inversely proportional to specificity, therefore, there is a trade-off between them. Now, for a given application whether high value for sensitivity or high value for specificity is of prime importance must be decided to optimize the model. The increase in threshold in tern increases number of negatives values, thus higher value of specificity and lower sensitivity will be achieved.

- Dice coefficient
  This computes the hormonic mean sensitivity and precision and its outcome is similar to F1 score.

- Matthews Correlation Coefficient (MCC) [14]
  MCC is defined as a correlation co-efficient between the target values and predicted values. This is a reliable statistical test, and its score will be high only when all the parameters of CM produce high scores. The value of MCC lies in the range of (+1 and −1).

MCC is more reliable compared to F1 score in classification. MCC is also known as phi-co-efficient. MCC is computed using below formula (6)

$$MCC = \frac{TP \times TN - FP \times FN}{\sqrt{(TP + FP)(TP + FN)(TN + FP)(TN + FN)}} \tag{6}$$

The value of MCC changes with change in any of the terms in the equation and the value returned varies in the range of (+1 and −1). The value +1 indicates perfect classification accuracy and −1 indicates no correlational between target and predicted values.

When any of product terms in the denominator is zero, the denominator becomes 1 and MCC becomes 0. If the values for FP and FN are equal 0, this leads MCC to 1 and this is the perfect classification/prediction. Conversely, if the values of TP and TN are equal to 0, then MCC value becomes −1, thus represents the case of worst performance of the model.

The qualitative analysis of CNNs performance in image analysis offers following advantages,

- **Interpretability:** CNN models suffer from lack of visibility; therefore, interpretation of the model is needed for better understanding of the model. The Gradient-weighted-Class-Activation Mapping (Grad-CAM) is a tool for visualizing the model using which image region heatmaps are generated which gives an insight into how predictions are made by the model.
- **Trustworthiness:** The heatmaps highlight potential regions in images which are the focus of CNNs to make classification. This also helps in authenticating the functionality of the system by a radiologist.

3 Methodology to Design a CNNs for COVID-19 Infection Analysis

This section discusses the methodology for design and development CNNs for COVID-19 detection using X-ray and CT scans. The various steps in the and development of CNN models are,

- Acquire COVID-19 chest images
- Image augmentation
- Synthetic image generation
- Pre-processing of images
- Design CNN model/ selection a pre-trained deep CNN model
- Training/Testing the model
- Evaluation of model performance (qualitative and quantitative)

3.1 Acquire COVID-19 Image Dataset

The COVID-19 datasets that are available in various repositories are to be gathered for preparation of a database. The CNN model is trained and tested on the database. The various databases publicly available for evaluation of CNN models are list in the Table 2.

3.2 Image Augmentation

Image augmentation is useful to increase the dataset and counter the issue of imbalanced COVID 19 dataset. The augmentation enhances inter-class variance between the image samples and therefore helps in avoiding model overfit [25]. This technique also helps in balancing image set which is used for CNN model training. There are two ways of image augmentation, they are offline augmentation, and another is augmentation of images on fly during training.

The traditional methods for augmentation of COVID-19 images include [26, 27]

- Image mirroring
- Addition of Gaussian noise
- Shearing images by applying affine transforms
- Decreasing/increasing the brightness of images
- Image translation
- Vertica flip, horizontal flip, and vertical plus horizontal flips
- Rotation by −90, 60, 90 and 150° , and
- Zooming by 20

Traditional methods of image augmentation are supported by Keras Library. During training images are selected randomly from dataset and specified operations will be carried out.

3.3 Synthetic Image Generation

The database of medical images for rare diseases often contain a few image samples, on the contrary, deep learning models need enormous number of images to offer better classification accuracy. One of the methods discussed is image augmented technique to enlarge the dataset which has a limitation on number of additional images that can be generated. Thus, there is a need for exploring synthetic images. In this context Generative Adversarial Networks (GANs) are useful in generating synthetic X-ray and CT-scan images pertaining to COVID-19 infection [28]. A variant of GAN referred as Auxiliary Classifier GAN (AC-GAN) [29] and Conditional Generative

Table 2 Dataset for COVID-19 images

Paper	Database
Linda Wang, Zhong Qiu Lin and Alexander Wong [15] [2020]	COVIDx is a benchmark database with highest number of COVID positive case samples. It consists of a total number of 13,975 chest X-rays collected from 13,870 number of patients This database integrates samples from five different datasets. They are, ActualMed COVID-19 Chest X-ray Dataset-Initiative, COVID-Image Data-Collection and Chest X-ray Dataset-Initiative, RSNA and COVID-19 radiography dataset
Mohammad Rahimzadeh, Abolfazl Attar, Seyed Mohammad Sakhaei [16] [2020]	COVID-CTSet: This dataset has total 63,849 number of CT images gathered from 377 patients. There are 15,587 number of samples, 95 samples are with corona virus patients, and 48,260, CT-images from 282 normal persons
Gianluca Maguolo, Loris Nanni [17] [2020]	NIH dataset/Chestx-ray8 database: This database has of 108,948 X-rays which are collected from 32,717 patients, and these X-rays are categorized into 8 varieties of infection CHE dataset: This dataset consists of 224,316 number of X-rays gathered from 65,240 number of patients which are classified into 14 classes KAG dataset: There are total 5,863 pediatric images. COV dataset: There are 144 number of X-rays of corona virus positive patients
Chiou-Jye Huang, Yung-Hsiang Chen, Ping-Huan Kuo, Yuxuan and Ma [18] [2020]	Surging News Network and World Health Organization (WHO): This database is formed from Hubei Province dataset, Zhejiang, and Guangdong Province cities

(continued)

Table 2 (continued)

Paper	Database
Enzo Tartaglione, Carlo Alberto Barbano, Marco Calandri, Claudio Berzovini and Marco Grangetto [19] [2020]	COVID-Chest XRay: there are total 287 Chest-X-ray and CT scans of COVID-19 and other types are of pneumonia infection. Out of total 137 CXRs, 108 are COVID-19 (+) cases and 29 are COVID (−) CORDA: Collected from websites and publications It contains total 447 chest X-rays collected from 386 patients, 150 samples are COVID-19 negative cases and 297 positive cases Montgomery-County X ray Set: This dataset with 138 X-ray images which are collected from Tuberculosis Control Program, 80 number of images are collected from normal patients, 58 from diseased patients Shenzhen Hospital X-ray Set: This database incorporates 662 X-rays, 326 X-rays are from healthy patients and 336 X-rays are of abnormal patients ChestXRay: This dataset consists of total 1583 normal patients, 2780 patients are bacterial pneumonia, 1493 by viral- pneumonia RSNA: This dataset incorporates of total 20,672 number of normal chest X-rays and 6012 pneumonia case
MohammadRahimzadeha, Abolfazl Attarb [20] [2020]	Covid chestxray consists of 180 number of X-rays from COVID-19 virus positive cases, 42 - pneumonia Rsna Pneumonia Detection Challenge: cit has 6012 number of pneumonia X-rays and 8852-number of X-rays of normal cases
Sakshi Ahuja, B.K.Panigrahi, Nilanjan Dey, Venkatesan Rajinikanth, Tapan Gandhi [21] [2020]	Covid-ct-dataset, UCSD-AI4H/COVID-CT, https://www.medrxiv.org/, https://www.biorxiv.org/: This is the collection of 349 CT scans for COVID-19 positive cases corresponding to 216-patients and 397-CT images collected from Non-COVID patients by integrating the dataset from Covid-ct-database, samples from medrxiv.org combined withbiorxiv.org website

(continued)

Table 2 (continued)

Paper	Database
Ali Narin, Ceren Kaya, Ziynet Pamuk [22] [2020]	Dr. Joseph dataset: This database collects Chest X-ray and CT-images from the patients having acute respiratory-distress-syndrome (ARDS), COVID-19, Middle-East-respiratory-syndrome (MERS), pneumonia, severe acute-respiratory syndrome (SARS). 50-corona virus positive samples and 50-normal Chest X-Ray Images from Kaggle database
Asmaa Abbas · Mohammed M. Abdelsamea · Mohamed Medhat Gaber [23] [2020]	JSRT dataset and Dr. Joseph Cohen's chest X-ray image dataset: It contains 80 number of chest X-rays of normal patients in Japanese Society of Radiological Technology Dr. Joseph Cohen's dataset: it has 105 total number of images with 11 images corresponding to COVID-19 and virus-SARS
Umut Ozkaya, Saban Ozturk, Mucahid Barstugan [24] [2020]	Societa Italiana di Radiologia Medica e Interventistica: Contains total number of 53 CT images gathered from COVID_19 positive patients

Adversarial Network (CGAN) [30] are used in generation of high-resolution images corresponding to COVID-19 positive cases.

3.4 *Pre-processing of COVID-19 images*

Typical pre-processing operations include image resizing, histogram equalization and wavelet decomposition etcetera.

- Image resizing- most of the pre-trained deep convolutional neural network models are trained on a particular size of images. Therefore, to match the input image size of pre-trained models, image resizing becomes one of the important steps.
- Histogram equalization- the quality of chest X-ray, chest CT scan is enhanced by histogram equalization. Even though medical images are acquired in a controlled environment, still the application of histogram enhancement techniques helps in further improvement of image quality.
- Wavelet decomposition- in some methods wavelet decomposition is initially applied on images dataset, then augmentation techniques are applied on LL, LH, HL and HH components to increase the number of samples [21]. Typically, Haar and Daubechies wavelet decomposition is selected, however, other wavelets can be also explored.

4 Pre-trained Deep CNN Models

There are various trained models which are trained on ImageNet database that comprises of millions of natural images. These pre-trained models are very much useful in building novel models for COVID-19 image classification with a slightly varied image dataset due their pre-acquired knowledge of the features of ImageNet dataset [31–33]. However, they have limitations in dealing with COVID-19 X-ray and CT- scan images, since these images are different from natural images to a great extent. Therefore, existing models such as VGG16, AlexNet and ResNet etc., cannot be readily used for COVID-19 image classification.

The primitive features of images remain same for any category and they differ only in high level features, therefore the knowledge of pretrained models can be exploited to derive features like edges, corner and curves etc., from COVID-19 dataset and models can be fine-tuned further to extract high level features. The advantages of pre-trained model are less complexity in training the models and encourages reuse of architecture. The techniques of transfer learning which are useful to customize pre-trained models in COVID-19 dataset classification are discussed in the next section.

4.1 Transfer Learning (TL)

This technique helps in repurposing pre-trained model for the classification of new dataset. The advantage of TL is that the model can be trained on new dataset comparatively fast and it is possible to tap the knowledge acquired by these models. Generally, in TL the earlier layers in the architecture are not trained on new data, rather the later part of the architecture will be trained, therefore it reduces the training time. In convolutional neural networks the earlier convolution layers extract more abstract features like edges and corners etc., from input images and the later layers extract the features which are more specific to given image dataset. Thus, these layers need to be retrained on the new dataset [34]. A CNN is visualized to contain two parts when transfer learning is to be applied. They are,

- Convolutional base
  This consists of a sequence of convolutional and pooling layers. Convolutional base extracts potential features from the image.
- Classifier
  This is a FC layer, which classifies the images based on the features from convolutional base.

The CNN models with suitable architecture and hyperparameters will automatically learn hierarchical representation of image features that are required for the classification. Features extracted by the first layer are general and therefore can be reused in varied problem domains. Features extracted by the last layer are dependent on specific dataset and needs fine-tuning. The pre-trained model is tuned on the new dataset through different training strategies [35].

- Train the model from the scratch, here only the pre-trained model architecture is reused, perhaps all the model layers are trained from the scratch. This indicates the case of reuse of architecture. This strategy needs high computation power for training. This strategy is adopted where the number of image samples are less and their nature is completely different from natural images, for example, medical images are different from natural-scene image. This strategy needs careful selection of learning rates as new weights must be calculated for all the layers.
- Train later layers and leave the earlier frozen, here, the difficulty is in determining how many layers should be kept frozen and how many need to be trained fresh. There is no thumb rule for it, rather it must be experimentally determined. The wrong decision in the selection of appropriate number of layers leads to model overfit. This strategy also needs careful selection of learning rate.
- Keep the convolutional base undisturbed and change only the classifier. This is suitable when the problem to be solved is similar in nature as that of pre- trained model and the number of image samples are smaller.

Transfer Learning is the most sought-after concept to solve new problems like COVID- 19 image analysis. Due to unavailability of sufficient data on COVID-19, it challenging to use complex deep CNNs for COVID-19 image classification. The selection of an appropriate pre-trained model like VGG16/19, ResNet50, DenseNet, AlexNet and SqueezeNet etc. is an important step in repurposing the model for COVID-19 image classification.

In the proposed chapter, the focus is not on general popular pretarined models like VGG, GoogleNet, AlexNet and MobileNet etc., as these models have very deep architecture and are not very good at the analysis medical images as they are trained on natural image dataset. The concept of transfer learning dictates that training the model on the relevant dataset is most important requirement in building a robust model.

From the literature it is understood that models which are developed using DenseNet, ResNet and SqueezeNet gave promising results in analysis of medical images such as chest-X-ray images and chest CT scans and there exists several benchmark models which are publicly available. Therefore, the focus of this chapter is to highlight the suitability of these models in the design of models for coronavirus infection diagnosis.

The focus of the chapter is on how progressive resizing concept is applied during the training of model, concept of training the model in multiple stages, in each stage the model is trained on one particular class of images and how to utilize ensemble models, that deploy multiple models and to arrive at the result combine the outcomes of individual models.

The suitable values for hyperparameters of convolutional neural networks like learning rate, drop out value, momentum, epochs, and batch size etc. need to be determined experimentally and there are no standard guidelines for selection of these parameters.

4.2 Repurposing DenseNet Architecture for COVID-19 Diagnosis

The architecture of pre-trained model DenseNet is popularly used in the design of Pneumonia diagnosis CNN models. The performance of these models is satisfactory to a large extent and even concurred with expert opinions, and therefore, these are deployed as assisting tools in AI supported health care units [36, 37]. In the present scenario. The knowledge of DenseNet based models is being exploited in the development of COVID-19 diagnosis systems. Following sections highlights the characteristics of a basic DenseNet architecture and discusses the design of a DenseNet based novel model for COVID-19 infection predication.

4.2.1 DenseNet Architecture [38]

The underlying concept of a dense network is to forward the residue of layers to all the layers that are ahead of them, e.g., layer-1 forwards its residue to all its

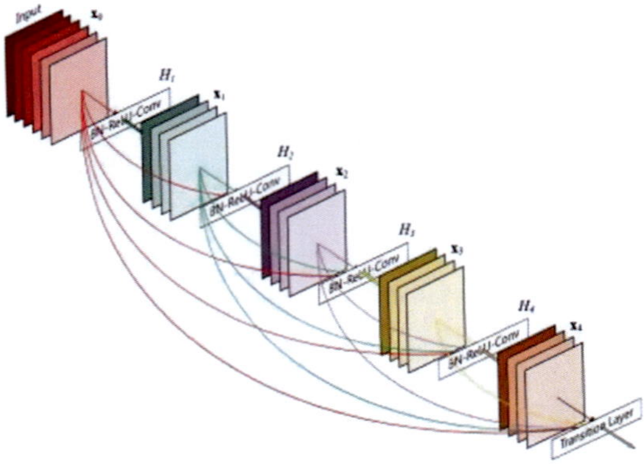

Fig. 3 DenseNet architecture [38]

forthcoming layers. This kind of connections between the layers improves flow of feature maps and hence the necessary information [38]. The connections between the layers in DenseNet is shown in the Fig. 3. The layers that between the dense blocks are the transition layers. These transition layers comprise of convolutional layers and pooling layers which transforms the feature map received by previous dense block. The DenseNet is named due to many block layers in the network. If there are 121 dense blocks, then it is DenseNet121 model and so. Another advantage of DenseNet is that dense connection provides regularization effect on model and thus reduces over-fitting when trained with smaller data.

DenseNet eliminates vanishing gradient issue and hence supports enhancing reuse of feature, and parameters efficiency.

4.2.2 Method 1: COVID-DenseNet Model [39]

The novel model COVID-DenseNet is designed using a pre-trained model CheXNet [36] which is used for pneumonia prediction from chest X-rays. This model is repurposed using transfer learning for COVID-19 positive prediction. The model CheXNet classifies chest X-ray images into 14 classes which corresponds to 14 different chest infections. The model CheXNet is developed by fine tuning DenseNet121 deep convolutional neural network on ChestX-ray14 database.

In this method, the reason for selecting CheXNet against other pre-trained model from Kaggle is CheXNet is trained on chest X-ray images not on natural images and therefore, the knowledge acquired by this model is more relevant to COVID-19 image set. Therefore, knowledge acquired this model will be of significant use to repurpose it for detection of COVID-19 diagnosis. Also, the performance of CheXNet is on par with radiologist's opinion. This model is used in healthcare units for Chest X-Ray annotation. The CheXNet model is retrained on COVIDx database keeping initial weights of Chest X-ray14 database.

This method uses COVIDx, which is a small and highly skewed dataset due to smaller number of images of COVID-19 positive patients. Therefore, this dataset in its state is not suitable for training a deep learning model. To counter this effect image augmentation is applied on COVID-19 images during training, this results into increase in the number of images from 238 to 11,416.

DenseNet-121 has total 121 densely connected convolution layers plus a fully connected layer with 1000 neurons in the final output layer. This FC layer is substituted with new one with two neurons and three neurons for two-class and three-class classification, for COVID 19 pneumonia classification.

The transfer learning technique applied on DenseNet121 to tune it for COVID-19 classification is discussed here.

The DenseNet-121 model with initial weights from ChexNet is retained end-to-end with hyperparameters Adamax optimizer, learning rate = 0.00001 and categorical cross-entropy loss function. The resulting model offers an overall good performance with accuracy = 93% and precision = recall = F1-score = 0.92 on three class classification, however, precision of class COVID-19 is 0.77 which is not satisfactory. However, the model offers overall performance with accuracy = 93% and precision = recall = F1-score = 0.96 on two class classification, with precision = recall = f1-score = 0.96 which is quite satisfactory.

4.3 ResNet with Transfer Learning for COVID-19 Image Analysis

The specialty of Residual Networks is that they attain training convergence quickly and offers satisfactory performance on image classification task [40]. Application of some of the ResNet models for COVID-19 diagnosis have shown appreciable performance. Therefore, it is necessary to understand the suitability of ResNets by fine-tuning to detect COVID-19 infections using various transfer learning techniques. The following sections discuss ResNet architecture in brief and provides an insight into a few ResNet architecture-based models that have offered satisfactory performance for corona virus infection prediction.

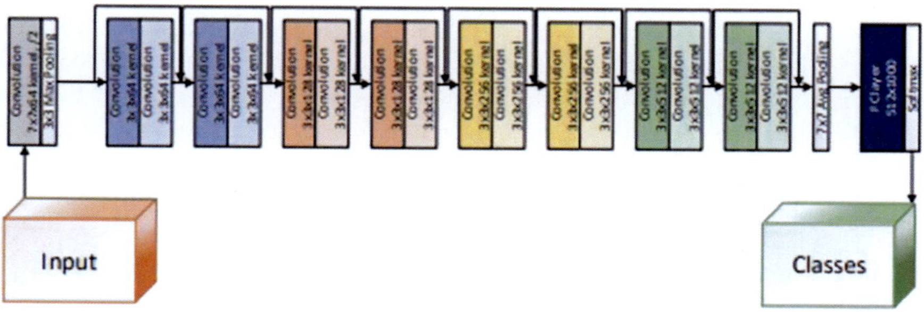

Fig. 4 The architecture of ResNet18 model [40]

4.3.1 Method 1: ResNet Architecture

ResNet offers advantages of easier gradient flow and efficient training, this model incorporates identity short connections to skip one or more number of convolutional layers in the architecture. This skip connection provides direct connection to every initial layer in the network thus helps in ease update of gradients. Figure 4 Shows the abstract block diagram of ResNet18, this model has 18-layers and resNet50 is having 50 layers.

Residual-Networks give a best performance with minimum number of parameters and faster training. Using residual networks, it is possible to input image of different sizes. This is very important for model training to achieve a high model performance.

4.3.2 Method 1: COVID-ResNet model [41]

A pre-trained ResNet-50 architecture is fine tuned for COVID-19 diagnosis model named as COVID-ResNet. This method uses the techniques of progressive resizing, cyclical learning in addition to discriminative learning to achieve faster training on residual neural networks.

In this method the network is finetuned with small images and then size of images is gradually increased during training process progressively in steps $128 \times 128 \times 3$, $224 \times 224 \times 3$, and $229 \times 229 \times 3$ pixels. Initially, ResNet50 is initialized with weights of ImageNet dataset. Images normalization is done using the mean and standard deviation of ImageNet dataset. For transfer learning, new head containing a sequence of Adaptive -average/max pooling is added, others layers added are i) batch normalization, ii) drop-out and, iii) linear layers. Training is performed in three stages.

Stage 1. The input images are resized to $128 \times 128 \times 3$, COVID-ResNet model is tuned in two steps, i. only the newly added head node of the model is trained, and ImageNet weights are retained for remaining layers. The learning rate chosen is 1e−3 and the 3 epochs, and ii. The entire model is tuned with discriminative-learning rate with 5 epochs. Stage 2. The head of the model from stage-I is fine-tuned further with images of size $-224 \times 224 \times 3$ and learning rate of 1e−4 for 3 epochs. Now, the entire model is further fine-tuned in 5 epochs with discriminative-learning rate. Stage 3. The entire is tuned for images of size $229 \times 229 \times 3$ in 25 epochs. The earlier layer was trained with a learning rate of 1e−6, last layer with 1e−4 and layers that in the middle of network were trained using equidistance learning rates, Adam optimizer is used and batch size selected is 32.

This model gives an accuracy of 96.23% and it is 13% more the COVID-Net model which gave an accuracy of 83.5%. In comparison with COVID-Net, it achieves 4.5 times lower parameters with a model size of 25.6 M against 116.6 M COVID-Net [50].

4.3.3 Method 2: Ensemble Fine-Tuned RsNet18 Model [42]

The ResNet18 model is used in this method, it has shallow architecture and takes less training time and offers good performance. The ResNet18 model consists of one 7×7 convolution layers, 2-pooling layers, 5-residual blocks and one fully connected layer. The residual block consists of two 3×3 convolution layers, batch-normalization and ReLU layers. The architecture is modified by skipping these two convolutional layers and feeding the input to final ReLu layer.

Datasets used for evaluation of model include, the RSNA-Pneumonia-Detection Challenge dataset, the COVID-19 image-data collection, and COVID-19 X-rays. Image augmentation is carried out through random cropping, horizontal flip, and rotation, and oversampling.

Four datasets generated for experimentation they are, Dataset_A: normal and diseased cases chest X-rays, Dataset_B: pneumonia and non-pneumonia cases chest X-rays, Dataset_C: COVID-19 and non_COVID-19 cases chest X-ray, Dataset_D: normal, pneumonia and COVID-19 chest X-rays, four version of ResNet18 are generated they are, model 'a' is trained the model on Dataset_A, model 'b' is trained trained on Dataset_B and model 'c' is trained on Dataset_C.

New versions of ResN18 model are obtained by additional training, Model 'A' is obtained by training on Dataset_D, Model 'Bab' is obtained by combining models: Model 'a' and Model 'b', Model 'Cac' is an ensemble obtained by combining models: Model 'a' and Model 'c'. Model 'Dbc' is an ensemble model obtained by combining models: Model 'b' and Model 'c'. Model 'Eabc' is an ensemble model obtained by combining models: Model 'a' and Model 'b' and Model 'c'.

Finally, all these models are trained on Dataset_D which is a super set of images, and weights of Model 'a', Model 'b' and Model 'c' are freeze.

In ensemble method every model contributes to the classification accuracy and inaccuracies in the model will be eliminated. To this ensemble model concatenation layer and classification layers are added to get final model. Now, this ensemble model is trained on Dataset_D and is tested for classification of dataset into normal, pneumonia and COVID-19. It is observed that Eabc model offers best performance.

In this method the focus on the way the model is trained stepwise on different class of images from the dataset and combine the outcome of these intermediate results.

4.4 SqueezeNet Model with Transfer Learning for COVID-19 Image Analysis

The SqueezeNet is a popular pre-trained CNN with a shallow architecture. SqueezeNet deep convolutional network is a light-weight model, and its performance is on par with Alex Net model with 50× fewer number of model parameters. The model is compressed to less than 0.5 MB, therefore which is suitable for deployment

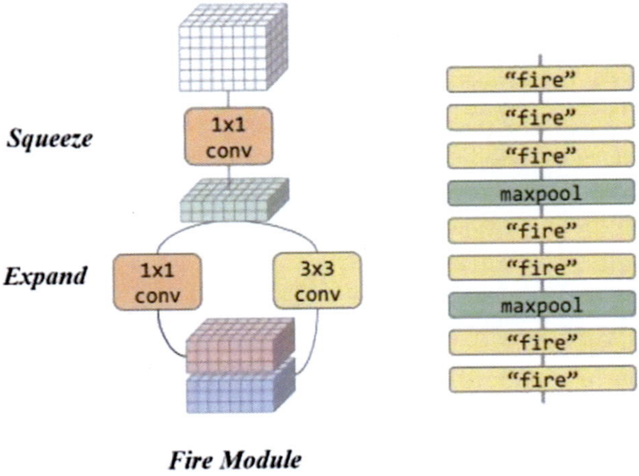

Fig. 5 The architecture of SqueezeNet [44]

of deep learning models in embedded systems [43]. There is a need to design portable embedded systems for COVID-19 diagnosis, and therefore SquezeNet is the most suitable pre-trained model for such system. SqueezeNet is utilized by fine tuning for the design COVID-19 diagnosis model. In the following sections architecture of a basic SqueezeNet model is given for quick reference and a few models that are implemented based on SqueezeNet model are discussed.

4.4.1 SqueezeNet Architecture [44]

SqueezeNet introduces a novel block in CNN architecture known as 'fire' module, it comprises of two main blocks, they are squeeze and expand. The squeeze block consists of 3, 1 × 1 convolutional filters/kernels, the expand block consists of 3, 1 × 1 filters/kernels and 4, 3 × 3 convolutional filters/kernels. The SqueezeNet contains total 9 number of fire modules. This model squeezes the input image in vertical dimension by 1 × 1 convolutional layer and expands the depth again by a double branch layers with kernel size 3 × 3 and 1 × 1 filters. Figure 5 shows the architecture of a basic SqueezeNet.

There are several methods in which explores effectiveness of SqueezeNet for the classification COVID-19 image and some of them are describe in this section.

4.4.2 Method1: Fine-Tuned SqueezeNet Model [45]

In this method the architecture of basic SqueezeNet model is modified further to optimize the computational complexity. To achieve fast training-convergence and reduce overfitting, batch normalization layer is added between the squeeze convolution layer and ReLu.

The Relu layers are substituted with Elu layers as they give better performance in comparison with Relu activation function. The Fig. 6(a) shows basic SqueezeNet fire module and (b) shows modified fire module.

The architecture of a basic SqueezeNet comprises of 8 fire modules, and it has two configurations. Simple bypass connection: it has 4-skip connections between Fire Modules; 2 & 3, 4 & 5, 6 & 7 and, finally, between 8 & 9. In complex bypass configuration there are another 4 skip connections between same pairs of fire Modules, a convolution layer with filter of size 1 × 1.

In this method SqueezeNet with simple bypass-connections is used. Here, a transpose-convolution layer is added to the last-custom Fire-Module. This arrangement multiplies the width and height of feature maps four times which are combined in depth with the feature maps from the second Custom-Fire-Module via a skip-connection.

Weighted sum is obtained through a convolutional layer consisting of 128-filters with size 1 × 1. Finally, all the feature-maps are joined and averaged with a Global Average Pooling layer. This design helps in to combining spatial information and features. Bayesian optimization is used for selection of initial learning rate, Momentum, L2-Regularization.

The SqueezeNet model is trained in four different schemes, namely SqueezeNet without bypass and transfer learning, SqueezeNet with simple bypass but without transfer learning. These schemes are compared with the modified model. The resulting architecture offers an accuracy of 83%, and sensitivity, specificity, precision and F1-score of 0.85, 0.81, 0.873, 0.83333 respectively.

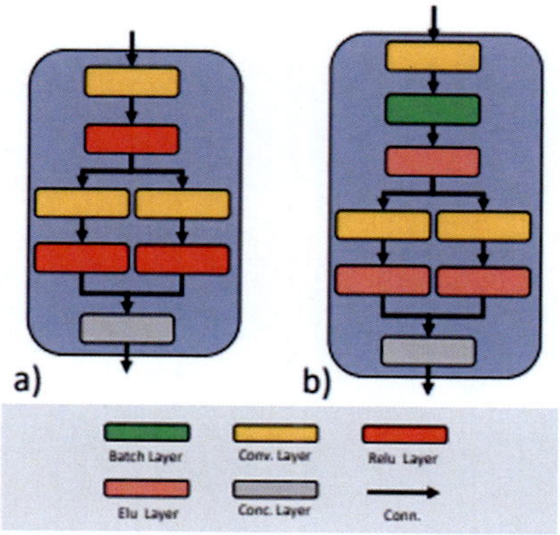

Fig. 6 The classic Fire-Module of the basic SqueezeNet **a**. modified Fire Module **b**

4.4.3 Method 2: COVIDiagnosis-Net Model [46]

The method builds a model known as Deep Bayes-SqueezeNet which is consists of three stages: offline augmentation, optimize SqueezeNet using Bayesian optimization technique and train the model. The proposed does three-class classification, Normal (no infection), Pneumonia and COVID. SqueezeNet model is repurposed for classification of COVIDx classes. To achieve the best prediction model, the CNN model is fine-tuned on hyperparameters using Bayesian method.

4.5 Comparison of DenseNet, ResNet and SqueezeNet Based Models

The previous sections presented a few CNN models that are designed by exploiting the architectures of very efficient pre-trained deep learning models through transfer learning. The Table 3 give the performance comparison of DenseNet, ResNet and SqueezeNet based models that are fine-tuned to adopt for COVID-19 infection diagnosis.

From the above table following observations are drawn:

i. The COVID-ResNet model offers excellent performance with an accuracy of 96.23%, and the model ensures balanced classification, precision = recall =

Table 3 Comparison of DenseNet, ResNet and SqueezeNet based models

		Accuracy	Precision/Positive predictive value	Recall/Sensitivity	F-score
COVID-DenseNet	COVID-19	90.33	0.77	0.85	0.81
	Pneumonia		0.93	0.91	0.92
	Normal		0.94	0.94	0.94
COVID-DenseNet [39]	COVID-19	95.00	0.90	0.94	0.92
	non-COVID-19		0.97	0.97	0.97
COVID-ResNet [41]	COVID-19	96.23	1	1	1
	Bacterial		0.95	0.97	0.96
	Viral		0.92	0.93	0.93
Multi-channel Resnet18 architecture [42]	COVID-19	90.00	0.83	1	0.91
	Pneumonia		0.98	0.88	0.94
	Normal		0.77	0.94	0.84
SqueezeNet variant		83.0	0.81	0.85	-
COVIDiagnosis-Net [46]		98.3	0.98	0.98	0.98

F1-score = 1 corresponding to COVID-19. This indicates that there will be no misclassification

COVID-19 positive cases, which helps in avoiding infection spread. Also, for other class samples all the metrics values are above 90 which is very much satisfactory.

ii. The COVIDiagnosis-Net offers high performance with an accuracy of 98.3%. The values for other metrics are, precision = recall = F1-score = 0.98 which are near 1 indicates the classification error can be ignored and this model can be used in assisting radiologists.

iii. The COVIDiagnosis_Net offers additional benefits of being a lightweight model as compared to COVID-ResNet and requires less memory and computational power. Therefore, can be used in the design of portable COVID-19 diagnostics model.

Performance of CNN models depend upon the strategies used in fine tuning hyper parameters. These include initial learning rate, Momentum and L1/L2-Regularization, initial weights for convolution filters etc. Others, factors are dropouts, type of pooling layer and batch normalization. There is a scope for evaluation of results by optimizing the architectures of Pre-trained models. In ensemble methods best performance metric like Matthew's Coefficient must be used for selection of CNN models for ensemble. It is good criteria to select pre-trained models which are prior trained on medical images. The models based on DenseNet, ResNet and SqueezeNet shows satisfactory results. However, since SqueezeNet is a lightweight model it has an added advantage.

The focus of the work was highlighting the development of COVID-19 diagnosis models using DenseNet, ResNet and SqueezeNet which gave promising results on medical images, in particular chest images. This chapter identifies these models from a pool of pretrained models, which may be helpful to the researchers.

The performance of convolutional neural networks or pretrained models change with the change in the hyper parameters of model and the training parameters, these parameters are to be determined experimentally as the there is no general guideline for selection of hyperparameters, and the performance of the model depends upon the combinational values and therefore, it is difficult to comment on which parameter contributes to the performance. Therefore, results are not shown for different values of parameters.

4.6 Ensemble Methods

The following section describes the ensemble methods used in the detection of COVID-19 from medical images.

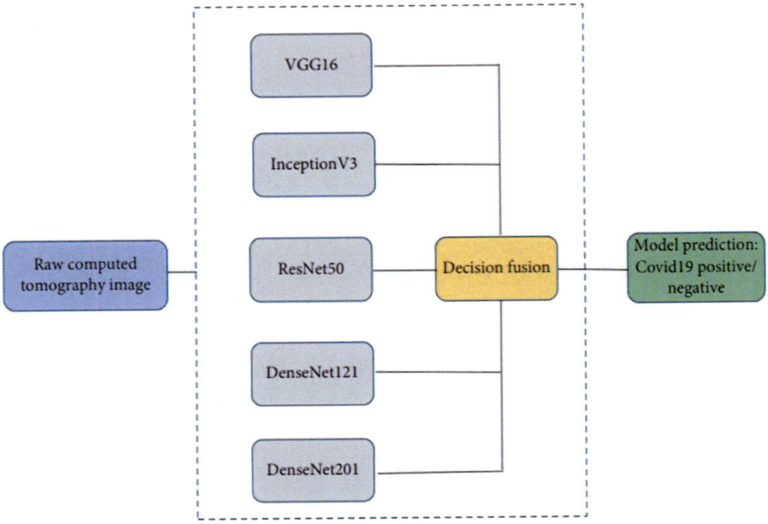

Fig. 7 Fusion based prediction model

4.6.1 Method 1 [47]

In this method, various Deep CNNs are used for detection of COVID19 from chest CT- images. The predictions obtained from individual models are fused to arrive at final prediction. The diagrammatic representation of this method is given in Fig. 7.

Experimental results indicate that the fusion-based technique is gives an accuracy of 86 with average-AUROC and F1-Score 0.883 and 0.867, respectively.

This method uses multiple pre-trained models for the infection diagnosis and combines their decision. This method has a disadvantage of high computational complexity and dependency on multiple pre-trained models which becomes more complex in tuning then for new dataset.

4.6.2 Method 2 [48]

In this method shallow network architectures namely MobileNetv2, ResNet18 and SqueezeNet plus models: ResNet101, Inceptionv3, VGG19, CheXNet, and DenseNet201 are tested for COVID-19 pneumonia classification wherein each individual model prediction is fused to compute the final prediction. The block diagram in Fig. 8 shows this technique. These pre-trained CNN networks are trained with stochastic Gradient-Descent (SGD), momentum optimizer for learning rate, batch with size16 and 20 epochs.

Model performance of COVID-19 prediction is evaluated on ROC curve and confusion metrics derived from it. Figure 8 illustrates the overall system diagram with the three- class image classification problem. The performance results are given in

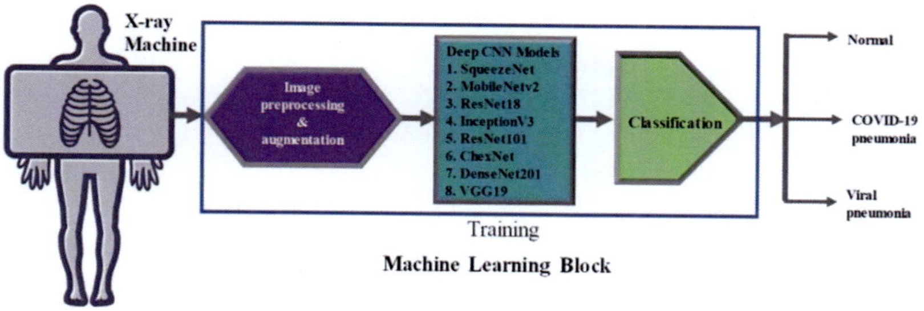

Fig. 8 Block diagram of the COVID-19 prediction system [48]

Table 4. Performance metrics for deep learning models for two-class prediction [48]

Schemes	Models	Accuracy	Precision (PPV)	Sensitivity (Recall)	F1 Scores	Specificity
Without image augmentation	SqueezeNet	99.29	99.3	99.29	99.29	99.29
	MobileNetv2	99.4	99.41	99.4	99.41	99.4
	ResNet18	**99.41**	**99.42**	**99.41**	**99.41**	**99.41**
	InceptionV3	99.41	100	98.81	99.4	100
	ResNet101	99.05	99.08	99.05	99.07	99.05
	CheXNet	**99.41**	**99.42**	**99.41**	**99.41**	**99.41**
	DenseNet201	99.3	99.4	97	97.8	99.75
	VGG19	99.41	99.76	99.05	99.4	99.76
With image augmentation	SqueezeNet	99.40	99.40	99.40	99.40	98.84
	MobileNetv2	99.65	99.65	99.65	99.65	99.26
	ResNet18	99.60	99.60	99.60	99.60	99.31
	InceptionV3	99.40	98.80	98.33	98.56	99.70
	ResNet101	99.60	99.60	99.60	99.60	99.31
	CheXNet	**99.69**	**99.69**	**99.69**	**99.69**	**99.23**
	DenseNet201	**99.70**	**99.70**	**99.70**	**99.70**	**99.55**
	VGG19	99.60	99.20	98.60	98.90	99.80

Table 4. It is clear from the results that ResNet18 and CheXNet give good predictions as compared to the other models. Therefore, these predictions can be combined to obtain the results.

The performance of model in coronavirus infection diagnosis depends upon the type of the dataset. in most of the cases it is imbalanced which gives erroneous results. Therefore, additional images are generated using image augmentation technique. Again, since there is no control on the class of images generated by keras library, this method at a time shows improved results and sometimes brings down the performance as shown in the Table 4, especially in case case of SqueezeNet. Therefore, it wise to generate additional images using GAN where it is possible to specify the class of images that the network must generate.

The models are justified on their performance based on F1 score, accuracy, sensitivity and specificity s for two class as well as three class classification problem. Other parameters are to be determined by experimentally. The keras tuner also provides option of leaving the system compute optimal combination for hyperparameters. In some cases, results are good with augmentation, however, augmentation does not contribute to results in all cases.

4.6.3 Method 3 [49]

This method is an ensemble model called as DeQueezeNet for COVID-19 infection inferences using X-ray. This model combines the outcome of two models namely DenseNet121 and SqueezeNet. The resulting model is trained on chest X-Ray images taken from the database COVID-19 data collection. Ensemble Learning is taken to the advantage for training multiple networks on the same image data and the model-predictions are averaged to arrive at final prediction. The predictions from various models can be combined using average, mean and mode. The models are selected by testing pre-trained models like AlexNet, ResNet50, DenseNet121, SqueezeNet and VGG16 on common COVID-19 dataset with training, validation, and testing configuration. The experimental results showed that DenseNet121 and SqueezeNet indicate the best results for precision, recall, and accuracy. Therefore, in this method DeQueezeNet is implemented using SqueezeNet and DenseNet121 models. The results attained are, accuracy = 94.52%, precision = 90.48% and recall = 96.15%. These results indicate that this model outperforms many state-of-the-art models.

5 Research Directions

The possible research challenges in the design of COVID-19 diagnosis deep learning models are listed below.

- Integrating the COVID-19 databases from all repositories to increase dataset size. This is very much essential for training a deep learning model to a best accuracy. Also, incorporate synthetic image generation models.
- Develop novel methods for transfer learning methods and hyperparameters selection, it is required to exploit the benefits of pretrained model. This eliminates the need for developing a model which incurs less computational overhead.
- There is a demand for developing handheld machines for COVID-19 diagnosis for rural area. Therefore, there is need for developing a shallow architecture model.
- State-of-the-art models show that ensemble methods give encouraging results. Therefore, there is a need to explore different ensemble models.

6 Conclusion

This chapter gives a thorough discussion on the role of deep learning in diagnosis of COVID-19 infection. The fundamentals of convolutional networks are elaborated in length which includes the conventional set of layers in the design of convolutional neural networks. This chapter also enlightens with important concepts such as augmentation, transfer learning and various metrics that must be used in evaluation of deep learning models as applicable to corona virus infection diagnosis. This chapter provides a good reference to various COVID-19 dataset is given and details are discussed which helps any researchers to understand the possible publicly available datasets. In this the major focus is on discussion of convolutional neural network models that gave promising results. These models include DenseNet based models, ResNet based models and SqueezeNet based models. The chapter emphases on exploring ensemble methods and gives the guidelines on selection of pretrained models for devising ensemble methods. From the comparison of results provided in this chapter it is evident that SqueezeNet based models give better results, and since SqueezeNet is a lightweight model, this model is used in the design of handheld diagnosis tools. The major contribution of this chapter is the research directions, in which all possible issues are discussed that need to be addressed.

References

1. H. Panwar, P.K. Gupta, M.K. Siddiqui, R. Morales-Menendez, V. Singh, Application of deep learning for fast detection of COVID-19 in X-Rays using nCOVnet. Chaos Solitons Fractals **138**, 109944 (2020)
2. M. Alazab, A. Awajan, A. Mesleh, A. Abraham, V. Jatana, S. Alhyari, COVID-19 prediction and detection using DeepLearning. Int. J. Comput. Inf. Syst. Ind. Manage. Appl. **12**, 168–181 (2020)
3. https://www.itnonline.com/content/infervision-launches-solution-fight-against-coronavirus. Accessed 26 Nov 2020
4. https://spectrum.ieee.org/the-human-os/biomedical/imaging/hospitals-deploy-ai-tools-detect-covid19-chest-scans. Accessed 26 Nov 2020
5. R. Yamashita, M. Nishio, R.K.G. Do et al., Convolutional Neural Networks: an overview and application in radiology. Insights Imaging **9**, 611–629 (2018)
6. M. Ramprasath, M. Vijay Anand, S. Hariharan, Image classification using convolutional neural networks. Int. J. Pure Appl. Math. **119**(17), 1307–1319 (2018)
7. V. Nair, G.E. Hinton, Rectified linear units improve restricted Boltzmann machines, in *ICML* (Omnipress, 2010), pp. 807–814
8. https://towardsdatascience.com/a-walkthrough-of-convolutional-neural-network-7f474f91d7bd. Accessed 28 Nov 2020
9. R.J. Borgli, H. Kvale Stensland, M.A. Riegler, P. Halvorsen, Automatic hyperparameter optimization for transfer learning on medical image datasets using bayesian optimization, in *2019 13th International Symposium on Medical Information and Communication Technology (ISMICT)*, Oslo, Norway (2019), pp. 1–6. https://doi.org/10.1109/ISMICT.2019.8743779
10. N. Srivastava, G. Hinton, A. Krizhevsky, I. Sutskever, R. Salakhutdinov, Dropout: a simple way to prevent neural networks from overfitting. J. Mach. Learn. Res. **15**(1), 1929–1958 (2014)
11. R.R. Selvaraju, M. Cogswell, A. Das, R. Vedantam, D. Parikh, D. Batra, Grad-CAM: visual explanations from deep networks via gradient-based localization. CoRR, abs/1610.02391 (2016), https://arxiv.org/abs/1610.02391
12. K. Simonyan, A. Vedaldi, A. Zisserman, Deep inside convolutional networks: visualising image classification models and saliency maps. arXiv:1312.6034 (2013)

13. C.M. Florkowski, Sensitivity, specificity, receiver-operating characteristic (ROC) curves and likelihood ratios: communicating the performance of diagnostic tests. Clin. Biochem. Rev. **9**(Suppl 1), S83–S87 (2008)
14. D. Chicco, G. Jurman, The advantages of the Matthews correlation coefficient (MCC) over f1 score and accuracy in binary classification evaluation. BMC Genomics **21**(1), 6 (2020)
15. L. Wang, A. Wong, COVID-Net: a tailored deep convolutional neural network design for detection of COVID-19 cases from chest radiography images. arXiv:2003.09871 (2020)
16. M. Rahimzadeh, A. Attar, S.M. Sakhaei, A fully automated deep learning-based network for detecting COVID-19 from a new and large lung CT scan dataset (in press). https://doi.org/10.1101/2020.06.08.20121541
17. G. Maguolo, L. Nanni, A critic evaluation of methods for COVID-19 automatic detection from X-ray images. arXiv:2004.12823 (2020)
18. C.-J. Huang, Y.-H. Chen, Y. Ma, P.-H. Kuo, Multiple-input deep convolutional neural network model for COVID-19 forecasting in China (2020). https://doi.org/10.1101/2020.03.23.20041608
19. E. Tartaglione, C.A. Barbano, C. Berzovini, M. Calandri, M. Grangetto Unveiling COVID-19 from chest x-ray with deep learning: a hurdles race with small data. arXiv:2004.05405 (2020)
20. M. Rahimzadeh, A. Attar, S. Sakhaei, A fully automated deep learning-based network for detecting COVID-19 from a new and large lung CT scan dataset (2020). https://doi.org/10.13140/RG.2.2.10063.92320/2
21. S. Ahuja, B.K. Panigrahi, N. Dey et al., Deep transfer learning-based automated detection of COVID-19 from lung CT scan slices. Appl. Intell. **51**, 571–585 (2020). https://doi.org/10.1007/s10489-020-01826-w
22. C.K. Narin, Z. Pamuk, Automatic detection of coronavirus disease (COVID-19) using X-ray images and deep convolutional neural networks. arXiv:2003.10849 (2020)
23. A. Abbas, M.M. Abdelsamea, M.M. Gaber, Classification of COVID-19 in chest x-ray images using DeTraC deep convolutional neural network. arXiv preprint arXiv:2003.13815 (2020)
24. Ş. Öztürk, U. Özkaya, M. Barstuğan, Classification of coronavirus images using shrunken features (2020). https://doi.org/10.1101/2020.04.03.20048868
25. F. Ucar, D. Korkmaz, COVIDiagnosis-Net: Deep Bayes-SqueezeNet based diagnosis of the coronavirus disease 2019 (COVID-19) from X-ray images. Med. Hypotheses **140**, 109761 (2020). https://doi.org/10.1016/j.mehy.2020.109761
26. R. Hu, G. Ruan, S. Xiang, M. Huang, Q. Liang, J. Li, Automated diagnosis of COVID-19 using deep learning and data augmentation on chest CT (2020). https://doi.org/10.1101/2020.04.24.20078998
27. O. Kim, M. McCourt, L. Wang, Parametrizing data augmentation in COVID-Net, advanced optimization techniques, convolutional neural networks, data augmentation, deep learning, healthcare, multimeric optimization, https://sigopt.com/blog/parametrizing-data-augmentation-in-covid-net-development/
28. S. Albahli, Efficient GAN-based Chest Radiographs (CXR) augmentation to diagnose coronavirus disease pneumonia. Int. J. Med. Sci. **17**(10), 1439–1448. https://doi.org/10.7150/ijms.46684. PMID: 32624700; PMCID: PMC7330663
29. A. Waheed, M. Goyal, D. Gupta, A. Khanna, F. Al-Turjman, P.R. Pinheiro, CovidGAN: data augmentation using auxiliary classifier GAN for improved Covid-19 detection. IEEE Access **8**, 91916–91923 (2020). https://doi.org/10.1109/ACCESS.2020.2994762
30. M. Loey, F. Smarandache, N.E.M. Khalifa, Within the lack of chest COVID-19 X-ray dataset: a novel detection model based on GAN and deep transfer learning. Symmetry **12**(4), 651 (2020). https://doi.org/10.3390/sym12040651
31. A. Jaiswal, N. Gianchandani, D. Singh, V. Kumar, M. Kaur, Classification of the COVID-19 infected patients using DenseNet201 based deep transfer learning. J. Biomol. Struct. Dyn. (2020). https://doi.org/10.1080/07391102.2020.1788642
32. K. El Asnaoui, Y. Chawki, Using X-ray images and deep learning for automated detection of coronavirus disease. J. Biomol. Struct. Dyn., 1–12 (2020). https://doi.org/10.1080/07391102.2020.1767212. Epub ahead of print. PMID: 32397844; PMCID: PMC7256347

33. H. Ko, H. Chung, W.S. Kang, K.W. Kim, Y. Shin, S.J. Kang, J.H. Lee, Y.J. Kim, N.Y. Kim, H. Jung, J. Lee, COVID-19 Pneumonia diagnosis using a simple 2D deep learning framework with a single chest CT image: model development and validation. J. Med. Internet Res. **22**(6), e19569 (2020). https://doi.org/10.2196/19569, https://www.jmir.org/2020/6/e19569. PMID: 32568730; PMCID: 7332254
34. M. Hussain, J.J. Bird, D.R. Faria, A study on CNN transfer learning for image classification, in *UKCI 2018: Advances in Computational Intelligence Systems*, ed. by A. Lotfi, H. Bouchachia, A. Gegov, C. Langensiepen, M. McGinnity. Advances in Intelligent Systems and Computing, vol. 840 (Springer, Cham, 2019). https://doi.org/10.1007/978-3-319-97982-3_16
35. https://towardsdatascience.com/a-comprehensive-hands-on-guide-to-transfer-learning-with-real-world-applications-in-deep-learning-212bf3b2f27a. Accessed 28 Nov 2020
36. P. Rajpurkar, J. Irvin, K. Zhu, B. Yang, H. Mehta, T. Duan, D. Ding, A Bagul, C. Langlotz, K. Shpanskaya, M. P Lungren, A.Y. Ng, CheXNet: radiologist- level pneumonia detection on chest X-rays with deep learning. arXiv: arXiv:1711.05225 [cs, stat] (2017)
37. J. Irvin, P. Rajpurkar, M. Ko, Y. Yu, S. Ciurea-Ilcus, C. Chute, H. Marklund, B. Haghgoo, R. Ball, K. Shpanskaya, J. Seekins, D.A. Mong, S.S. Halabi, J.K. Sandberg, R. Jones, D.B. Larson, C.P. Langlotz, B.N. Patel, M.P. Lungren, A.Y. Ng, CheXpert: a large chest radiograph dataset with uncertainty labels and expert comparison, in *AAAI Conference on Artificial Intelligence*, vol. 1 (2019)
38. G. Huang, Z. Liu, G. Pleiss, L. Van Der Maaten, K. Weinberger, Convolutional networks with dense connectivity. IEEE Trans. Pattern Anal. Mach. Intell. (2019) https://doi.org/10.1109/TPAMI.2019.2918284. Epub ahead of print. PMID: 31135351
39. L. Sarker, M.M. Islam, T. Hannan, Z. Ahmed, COVID-DenseNet: a deep learning architecture to detect COVID-19 from chest radiology images. Preprints **2020**, 2020050151 (2020). https://doi.org/10.20944/preprints202005.0151.v1
40. M.S. Ebrahimi, H.K. Abadi, Study of residual networks for image recognition. arXiv preprint arXiv:1805.00325 (2018)
41. M. Farooq, A. Hafeez, COVID-ResNet: a deep learning framework for screening of COVID19 from radiograph, arXiv:2003.14395 (2020)
42. S. Misra, S. Jeon, S. Lee, R. Managuli, I.-S. Jang, C. Kim, Multi-channel transfer learning of chest X-ray images for screening of COVID-19. Electronics **9**, 1388 (2020)
43. F. Iandola, M. Moskewicz, K. Ashraf, S. Han, W. Dally, K. Kurt, SqueezeNet: AlexNet-level accuracy with 50x fewer parameters and <0.5 MB model size. arXiv preprint arXiv:1602.07360 (2016)
44. https://codelabs.developers.google.com/codelabs/keras-flowerssqueezenet/. Accessed 27 Nov 2020
45. M. Polsinelli, L Cinque, G. Placidi, A light CNN for detecting COVID-19 from CT scans of the chest. arXiv preprint. https://doi.org/10.1101/2020.02.14.20023028
46. F. Ucar, D. Korkmaz, COVIDiagnosis-Net: eep Bayes-SqueezeNet based diagnosis of the coronavirus disease 2019 (COVID-19) from X-ray images. Med. Hypotheses **140**, 109761 (2020). https://doi.org/10.1016/j.mehy.2020.109761
47. A.K. Mishra, S.K. Das, P. Roy, S. Bandyopadhyay, Identifying COVID19 from chest CT images: a deep convolutional neural networks based approach. Hindawi J. Healthc. Eng. **2020**, 7 (2020). https://doi.org/10.1155/2020/8843664. Article ID 8843664
48. M. Chowdhury, T. Rahman, A. Khandakar, et al. Can AI help in screening viral and COVID-19 pneumonia? arXiv preprint arXiv:2003.13145 (2020)
49. S. Kumar, S. Mishra, S.K. Singh, Deep Transfer Learning-based COVID-19 prediction using Chest X-rays. medRxiv preprint. https://doi.org/10.1101/2020.05.12.20099937
50. L. Wang, A. Wong, OVID-Net: a tailored deep convolutional neural network design for detection of COVID-19 cases from Chest radiography images. arXiv:2003.09871 (2020)

Deep Learning for COVID-19

B. S. Prashanth, M. V. Manoj Kumar, Likewin Thomas, M. A. Ajay Kumar, Dinghao Wu, B. Annappa, Anirudh Hebbar, and Y. V. Srinivasa Murthy

Abstract Ever since the outbreak in Wuhan, China, a variant of Coronavirus named "COVID 19" has taken human lives in millions all around the world. The detection of the infection is quite tedious since it takes 3–14 days for the symptoms to surface in patients. Early detection of the infection and prohibiting it would limit the spread to only to Local Transmission. Deep learning techniques can be used to gain insights on the early detection of infection on the medical image data such as Computed Tomography (CT images), Magnetic resonance Imaging (MRI images), and X-Ray images collected from the infected patients provided by the Medical institution or from the publicly available databases. The same techniques can be applied to do the analysis of infection rates and do predictions for the coming days. A wide range of open-source pre-trained models that are trained for general classification or segmentation is available for the proposed study. Using these models with the concept of transfer learning, obtained resultant models when applied to the medical image datasets would draw much more insights into the COVID-19 detection and prediction process. Innumerable works have been done by researchers all over the world on the publicly available COVID-19 datasets and were successful in deriving

B. S. Prashanth (✉) · M. V. Manoj Kumar · A. Hebbar
Department of Information Science and Engineering, Nitte Meenakshi Institute of Technology, Yelahanka, Bengaluru 560064, Karnataka, India

L. Thomas
Department of Computer Science and Engineering, PES Institute of Technology and Management, Shivamogga 577204, Karnataka, India

M. A. Ajay Kumar · D. Wu
College of Information Sciences and Technology, Pennsylvania State University, University Park, PA 16802, USA
e-mail: duw12@psu.edu

B. Annappa
Department of Computer Science and Engineering, National Institute of Technology Karnataka, Surathkal, Mangalore 575025, Karnataka, India
e-mail: annappa@ieee.org

Y. V. Srinivasa Murthy
Department of Computational Intelligence, School of Computer Science and Engineering, Vellore Institute of Technology (VIT), Vellore 632014, Tamil Nadu, India

J. Nayak et al. (eds.), *Understanding COVID-19: The Role of Computational Intelligence*, Studies in Computational Intelligence 963, https://doi.org/10.1007/978-3-030-74761-9_23

good results. Visualizing the results and presenting the summarized data of prediction in a cleaner, unambiguous way to the doctors would also facilitate the early detection and prevention of COVID-19 Infection.

Keywords Data visualization · Deep learning · CNN · DNN · Transfer learning · Medical imaging · COVID-19

1 Introduction

COVID-19 is a novel strain of coronavirus that has been observed in humans. The COVID-19 had its first outbreak in Wuhan, Hubei province, China [1]. To date, there is no clinically approved medicine that can prevent or cure COVID-19, though the medical practitioners and pharmaceutical agencies are striving hard to develop medicine to cure or to prevent the spread of COVID-19. There are medicines that are being administered to alleviate the progression of fever, cold, and pneumonia-like COVID-19 symptoms. The medical fraternity and the government agencies from various countries have invested huge amounts of revenue and time to invent a medication that can successfully suppress the COVID-19 once and for all. In this appalling time, efforts were made to use Artificial Intelligence and deep learning methodologies to combat the non-medical aspects such as tracking and forecasting outbreaks, detecting the non-compliance of infected patients, containing the outbreak, identification of the hot zones which will help to contain the infection spread. Even in the medical image analysis of the dataset such as CT, MRI, X-Ray images, deep learning, and AI played a pivotal role in the detection, classification, and identification of the infection in a swift manner. In the field of healthcare, the role of AI and deep learning was quintessential and helped the medical institutions and hospitals to provide faster diagnoses in a systematic manner.

Due to the widespread essence of the coronavirus, patients are being admitted to health care in batches. This had pushed the government and medical agencies to their edge to accommodate the higher number of admissions to the medical facility. Setting up an atmosphere where the patient could get quick treatment and in a swift manner is a daunting task. Rapid diagnosis is quintessential and proven to be effective to contain the widespread of the COVID-19 virus. The mortality rate keeps on rising all over the world and it is evident when WHO (World Health Organization) decided to put nCoV as an epidemic disease on February 11, 2020, coiling the term COVID-19 which stands for Coronavirus Disease 2019.

Deep learning techniques were helpful in combating the socio-economic problems that took birth due to the COVID-19 pandemic as well. The tools such as Dashboards are being used by many countries are still helpful for the common people to get information about the precautions to be taken, or the infection rate, fatality rate, etc. These dashboards are internally using various AI and deep learning models for processes such as data gathering or retrieval, assimilation of data, identification of the useful insights from the gathered data from the various sources, and provide meaningful insights based on it in the dashboards. The World Health Organization does use these dashboards extensively to provide all the information about the pandemic to the world [4].

2 Overview of the COVID-19 Pandemic

Coronaviruses are the class of viruses that are known to cause illness such as the common cold, cough, fever, etc. [1, 6]. According to the stats provided by the World Health Organization (WHO) dated Aug 28, there are a total of 24,021,218 positive cases all over the world with the mortality rate is around 821,462 as shown in Fig. 1 [4].

Though on the outset, individuals from all age groups are susceptible to the infection of COVID-19, the probability of the disease being fatal is more around the people aged around 60 and above. The disease had also become very infectious to those who have respiratory disorders and those with chronic medical conditions [5].

The virus spreads through the air medium the most, it spreads when an infected person coughs or sneezes. The droplets sprayed via cough can spread up to 6 feet away. If a healthy person breathes them or swallows them, the virus starts to incubate in that new human body. Some of the infected patients exhibit asymptotic behavior without any symptoms but they still can become a carrier. The virus can also spread through surfaces, so touching a surface with a virus and then followed by touching the mouth, nose or eyes could also lead to the spread of coronavirus [6, 7].

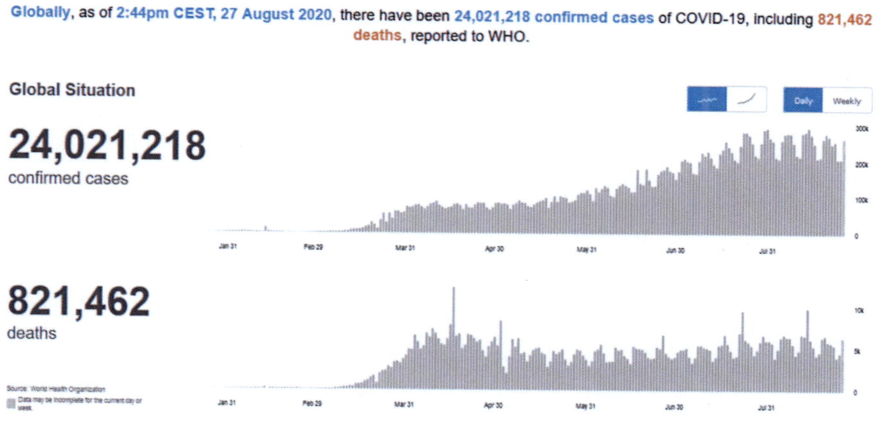

Fig. 1 WHO dashboard summarizing the confirmation cases and deaths due to COVID-19 [4]

3 Introduction to Neural Networks and Deep Learning

To understand deep learning, one needs to understand the difference between rule-based models and data-based models. The complex problems can be solved by imposing a set of mathematically proven rules fed to the machines using which the machines are able to understand and interpret the input data and learn. This is the fundamental idea behind Artificial Intelligence. Contrary to that, machine learning algorithms rely on the input data and they learn the inference based on the quality of the input data. The higher the quality of the data, the higher is the learning. The traditional machine learning algorithms work exceptionally well if there are no well-defined rules. Machine learning algorithms work well for statistical data, or text, numerical data, but they do not yield better results for the other kinds of data such as audio, video, unstructured data, etc. Though perception wise, often Artificial Intelligence and Machine learning are used synonymously, but in fact, Artificial Intelligence is a broader subject that enables and induces smartness to machines and Machine learning is a technique to enable them to be smart. Deep learning is one sub-branch of Machine Learning which comprises Algorithms which is based on mimicking the activities of the human brain.

3.1 Neural Networks

Neural Networks or otherwise known as 'Artificial Neural Networks', are a collection of neurons that identify the complex relationship between input and output. They are loosely inspired by the operations involved in biological neurons of a human brain. The idea of Neural Networks was first started by a psychologist named Frank Rosenblatt, who initially coined the term 'Perceptron' in the year 1958. He observed the similarities between a biological neuron as well as an artificially created neuron. In the next section let us see the architecture of neural networks in detail.

3.1.1 Architecture of Artificial Neural Networks

A typical Neural Network architecture generally constitutes Input Layer, Hidden Layer, Output Layer. Figure 2 depicts a simple Neural Network architecture.

The order of the layers is sequential in nature i.e., the input layer always comes first, then the hidden layers, and finally the output layer. Each layer consists of neurons where complex operations take place which helps in the prediction. The input layer as the name suggests takes in the input of various forms such as numerical values, images, texts, etc. And pass it onto the next layer that is the hidden layer. No mathematical operation takes place in the input layer and a neural network contains only one input layer, but it can have any number of nodes in that layer. The hidden layer is the most important layer of a Neural Network. In this layer, the complex

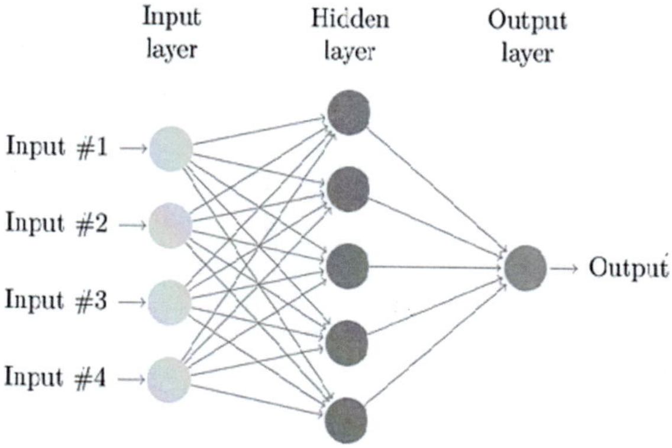

Fig. 2 Simple architecture of Artificial Neural Networks

relationship between the input values and the output value are identified. A prominent step called 'Feature Extraction' takes place in these layers where relevant and the most important features from the input values are identified and extracted such that it can be helpful for the network to train on the data and predict accurately for the future, unknown values. Unlike the input layer, the hidden layer can consist of any number of layers along with any number of nodes. The relationship between the input and the output values are generally in a non-linear fashion. To help the Neural Networks to train on the data, special functions known as 'Activation Function' are present which introduces the non-linearity in the network. There are many flavors of activation functions such as 'Sigmoid', 'Tanh', 'ReLU', 'SoftMax' etc. More details about the activation function will be given in the upcoming section. Finally, in the output layer, the model gives out its prediction based on the extracted features from the data. Just like the input layer, even the neural network contains only one output layer but can have any number of nodes based on the type of problem that is being solved. The process of flow of data starting from the input layer, traversing through the hidden layer, and finally to the output layer is known as the 'Forward Propagation'. The following section examines the various activation functions which are commonly used in Neural & Deep Neural Networks.

3.1.2 Activation Functions

The activation functions in neural networks introduce non-linearity into the model and aid the network to identify the complex relationship between the input and output. After the computation of input with activation function, the network decides which neurons to fire and in each layer of the hidden layer, thus influencing the final prediction. A few of the prominent activation functions are discussed in this section.

3.1.2.1 Sigmoid Activation Function

This activation is preferred in the output layer when the problem type of binary classification (disease/no disease, fraud/not fraud), since the input from the hidden

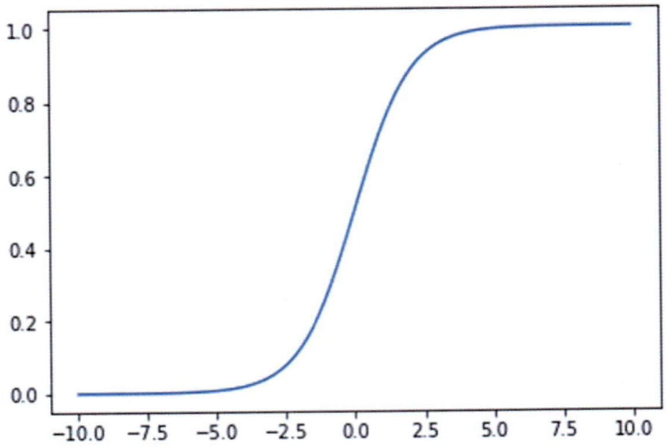

Fig. 3 Sigmoid activation function

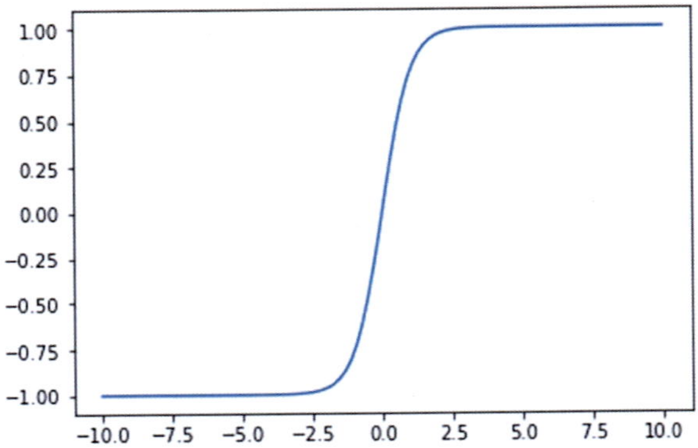

Fig. 4 Tanh activation function

layer, are squashed to the range between 0 and 1 i.e., the probability score for the prediction will lie within these two values. Figure 3 represents the sigmoid activation function.

Equation for sigmoid activation function is given as:

$$\mathrm{y(x)} = 1/e^{(-\mathrm{x})} \tag{1}$$

Where,
y(x) = prediction y is generated for every value of x.

3.1.2.2 Tanh Activation Function

Tanh activation is another popular activation used while training a network. Unlike Sigmoid activation which has a range between (0 to 1), the Tanh activation function has the range between (−1 to1). This centers the data close to zero, thus allowing the layers to learn the features in a better manner. Figure 4 represents the Tanh activation function.

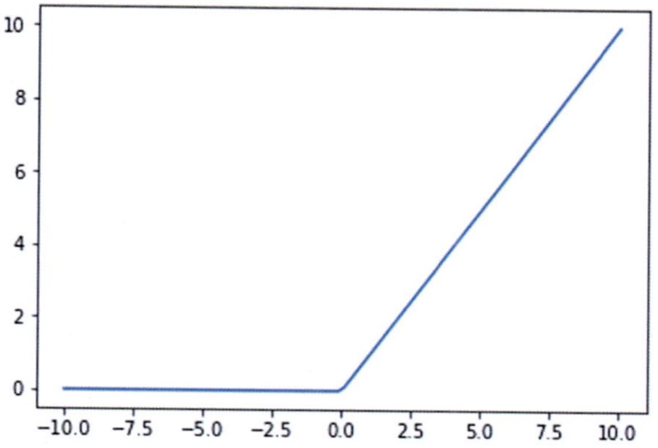

Fig. 5 ReLU activation function

Equation for tanh activation function is given by:

$$y(x) = e^{x} - e^{(-x)}/e^{x} + e^{(-x)} \quad (2)$$

3.1.2.3 ReLU Activation Function

This is perhaps the most popular and frequently used activation function in neural networks and more specifically Convolutional Neural Network. Since ReLU converts all negative pixels in the images to zero and extracts the maximum pixel values, they are helpful in extracting useful features from an image and ignore the noise. ReLU is computationally more feasible compared to sigmoid or tanh and hence faster compared to the other two activation functions. Figure 5 represents the ReLU activation function.

Equation for the ReLU activation function is given by:

$$y(x) = \max(0, x) \quad (3)$$

3.1.2.4 SoftMax Activation Function

SoftMax is one such activation function that is used in the output layer, unlike other activations that can be used both in the hidden layers as well as the output layer. In this activation, the probability scores of each class are given in the output nodes, such that the sum of the probability score is equal to one. Figure 6 represents the SoftMax activation function.

The equation for the SoftMax activation function is given by,

$$P\left(y = j \middle| \theta^{(i)}\right) = e^{\theta(i)} / \sum_{j=0}^{k} e_{k}^{\theta(i)} \quad (4)$$

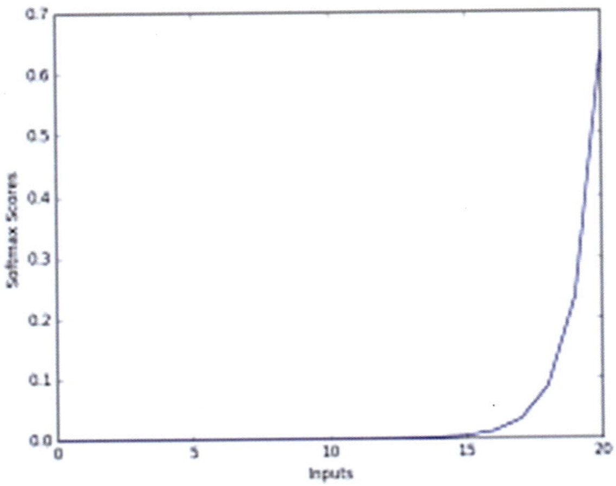

Fig. 6 SoftMax activation function

3.1.3 Advantages and Disadvantages of a Neural Network

Undoubtedly, Artificial Neural Networks (ANN) have offered a robust classification method, this section briefs some of the noteworthy and inherent advantages and disadvantages of ANN.

Advantages of a Neural Network

Neural networks can train by example. They can train with a set of data and can predict the unseen data with ease. Neural networks can generalize. They have the power to predict a new variety of data-based previous trends of data. Neural networks are robust and fault tolerant. They can predict accurately even if the input data has noise or incomplete. Neural networks can understand the complex and non-linear relationship between the input and the output.

Disadvantages of a Neural Network

Neural networks are a black-box model. It means that although accurate predictions can be obtained, it is hard to explain the reason behind that prediction since the relationship between the input and the output is complex and non-linear. Training a neural network is computationally intensive when it is run on a local machine. It mainly depends on the complexity of the network and the amount of data used but does take a lot of time and resources compared to conventional machine learning algorithms. To achieve good results with a neural network a lot more data is required. To develop a custom neural network, a lot of expertise and skill set is required. Thanks to several open-source libraries such as Keras and PyTorch, building a network of certain complexity has been made possible. But these cannot always be relied on and there might arrive a point where the model must be built from scratch. Certain neural network architectures such as Artificial Neural Networks, Convolutional Neural Networks, and Recurrent Neural Networks rely on supervised learning methodology (structured data) to train and predict on any given data and fail to perform well on

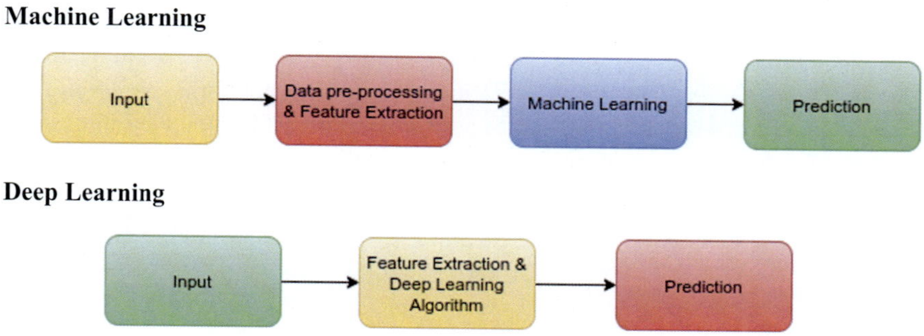

Fig. 7 Notable differences between machine learning and deep learning

those data which do not have labels with them (unsupervised learning). So, to overcome the training of a model via unsupervised methods, there exists another subset of machine learning, known as 'Deep Learning', which can handle both structured as well as unstructured data. This will be discussed in the upcoming section.

3.2 Deep Learning

Deep Learning is a subset of Machine Learning which can be used to train in supervised, semi-supervised, and unsupervised models. This is inspired by Artificial Neural Networks. Deep Learning can extract features from the data in a hierarchical fashion i.e., at first low-level features are extracted, then mid-level features, and finally high-level features are extracted.

3.2.1 What Is Deep Learning?

The main difference between machine learning and deep learning is that in machine learning, the dataset should be pre-processed thoroughly before using it to train the model. In deep learning, the data pre-processing step is not mandatory since the models are robust to noise and missing data, little to no data preprocessing is required. Figure 7 depicts the differences in a clear manner.

In the current scenario, both deep learning and neural networks are so entwined to each other that the terms are used interchangeably, although there are some key differences between the two terminologies which will be discussed in the next section.

3.2.2 Differences Between Deep Learning and Neural Network

Some of the key differences between deep learning and neural networks are. Deep learning is a subset of machine learning which allows the computer systems to improve with experience and data. Neural Networks are inspired by the working mechanism of the nervous system. Deep learning is a type of machine learning method which represents artificial networks. Neural networks are biologically inspired networks that are assigned to perform certain tasks. Deep learning is associated with data transformation and feature extraction, whereas neural networks use neurons to obtain input values, perform operations and provide the output value. Some of the famous architectures in neural networks are Perceptron, Feed-Forward Network, Hopfield Network, Boltzmann machine, etc.

3.2.3 Architectures in Deep Learning

During the last two decades advancement in learning algorithms have given us various deep learning architectures, this helped us to expand the number and type of problems that neural networks could solve. Deep learning is not a stand-alone approach, but it is a range of classification algorithms that one can apply to range of problems.

3.2.3.1 Convolutional Neural Network

CNN is a Deep-Learning algorithm that is used to deal with image data. They are commonly used in image classification, object detection, etc. They extract important features from the data and based on the information obtained they perform prediction. Following are the steps involved in CNN: *Convolution operation.* To achieve this kernel/filter is used over the input image where the filter multiplies its overlapping values by traversing over each pixel of an input image and outputs its sum as a single value until the entire image is traversed. With more added layers it can extract high-level features and understand the data. *Pooling* – This step is similar to Convolution operation i.e.; it helps in reducing the spatial size of the convolved feature. These are also helpful since it extracts the most important features irrespective of rotational and positional variance.

3.2.3.2 Recurrent Neural Network

Recurrent Neural Network is a class of deep learning which specializes in handling sequential data. This feature makes this architecture unique compared to other deep learning architecture since they are not well equipped to handle sequential data. In certain architectures such as feed-forward networks, the prediction is made based on current input, whereas in RNNs, the prediction is based on both current input as well as the information it has learned from the previous input, thus RNN architecture is said to have a short-term memory. RNNs are used in applications such as Text-to-speech conversion, stock price prediction, etc.

3.2.3.3 Deep Belief Networks

Deep Belief Networks (DBNs) are another class of deep learning algorithm that use probabilities and unsupervised learning to give predictions. Unlike other architectures, in DBNs, each layer gains knowledge from the input and works globally and regulates each layer. In this type of network, every node is connected to their prior and their subsequent layers. While their initial nodes are undirected towards the hidden layers, subsequent layers are having directed connections. In deep belief networks, a "greedy algorithm" is implemented to pre-train the model, thus allowing the model to find the global optimum in a seamless manner.

3.2.3.4 Autoencoders

Autoencoders belong to a unique class of deep learning architectures. They are unique because the output obtained in this network is the same as input, i.e., the dimension of output is equal to that of input. Although the dimensions are the same, in the hidden layers which have a reduced number of nodes compared to the input layer and output layer, the input is reduced to its smaller representation before reconstructing the same in the output layer. Autoencoders consist of three components: Encoder, Code, and Decoder. An encoder is a feed-forward network which reduces the dimension of the input. This reduced representation is stored in the layer called code. This reduced representation is fed to the decoder which has a similar structure to that of the encoder. In this section, the reduced representation is reconstructed back to its original format. Some of the types of autoencoders are Convolutional Encoder, Variational Autoencoder, Denoising Autoencoders, and Deep Autoencoders. The following section explores some of the recent works on COVID-19 using Deep Learning.

3.2.4 Evaluation Metrics in Deep Learning

The performance of the CNN model has to be measured and should be compared against the CNN models that solves the problems of similar kind, In order to measure the performance of the CNN models, the following are the parameters used.

1. *Accuracy*- The accuracy of the model states how good the model is able to learn from the dataset supplied to it. The accuracy term is always expressed in terms of percentage. There are two types of accuracy when it comes to CNN models, they are

   a. *Test accuracy* – which gives the accuracy of the model during training phase
   b. *Validation accuracy* - which gives the accuracy of the model when a new dataset is provided for the classification or detection

2. *Loss* – The loss metrics is the converse of the Accuracy parameter; it gives the difference between the actual prediction of the model with expected prediction. The loss can be defined on the same terms of accuracy, i.e., Validation loss and testing loss that is the loss incurred during testing and validation phases of model training
3. *Precision* – It measures how many observations predicted by the CNN model as positive are in fact positive, it is also called as Positive Predictive value (PPV)
4. *Sensitivity* – It measures true positive rate.
5. *F-Score* – It is a weighted value of precision and recall.

4 Recent Works on COVID-19 Using Deep Learning

The following section summarizes some of the recent literature done on COVID-19 classification using novel or pre-trained convolutional neural networks. The following section examines a plethora of CNN architectures which are relevant to understand the recent research work done by various researchers around the globe. Some of these architectures such as [8, 16–18] are the CNN models which are custom developed and having implementation for the COVID-19 detection and classification. The other architectures used Transfer learning paradigm wherein, a pretrained model used for the similar task is reused again with tweaks on the layers to suit it to the COVID-19 Identification and Classification, these works can be referred here [10, 13, 20]. There are other architectures which are discussed further in the section which sheds some insights on the hybrid models which use a different approach by cascading architectures and techniques such as in [21, 23]. The surveyed works are summarized based on the following factors/parameters for COVID-19 detection and classification.

1. Type of Dataset
2. Pre-processing methodology,
3. Novel CNN architectures,
4. Architectures using Transfer Learning,
5. Binary or Multiclass Classification,
6. Other CNN architecture

4.1 Type of Dataset

From the radiological perspective, most of the work that has been carried out by various researchers use radiology datasets such as chest X-ray, chest CT scans, etc. for their experiments using Deep learning. The work carried out by [8] used a chest-X-ray dataset to train the modified DarkCOVIDnet. The work presented by [10] proposed a novel CNN architecture called CovXNet which is trained on the Chest-X-ray images which were obtained from the Guangzhou Medical Center, China [11] which comprises of chest X-ray images of COVID, Normal, Pneumonia (viral), pneumonia (bacterial) patients and is signed by the radiologists. COVIDiagnosis-Net [15] is another novel work that uses Deep Bayes-SqueezeNet based technique to provide diagnostics about coronavirus using the COVIDX Dataset which comprises posteroanterior chest radiography images for training the model. Though there is a wide range of open-source datasets available for radiology datasets for chest CT scans and X-Ray, many of the images in the datasets are collected raw and need additional steps of preprocessing before usage. These datasets are constantly updated from various sources regularly and the updated datasets can be used for further tuning of various models. Out of all the datasets, the COVIDX dataset is the biggest one in the numbers of COVID-19 positive cases.

4.2 Pre-processing Methodology

In order to train a custom CNN or to use an existing pre-trained model either using it as a stock or for fine-tuning, the input data to the CNN's must to be preprocessed such that it complies to the existing pre-trained architecture guidelines. Some of the techniques involve the conversion of RGB images into Gray-Scale images which helps to speed up the processing of images in CNN adding to the advantages such as ease of visualization in grayscale, lesser space requirement, etc. The other form of preprocessing methods involves applying Fourier or Discrete Wavelet Transforms on the images and feeding the processed images for training in CNN [12]. Each of the CNN mandates the range of input image size to be used for their respective architecture. Theoretical understanding of the pre-trained architectures is needed to decide on the exact image size to which the input images need to be resized. For example, the publicly available dataset presented in the chest X-ray image dataset [9] consists of a wide range of images at different sizes and resolutions. The CNN requires the image to be resized to fit into CNN for training. The work presented in [8] uses resize operation on the dataset using python tools and API's such as OpenCV [9] and resizes the image to 256×256 size and used for training the model. Similarly, different models need the images to be fed at a specified size such as the work [10] preprocess the images in the dataset to 128×128. The other notable preprocessed image size is 224×224 which is used in [15]. The uniformity of the image size is quite pivotal in the CNN model performance. The method of generating additional images for training can be achieved with the concept of augmentation where the images are rescaled, sheared, and moved randomly in all directions and considered as the additional dataset for processing. These kinds of operations are vital in training CNN with limited datasets. The work presented in [15] used data- augmentation approach to generate additional data for the training and validation. The following Fig. 8 depicts some of the preprocessing operations employed in CNN.

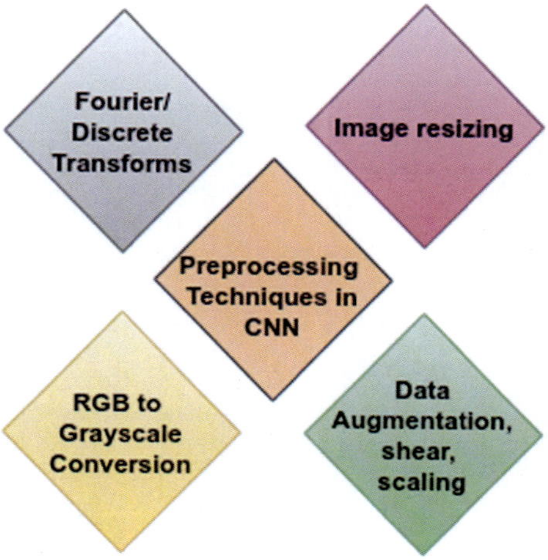

Fig. 8 Data preprocessing techniques in CNN

4.3 *Novel CNN Architectures*

The following subsection explores some of the novel architecture which is used for COVID-19 detection and classification.

4.3.1 DarkCOVIDNet

The architecture presented in [8] is called DarkCOVIDNet. The model designed yielded 98.08% accuracy for binary classes and 87.02% accuracy for multi-class classification. The dataset used for the model is taken from the combined dataset accumulated by Cohen JP [9] using various opens access sources. The work uses the DarkNet-19 pre-trained model which is used in a real-time object detection system named YOLO (you only look once), basically used for classification. The proposed architecture for basic classification is as shown in Fig. 10.

The model was initially trained for three classifications, which are COVID1, No-findings, Pneumonia and then later the trained model was tested against two cases COVID, No-findings. The model performed pretty good for the dataset and was able to attain an accuracy of 98.08%. The findings are plotted in the graph and can be seen in Fig. 9. Though the accuracy seems valid, the nature of the model learning is erratic, an apparent steep fall in the accuracy value at certain points in the plot is evident from Fig. 10. Further tune-up of the model is possible which might smoothen the learning of the model about COVID-19 features.

The heart of the architecture is the usage of object detection architecture such as DarkNet-19 as a base architecture. The key takeaway from the work is the fact that prebuilt object detection CNN's can also be used for Detection of COVID-19 from chest X-ray images assuming that the pattern of infection in the chest X-ray images can be treated as objects.

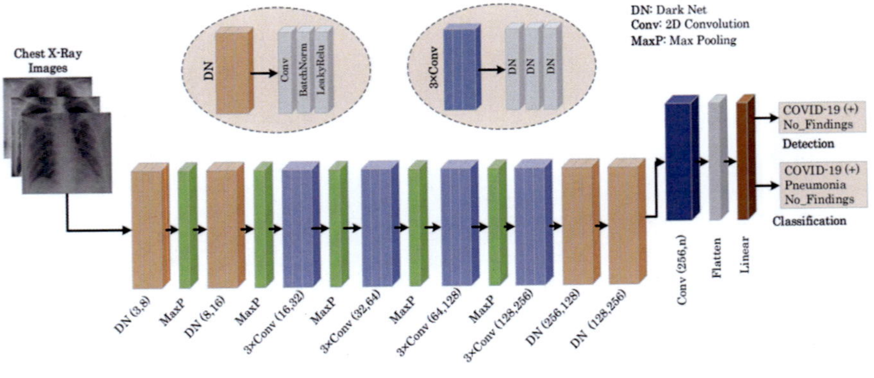

Fig. 9 The architecture of the DarkCOVIDNet [8]

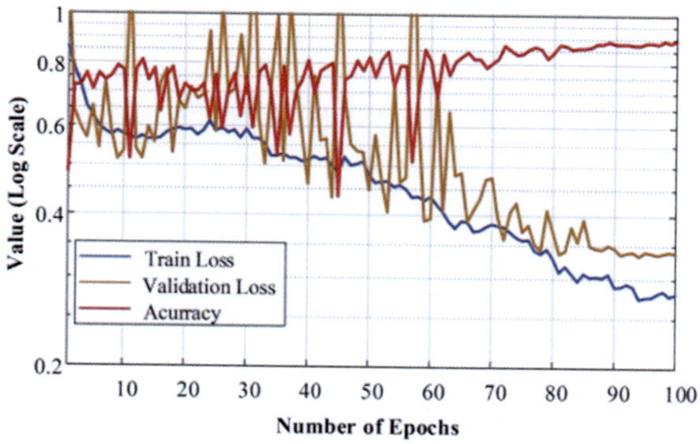

Fig. 10 Results obtained by the DarkCOVIDNet [8]

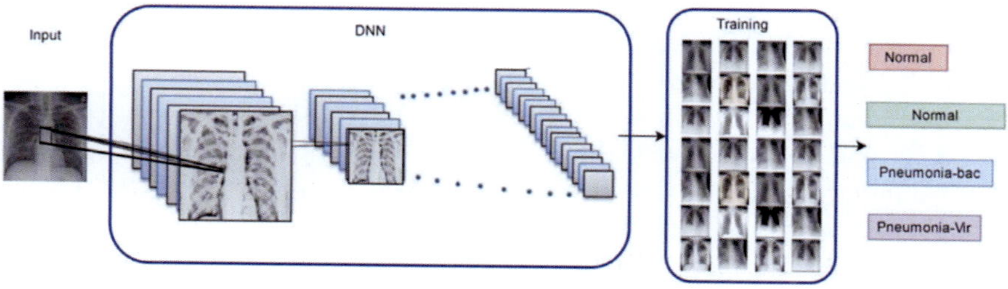

Fig. 11 The architecture of CoroNet [14]

4.3.2 CoroNet

The work presented in [14] introduces a new CNN architecture which is based on the Xception architecture, which is a classification model, pre-trained to process the general images from Imagenet. The model is applied with COVID-19, chest pneumonia X-ray images from publicly available databases [9].

It is a 4-class classification that is COVID/pneumonia bacteria/pneumonia viral/Normal) and they achieved an accuracy of 89.6%. For three-class classification (COVID/Pneumonia/normal) they attained the accuracy of up to 95%. The architecture of the model is as shown in Fig. 11.

CoroNet is an architecture that is custom tweaked for the identification and classification of COVID-19. The architecture uses Xception CNN architecture as its base model [20] and it comprises 71 layers deep CNN model which is trained using the ImageNet image dataset [10]. The Xception model uses depthwise separation of the convolutional layers and it uses residual connections in place of regular convolutional layers. This reduces the number of operations required for the classification since the depth wise separation converts $n \times n \times k$ convolutional operations into $1 \times 1 \times k$ operations which in turn reduces the convolutional operations by 1/k. The residual

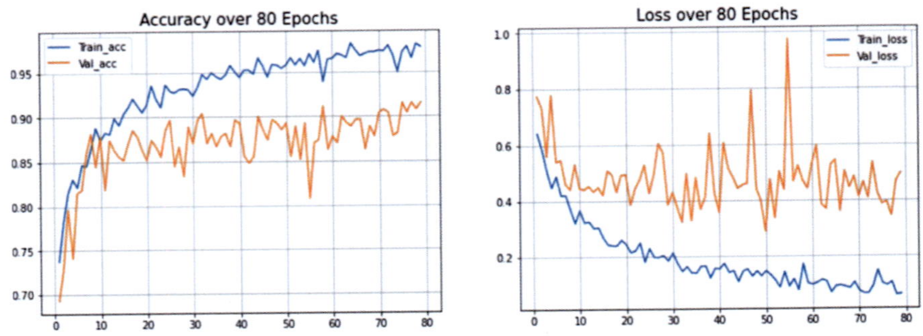

Fig. 12 CoroNet performance in terms of accuracy and loss for epoch = 80 [14]

connections are those connections that allow certain convolutional layer output to skip over a few layers. The networks which employ these are called residual networks.

The idea behind the usage of residual networks is to overcome the problem of vanishing gradient. The results obtained by the model are as shown in Fig. 12 and the dataset used for the training of this model is taken from [9].

With the residual networks in place, the CNN designed used very few layers for the initial learning of COVID-19 features from the dataset. This had speed up the overall learning time needed for the Model to learn the patterns of COVID-19 features indeed. The model also sheds some insights about the execution time which is comparatively less due to the use of Residual networks.

4.3.3 COVIDiagnosis-Net

The work in [15] introduces a COVID-19 detection model based on Artificial Intelligence. The model uses deep SqueezeNet with Bayes optimization techniques for attaining higher accuracy. The dataset used for training is from [16] which is a COVIDx dataset. The architecture addresses the irregularities or noises in the images in the dataset by performing offline data augmentation and then the processed images are fed into the training steps of CNN. The offline augmentation enables the model to be used to build apps on handheld devices or embedded systems as well. The Augmented data is then passed through the standard SqueezeNet pretrained Model. SqueezeNet architectures are designed to have very few layers with lesser datasets which helps to easily fit into computer memory that can be transmitted over the networks. The work also uses hyperparameter tuning as a feedback network and adjusts the weight accordingly with the help of the Bayesian optimization method. The detailed architecture of the COVIDiagnosis-Net is as shown in Fig. 13.

The model is trained with a dataset [16] and the model yielded an accuracy of 98.3% with the F1 score of 0.983. The model uses offline methods to do the preprocessing of the images which reduce the overall workload of the designed architecture. The core of this architecture is the offline augmentation method combined with

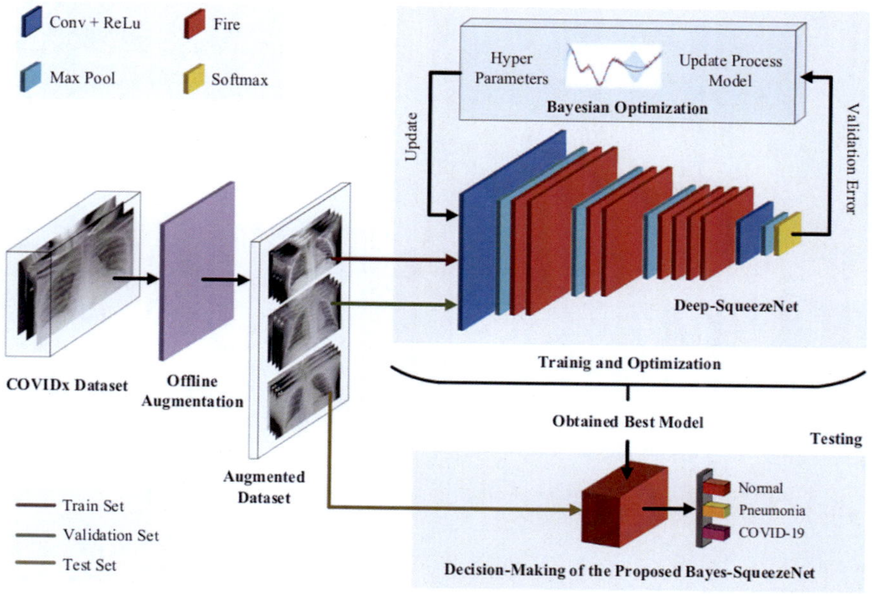

Fig. 13 COVIDiagnosis-Net architecture [15]

Bayesian optimization which is used to optimize expensive to evaluate functions which requires further exploration to be used for new architecture designs.

4.3.4 COVID-Net

Another important architecture that was trained on the COVIDx dataset as presented in [16]. The model is custom-tailored for COVID-19 Identification and Classification.

The model is designed for 3 Class Classifications, i.e. No infection (Normal), Non-COVID (viral/bacterial), and COVID infected. The detailed model of the COVID-Net architecture is as shown in Fig. 14. The model is compared against the VGG-19 and ResNet-50 architecture and it outperformed both in terms of accuracy. The model yielded 90.5,91.3 and 98.9 respectively for normal, non-COVID, and COVID infected.

The COVID-Net architecture is pre-trained with the ImageNet dataset and then was trained on COVIDx data. The architecture varies the learning rate over a period and uses Adam optimizer for learning. If the learning rate is stagnant for a certain period, the learning rate is decreased. The architecture also uses augmentation, residual networks, and hyperparameter tuning methods to attain the results. The key takeaway from exploring this architecture is the hyper-parameter tuning which can also be an effective method to improvise the accuracy and performance of the model. Some of the hyperparameters which can be tuned are, size of the network,

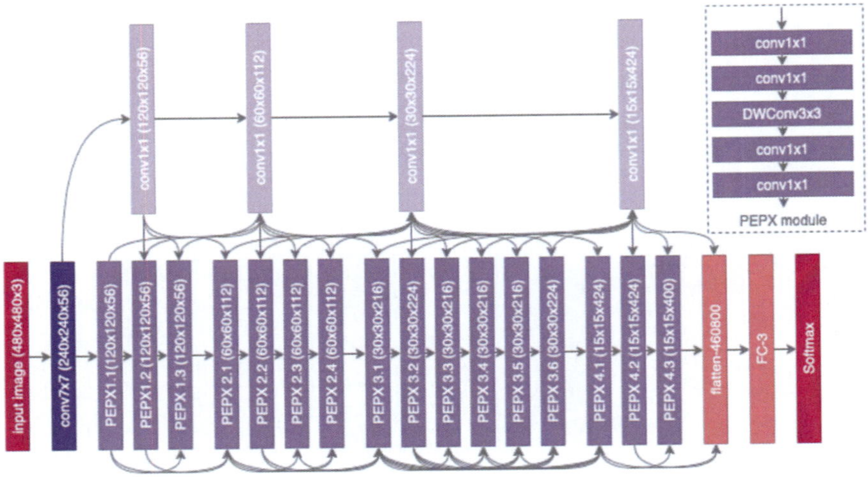

Fig. 14 COVID-Net architecture [16]

number of layers, dropout rate, learning rate, kernel size etc. By varying these parameters after the initial learning, the performance can be improvised over time which is quite evident from the COVID-Net.

4.4 CNN Architectures Based on Transfer Learning

Transfer learning is a paradigm using which the knowledge and learning gained from the one model is reused or transferred to another model for which the features and dataset being used for learning are similar. The approach is widely used because it is tedious to develop a model well suited for the application cause from scratch and it makes sense that if there is a pre-trained model which has been designed to solve a similar kind of problem earlier to be reused. This effectively reduces the time and cost involved in designing the model to solve the problem using the Deep learning approach.

When it comes to the reusability of the learning, there are many approaches, it depends on the type of the problem, dataset and its features, and the approach itself. Some of the approaches are, ***Classifier, Standalone Feature Extractor, Integrated Feature Extractor, Weight Initialization and update***. The following subsections explore some of the architectures which are used for COVID-19 detection and classification using the transfer learning approach.

4.4.1 CovXNet

The work proposes a novel CNN architecture termed as CovXNet which uses dilation rate and depth wise convolution for extracting the rich features out of the chest X-ray [10] as shown in Fig. 15. COVID-19 does cause pneumonia and the symptoms of it are kind of similar to normal pneumonia, the model is trained first with X-ray images of the (viral/bacterial) pneumonia. The trained model weights are then transferred using the concept of Transfer learning and several layers are fine-tuned to process the X-ray images. The model is trained with X-ray images of infected/non-infected patients of pneumonia for various resolutions and each of the learning was accumulated in the end for the meta-learning process with the help of the Stacking Algorithm. The results obtained were having an accuracy of 97.4% for COVID.

The detailed architecture is as shown in Fig. 16. The study also experimented with the models which are trained with resized images at various levels and used the learning from each model to perform output prediction. The process is termed as a Meta-learning phase where the inferences from the models are accumulated to

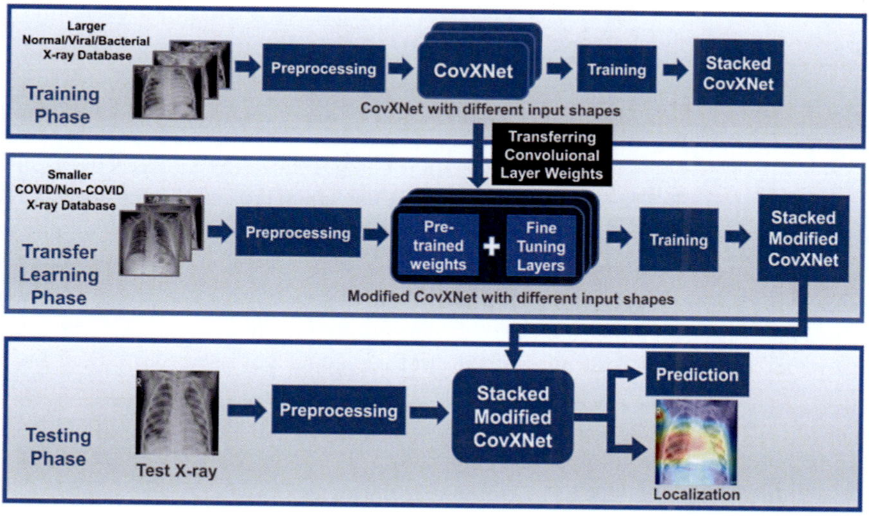

Fig. 15 The workflow of CovXNet [10]

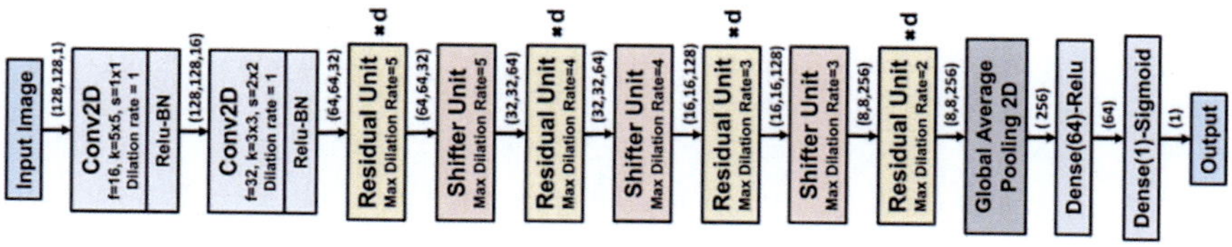

Fig. 16 The architecture of CovXNet [10]

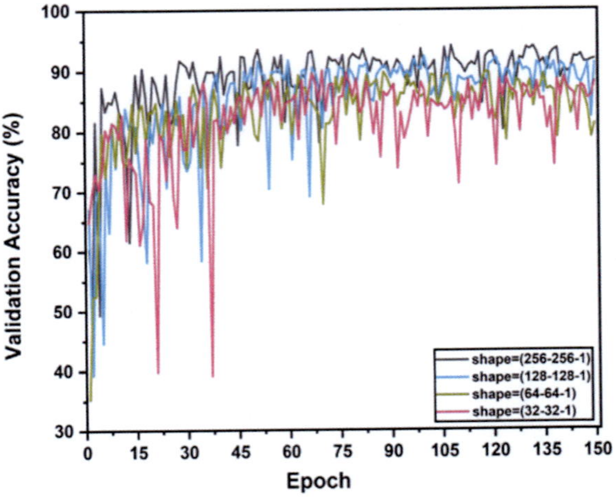

Fig. 17 Epoch vs accuracy for stacked meta learner architecture [10]

a final summative learning model called Meta Learner which does the predictions. The results obtained are summarized below in Fig. 17.

4.4.2 nCOVnet

The work proposed in [13] introduces a new CNN architecture that uses the Transfer learning approach. The Model is called nCOVnet which is used to detect the COVID-19 infection based on the chest X-ray images. The model uses a dataset compiled by Cohen et al. [9] for training and validation. The model uses VGG-16 as base model to which custom 5 layers are added and the modified VGG16 model was trained on the dataset [9]. The input images of the model are of size $224 \times 224 \times 3$ which is of depth 3 and an overall image dataset of 337 images is used to train the model. The model provided an accuracy of 97% for COVID-19 positive patients and an overall accuracy of 88%. The model was trained for 80 epochs with the learning rate to be 0.0001. The training accuracy yielded was around 93–97%. Once the model was trained and weights were computed, the model boasts to take a mere 5 sec for COVID-19 classification on the supplied new images. The results obtained are presented in Fig. 18.

4.4.3 Automated Deep Transfer Learning Approach for COVID-19 Detection

The work presented in [17] used the Xception model as a base model for feature extraction and the learning was transferred to the custom model for classification of COVID-19. The overall model is depicted in the following Fig. 19.

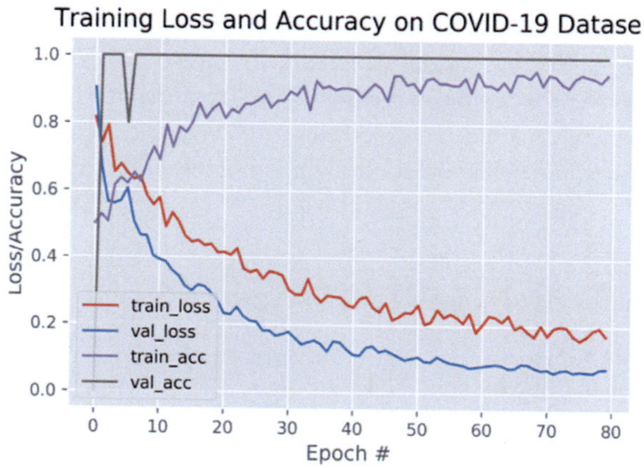

Fig. 18 Loss/accuracy vs epoch stats of nCOVnet [13]

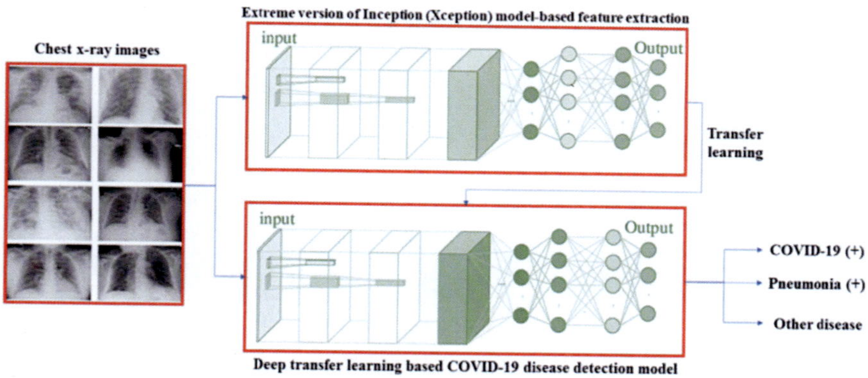

Fig. 19 The architecture used for the classification of COVID-19 [17]

The model was trained on the chest image X-ray dataset [8, 9]. The model uses an additional logistic regression technique to perform the binary classification i.e., COVID-19, and non-COVID. The model obtained an overall accuracy of 99.2% in training and 97% accuracy during validation.

The concept of Transfer Learning is a very effective approach because it saves time of building the model from the scratch for COVID-19 detection and classification. There is plethora of algorithms which have worked effectively for the general image classification whose learning can be reused to solve the problems related to COVID-19 classification. The transfer learning is actually used as either weight initialization approach or as a feature extractor. The architecture of Modified CovXnet embodies the first approach of weight transfer which is an executive summary of the learning from the pre-trained model called as CovXnet. The architecture called nCovNet [13] also uses the weight transfer paradigm for transferring the learning of the model for COVID-19 detection and Classification. The work proposed in [18] demonstrates the feature transfer from the final Neural Network output and input it as another input to the layers of the Second Model which is to be trained on COVID-19 dataset. The approach seems to be working as it was evident from the performance parameters of [18]. The approach has to be examined carefully, both the approaches might work or might not work for various architecture which is another research horizon to explore.

4.5 Binary v/s Multiclass Classification

Another horizon using which the literature survey was conducted is based on whether the model is doing Binary classification or Multiclass classification. While examining the Radiology data such as CT images, X-ray images of COVID-19 datasets from the various public sources for COVID-19 identification and classification, the features that were under focus for being examined by the various CNN models were similar to that of pneumonia disease which was induced by the other viruses or bacteria. Moreover, COVID-19 infection does also exhibit pneumonia-like features when examined using X-ray and CT images of COVID-19 infected patients. There were CNN models that were developed to address the false positive cases as well and it is summarized in Table 1 below.

Table 1 Classification of Deep learning models based on the mode of classification

Model name	Binary/Multiclass	Classes	Literature reference
DarkCOVIDNet	Binary & Multiclass	COVID/non-COVID (Multiclass) COVID/non-COVID/Pneumonia (Multiclass)	[8]
CoroNet	Multiclass	COVID/pneumonia(bacteria)/pneumonia (viral)/Normal (4-class) COVID/Pneumonia/normal (3-class)	[14]
COVIDiagnosis-Net	Multiclass	COVID/Pneumonia/normal (3-class)	[15]
COVID-Net	Multiclass	COVID/Pneumonia/normal (3-class)	[16]
CovXNet	Binary/Multiclass	COVID/non-COVID (Multiclass) COVID/non-COVID/Pneumonia (Multiclass)	[10]
nCOVnet	Binary	COVID/non-COVID (Multiclass)	[13]
Automated Deep Transfer Learning Approach for COVID-19 detection	Multiclass	COVID/Pneumonia/normal (3-class)	[17]

4.6 *Other Notable Architectures Used for COVID-19 Classification*

This chapter has presented a range of Deep learning methods for Covid 19 related classification problems. In this section we present some of the noteworthy Deep learning model which also concentrate on learning for Covid 19 related tasks.

4.6.1 Convolutional CapsNet

The work [18] proposes a novel approach called "capsnet" for the detection and classification of the COVID-19 pandemic. The core idea behind the work is the use of the capsule networks which provide fast and accurate diagnostics of the COVID-19. The model is tested for the binary classification as well as multiclass classification and achieved an accuracy of 97.24 and 84.22% respectively. The method can be used along with the regular diagnosis of the medical practitioners for fast screening. The architecture of the CapsNet is shown in Fig. 20.

Capsule networks are a type of Artificial Neural networks which are used to model the hierarchical relationship that exists in the training images. The idea behind the use of this network is to add structures called "capsules" with the regular layers of CNN and reuse the learning of these capsules to obtain the final learning or higher learning. The output of these capsules is a vector comprising the probability of observation and a pose for that observation [19]. In the convolutional neural network, the learned features are pooled in the pooling layers before transmitting to the next layer. The

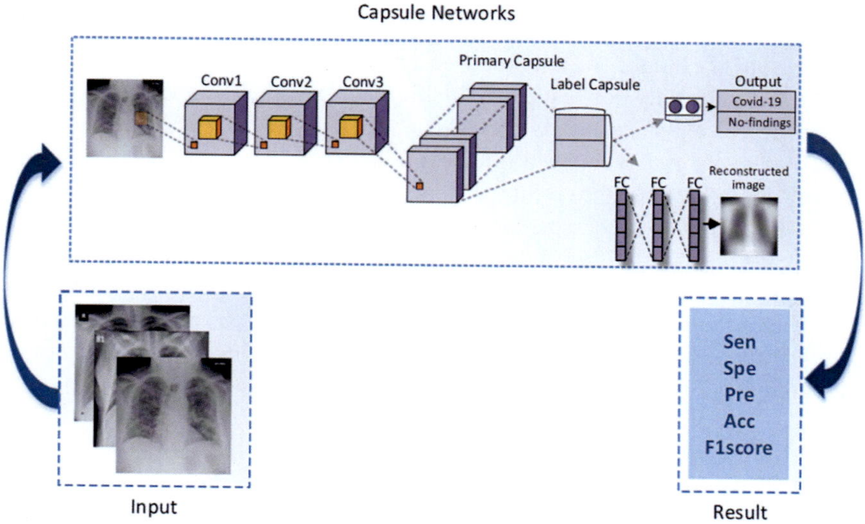

Fig. 20 Flowchart of the CapsNet [18]

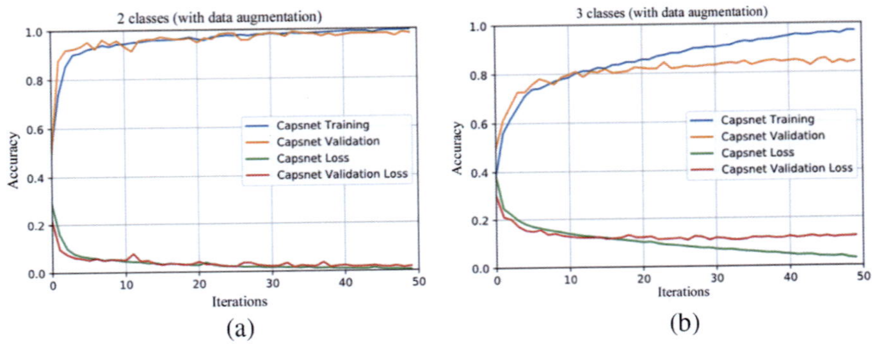

Fig. 21 **a** Accuracy vs epoch for 2 Class classification using Capsnet (with augmentation) (left), **b** Accuracy vs epoch for 3 Class classification using Capsnet (with augmentation) (right)

capsules can be used to trap this information before going to the fully connected layers and can be combined in the later stages to yield better accuracy. The model was trained on the X-ray dataset which was accumulated using [9] and other open source databases and it consists of 1050-COVID-19 images with 1050 non-COVID chest X-rays images. The approach uses a data augmentation approach for the data preprocessing to improvise the accuracy. The results are summarized in Fig. 21(a) and (b).

4.6.2 CNN Architecture by Combining Standard Models

The Table 2 summarizes the performance of the various CNN architectures used for COVID-19 Detection and classification.

The work presented in [20] proposes a different approach wherein the output of two CNNs are combined and passed on to the custom CNN model for COVID-19 detection and classification as shown in Fig. 22.

The concept itself is interesting to be further explored since the learning from two models might increase the performance of the output, but still careful considerations needs to be done before deciding on the architectures to be used. The hinderance of this approach is the time taken for the model design and parameters since the model requires the two models of choice to complete the learning before passing the features to the custom CNN architecture for the output. The various architecture performances with COVID-19 classification (Binary/Multiclass) is summarized in the following Fig. 23 , and Fig. 24.

Table 2 Architectures for COVID-19 detection and classification

Work	Dataset used	Transfer learning	Architecture name	Binary/Multiclass	Accuracy
[8]	[9]	NO	DarkCOVIDNet	Binary/Multiclass	Binary: 98.08%, Multiclass: 87.02%
[14]	[9]	NO	CoroNet	Multiclass	Multiclass: 89.6 (4 class)–95% (3 class)
[15]	[16]	NO	COVIDiagnosis-Net	Multiclass	Multiclass: 98.3%
[16]	[16]	YES	COVID-Net	Multiclass	Multiclass: 91.3%
[10]	[10]	YES	CovXNet	Binary/Multiclass	Multiclass: 90.2%
[13]	[9]	YES	nCOVnet	Binary	Binary: 93%
[17]	[8, 9]	YES	Automated Deep Transfer Learning	Multiclass	Multiclass: 97%
[18]	[9, 16]	NO	CapsNet	Binary/Multiclass	Binary: 97.24%, Multiclass: 84.22%
[20]	[9]	NO	Hybrid Architecture	Multiclass	Multiclass: 99.5%

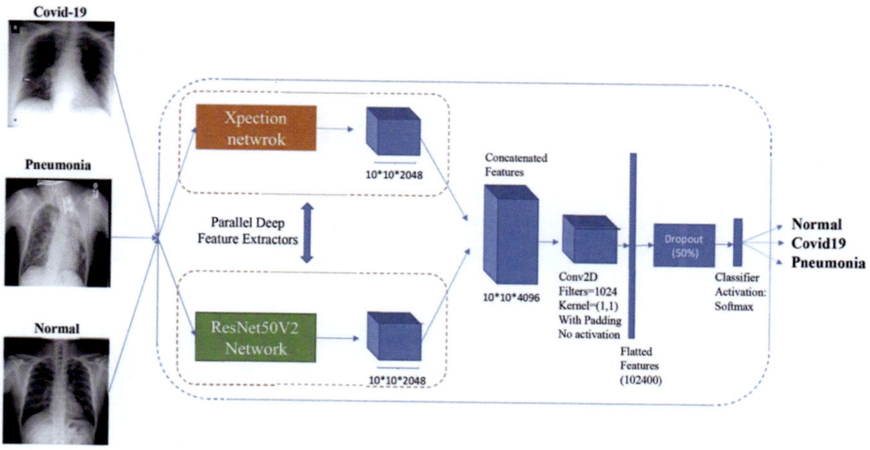

Fig. 22 The architecture of the Hybrid Approach using Xception and ResNet50V2 [20]

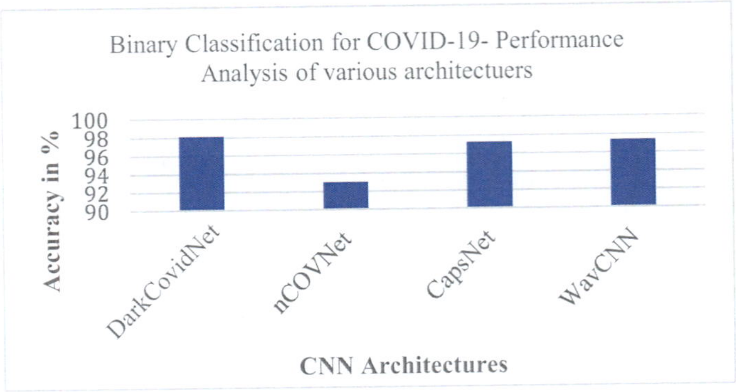

Fig. 23 Binary classification performance analysis of various architectures for COVID-19 classification

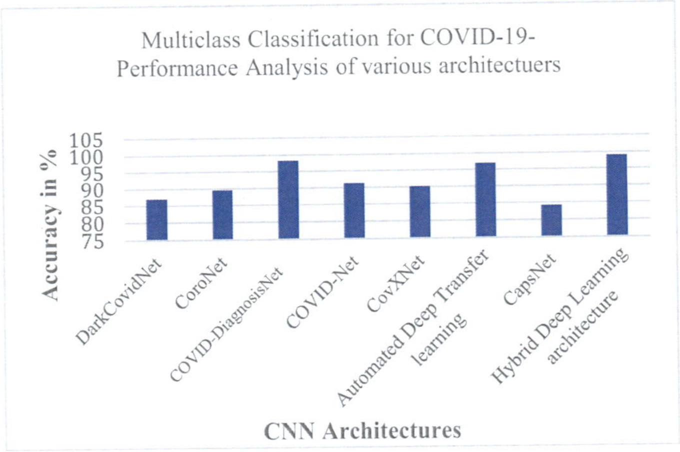

Fig. 24 Multiclass classification performance analysis of various CNN architectures for COVID-19 classification

5 Sources of COVID-19 Datasets

The following section examines the various datasets that are used by various researchers all over the world for the COVID-19 detection and classification. An extensive detailed survey of various COVID-19 datasets was conducted by the researchers covering the various data sources and the following table summarizes the various datasets, its sources, and applications [21] as shown in Table 3.

The datasets quoted above are all available as public for the usages of the researchers, some of these datasets are compiled using the existing datasets. In terms of the number of samples for chest-X-ray COVIDNet stands out with 13,800 chest

Table 3. COVID-19 X-ray and CT image dataset (chest/lungs)

Dataset name	Type	Description	Learning architectures	Learning goals	Source
COVID-chest X-ray-dataset	Chest X-ray	(Total Samples-654) COVID-19 Chest X-Ray (203)	Miscellaneous	Data collection	[9]
COVIDx	Chest X-ray	Normal (8066), pneumonia (5575), COVID-19 (617), Total - 14258	COVIDNet,	COVID-19 identification and classification	[27]
Extensive COVID-19 X-Ray and CT chest images dataset	Chest X-ray/CT images	X-ray images (5500-Normal, 4044-COVID-19), CT images (2628-Normal, 5427-COVID-19 infected)	Miscellaneous	Data collection	[26]
COVID-19 Radiography database	Chest X-ray	COVID-19 (1143), Normal (1341), Viral Pnemonia (1345)	Miscellaneous	Data collection	[28]
Miscellaneous	Chest CT images	349 CT images COVID	NA	COVID-19 identification and classification	[2]
COVID-19 SIRM database	Chest X-ray	COVID-19 (68 images)	Miscellaneous	Data collection	[29]

X-ray images which is an amalgamation of many image dataset from publicly available datasets. As far as the Chest CT image set is concerned, the CT image dataset that is used in [2] stands out with a greater number of samples. The image quality of the datasets specified in the table are substantial with standard quality and can be used by the researchers to carry out their independent research work.

6 Discrete Wavelet Transforms and CNN

The following section examines the impact of combining traditional image processing methods such as Discrete Wavelet Transforms (DWT) with CNN's using pretrained models as a base. The section also examines the concept of DWT in detail with relevant examples. The outcome of this study is to examine the impact of DWT when combined with CNN's for solving classification problem. Though efforts have been done to use DWT along with CNN's for modelling problem such as image reconstruction, texture mapping, frequency observation, etc., not much substantial work is done on the usage of DWT in addressing classification problem.

The two approaches of transformations that can be applied on the images for image processing paradigm such as classification, object detection, tracking, compression, restoration, segmentation, etc. are, Spatial Approach, Spectral Approach.

6.1 *Frequency Domain Processing*

The term "spectral" means frequency, the spectral Approach relies on the modified spatial images which are obtained upon applying Fourier transforms on it. Spectral Image processing is very vital in technology such as Image reconstructions, object tracking, Automated target detection [24], Filtering and Compression etc. Some of the spectral approaches use either Fourier or wavelet transforms over spatial images. The following subsection will explore each of them in detail.

6.1.1 Discrete Wavelet Transform on Images

Wavelets (small waves) are the functions that are concentrated in time as well as in frequency around a certain point in the image. Wavelet transforms rely on the low and high pass filter extensively to gain insights of the various frequency levels along with their locality. The method is called Discrete Wavelet Transform because the method is focused on a range of frequencies in the image rather than the entire set of frequencies.

In [25], Parida, P., and Bhoi, N. discuss the discrete wavelet transformation method as shown in Fig. 25. Which gives a view of how DWT extracts the features by applying low and high pass filters at the cost of resolution due to subsampling.

The filters are applied column-wise along the image, upon the application of the filter, there will be redundant samples coming out of the low pass and high pass filter.

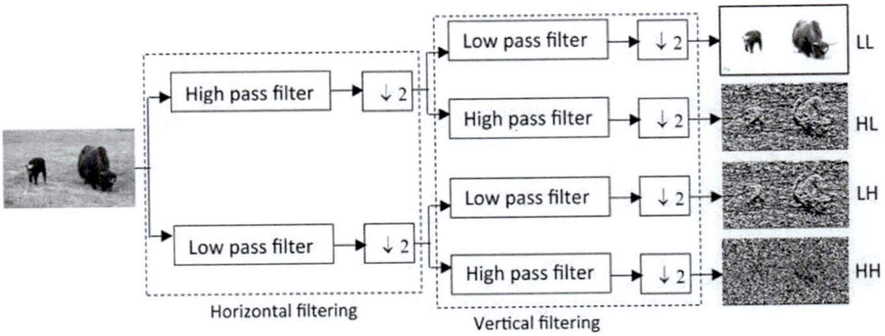

Fig. 25 Discrete wavelet transform applied to a sample image [25]

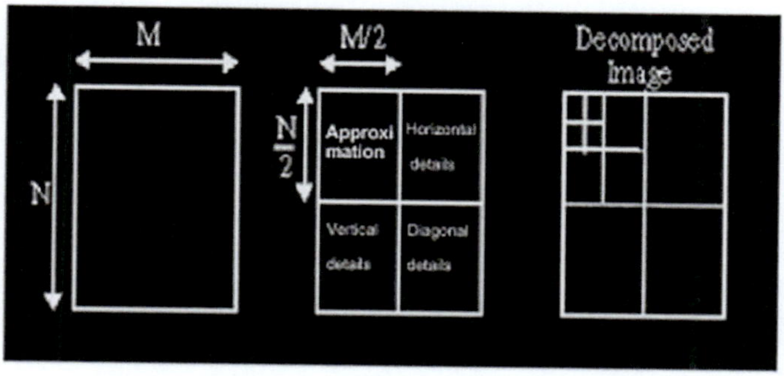

Fig. 26 Level 1 decomposition [25]

Since losing half of the samples would not let one lose information, Down sampling operation is done on the image by 2 every iteration (Nyquist Algorithm). The output of the low pass filter and high pass filter after sub-sampling is fed to the low pass and high pass filter again followed by a subsampling by 2, the difference lies in the fact that the operation is applied row-wise on the image. The output obtained at the end of this iteration will be an image that passed through LL (2 low pass), LH (Low pass followed with high pass), HL (High pass followed with Low pass) and HH (2 High pass filter). One important concept to be aware of is that High pass filter extract the edges and low pass filter does the approximation of the image. The output after this is represented as Fig. 26 shown below.

The LL image gives the Approximation, HH gives diagonal edges (since it passes through two High pass filters, first with column and with the row), LH gives the horizontal edges of the image and HL gives the vertical edges of the image. Some of the important observations from the study are, in this process, the images are sampled twice, once for row and once for column incurring sampling rate of 1/4th the image. The resolution of the image also decreases at the same rate. The overall purpose of this DWT is finding the localization of frequency as well as locality, the study helps us to understand that in a particular location whether we have higher or lower frequency which will help us to understand the distribution of frequency over a specified portion of the image.

a. Experimental Testbed

In order to study the behavior of the CNN models with DWT in place, a pre-trained model named 'AlexNet' which is pretrained for general image classification using a subset of dataset from ImageNet dataset (~1.2 Million Images-subset) and it has classified objects among 10,000 classes approximately. The dataset comprises of 349 CT images found from clinical findings of COVID-19 from 216 patients and 397 Non-COVIID CT images.

The overall objective of the study to train the AlexNet to learn the frequency distribution of the infection from the Chest CT images of COVID/Non-COVID image set. The following diagram shows the progression of the COVID-19 infection 12 days after the contact of the virus as shown in the Fig. 27a, the Fig. 27b shows the CT

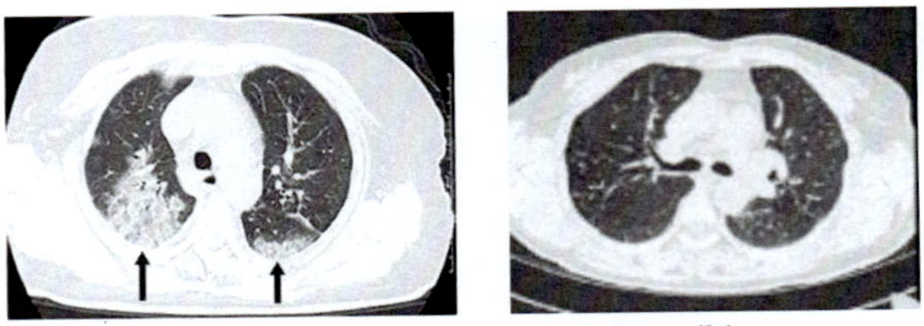

Fig. 27 **a** - chest CT view of the COVID-infected patient. **b** - normal chest CT view of Non-COVID patient

Parameters	AlexNet	
Layer 1 (CL)	Fliters	96
	Kernel Siz	11*11
	Padding	valid
	Stride	4
	Activatio	Relu
Max Pooling	Poolsize	2*2
	Stride	2
	Padding	valid
Batch normalization		
Layer 2(CL)	Fliters	256
	Kernel Siz	11*11
	Padding	valid
	Stride	4
	Activatio	Relu
Max Pooling	Poolsize	2*2
	Stride	1
	Padding	valid
Batch normalization		

Layer 3(CL)	Fliters	384
	Kernel Siz	3*3
	Padding	valid/1
	Stride	1
	Activatio	Relu
Batch normalization		
Layer 4(CL)	Fliters	384
	Kernel Siz	3*3
	Padding	valid/1
	Stride	1
	Activatio	Relu
Batch normalization		
Layer 5(CL)	Fliters	256
	Kernel Siz	3*3
	Padding	valid/1
	Stride	1
	Activatio	Relu
Max Pooling	Poolsize	3*3
	Stride	2
	Padding	valid
Batch normalization		

Basic Parameters		
Input shape	224X224X3	
Output activation functio	Softmax	Sigmoid

Other parameters		
loss model(crossentropy	categorical	Binary
optimizer	Adam	
Metrics	Accuracy	
Batch Size	10	30

Fig. 28 Model parameters of the AlexNet architecture (implemented)

image of the non-infected patient, and the distribution of the infection is pointed out in the Fig. 26a.

The experiments were carried out in two phases, Phase-I—Use the stock AlexNet Architecture without any preprocessing step apart from the resize operation and record the behavior, Phase-II Use the DWT preprocessing and feed the AlexNet with the preprocessed images on various layers with various inputs passing through the filters such as LL/LH/HL/HH.

6.1.2 Phase-I Implementation and Result

The Stock architecture of AlexNet [3] is fed with the input dataset comprising the 349/397 COVID/Non-COVID images from [2].

The model parameters used for the experimentation is as shown in the below Fig. 28. The architecture uses the stock parameters for the model design, and it

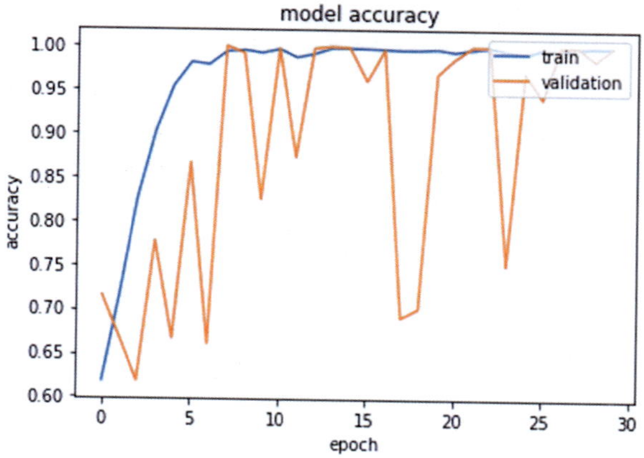

Fig. 29 Accuracy v/s Epoch Stock AlexNet architecture

is implemented using Keras Library. The result of the Experiment shows the erratic nature of the model during validation. The graph of Accuracy v/s epoch for the Stock AlexNet architecture shows the same on run. The Fig. 29 shows the graph. From the experiment, the AlexNet appears to have an erratic learning when it comes to the medical images especially the COVID-19 images and its quite evident in the accuracy value during validation. This is due to the nature of the input; the medical images are rich in information and are taken under the restricted environment. More ever additional model tweaking of the parameters such as Optimizers, Pooling Layers, loss models, or increasing dense layers did not induce the smoothness in the validation curve in the Graph.

6.1.3 Phase-II Implementation and Result

The following Fig. 30 shows the model design for the AlexNet with Discrete Wavelet Transforms as preprocessing method, the wavelet used for the transforms is biorthogonal wavelet.

The idea here is to make the AlexNet remember these distributions of the frequencies from the COVID-19 dataset and use the learning for the Classification of the new input images. Though the preprocessing steps involve the four-layer output, the HL and HH layer output is not of significant importance, so the Model is focused on the approximation of the original image (LL) and Horizontal edges (LH) processed images. The output of the DWT is as shown below Table 4. The AlexNet is first trained with the LL images (Preprocessed) with the stock architecture under standard settings, the results obtained is as shown in the Fig. 31. The state-of-the-art performance of 98% is achieved within first 15 epoch and is evident from the diagram. The inference drawn out of this is the fact that the AlexNet was able to learn the features using the approximation of the images of Chest CT images (for COVID-19 infected) obtained by the LL preprocessing method.

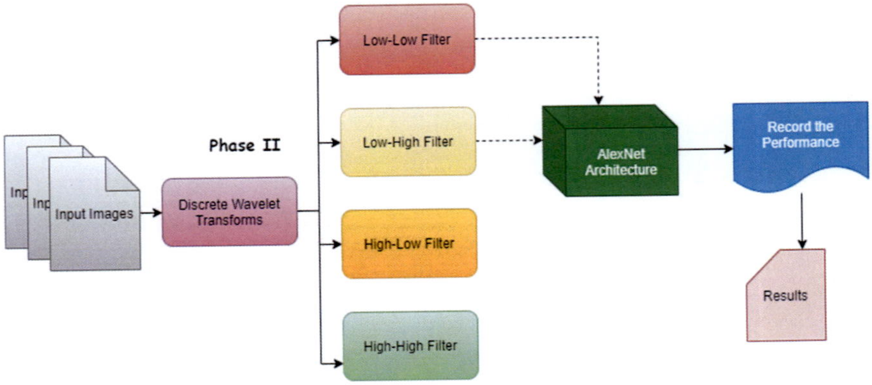

Fig. 30 Proposed architecture for experimentation

Table 4 DWT in action (chest CT image of COVID-19 infected patient)

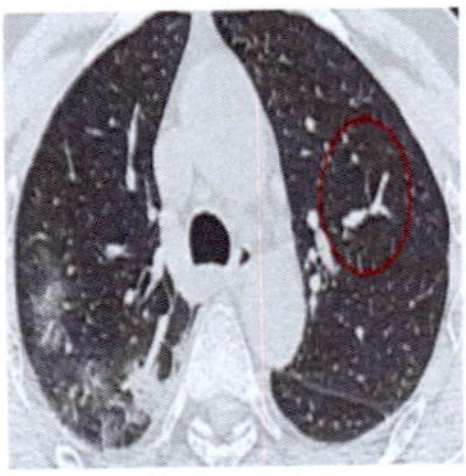	Low pass – Low Pass filter (LL)	
	Low Pass – High Pass Filter (LH)	
	High Pass – Low Pass filter (HL)	
	High Pass – High Pass filter (HH)	

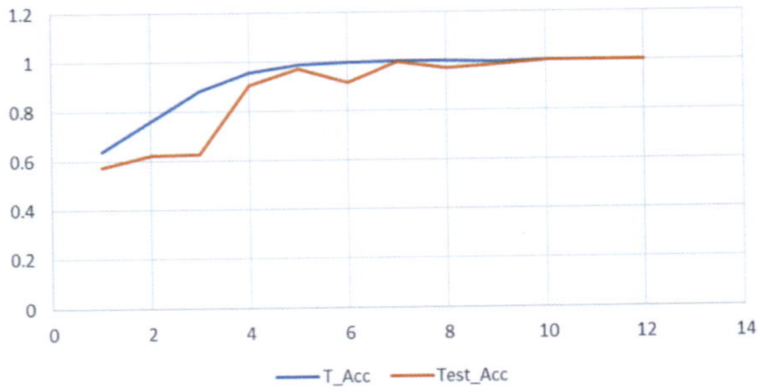

Fig. 31 AlexNet with DWT performance (accuracy vs epoch)

Though the features and image size were less, the validation curve appears to be smooth compared the erratic behavior exhibited by the AlexNet without DWT preprocessing. Upon applying the preprocessing of the images using LL1 the image size is reduced by 1/4th of the volume and feature loss is also evident. But with less features to learn, the AlexNet had performed well for the training and testing image sets.

The output of the LH1 though on the other hand did not provide significant learning for the model and it failed as the horizontal features were too minimal for the model to learn. The output was furthered smoothened with the additional resize operation on the LL images as shown in the Fig. 31, The state-of-art results are obtained with minimum epochs which is evident in the performance graphs shown in Fig. 30 and 32.

Though the DWT preprocessing step has given the ability to the AlexNet to perform better in terms of accuracy and loss, the method involves subsampling of the images on every level of decomposition which might incur feature loss. The loss of features might also hinder the learning of models when more than two level of decomposition using DWT is done. Another important point to note is the fact that more than two level decomposition will not yield any results for any pretrained models. Though there are architectures in place for the COVID-19 detection which has yielded better results and some of which have gone through the radiological trials, the proposed architecture opens up a new horizon of building a hybrid model which combines the powerful techniques of the traditional image processing methods with the Convolutional Neural Networks and evidently it had made a significant impact on the performance of the model in terms of accuracy and loss metrics.

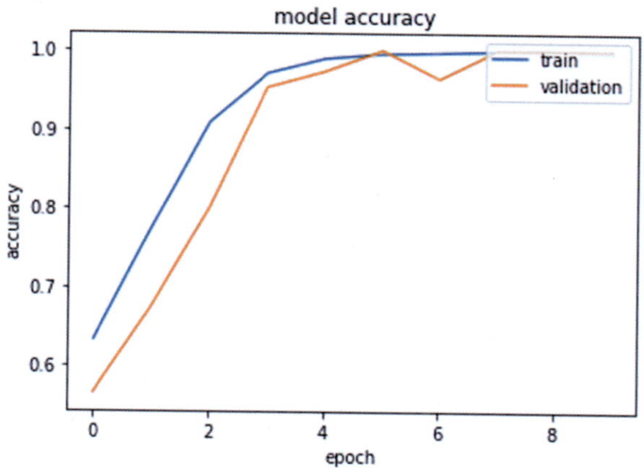

Fig. 32 AlexNet with DWT performance (accuracy vs epoch) with resize operation after the DWT step

Fig. 33 Research prospects

7 COVID-19 Medical Data Visualization

Medical information systems and medical devices are recording tons and tons of data related to treatment, disease progression, patient information, pace of disease spread, primary and secondary contracts etc. It becomes at most important to comprehend this data in an understandable way. The large quantity of data can be summarized with the help of data summarization and visualization tools. The summary and the visualization graphs generated will give the clear insight of what is happening on the ground, in turn helps in taking the informed decisions. It immensely helps the administration to act quickly, and reason the decision. There are several tools which could help in the process of summary generation and visualization. Following are some of the software tools that could benefit the healthcare community for visualizing and summarizing the humongous quantity of data currently being recorded in healthcare information systems.

Few of the noteworthy visualization of COVID 19 data include John Hopkins University Dashboard, WHO COVID-19 Dashboard, The University of Washington Dashboard, South China Morning Post Dashboard, CDC Dashboard, The COVID tracking project (COVID Tracking), ECDC Dashboard University of Virginia COVID-19 interpret and Few other attempts are COVID19-SEIR, COVID Trends, 91-DIVOC Rt Covid-19 Dashboard, IHME COVID-19 Dashboard, WORLDOMETER, COVID ACT Now Dashboard COVID-19 Measures Oxford, Carnegie Mellon Dashboard, Stanford COVID-19 Dashboard, and 91-DIVOC.

For in order to develop novel data visualization method for COVID 19, it is at most important to understand the pandemic demographic information, the disease trait and dynamics of the parameters. The new visualization method should not be generalized. From our observations at the outputs generated by the various visualization tools referred in the above paragraph, it is relatable to fact that the tools cannot be generalized to summarize/visualize from one population to other population, or in other words from one ethnic group to another ethnic group. A variable approach for the visualization of data needs to be implemented based on disease dynamic at any given place. We would surely recommend exploring this problem of visualization on the basis of demographics and ethnicity, our future work will definitely focus on this problem.

One other aspect visualization can explore is changing behavior of virus, depending on the weather pattern in different patterns across the world, one typical example could be - the European countries have gone to second lockdown due to high mortality rates, whilst in Asian countries (for example India), even though the infection rate is not under control, the mortality rate is significantly coming down. This contextual information about the disease dynamics needs to be considered rather mare generalizing the visualization approach. The visualization method can embody 360-degree attributes of the pandemic.

There are various aspects through which data can be visualized, but it is at most important to make sure that the visualization tool is not biased due to irrelevant assumption and generalization.

8 Factors that Indicate the COVID-19 Infection Progression (Radiology Perspective)

Among the radiology data such X-ray and chest CT images considered by the various researchers to develop CNN classification models, X-ray images tend to provide rich information related to the COVID-19 infection. The presence of patches or bands like ground glass opacity in mid-to-lower or peripheral lung zone in chest radiography is highly suggestive of COVID-19 infection. This observation along with the clinical observation can be collaborated to make final confirmation of COVID-19 infection [22]. Initial CT findings in COVID-19 show a similar kind of presence of patches in the lung zone of CT. The appearance of these patches is more in the lower lobes and is less frequent in the middle lobe. Some of the uncommon but observed findings are pericardial effusion, lymphadenopathy, cavitation & CT halo sign with the disease progression which is evident from the Chest CT scan images [23].

9 Framework and Research Methodology

The following Table 5 summarizes the survey framework used for the study.

Table 5 Survey framework

Objectives	1. Literature collection on the recent works done for COVID-19 detection 2. Literature survey on the dataset discovery and aggregation 3. Explore the CNN architectures, machine learning algorithms for the techniques for COVID-19 detection and classification 4. Explore the architectural designs and patterns for basing the proposed work
Survey methods	Systematic Exploration of articles, white papers, journals and medical databases/journals
Databases explored	IEEE-Xplore, Elsevier Journals, Radiopedia.org, Springer databases, acs.org, GitHub databases
Datasets accumulated	[2, 9, 16]
Keywords	Deep learning for COVID-19, COVID-19 classification and detection, Convolutional Neural Networks, Transfer Learning, Ensemble Learning
Inclusion criteria	COVID-19 Classification using CNN's
Exclusion criteria	Machine Learning for COVID-19 and its relevant works
#papers collected	127
#papers shortlisted	22
Outcomes achieved	1. Architectures summary presented in Table 2 2. Datasets examined under survey is presented in Table 3 3. Analysis of each architecture is presented under the architecture name in Sect. 4

10 Future Research Prospects

COVID 19 has posed several challenges and opportunities at a short span. Solving these challenges will yield greater benefit for the research fraternity at large. This section briefs some of the pressing challenges surfaced due to the pandemic situation, any researchers willing to take up COVID 19 related research topic can further explore the pointers given in this section. Each one of the pointers given in the rest of the sections can be explored in depth and suitable solutions can be proposed.

Data Science related areas could actively contribute to find out the solutions for threats developed due to infection, severity, and outcome risk. Infection risk is related to groups of people and the most susceptible set of individuals in that group. Severity risk is related to a specific or group of individuals developing the severe disease symptoms and complications. Outcome risk is related to the treatments and its outcome.

There are various realms from which the challenges related to COVID 19 looks challenging, if effectively solved could serve medical eternity in a larger extent.

The aspects to be considered for enhancement range from time complexity, space complexity, data quality, recording, record keeping, transfer learning, and in the aspect of transition networks. All these aspects need to be critically brought under the scrutiny of future research exploration.

There are some other aspects too from which the problems need to be solved, there needs to be methods and techniques which facilitate the way for faster development and discovery of drugs. It is more important to find out if the already available drug can cure the new disease. For this to happen, one should know the ontology of disease and the characteristics of the new disease. There are other dimensions which involve understanding the path of the virus, in other words, or to trace the virus entry point. Some other important aspects are Screening patients using face scans. Building biomedical knowledge graphs, predicting drug-target interactions, Predicting the spread of infectious disease using social networks, understanding viruses through proteins, Predicting viral-host protein–protein interactions.

11 Conclusion

It is apparent from the pandemic situation that, no country on this planet was prepared enough to handle the disaster. Unimaginable number of casualties have happened in such a short span of time, the effect of the disease spread has affected every individual in one or the other way. The economy is struggling, many have lost their livelihood, borders are closed, economic activities have come to halt, healthcare facilities and infrastructures are in a loan that have never been experienced in any time before.

It is a very challenging time that data science can come into action and solve some of the pressing issues. This paper presented an effort to consolidate the novel Deep Learning architectures for solving COVID 19 related problems. Several Deep Learning approaches have been proposed to address the problem of disease identification, progression, and diagnosis. This work also focused on comparing the several Deep Learning architectures based on several key performance factors. The attempt also was made to compare the different datasets used for evaluating Deep Learning architectures. The chapter also presented the need of Deep Learning instead of conventional neural network architectures. The concluding sections of the paper focused on the tools and techniques for visualizing and summarizing COVID 19 related data. Visualization and Summarization greatly helps in understanding the trend which facilitates informed decision making. There are several pressing issues in effectively handling the disease spread, diagnosis, and management - noteworthy issues related to pandemic which requires further investigation and research are also presented in the final sections of this chapter. The systematic attempt has been made to consolidate, compare, and evaluate the existing noteworthy Deep Learning architectures for handling various scenarios of COVID 19.

References

1. What Is COVID-19? | coronavirus. (n.d.), https://coronavirus.dc.gov/page/what-COVID-19. Accessed 28 Aug 2020

2. X. Yang, J. Zhao, Y. Zhang, S. Zhang, P. Xie, COVID-CT-Dataset: A CT Image Dataset about COVID-19 (n.d.), https://www.medrxiv.org/. Accessed 10 Nov 2020
3. A. Krizhevsky, I. Sutskever, G.E. Hinton, ImageNet classification with deep convolutional neural networks. Commun. ACM **60**(6), 84–90 (2017)
4. WHO Coronavirus Disease (COVID-19) Dashboard | WHO Coronavirus Disease (COVID-19) Dashboard (n.d.), https://COVID19.who.int/. Accessed 28 Aug 2020
5. Overview | Coronavirus Disease COVID-19 (n.d.), https://www.COVID19treatmentguidelines.nih.gov/overview/. Accessed 28 Aug 2020
6. COVID-19 basics - Harvard Health (n.d.). https://www.health.harvard.edu/diseases-and-conditions/COVID-19-basics. Accessed 28 Aug 2020
7. Coronavirus & COVID-19 Overview: Symptoms, Risks, Prevention, Treatment & More. (n.d.), https://www.webmd.com/lung/coronavirus. Accessed 28 Aug 2020
8. T. Ozturk, M. Talo, E.A. Yildirim, U.B. Baloglu, U. Ozal Yildirim, R. Acharya, Automated detection of COVID-19 cases using deep neural networks with X-ray images. Comput. Biol. Med. **121**, 103792 (2020)
9. J.P. Cohen, COVID-19 image data collection (2020), https://github.com/ieee8023/COVID-chestxray-dataset
10. T. Mahmud, M.A. Rahman, S.A. Fattah, CovXNet: a multi-dilation convolutional neural network for automatic COVID-19 and other pneumonia detection from chest X-ray images with transferable multi-receptive feature optimization. Comput. Biol. Med. **122**, 103869 (2020)
11. D. Kermany, K. Zhang, M. Goldbaum, Labeled optical coherence tomography (OCT) and chest X-ray images for classification, version 2 (Mendeley Data, 2018)
12. S. Fujieda, K. Takayama, T. Hachisuka, Wavelet convolutional neural networks, arXiv preprint arXiv:1805.08620 (2018)
13. H. Panwar, P.K. Gupta, M.K. Siddiqui, R. Morales-Menendez, V. Singh, Application of deep learning for fast detection of COVID-19 in X-rays using nCOVnet. Chaos Solitons Fractals **138**, 109944 (2020)
14. A.I. Khan, J.L. Shah, M.M. Bhat, Coronet: a deep neural network for detection and diagnosis of COVID-19 from chest x-ray images. Comput. Meth. Programs Biomed. **196**, 105581 (2020)
15. F. Ucar, D. Korkmaz, COVIDiagnosis-Net: Deep Bayes-SqueezeNet based diagnosis of the coronavirus disease 2019 (COVID-19) from X-ray images. Med. Hypotheses **140**, 109761 (2020)
16. L. Wang, A. Wong, COVID-Net: a tailored deep convolutional neural network design for detection of COVID-19 cases from chest X-Ray images, arXiv preprint arXiv:2003.09871 (2020)
17. N.N. Das, N. Kumar, M. Kaur, V. Kumar, D. Singh, Automated deep transfer learning-based approach for detection of COVID-19 infection in chest X-rays. IRBM (2020)
18. S. Toraman, T.B. Alakuş, İ Türkoğlu, Convolutional capsnet: a novel artificial neural network approach to detect COVID-19 disease from X-ray images using capsule networks. Chaos Solitons Fractals **140**, 110122 (2020)
19. Capsule neural network - Wikipedia (n.d.), https://en.wikipedia.org/wiki/Capsule_neural_network. Accessed 6 Sept 2020
20. M. Rahimzadeh, A. Attar, A modified deep convolutional neural network for detecting COVID-19 and pneumonia from chest X-ray images based on the concatenation of Xception and ResNet50V2. Inf. Med. Unlocked **19**, 100360 (2020)
21. J. Shuja, E. Alanazi, W. Alasmary, A. Alashaikh COVID-19 datasets: a survey and future challenges (medRxiv 2020)
22. Radiologists find chest X-rays highly predictive of COVID-19 (n.d.) https://medicalxpress.com/news/2020-09-radiologists-chest-x-rays-highly-COVID-.html. Accessed 7 Sept 2020
23. COVID-19 | Radiology Reference Article | Radiopaedia.org. (n.d.). https://radiopaedia.org/articles/COVID-19-4?lang=us. Accessed 7 Sept 2020
24. K. Krishnamurthy, Spectral Image Processing Theory and Methods: Reconstruction, Target Detection, and Fundamental Performance Bounds. Dissertation, Duke University (2011). https://hdl.handle.net/10161/3945

25. P. Parida, N. Bhoi, Wavelet based transition region extraction for image segmentation. Fut. Comput. Inf. J. **2**(2), 65–78 (2017)
26. W. El-Shafai, F.A. El-Samie, Extensive COVID-19 X-ray and CT chest images dataset, version 3 (Mendeley Data 2020). https://doi.org/10.17632/8h65ywd2jr.3
27. COVID-Net/COVIDx.md at master, lindawangg/COVID-Net, GitHub (n.d.), https://github.com/lindawangg/COVID-Net/blob/master/docs/COVIDx.md. Accessed 24 Dec 2020
28. COVID-19 Radiography Database | Kaggle. (n.d.), https://www.kaggle.com/tawsifurrahman/covid19-radiography-database. Accessed 24 Dec 2020
29. COVID-19 DATABASE | SIRM. (n.d.). https://www.sirm.org/en/category/articles/covid-19-database/. Accessed 24 Dec 2020

The manufacturer's authorised representative in the EU is Springer
Nature Customer Service Centre GmbH, Europaplatz 3, 69115 Heidelberg,
Germany. If you have any concerns regarding our products, please
contact ProductSafety@springernature.com

Printed and bound by CPI Group (UK) Ltd, Croydon, CR0 4YY
07/10/2026
02229168-0004